RESEARCH HANDBOOKS IN CLIMATE LAW

Series Editor: Jonathan Verschuuren, *Tilburg University, the Netherlands*

This important and timely series brings together critical and thought-provoking contributions on the most pressing topics and issues within the field of climate law. The volumes in this significant series cover a wide array of the effects of climate change on such diverse fields as trade, human rights, energy, disasters, finance, and migration.

Each *Research Handbook* comprises specially-commissioned chapters from leading academics, and practitioners, as well as those with an emerging reputation and is written with a global readership in mind. Equally useful as reference tools or high-level introductions to specific topics, issues and debates, these *Handbooks* will be used by academic researchers, postgraduate students, practising lawyers and lawyers in policy circles.

Titles in this series include:

Research Handbook on REDD+ and International Law
Edited by Christina Voigt

Research Handbook on Emissions Trading
Edited by Stefan E. Weishaar

Research Handbook on Climate Change and Trade Law
Edited by Panagiotis Delimatsis

Research Handbook on Climate Change and Agricultural Law
Edited by Mary Jane Angelo and Anél Du Plessis

Research Handbook on Climate Change and Agricultural Law

Edited by

Mary Jane Angelo

Professor of Law and Director, Environmental and Land Use Law Program, University of Florida Fredric G. Levin College of Law, USA

Anél Du Plessis

Professor of Law, Faculty of Law, North-West University, South Africa

RESEARCH HANDBOOKS IN CLIMATE LAW

Edward Elgar PUBLISHING

Cheltenham, UK • Northampton, MA, USA

Published by
Edward Elgar Publishing Limited
The Lypiatts
15 Lansdown Road
Cheltenham
Glos GL50 2JA
UK

Edward Elgar Publishing, Inc.
William Pratt House
9 Dewey Court
Northampton
Massachusetts 01060
USA

A catalogue record for this book
is available from the British Library

Library of Congress Control Number: 2016953944

This book is available electronically in the **Elgar**online
Law subject collection
DOI 10.4337/9781784710644

FSC
www.fsc.org
MIX
Paper from
responsible sources
FSC® C013056

ISBN 978 1 78471 063 7 (cased)
ISBN 978 1 78471 064 4 (eBook)

Typeset by Servis Filmsetting Ltd, Stockport, Cheshire
Printed and bound in Great Britain by TJ International Ltd, Padstow, Cornwall

Contents

List of contributors vii
Preface and acknowledgements xii

1 Untangling the climate-food web: achieving food security and agricultural climate-resilience 1
Mary Jane Angelo

2 Climate change and agriculture under the United Nations Framework Convention on Climate Change and related documents 21
Jonathan Verschuuren

3 Resilience and transformation of agro-ecosystems in a changing climate 47
Lance H Gunderson

4 Climate change, complexity, agriculture and challenged governance 74
Paul Martin

5 Climate change and water law in agriculture 103
Robert W Adler

6 Climate change and invasive species law in agriculture: a southern hemisphere view of key legal and institutional issues in Australia 135
Elodie Le Gal

7 Genetically modified food crops in an era of climate change: an answer in search of a problem 167
Rebecca M Bratspies

8 Adapting agriculture through land use controls 196
Keith H Hirokawa

9 Much ado about methane: intensive animal agriculture and greenhouse gas emissions 228
Michelle Nowlin and Emily Spiegel

10 Adaptation of agricultural trade and investment rules to climate change 274
Christian Häberli

11 Agriculture and payments for ecosystem services in the era of climate change 315
JB Ruhl

12 Climate change and agrofuels: Brazilian ethanol and the Cerrado biome 331
Sérgio Sauer, José Paulo Pietrafesa and Pedro Araujo Pietrafesa

13 Adaptation to climate change in smallholder agriculture in Kenya: the role of law 367
Robert Kibugi

14 Agricultural land use conflict in the context of climate change: an Australian case study 385
Amanda Kennedy and Amy Cosby

15 Climate change and land grabbing 423
Akachi Odoemene

Index 449

Contributors

Robert W Adler is the Jefferson B and Rita E Fordham Presidential Dean and Distinguished Professor of Law at the University of Utah, SJ Quinney College of Law. His books include *Modern Water Law: Private Property, Public Rights and Environmental Protection* (Foundation Press, 2013, with Craig and Hall); *Restoring Colorado River Ecosystems: A Troubled Sense of Immensity* (2007); *Environmental Law: A Conceptual and Pragmatic Approach* (with Driesen and Engel, 3rd edn 2015); and *The Clean Water Act: Twenty Years Later* (with Landman and Cameron, 1993). He has represented Utah and national environmental groups in pro bono litigation, and served on the boards of local and national organizations and advisory groups, including FRIENDS of Great Salt Lake, Hawkwatch International, EPA Federal Advisory Committees, and a Taskforce on Unfunded Federal Mandates for the US Advisory Commission on Intergovernmental Relations.

Mary Jane Angelo is a Professor of Law and Director of the Environmental and Land Use Law Program at the University of Florida Levin College of Law. Professor Angelo has extensive publications on a variety of environmental law topics including the recent books *Food, Agriculture and Environmental Law* (with Czarnezki and Eubanks) and *The Law and Ecology of Pesticides and Pest Management*. She has served on National Academy of Sciences committees on Everglades Restoration and on Ecological Risk Assessment under FIFRA and the Endangered Species Act. Professor Angelo joined the UF faculty in 2004 after many years of environmental law practice, including serving as an attorney for the US Environmental Protection Agency in Washington, DC, and for the St. Johns River Water Management District in Florida. Professor Angelo received her BS in Biological Sciences from Rutgers University, and both her MS, in Entomology, and JD from the University of Florida.

Rebecca M Bratspies is a Professor of Law and the director of the CUNY Center for Urban Environmental Reform (CUER). She has published widely on regulatory policy – with a focus on environmental democracy, regulating new technologies, and corporate responsibility. Her recent scholarship explores questions of urban sustainability, and the intersection of human rights and environmental regulation. She is a scholar with the Center for Progressive Reform, the Environmental Law Collective and has served as an appointed member of the ABA Standing Committee on Environmental Law, a member of the Executive Committee of the American Association of Law Schools Section on the Environment, and an advisor to the Consultative Group on Agricultural Research.

Amy Cosby is part of the research team at the Australian Centre for Agriculture and Law at the University of New England (UNE). She holds a Bachelor of Agriculture/Bachelor of Laws (Hons) and a PhD in the area of precision agriculture, also from UNE. Her research interests are in the areas of land use governance and global food security.

Anél Du Plessis (BA Law, LLB, LLM, LLD) is Professor of Law at the North-West University (NWU) (Potchefstroom Campus) and heads its LLM programme in Environmental Law and Governance. Her published research focuses on a combination of environmental-, local government and different aspects of human rights law and environmental governance. Anél is an elected member of the South African Young Academy of Science, a rated researcher with the National Research Foundation of South Africa and an alumnus of the DAAD, Fulbright and Alexander von Humboldt Scholarship programmes. She is a member of the Executive Committee of the Environmental Law Association of South Africa and Assistant Editor of the *Potchefstroom Electronic Law Journal*.

Lance H Gunderson is a Professor of Environmental Sciences at Emory University, Atlanta, GA, USA. He is a systems ecologist who is interested in how people understand, assess, and manage large social ecological systems. He is Co-editor in Chief of Ecology and Society and Chair of the Resilience Alliance. He is a Fellow with the Beijer International Institute for Ecological Economics, Swedish Royal Academy of Sciences.

Christian Häberli works as a Senior Research Fellow at the World Trade Institute of Bern University on various trade, agriculture, employment and development policy issues. His main research topic is food security from a trade and investment perspective. He also works as a lecturer and policy adviser in Europe and in the Global South. Graduated in 1977 at Basel University with a PhD on the subject of African Investment Law. Degrees for Development Sciences in Geneva (1975) and for Theology in Bern (2009). Professional career for the International Labour Organization (ILO) and for the Swiss Government with assignments in Madagascar, Thailand, Nepal and Switzerland.

Keith H Hirokawa teaches environmental law and natural resources law at Albany Law School. His scholarship explores convergences in ecology, ethics, economics, and law, with particular attention given to local environmental law, ecosystem services policy, watershed management, and environmental impact analysis. Prior to joining the faculty at Albany Law, Professor Hirokawa taught at Texas Wesleyan University School of Law and the University of Oregon School of Law. His books include *Greening Local Government* (2012, with Patricia Salkin), *Environmental Law and Contrasting Ideas of Nature* (2014), and *Rethinking Sustainability to Meet the Climate Change Challenge* (2015, with Jessica Owley).

Amanda Kennedy is the Deputy Director of the Australian Centre for Agriculture and Law at the University of New England in Australia (UNE). Her research interests include instruments and institutions for effective land use governance, and environmental justice. She was the recipient of an Australian Research Council (ARC) Discovery Early Career Researcher Award between 2012 and 2015 to examine systems for managing land use conflict.

Robert Kibugi teaches law at the University of Nairobi, and is a specialist on climate change, water services, land law, land use law and energy. He holds a doctorate in law from the University of Ottawa. He has extensive public policy and legislative experience in Kenya, including design and draft of laws and policies on climate change, water services and land degradation.

Elodie Le Gal is a French-Australian academic lawyer and researcher who specialises in legal and institutional analysis. Her research, and research-led teaching expertise, which supports team-work and strategic community-led applied research in Australia and overseas through strong research collaborations with international universities, the private sector and the IUCN Academy of Environmental Law, focuses on the sustainable governance of natural resources for improving people's livelihoods and the promotion of fair economic development. She obtained her PhD at the University of New England in 2013. She also has a Master of Laws (Diplôme d'Etudes Approfondies) in Private International Law/International Business Law from the Pantheon-Assas Faculty of Law, Paris.

Paul Martin specialises in research on the effectiveness of environmental governance, with a particular emphasis on the rules and organisations that affect rural sustainability. His background is in law and corporate strategy, and for many years he led environmental organisations, whilst being a corporate director and advisor. His team in the AgLaw Centre uses multidisciplinary and empirical approaches to understand the performance in practice of legal arrangements and to propose law and policy reforms.

Michelle Nowlin is a Supervising Attorney with the Duke Environmental Law and Policy Clinic and Clinical Professor of Law at Duke University

School of Law, with a secondary appointment at the Nicholas School of the Environment. Her practice and scholarship focus on the intersection of food and agricultural production and environmental protection. Prior to Duke, she worked as a Senior Attorney with the Southern Environmental Law Center, where she developed and led the organization's Hog and Poultry Industry Project and litigated cases involving the Clean Water Act, Endangered Species Act and National Environmental Policy Act. She earned her BA, with Highest Honors, from the University of Florida, and earned a dual JD/MA in Environment and Natural Resources Economics and Policy from Duke University.

Akachi Odoemene's research focuses on African Social and Cultural History, Peace and Conflict Studies, Urban History, Ethnic Studies and Development Studies. His work has appeared in several edited volumes, as well as the *Journal of Historical Sociology*, *Armed Forces and Society*, *African Development*, *African Economic History*, *African Identities* and many others. He was an African Humanities Program (AHP) Fellow of the American Council of Learned Societies (ACLS) in 2009, and a Hewlett Visiting Scholar at PSTC, Brown University, Providence, USA in 2012. Dr Odoemene is currently an Oxford-Princeton Global Leaders Post-doctoral Fellow at the Woodrow Wilson School of Public and International Affairs, Princeton University, USA and a Senior Lecturer in the Department of History and International Studies, Federal University Otuoke, Bayelsa State, Nigeria.

José Paulo Pietrafesa has a PhD in Sociology and is Professor at the Federal University of Goiás, researching in themes such as the expansion of sugarcane production and its social (education and working conditions) and environmental consequences in the Goiás State.

Pedro Araujo Pietrafesa has a PhD in Social Studies (University of Brasilia) and is a Professor at the Pontifical Catholic University of Goiás (PUC-GO).

JB Ruhl joined Vanderbilt's law faculty as the David Daniels Allen Distinguished Chair of Law in 2011. He is an expert in environmental, natural resources, and property law, and also studies the legal industry and legal technology. He was named director of Vanderbilt's Program on Law and Innovation in 2014 and serves as the Co-director of the Energy, Environment and Land Use Program. He received a BA in Economics and Juris Doctorate from the University of Virginia, a Masters of Law from George Washington University, and a PhD in Geography from Southern Illinois University.

Sérgio Sauer has a PhD in Sociology, is a professor at the University of Brasília (UnB) and a researcher of CNPq, studying agrarian social movements, land struggle and the expansion of agribusiness in the Amazon region.

Emily Spiegel is a Law Fellow in the Duke Environmental Law and Policy Clinic. She previously worked on agricultural and environmental issues at the Food and Agriculture Organization, the Environmental Law Institute, the International Fund for Agricultural Development, and the US Department of Agriculture. She also served as a Peace Corps volunteer in Jordan. Spiegel received her BS from Georgetown University and her JD from Duke University School of Law. Her professional interests lie in the areas of agricultural law, law and development and conservation law.

Jonathan Verschuuren is a Professor of International and European Environmental Law at Tilburg University and Marie Sklodowska Curie fellow at the University of Sydney. His research mainly focuses on various legal aspects of climate change, such as coastal adaptation and climate change and agriculture. Verschuuren has written more than 300 publications in the field of environmental law, including several books and many articles in outstanding refereed journals throughout the world. Some of his recent publications include *Research Handbook on Climate Change Adaptation Law* (Edward Elgar 2013), 'Hydraulic Fracturing and Environmental Concerns: The Role of Local Government' *Journal of Environmental Law* (2015), and 'The Paris Agreement on Climate Change: Agriculture and Food Security' *European Journal of Risk Regulation* (2016).

Preface and acknowledgements

Many of the scientific aspects around climate change are now much better understood than a few decades ago. Very little uncertainty exists about the causes of climate change and increasingly we understand better the complexity of human and systems adaptation. Yet, an acute need remains to untangle critical nexuses, one of which is the nexus between the phenomenon of climate change, the pursuit of sustainable development and the effort to improve food security in all parts of the world. The law has a direct influence on how we approach and understand this nexus because of the fact that legal rules, principles and instrumentation determine and regulate the relationships between the actors involved at multiple levels of climate and agricultural governance.

Climate change and agriculture are interrelated in many ways on multiple scales. Average temperature fluctuations, rainfall and climate extremes, changes in the features and occurrence of pests and diseases among plants and animals, changes in the nutritional properties of some foods and sea-level changes have a direct bearing on traditional approaches to crop production and other agricultural processes. On the other hand, agriculture contributes to climate change via many means including anthropogenic emissions of greenhouse gases and the conversion of non-agricultural land (for example, natural forests and environmentally sensitive areas) into agricultural land.

This book is the first attempt to consolidate scholarly understanding and recent developments in the Northern and Southern hemispheres on the legally relevant complexities underlying the combating of, and adaptation to, climate change in an era of growing pressure on agriculture and food security.

The *Research Handbook on Climate Change and Agricultural Law* is the latest edition in the Edward Elgar series on 'Research Handbooks in Climate Law'. It has brought together scholars from across the globe in an attempt to untangle the climate-food web and to understand the nexuses on which food security and climate-resilient development depend. Together, the authors address the following key questions:

- to what extent do domestic agricultural laws and policies in the northern and southern hemispheres address climate mitigation and adaptation?
- are the policies and legal tools in existing national agricultural law frameworks sufficient for human and agro-ecosystem adaptation?

- what are the legally relevant challenges to marrying climate mitigation and adaptation with agricultural development, economic growth, food security objectives and sustainable livelihoods?
- how do individual countries' agricultural law and policy provisions for mitigation and adaptation measure up against the expectations of international and regional climate law and policy?
- what are the weaknesses, strengths, gaps and innovations in existing agricultural laws and policies?
- what new and innovative tools can be employed to address climate change mitigation and adaptation in the agricultural sector? and
- what are the remaining research needs?

These questions are approached from the perspective of the legal systems of the United States, South Africa, Kenya, Australia, Brazil, and other countries, with the point of departure being that agriculture is at the same time both a significant contributor to global climate change, and a segment of the economy that will experience some of the most serious impacts of climate change. While this book focuses on the sectors of climate change and agriculture it also serves to highlight and open debate about the limits of the role of law and policy in effectively addressing the challenges we face. It challenges us to rethink the governance instruments we use to regulate behaviour and to question global forces in how we do things in poor and wealthy national and local contexts.

This book is the result of cooperative efforts of each and every contributing author, the editors, the publisher and those colleagues in our academic circles who continue to shape our collective understanding of the intricacies in the fields of climate change, agricultural law, environmental law and sustainability. The editors would also like to thank Rachael Vinyard, Jennifer Lomberk, Italo Lenta, and Gennaro Scibelli for outstanding research and editorial assistance.

Mary Jane Angelo
Anél du Plessis
August 2016

1. Untangling the climate-food web: achieving food security and agricultural climate-resilience

Mary Jane Angelo

1. INTRODUCTION

Any lingering doubts that climate change, if unchecked, will cause significant global harm were put to rest at the United Nations Climate Change Conference (Conference of the Parties (COP21)) in Paris in late 2015. For the first time, it appears that the international community has taken seriously the threat of climate change and the urgency to act. In what has been hailed as 'an extraordinary effort at international diplomacy',[1] 195 nation states rallied around the call to ensure that the globe does not warm by more than 2 degrees C to avoid the catastrophic impacts that would likely occur with any greater increase in temperature.

Many believe that the agreement that grew out of COP21, referred to as the 'Paris Agreement', represents the last best hope for avoiding catastrophic impacts from climate change. In the agricultural arena, a failure to fulfill the commitments made in the Paris Agreement could result in dramatic increases in hunger and poverty throughout the world. Even if the promises of the Paris agreement are realized and countries throughout the world succeed in substantially reducing GHG emissions, climate change impacts will occur. Without substantial GHG emission reductions to slow the pace of change, these impacts could be severe and could dramatically threaten food security.

Although the impacts of climate change on agriculture will vary based on geologic location and other factors, there is widespread agreement that climate change will likely result in significantly decreased crop yields, increased food prices, and decreased food security for millions of people, particularly in the developing world. In its 2014 report, the IPCC concluded, that while projected impacts vary across crops and regions,

[1] Coral Davenport, 'Nations Approve Landmark Climate Accord in Paris' *The New York Times* (Le Bourget, France, 12 Dec 2015), accessed 10 October 2016 at http://www.nytimes.com/2015/12/13/world/europe/climate-change-accord-paris.html?smprod=nytcore-ipad&smid=nytcore-ipad-share&_r=.

crop yields could decrease by more than 25 percent.[2] These changes will take place during the same timeframe that a larger and generally wealthier population will raise the demand for agricultural production by approximately 60 percent.

Never in history has there been an anthropogenic environmental problem as dramatic as the climate change crisis.[3] As an ever-increasing body of scientific research continues to demonstrate the link between anthropogenic carbon emissions and the continued increases in global atmospheric temperatures, the vast majority of the world's scientists concur that climate change is already occurring and will continue to occur far into the future.[4] Scientists project that, in the absence of substantial reductions in global GHG emissions, the expected increase in global temperature will produce significant climatic changes that will lead to a multitude of environmental harms, including continued warming,[5] sea level rise,[6] salt water intrusion, increased precipitation and flooding in some areas,[7] increased incidence of drought in many geographic locales,[8] increased frequency of heat waves,[9]

---

2 See IPCC 'Climate Change 2014: Synthesis Report Summary for Policy Makers' (2014).

3 See Raymond B Ludwiszewski and Charles H Haake, *Climate Change: A Heat Wave of New Federal Regulation and Legislation*, FED. LAW, June 2009, at 32 (explaining that global climate change is currently the top environmental concern). See also see IPCC 'Climate Change 2014: Synthesis Report Summary for Policy Makers' (2014).

4 Intergovernmental Panel on Climate Change, 'Climate Change 2013: The Physical Science Basis, Summary For Policymakers' (2013) 17 (stating that 'It is extremely likely that more than half of the observed increase in global average surface temperature from 1951 to 2010 was caused by the anthropogenic increase in greenhouse gas concentrations and other anthropogenic forces together'), ('IPCC Report'), accessed 10 October 2016 at http://ipcc.ch/pdf/assessment-report/ar5/wg1/WG1AR5_SPM_FINAL.pdf; Endangerment and Cause or Contribute Findings for Greenhouse Gases Under Section 202(a) of the Clean Air Act; Final Rule, 74 Fed. Reg. 66496-66546 (15 December 2009).

5 IPCC Report, above note 4 at 7 states that it is 'virtually certain' (>99% probability of occurrence) that future warming will occur. For an explanation of the probability terminology used in the IPCC report, see IPCC 'Climate Change 2014: Synthesis Report Summary for Policy Makers' (2014) at 2.

6 IPCC Report, above note 4 at 7, It is 'very likely' (99–100% probability of occurrence) that there will be increased incidents of high sea level.

7 It is 'likely over many areas' (66–100% probability of occurrence) that there will be increased heavy precipitation in some areas of the globe. *Id.*

8 It is 'likely' (66–100% probability of occurrence) that there will be an increased number of droughts. *Id.*

9 It is 'very likely' (99–100% probability of occurrence) that there will be an increased number of heat waves. *Id.*

and more intense severe weather events.[10] These impacts will have significant consequences for agriculture and food security. Achieving the goal of global food security in the face of these anticipated dramatic impacts will require innovative and transformative approaches that go well beyond the COP21 commitments to reduce GHG emissions, to create a more climate-resilient agricultural system.

This volume, which focuses on agriculture and climate change, is part of a broader series on climate change. The theme of this volume centers on the goal of achieving a low-carbon climate resilient agricultural system that provides for future food security throughout the developed and developing regions of the globe. We invited top legal, science and social science scholars from around the world to contribute chapters to explore issues related to the types of domestic and international agricultural and environmental laws and policies that address climate change mitigation and adaptation and the challenges of attempting to reconcile these laws and policies with agricultural development, economic growth, food security objectives and sustainable livelihoods. This volume is intentionally interdisciplinary and international in scope and includes contributions representing a range of geographic perspectives, representing both the northern and southern hemispheres.

The contributions to this volume seek to address following issues:

(1) To what extent do domestic agricultural laws and policies in the northern and southern hemispheres address climate mitigation and adaptation?
(2) Are the policies and legal tools in existing national agricultural law frameworks sufficient for human and agro-ecosystem adaptation?
(3) What are the legally relevant challenges to marrying climate mitigation and adaptation with agricultural development, economic growth, food security objectives and sustainable livelihoods?
(4) How do individual country's agricultural law and policy provisions for mitigation and adaptation 'measure up' against the expectations of international and/or regional climate law and policy?
(5) What are the weaknesses, strengths, gaps and innovations in existing agricultural laws and policies?
(6) What new and innovative tools can be employed to address climate change mitigation and adaptation in the agricultural sector?
(7) What are the remaining research needs?

---

10 It is 'more likely than not' (50–100% probability of occurrence) that there will be more intense tropical cyclone events. *Id.*

Following this introductory chapter, Chapter 2 provides Jonathan Verschuuren's broad overview of international law on climate change as it relates to agriculture. In Chapter 3 Lance Gunderson provides a scientific perspective on the complex and dynamic nature of agroecosystems in a broader 'systems' ecology framework. In Chapter 4 Paul Martin then takes the general concept of 'systems thinking' and proposes extending it as a model for climate change governance. The next several chapters address a variety of legal tools for climate change adaptation and mitigation in the context of agriculture: water law (Chapter 5, Robert Adler); invasive species law (Chapter 6, Elodie Le Gal); genetically modified organism law (Chapter 7, Rebecca Bratspies); land use law (Chapter 8, Keith Hirokawa); and intensive animal farming law (Chapter 9, Michelle Nowlin and Emily Spiegel). The volume then turns to a series of chapters related to the roles of trade, finance, markets and government subsidies, including chapters addressing: international trade law (Chapter 10, Christian Häberli); payments for ecosystem services (Chapter 11, JB Ruhl); and government subsidies of biofuels (Chapter 12, Sérgio Sauer, José Paulo Pietrafesa and Pedro Araujo Pietrafesa). Finally in the last chapters the volume explores issues related to legal protection of agricultural community interests in the context of climate change adaptation, including chapters on: adaptation in smallholder agriculture (Chapter 13, Robert Kibugi); agriculture and land use conflicts (Chapter 14, Amanda Kennedy and Amy Cosby); and climate change and land grabbing (Chapter 15, Akachi Odoemene).

2. THE CLIMATE CHANGE–FOOD SECURITY LINK

There is no longer any doubt that human actions have set into motion a transformation of the Earth's climate in a way that humans have never done before. Although the imperative to mitigate the ongoing emissions of greenhouse gases that are at the root of this change continues, there is no longer any doubt that human societies throughout the world will be affected and will have to find ways to adapt to the changes. The anticipated changes are numerous and include temperature increase, carbon dioxide increase, increased frequency and extent of extreme weather conditions, increased drought in many areas and increased severe storm events and flooding in others, increased movement of pest and disease populations, increased pest and disease outbreaks, saltwater intrusion along the coasts and loss of land due to sea level rise loss. The aggregate effects of these changes are predicted to have severe impacts on food security due to significant decreases of agricultural production in many areas, significant food price increases, decreased nutritional value of some foods and an increased risk of health problems due to adulterated and unsafe foods. All of these changes will be occurring against a backdrop of a growing global population and increased food consumption and demands for different food in regions with significant economic growth. Although all aspects of society will be impacted to some extent, perhaps the greatest challenge of these changes will be feeding the growing global population.

The United Nations Food and Agriculture Organization (FAO) defines 'food security' as existing when 'all people at all times have physical or economic access to sufficient safe and nutritious food to meet their dietary needs and food preferences for an active and healthy life'.[11] According to the FAO the four dimensions of food security are: food availability, food accessibility, food utilization and food systems stability. The adverse impacts of climate change, such as increased crop loss from pests, disease or drought, that decrease food production will reduce food availability. Increased food prices and reduced farm incomes will reduce food accessibility. Climate impacts that affect the safety and nutritional value of food will reduce food utilization. Finally climate changes that disrupt food systems due to extreme climate events will reduce food system stability.

Currently the global population exceeds 7 billion people[12] and that number is expected to increase by one-third by 2050.[13] The majority of population growth will occur in developing countries. Population growth coupled with growing income and consumption will result in a need for global food production to increase by 60 percent by 2050. To avoid a food security catastrophe we must find ways to transform our agricultural systems by both finding ways to reduce the substantial contributions of GHG emissions from agriculture to help to mitigate climate change impacts and by finding ways to adapt agricultural systems to the inevitable changes that are, and will continue, to occur regardless of the success of mitigation efforts. Achieving these mitigation and adaptation goals will require, among other things, preserving natural resources and vital ecosystem services, while at the same time transitioning to more productive agricultural production systems that have greater climate resiliency that will enable them to adapt to greater perturbations and long-term climate

---

[11] United Nations Food and Agriculture Organization, *Climate Change and Food Security: A Framework Document* (2008) 3, accessed 10 October 2016 at http://www.fao.org/forestry/15538-079b31d45081fe9c3dbc6ff34de4807e4.pdf.

[12] Sam Roberts, 'U.N. Says 7 Billion Now Share the World' *The New York Times* (31 October 2011), accessed 10 October 2016 at http://www.nytimes.com/2011/11/01/world/united-nations-reports-7-billion-humans-but-others-dont-count-on-it.html?_r=1.

[13] United Nations Food and Agriculture Organization, above note 11 at 42.

variability. Creating a more resilient agricultural system will require a major transformation in the way land, water, soil nutrients and genetic resources are managed, which in turn will require changes in national, local and international laws and policies as well as changes in financial mechanisms.

Climate change and agriculture are closely linked, as each can have dramatic impacts on the other.[14] Industrialized agriculture relies heavily on fossil fuel-intensive farming, processing and distribution systems, which emit greenhouse gases.[15] Throughout the developed world a significant portion of all fossil fuel use is for agricultural activities such as pesticide and fertilizer production and application, mechanized plowing and harvesting, mechanized processing and distribution of farm products via fossil fuel powered transportation systems.[16] The predicted impacts to agriculture due to increase in temperature, altered rainfall patterns and increased extreme weather events are expected to result in substantial reductions in global agricultural production and significant increases in food prices. Reduced food production and higher food prices will further compromise food security concerns in regions of the world which are already food insecure and may cause increased food security problems in other regions.

2.1 Agriculture's Contribution to Climate Change

Agricultural systems are significant contributors to climate change both in terms of GHG emissions and because of conversion of forestland, which can serve as a carbon sink, into cropland. Most of the developed world relies on fossil fuel-heavy industrialized agricultural practices to produce high yields. Chemical pesticides and fertilizers used in industrialized farming are derived from fossils fuels[17] and heavy equipment used

---

[14] Mary Jane Angelo, 'Corn, Carbon and Conservation: Rethinking US Agricultural Policy in a Changing Global Environment' (2010) 17 Geo. Mason L. Rev. 593.

[15] William S Eubanks II, 'A Rotten System: Subsidizing Environmental Degradation and Poor Public Health with Our Nation's Tax Dollars' (2009) 28 Stan. Envtl. L.J. 213, 269–70; Angelo, above note 14.

[16] *See Id.*; Peter Warshall, 'Tilth and Technology: The Industrial Redesign of Our Nation's Soils' in Andrew Kimbrell (ed.), *Fatal Harvest: The Tragedy of Industrial Agriculture* 221, 225 (Foundation for Deep Ecology/Island Press 2002). Jason J Czarnezki and Elisa K Prescott, 'Environmental and Climate Impacts of Food Production, Processing, Packaging, and Distribution' in Mary Jane Angelo, Jason J Czarnezki and William S Eubanks, II (eds), *Food, Agriculture and Environmental Law* (Environmental Law Institute 2013).

[17] Warshall, above note 16 at 221, 225.

for growing, harvesting, transportation and processing use significant amounts of gasoline and diesel fuel.[18] The food sector accounts for approximately 30 percent of the world's total energy consumption and over 20 percent of greenhouse gas emissions.[19]

A particular industrial agricultural practice, concentrated animal feeding operations, is responsible for the release of significant amounts of methane gas, a natural animal waste and one of the most powerful GHGs.[20] In these operations substantial amounts of methane gas are emitted by hundreds or thousands of animals confined to a small areas and fed large quantities of corn and other grains.[21] Methane gas is over 20 times more powerful than carbon dioxide in exerting a greenhouse effect.[22] It is estimated that GHG emissions from livestock account for almost two-thirds of all GHG emissions. Of those two-thirds, 78 percent is in the form of methane. According to FAO estimates, with the right policies and technologies, it is possible to reduce GHG emissions from livestock by about 30 percent.[23]

On another front, in our search for alternative renewable fuels, we may have unwittingly accelerated agriculture's contribution to climate change by substantially increasing corn ethanol production.[24] Between the 1980s and 2007 corn ethanol production in the US increased from approximately

---

[18] Czarnezki and Prescott, above note 16; William S. Eubanks II, 'The Sustainable Farm Bill: A Proposal for Permanent Environmental Change' (June 2009) 39 ELR 10493, 10504; Warshall, above note 16, at 225.

[19] Food and Agriculture Organization of the United Nations, 'Energy-Smart Food for People and Climate' (2011) 49, accessed 10 October 2016 at http://www.fao.org/docrep/014/i2454e/i2454e00.pdf.

[20] *See generally* Hannah MM Connor, 'The Industrialization of Animal Agriculture: Connecting a Model With Its Impacts on the Environment' in Angelo, Czarnezki and Eubanks, *Food, Agriculture and Environmental Law* (2013).

[21] H Steinfeld *et al*, Food and Agriculture Organization of the United Nations, 'Livestock's Long Shadow' (2006) 271.

[22] *Ibid* at xxi.

[23] Gerber, P.J., *et al.*, Food and Agriculture Organization of the United Nations, 'Tackling Climate Change Through Livestock: A Global Assessment of Emissions and Mitigation Opportunities' (2013) xiii, accessed 10 October 2016 at http://www.fao.org/3/i3437e.pdf.

[24] High Level Panel of Experts on Food Security and Nutrition, 'Biofuels and Food Security: A report by the High Level Panel of Experts on Food Security and Nutrition of the Committee on World Food Security' (2013) 50, accessed 10 October 2016 at http://www.fao.org/fileadmin/user_upload/hlpe/hlpe_documents/HLPE_Reports/HLPE-Report-5_Biofuels_and_food_security.pdf; Timothy D Searchinger *et al.*, 'Use of U.S. Croplands for Biofuels Increases Greenhouse Gases through Emissions from Land Use Change' (2008) 319 *Science* 5867, 1238–40.

175 million gallons to almost 6.5 billion gallons.[25] By 2011, 40 percent of all corn grown in the US went to ethanol production, up from less than 1 percent in 1980.[26] Although corn ethanol is considered to be a 'renewable' source of energy,[27] there is widespread doubt that its use will reduce GHG emissions in any meaningful way.[28] In the US, a long running series of federal subsidy programs have incentivized corn ethanol production.[29] In recent years, a renewable fuel standard mandated by the Energy Policy Act of 2005 and the Energy Independence and Security Act of 2007 created a strong market for corn ethanol.[30] These policies promote increased industrialized corn production via large-scale monoculture production. Similar incentives for biofuel production exist in other parts of the world. In Chapter 12, Sérgio Sauer, José Paulo Pietrafesa and Pedro Araujo Pietrafesa describe legal incentives and financial investments that have increased demand for sugarcane as an alternative energy source to replace petroleum. The authors describe GHG emissions from sugarcane biofuel production and use, and describe how this energy source results in loss

---

[25] James A Duffield, Irene M Xiarchos and Steve A Halbrook, 'Ethanol Policy: Past, Present, and Future' (2008) 53 S.D. L. Rev. 425, 425; Karl R Rabago, 'A Review of Barriers to Biofuel Market Development in the United States' (2008) 2 Envtl. & Energy L. & Pol'y J. 211, 212 (describing the remaining barriers to full commercial success for biofuels in the United States).

[26] 'Percentage of Corn Crop Used for Ethanol' accessed 13 November 2016 at http://www.ers.usda.gov/topics/crops/corn/policy/.

[27] 'Growth Energy, About Growth Energy', accessed 10 October 2016 at http://www.growthenergy.org/about-growth-energy/.

[28] Christopher Jensen, 'Caution Flags Raised Over Ethanol Industry's 15 percent Solution' *The New York Times* (10 May 2009).

[29] Wallace E Tyner, 'The US Ethanol and Biofuels Boom: Its Origins, Current Status, and Future Prospects' (2008) 58 *Bioscience* 646, 646; Robert W Hahn, 'Ethanol: Law, Economics, and Politics' (2008) 19 Stan. L. & Pol'y Rev. 434, 437–45 (describing how federal subsidies have driven the development of the ethanol fuel industry in the United States); Gary D. Libecap, 'Agricultural Programs with Dubious Environmental Benefits: The Political Economy of Ethanol' in Rodger E Meiners and Bruce Yandle (eds), *Agricultural Policy and the Environment* (Rowan & Littlefield Publishers 2003) (quoting David Pimentel) 89.

[30] Energy Policy Act of 2005, Pub. L. No. 109-58, 119 Stat. 594; Energy Independence and Security Act of 2007, Pub. L. No. 110-140, 121 Stat. 1492; Mark Holt and Carol Glover, 'Cong. Research Serv., Energy Policy Act of 2005: Summary and Analysis of Enacted Provisions 100' (2006), accessed 10 October 2016 at http://www.congressionalresearch.com/RL33302/document.php?study=Energy+Policy+Act+of+2005+Summary+and+Analysis+of+Enacted+Provisions; Fred Sissine, 'Cong. Research Serv., Energy Independence and Security Act of 2007: A Summary of Major Provisions 6' (2007), accessed 10 October 2016 at http://www1.eere.energy.gov/manufacturing/tech_assistance/pdfs/crs_report_energy_act_2007.pdf.

of biodiversity and deforestation in Brazil. Similarly Akachi Odoemene, in Chapter 15, describes how the push for biofuel production is a major contributor to the 'land grabbing' phenomenon that is occurring in Africa and other parts of the world.

The increased conversion of forestland to agriculture, particularly in the context of land grabbing in parts of the developing world, is another major contributor to climate change. As forested land is burned or otherwise cleared to make way for crops and pasture, carbon preciously contained in forest biomass and organic soils is released into the atmosphere. Although crop plants may absorb carbon, the carbon is typically held in the crop biomass for short periods of time. The land grabbing phenomenon described in Chapter 15 has become a significant driver of land conversion. As Odoemene explains, the 2007–2008 global food crisis brought on by a combination of extreme weather conditions and the global financial crisis, led to a dramatic rise in food prices sparking fears of food insecurity. These fears led countries with net food import balances to look for ways to achieve food security and independence, often by investing in large areas of land in developing countries. For example as China experienced rapid economic growth and its more affluent population demanded more and different foods, it became a net food importer. To guarantee food security and independence the Chinese government, along with private investors, began to purchase enormous areas of arable land in other countries, contributing to deforestation and thus increasing atmospheric carbon.

2.2 Climate Change Impacts on Agriculture

Not only is agriculture a significant contributor to global climate change, but it is also a segment of the economy that will experience some of the most serious impacts of climate change. The IPCC predicts that the most significant impacts of climate change will be to agriculture, particularly in the developing world.[31] The adverse effects of climate change on agriculture will vary greatly depending on geographic location and regional variation in agricultural systems.[32] The most dire concern over the effect of climate change on agriculture is regarding how it will alter crop yield, which is directly linked to food prices and food security. The anticipated impact on crop yield will result from a number of likely changes including

---

[31] Purnamita Dasgupta *et al.*, 'Rural Areas' in *Climate Change 2014: Impacts, Adaptation, and Vulnerability*. (Fifth Assessment Report of the Intergovernmental Panel on Climate Change 2014) 631.

[32] Susan Charles, 'Climate Change: Impacts On Food Safety' (2011) 26 Sum Nat. Resources & Env't 44.

changes in precipitation, in temperature, in sea level rise, in carbon dioxide levels and in crop disease and pest outbreaks.

One of the most significant risks to agriculture will involve water resources. Altered rainfall patterns are expected to cause droughts in some regions and flooding in others.[33] These alterations will likely worsen water shortages in many already water-stressed areas.[34] Increased temperatures will further exacerbate water scarcity, which in turn will result in decreased crop yields.[35] The most dramatic changes in precipitation and temperature are likely to occur in the poorest and most vulnerable regions in the developing world,[36] which are expected to experience disproportionate risks to food security. In Chapter 5 of this volume Robert Adler explores the increasing need for water to grow food for an expanding population with rising affluence and changing food preferences. He explores the impacts of climate change on the relationship between water and agriculture and the degree to which domestic and international water law will need to change to adapt to those impacts.

As the effects of climate change become more prevalent and widespread, the potential to greatly impact global food security will be more apparent.[37] Temperature increases, altered precipitation patterns and greater weather extremes will adversely affect global and regional food production, which will result in reduced food availability and food accessibility. The world's poorest and most vulnerable regions are likely to experience the most dramatic and harmful changes in precipitation and temperature,[38] resulting in disproportionate food security impacts. Exacerbated food insecurity in the poorest regions of the world will present increased challenges as to the ongoing struggle to meet a growing global population's demand for food.

---

33 *Ibid.*

34 *Ibid.*

35 *Ibid.*

36 *Ibid.*

37 The United Nations Food and Agriculture Organization defines 'food security' as existing when 'all people at all times have physical or economic access to sufficient safe and nutritious food to meet their dietary needs and food preferences for an active and healthy life'. The FAO definition encompasses four dimensions: food availability, food accessibility, food utilization, and food systems stability. FAO CC & Food Security, 3.

38 Charles, above note 32 at 44.

3. ACHIEVING FOOD SECURITY AND A CLIMATE RESILIENT AGRICULTURE THROUGH MITIGATION AND ADAPTATION

As the globe warms while simultaneously global populations grow larger and more affluent, agriculture faces three major and interconnected challenges: (1) food security; (2) adaptation to climate change; and (3) climate change mitigation.[39] To adequately address these challenges, radical changes to our food production systems will be necessary. To meet growing food needs in a sustainable manner, we will need to find ways to produce more food with more efficient use of natural resources. Food systems must become more resilient to both the gradual changes and extreme events caused by climate change.

Strategies for addressing climate change can generally be divided into two broad categories: mitigation strategies and adaptation strategies. Mitigation strategies are designed to avoid the unmanageable impacts of climate change by reducing GHG emissions. Agricultural mitigation strategies reduce GHG emissions and sequester carbon through less fossil fuel-dependent more sustainable agricultural practices.[40] Adaptation strategies are designed to manage the unavoidable effects of climate change. In the context of agriculture and food security, these strategies seek to build agricultural systems that can adapt to maintain adequate food production under changing climatic conditions.[41] Ideally sufficient mitigation strategies would be employed to avoid most impacts of climate change. To date most of the strategies that have been employed in the US and other developed countries have fallen under the category of mitigation.[42] Given the modest success of these strategies, research predicts that even if the atmospheric concentrations of GHGs could be stabilized at current levels, impacts on agricultural production

---

[39] Food and Agriculture Organization of the United Nations 2013, 'Climate-Smart Agriculture Source Book' (2013), accessed 10 October 2016 at http://www.fao.org/3/i3325e.pdf.

[40] *Ibid*; United Nations Food and Agriculture Organization, 'Climate Change and Food Security: A Framework Document' (2008) ('FAO Food Security') xi, accessed 10 October 2016 at http://www.fao.org/forestry/15538-079b31d45081fe9c3dbc6ff34de4807e4.pdf.

[41] *Ibid*.

[42] JB Ruhl, 'Climate Change Adaptation and the Structural Transformation of Environmental Law' (2010) 40 Envtl. L. 363, 374–5; Robin Kundis Craig, 'Stationarity is Dead' – Long Live Transformation: Five Principles for Climate Change Adaptation Law' (2010) 34 Harv. Envtl. L. Rev. 9, 18–19.

will continue well into the future.[43] Thus it is critical that policy solutions include both mitigation and adaptation strategies to ensure that agricultural production, and consequently food security, is not threatened by climate change.[44]

3.1 Mitigation Strategies

The US Global Change Research Program defines mitigation as 'options for limiting climate change by, for example, reducing heat-trapping emissions such as carbon dioxide, methane, nitrous oxide, and halocarbons, or removing some of the heat trapping gases from the atmosphere'.[45] Mitigation strategies seek to avoid the unmanageable consequences of climate change by reducing GHG emissions or finding ways to capture or sequester carbon to reduce atmospheric concentrations of GHGs.[46] Effective mitigation measures to keep global temperature increases below 2 degrees C are critical to avoiding the most catastrophic impacts of climate change, including substantial reduction in global food security.[47]

As described above agriculture is a significant contributor to climate change and thus is a logical target for mitigation strategies. Nevertheless as Jonathan Verschuuren describes in Chapter 2, agriculture has played a minor role in the UNFCCC. Neither the Paris Agreement nor the Kyoto Protocol treat agriculture's contribution to climate change in more than a cursory manner. For example under Kyoto only 2.5 percent of registered projects included in the Clean Development Mechanism were related to agriculture. Verschuuren also describes the modest role that agriculture played in the Paris Agreement, but concludes that despite the limited

---

43 Steven K Rose and Bruce A McCarl, American Agricultural Economics Association, 'Greenhouse Gas Emissions, Stabilization and the Inevitability of Adaption: Challenges for US Agriculture' (2008) 23 Choices 15, accessed 10 October 2016 at www.choicesmagazine.org/2008-1/theme/05.pdf.

44 IPCC Report, above note 4 at 17–18.

45 US Global Change Research Program, Global Climate Change Impacts in the United States 10-11 (2009) ('USGCRP, Impact Report'), accessed 10 October 2016 at http://downloads.globalchange.gov/usimpacts/pdfs/climate-impacts-report.pdf.

46 BS Fisher *et al.*, 'Issues Related to Mitigation in the Long Term Context' in B Metz, OR Davidson, PR Bosch, R Dave, LA Meyer (eds), *Climate Change 2007: Mitigation. Contribution of Working Group III to the Fourth Assessment Report of the Inter-governmental Panel on Climate Change 171* (2007), accessed 10 October 2016 at http://www.ipcc.ch/pdf/assessment-report/ar4/wg3/ar4-wg3-chapter3.pdf; FAO Food Security above note 40 at 8; Craig, above note 42, at 19.

47 FAO Food Security, above note 40 at 59.

attention paid to agriculture, ultimately the Paris Agreement will require dramatic reductions in GHG emissions for the agricultural sector to meet its objectives.

In the US, efforts to regulate greenhouse gas emissions generally have been through the implementation of Clean Air Act regulatory programs or other programs targeted at the energy and transportation sectors and have not focused on agricultural activities.[48] To achieve significant reductions in GHGs from agriculture, strategies will need to be implemented that are tailored to address the types of activities that take place in the agricultural world. Such strategies will not fit neatly into programs designed to reduce emissions from the energy, transportation and industrial sectors. Agriculture-targeted mitigation strategies include improved energy efficiency on the farm, reduced use of fossil fuel-derived synthetic fertilizers and pesticides, reduced use of fossil fuels to run heavy farm equipment, transportation and processing equipment, and the creation of local or regional food systems that decrease the distances that agricultural products are transported to their end users.

The FAO has argued in favor of global adoption of agricultural mitigation 'best practices'. These best practices would contribute to the FAO's articulated mitigation goals: reducing CO_2 emissions, reducing methane and nitrous oxide emissions and sequestering carbon.[49] Agricultural mitigation best practices geared toward reducing CO_2 emissions include reducing land conversion and deforestation rates, increasing energy efficiency in the commercial agriculture and agro-industrial sectors and improving control of wildfire and burning associated with agricultural fields.[50] Best practices aimed at reducing methane and nitrous oxide emissions include improving livestock nutrition, increasing management efficiency of wastes from livestock, increasing efficiency of nitrogen fertilizer applications in cultivation and reusing treated wastewater for irrigation.[51]

---

[48] Prevention of Significant Deterioration and Title V Greenhouse Gas Tailoring Rule, 75 Fed. Reg. 31514, 31520 (3 June 2010); Memorandum of 25 June 2013–Power Sector Carbon Pollution Standards 78 Fed. Reg. 126 (1 July 2013), accessed 10 October 2016 at http://www.gpo.gov/fdsys/pkg/FR-2013-07-01/html/2013-15941.htm; Carbon Pollution Emission Guidelines for Existing Stationary Sources: Electric Utility Generating Units 2015, 80 FR 64662. (Following the issuance of the regulation, the rule was challenged by a group of petitioners including electricity companies. On 9 February 2016, the Supreme Court stayed implementation of the Clean Power Plan pending judicial review.)

[49] *Ibid.*

[50] *Ibid.*

[51] *Ibid.*

To achieve the goal of sequestering carbon, FAO suggests best practices such as improved pasture and grazing management in grasslands, improved management of organic matter in soils, minimized mechanical soil disturbance, increased use of marginal lands for planted forests or cellulosic biomass, planting trees for carbon sequestration purposes and the 'introduction of integrated agroforestry systems that combine crops, grazing lands and trees in ecologically sustainable ways'.[52] According to the FAO, these practices could have the collateral benefit of strengthening global food security by increasing market stability, providing agricultural employment opportunities, and boosting the sustainability of vulnerable agricultural systems.[53]

Animal production holds promise for significant mitigation, as Michelle Nowlin and Emily Spiegel explain in Chapter 9, because intensive animal agriculture is a significant emitter of the most potent GHGs. This chapter presents several mitigation measures and regulatory tools that are emerging to reduce GHG contributions from large-scale animal protection.

Another mitigation strategy that has received widespread attention in recent years is a reduction of food waste and loss. Currently approximately one-third of food produced is lost or wasted. From an economic standpoint, this represents a loss of approximately 2.6 trillion dollars.[54] Moreover at a global scale food loss and waste contributes approximately 8 percent of all annual GHG emissions, accounting for approximately 3.3 gigatons of CO_2 equivalents.[55] In the United States in particular, 31 percent (133 billion pounds of the 430 billion pounds) of the available food supply in the United States in 2010 went uneaten.[56]

Reducing food waste can serve as an important mitigation strategy because it will result in a reduction in GHG emissions from producing, processing, transporting and disposing of the wasted food.

---

52 *Ibid.*

53 *Ibid.*

54 FAO, 'If We Had to Pay the Bill to Nature, What Would Food Waste Cost Us?' (FAO, 17 September 2014), accessed 10 October 2016 at http://www.fao.org/zhc/detail-events/en/c/243143/.

55 FAO, 'Food Wastage Footprint: Impacts on Natural Resources' (2013) 17, accessed 10 October 2016 at http://www.fao.org/docrep/018/i3347e/i3347e.pdf.

56 Jean C Buzby, *et al.*, United States Department of Agriculture, 'The Estimated Amount, Value, and Calories of Postharvest Food Losses at the Retail and Consumer Levels in the United States' (2014) 11, accessed 10 October 2016 at http://www.ers.usda.gov/media/1282296/eib121.pdf.

3.2 Adaptation Strategies

According to the US Global Change Research Program, adaptation 'refers to changes made to better respond to present or future climatic and other environmental conditions, thereby reducing harm or taking advantage of opportunity'.[57] Adaptation will be crucial even if significant mitigation strategies are implemented. Based on the assumption that at least some climate change impacts will occur even if atmospheric GHG concentrations are stabilized through mitigation, adaptation strategies focus on strengthening natural and human systems' capacity to withstand changing conditions.[58]

Adaptation strategies include both technological and institutional policy changes. Examples of technological adaptations include improved crop varieties that are resistant to disease, drought or other climate change-induced conditions;[59] more efficient water conservation measures;[60] improved irrigation systems;[61] and development of better weather forecasting systems.[62] Institutional and policy changes discussed typically involve increased governmental support, including in the form of subsidies and incentives for adaptive farming practices.[63]

The FAO has set forth three overarching goals of adaptation for agriculture:

(1) protecting food production systems against increasing variability and extremes in weather patterns;
(2) minimizing disruptive impacts to food supplies from temperature increases and altered precipitation patterns; and

---

57 USGCRP, Impact Report, above note 45.

58 Lindsay F Wiley, 'Mitigation/Adaptation and Health: Health Policymaking in the Global Response to Climate Change and Implications for Other Upstream Determinants' (2010) 38 J.L. Med. & Ethics 629, 631.

59 NV Fedoroff *et al.*, 'Radically Rethinking Agriculture for the 21st Century' (2010) 327 *Science* 833, 833.

60 Anita Wreford *et al.*, Organisation for Economic Co-operation and Development, 'Climate Change and Agriculture: Impacts, Adaptation and Mitigation' (2010) 20, 67.

61 World Bank, World Development Report 2008, 'Adaptation and Mitigation of Climate Change in Agriculture' 2 ('World Development Report), accessed 10 October 2016 at http://siteresources.worldbank.org/INTWDRS/Resources/477365-1327599046334/Brief_AdptMitClimateChng_web.pdf.

62 World Development Report, above note 61.

63 Shikui Dong, *et al.*, 'Vulnerability of Worldwide Pastoralism to Global Changes and Interdisciplinary Strategies for Sustainable Pastoralism' (2011) 16 *Ecology and Society* 10, 24.

(3) protecting natural systems through enhanced provision of environmental services.[64]

Agricultural adaptation practices that would address the first goal include research, development and dissemination of new adaptive crop varieties either through traditional breeding or genetic modification.[65] Practices to meet the second goal include more efficient water management, improved livestock management and increased energy efficiency within the agro-industrial sector.[66] As to the third goal ecosystem provisioning services could be enhanced through practices such as planting forests or other cellulosic biomass on degraded or marginalized land as carbon sinks, implementing watershed protection measures and planting on-farm refugia to promote biodiversity.[67]

Planning, including land use planning, will play an important role in adaption. In Chapter 8 Keith Hirokawa addresses the role of land use planning in adaptation. He describes how impacts of climate change are first experienced at a local level and emphasizes the significant role that local governments have played and continue to play in preparing for climate change. Hirokawa examines several local land use regulatory schemes that address predicted vulnerabilities related to food production and food security.

On a related topic, in Chapter 14 Amanda Kennedy and Amy Cosby address issues related to agricultural land use conflict in the context of climate change. Using a case study from Australia the authors explore the conflicts between agriculture and fossil fuel exploration created by increasing global energy demand. This chapter describes regulatory reforms that have been employed, albeit with modest success, to address these conflicts. The authors assert that existing regulatory instruments that disempower rural stakeholders by giving priority to the economic benefits of extractive industries exacerbate, rather than resolve, these conflicts. The authors conclude by presenting an alternative approach, which views land use conflict through the lens of environmental justice to ensure agricultural interests and the opportunity for meaningful participation in land use decision-making. To ensure food security in the face of climate change, it is critical that adaptation planning efforts focus on reforming agricultural systems on a local level. The United Nations Secretary General Special Report

---

64 FAO Food Security, above note 40 at 31.
65 *Ibid.*
66 *Ibid.*
67 *Ibid.*

concludes that our agricultural system must be reshaped to be regenerative and smallholder centered.[68] In Chapter 13 Robert Kibugi addresses the need for and challenges of reforming smallholder farms in poor rural areas in Africa.

4. CLIMATE RESILIENT AGRICULTURE AS BOTH A MITIGATION AND ADAPTATION STRATEGY

In Chapter 3 Lance Gunderson explores the ways in which agriculture can adapt to climate change using a 'systems' framework, which takes into account the relationships among many elements or components to explain patterns and properties of complex phenomena. Gunderson describes agro-ecosystems as complex adaptive and dynamic systems and explores the need for managing such systems for resilience and transformation. This chapter also discusses the factors that contribute to declining resilience including a decrease in functional diversity and erosion of various forms of capital and limitations on systems capacity for renewal and regeneration in the context of climate change. Building on the 'systems' concept, in Chapter 4 Paul Martin posits that the human challenge of climate change will force adaptation within the law, which will extend beyond the usual regulatory fixes. In this chapter Martin proposes a novel approach to address the complexity of climate change employing 'systems thinking' as a way to address climate governance.

Gunderson and Martin's focus on systems approaches highlight the complex interactions at play in agricultural production and food value chain systems and the need to capitalize on those interactions to build resiliency into those systems. Agricultural systems, as types of ecosystems, can benefit from increased ecological resilience. The concept of ecological resilience captures the strength of redundancies in ecosystems from reinforcing processes and substituting functions provided by the presence of a diversity of species. The redundancies inherent in a diverse ecosystem enable the system to absorb perturbations and persist despite the disruption.[69] In the agricultural context a resilient agro-ecosystem has the ability to continue to function and produce despite disruptions such as increased pest populations, disease or changed rainfall patterns.

---

[68] UN News Centre 'Small Farmers Can Be Major Actors in Reducing Agriculture's Carbon Footprint' (*UN News Centre*, 8 July 2015), accessed 10 October 2016 at http://www.un.org/apps/news/story.asp?NewsID=51358#.VleM2mY4hok.

[69] *Ibid* at 6.

Agricultural ecosystems with high levels of resilience are more likely to be able to withstand a greater range of climate change-induced perturbations than are those lacking resilience.

For an agricultural ecosystem to be resilient several factors must be present. One of the most significant factors is an increase in species richness. Individual species each perform specific limited ecosystem functions. Greater species richness translates into greater functional diversity in the ecosystem.[70] The presence of multiple species that can compensate for the loss of functions provided by other species imbues the systems with the ability to dampen the effects of perturbations, such as those resulting from climate change.[71] Consequently the creation of a resilient agriculture depends on sufficient redundancy built into the systems so that unexpected disturbances, whether anthropogenic or natural, can be absorbed without causing the system to collapse. Introducing a diverse array of species into the agricultural system is the best way to accomplish this goal.

One of the greatest challenges to promoting and enhancing biodiversity in agricultural systems is invasive species. Invasive species can reduce biodiversity by outcompeting numerous native species, thereby dramatically reducing the number and abundance of diverse species. In addition, invasive species that become crop pests may create the need for substantial chemical pesticide use, which can reduce biodiversity by killing the natural diversity of pest predators and parasites that exist in and near croplands. In Chapter 6 Elodie Le Gal provides an in-depth description of the complexity and uncertainty regarding the interactions between climate change and invasive species. This chapter explores the limited efficacy of invasive species law in climate adaptation to protect agriculture from increased numbers of invasive species.

In Chapter 10 Christian Häberli tackles the issue of resiliency from a different standpoint. Rather than look at the resiliency of agricultural production systems, Häberli addresses the resiliency of the food value chain under the present international regulatory framework and concludes that climate change mitigation and adaptation efforts must take into account the circumstances of weak states, small farmers and poor consumers with regard to trade liberalization and foreign investment. Häberli proposes a number of changes that are needed to build resiliency into the food value

---

[70] Garry Peterson, *Contagious Disturbance and Ecological Resilience* (PhD dissertation, University of Florida 1999) 209.

[71] Lance H Gunderson, CS Holling, Lowell Pritchard, Jr and Garry Peterson, 'Resilience of Large-Scale Resource Systems' 9, in Lance H Gunderson and Lowell Pritchard, Jr (eds), *Resilience and the Behavior of Large-Scale Systems* (Island Press 2002).

chain, including multilateral climate-smart best farming and processing practices which must be protected against legal challenges in the WTO. This chapter also highlights the fact that while poor consumers are particularly vulnerable to food prices increases, poor food producers rarely benefit from them. Climate action can yield the highest returns for agriculture as the primary source of economic activity for poor populations.

Current industrialized agricultural practices tend to reduce diversity and therefore are not resilient. In large-scale industrialized agriculture monoculture planting can cover hundreds, if not thousands, of acres. More resilient farming systems introduce and promote diversity both in number of crop types and types of crop varieties. Diverse crops and varieties are planted together in intercropping or rotated and natural refugia are provided on the farm to provide the biological diversity necessary to reduce vulnerability to change. Planting a variety of crop types can provide a safety net in case one crop is compromised by an outbreak of pests or disease. Moreover resilient agro-ecosystems maintain and protect natural populations of beneficial species such as predators and parasites of pest species, pollinators, soil microbes and a diverse array of other organisms.[72] This is accomplished by reducing pesticide inputs that kill beneficial parasites and predators and by providing on-farm conditions that allow a diverse array of species to thrive.

The importance of building resilience into agricultural systems is highlighted by the FAO's 'climate smart' sourcebook.[73] According to the FAO, climate smart agriculture integrates 'the three dimensions of sustainable development (economic, social and environmental) by jointly addressing food security and climate challenges'.[74] It is composed of three inter-related main pillars:

(1) Sustainably increasing agricultural productivity and incomes.
(2) Adapting and building resilience to climate change.
(3) Reducing and/or removing greenhouse gases emissions, where possible.

Goal number (3) can be achieved via the types of mitigation measures discussed in this volume. The adaptation strategies proposed in this volume with help to achieve goal number (1). To meet goal number (2) it will

---

[72] Robin Kundis Craig, 'Stationarity Dead'— Long Live Transformation: Five Principles for Climate Change Adaptation Law' 34 Harv. Envtl. L. Rev. 9 (2010) (describing some activities that promote resilience in agricultural systems).

[73] FAO 2013, 'Climate-Smart Agriculture Source Book' note 39 above.

[74] *Ibid.*

be necessary to integrate mitigation and adaption strategies to develop resilient agricultural systems.

5. CONCLUSION

This volume presents a diverse set of ideas related to climate change and agriculture representing a wide variety of perspectives in terms of field of study and geographic perspective.

Although a wide range of ideas and perspectives are presented, several common themes emerge. Climate change and agriculture are part of a complex web of science, law and policy, which extends from the global scale to the smallholder. Agriculture is a significant contributor to climate change and thus should be considered part of the solution, as well as part of the problem. Consequently changes to agricultural systems that reduce GHG emissions, sequester carbon or put land to use in ways that reduce overall atmospheric carbon can be important tools for climate change mitigation. Conversely agriculture in general, and food security in particular, will suffer serious adverse impacts from climate change even with mitigation measures in place. Accordingly agricultural adaptation strategies targeted at agricultural production will be critical to ensuring food security in the future. Because of the pervasive complexity and uncertainty regarding climate change impacts on agriculture, it will be important to ensure that any adaption efforts employ systems approaches aimed at building resiliency in agricultural production as well as in the entire agricultural value chain. In many cases resilient agricultural systems are comprised of both mitigation and adaptive elements. Thus building more resilient systems will have benefits in reducing the adverse effects of climate change as well as adapting to the inevitable effects that will occur. Although climate change will result in adverse impacts throughout the globe, disproportionate impacts will be felt by the poorest and most vulnerable populations. Regions of the developing world face the greatest threats to food security. Mitigation and adaptation strategies, including regulatory and financial policies must include measures to ensure greater food security for poor and vulnerable populations. This volume provides a number of proposals for climate change mitigation and adaptation aimed at providing food security for a growing population in an era of dramatic changes to the global environment.

2. Climate change and agriculture under the United Nations Framework Convention on Climate Change and related documents

Jonathan Verschuuren[1]

1. INTRODUCTION

Agriculture contributes to climate change to a considerable extent.[2] Agriculture is also among the sectors that will suffer the largest negative impacts of climate change, for which, consequently, huge adaptation efforts are needed.[3] At the same time this sector faces the challenge of feeding a growing world population that is getting richer, leading to a tremendous increase in demand for agricultural products between now and 2050.[4] Yet agriculture has only played a marginal role in the United Nations Framework Convention on Climate Change (UNFCCC)[5] and the

[1] This project has received funding from the European Union's Horizon 2020 research and innovation programme under the Marie Sklodowska-Curie grant agreement No 655565.

[2] In its Fifth Assessment Report, the IPCC's Working Group III concludes that the AFOLU sector (agriculture, forestry and other land use) is responsible for just under a quarter (~10–12 GtCO2eq/yr) of anthropogenic GHG emissions mainly from deforestation and agricultural emissions from livestock, soil and nutrient management. P Smith *et al.*, 'Agriculture, Forestry and Other Land Use (AFOLU)' in Ottmar Edenhofer *et al.* (eds), *Climate Change 2014: Mitigation of Climate Change. Contribution of Working Group III to the Fifth Assessment Report of the Intergovernmental Panel on Climate Change* (Cambridge University Press 2014) 816.

[3] JR Porter *et al.*, 'Food Security and Food Production Systems' in Christopher B Field *et al.* (eds), *Climate Change 2014: Impacts, Adaptation, and Vulnerability. Part A: Global and Sectoral Aspects. Contribution of Working Group II to the Fifth Assessment Report of the Intergovernmental Panel on Climate Change* (Cambridge University Press 2014) 513.

[4] According to the IPCC, all aspects of food security are potentially affected by climate change. Porter *et al.* (n 3) 488.

[5] United Nations Framework Convention on Climate Change (9 May 1992), (1992) 31 ILM 849, entered into force 21 March 1994.

Kyoto Protocol.[6] Because emissions from agriculture have been rising on a yearly basis since 1990,[7] and because the increase in demand for agricultural products (both for food and for biofuels) will cause emissions to rise further, agriculture can no longer be ignored in international negotiations.[8] The Paris Agreement, though, is not expected to drastically change the role of agriculture in the international legal framework on climate change mitigation and adaptation.

This chapter critically assesses the current and potential role of the UNFCCC and related documents, most importantly the Kyoto Protocol, in addressing the combined challenges of reducing greenhouse gas emissions and adapting to changing climate while increasing productivity for the agricultural sector. The chapter first focuses on mitigation, next addresses adaptation and then briefly looks ahead by reviewing the Paris Agreement, which was finalized at the UNFCCC's COP21 in December 2015.[9]

2. AGRICULTURE IN THE UNFCCC AND THE KYOTO PROTOCOL: MITIGATION

Greenhouse gas emissions attributed to the agricultural sector originate from a variety of sources. The UNFCCC and related documents do not deal with all of these emissions in an integrated, sector-specific way. First of all a distinction is made between CO_2 emissions and non-CO_2 emissions. Non-CO_2 emissions are methane and nitrous oxide emissions from livestock and the use of fertilizers, respectively. These can be fairly easily calculated or estimated at national levels (for instance by measuring methane emissions per animal or nitrous oxide emissions per unit of arable land),[10] provided

---

[6] Kyoto Protocol to the United Nations Framework Convention on Climate Change, UN Doc FCCC/CP/1997/7/Add. 1 (10 December 1997), (1998) 37 ILM 22, entered into force 16 February 2005.

[7] Smith *et al.* (n 2) 823.

[8] According to the IPCC's WGIII, 'lack of an international agreement that supports a wide implementation of AFOLU measures can become a major barrier for realizing the mitigation potential from the sector globally'. Smith *et al.* (n 2) 859.

[9] Paris Agreement under the United Nations Framework Convention on Climate Change, FCCC/CP/2015/L.9 (12 December 2015), not yet entered into force.

[10] Peter Wehrheim and Asger Strange Oleson, 'Land Use, Land Use Change and Forestry – How to Enter the Climate Impact of Managing Biospheres and Wood into the EU's Greenhouse Accounting' in G van Calster *et al.* (eds), *Research Handbook on Climate Change Mitigation Law* (Edward Elgar 2015) 304.

uniform emission factors are used.[11] At the individual farm level these emissions are much more difficult to measure, since a variety of factors determine the amount of emissions (such as the diet of individual animals, soil composition, weather systems of individual regions, the way in which fertilizer is applied, and so on).[12] CO_2 emissions from agriculture are even more difficult to estimate because both the emission and the removal of CO_2 affects crops and other vegetation, as it is absorbed from the air. Therefore for practical and political reasons, 'agricultural emissions' under the UNFCCC only includes methane and nitrous oxide emissions.[13] These are called non-land emissions; emissions associated with the application of fertilizer and manure on soils are the only soil-based emissions included in the UNFCCC process.[14] Greenhouse gas emissions from soils through tillage, cropland and grazing land management, or the conversion of land into agricultural land are not considered part of the agricultural sector but are instead part of the LULUCF sector (Land Use, Land Use Change and Forestry).[15] Because this approach risks fragmenting opinion on the contribution of agriculture to climate change, with double counting and inconsistencies, the IPCC in its 2006 guidelines decided to combine its assessments of emissions from agriculture and LULUCF.[16] The IPCC, now under AFOLU (Agriculture, Forestry and Other Land Use), reports on emissions from grassland, cropland, forest land, settlements, wetlands, other land, harvested wood products, liming and urea application, direct and indirect N_2O emissions from managed soils, emissions from biomass burning, enteric fermentation, manure management, rice cultivation and other agricultural activities.

---

11 Andrew Macintosh and Lauren Waugh, *An Introduction to the Carbon Farming Initiative: Key Principles and Concepts* (ANU 2012) 15.

12 Hugh Saddler and Helen King, *Agriculture and Emissions Trading: The Impossible Dream?* (Australia Institute 2008) 102.

13 *Ibid* at 307.

14 *Ibid.*

15 *Ibid.*

16 HS Eggleston *et al.* (eds), '2006 IPCC Guidelines for National Greenhouse Gas Inventories' (*IGES*, 2006), accessed 10 October 2016 at http://www.ipcc-nggip.iges.or.jp/public/2006gl.

3. MITIGATION OBLIGATIONS FOR AGRICULTURAL EMISSIONS UNDER THE UNFCCC AND THE KYOTO PROTOCOL

3.1 Mitigation Policy and Reporting Requirements

The basic commitments in the UNFCCC are not restricted to specific sectors and therefore apply to agriculture. All parties must have (and report to the COP) national inventories of anthropogenic greenhouse gas emissions and policies that address emissions by their sources and removal by sinks of all greenhouse gases not controlled by the Montreal Protocol.[17] The commitment to sustainably manage, conserve and enhance sinks and reservoirs specifically mentions biomass.[18] For Annex I parties, more stringent formulations apply. These states have to adopt national policies and measures on limiting greenhouse gas emissions and protecting sinks and reservoirs,[19] and they have to report

> . . . detailed information on its policies and measures . . . as well as on its resulting projected anthropogenic emissions by sources and removals by sinks of greenhouse gases . . . with the aim of returning individually or jointly to their 1990 levels these anthropogenic emissions of carbon dioxide and other greenhouse gases not controlled by the Montreal Protocol.[20]

The Kyoto Protocol further expands the reporting obligation for Annex I parties, to mandate estimates to be made of changes in carbon stocks since 1990.[21] Later, this was expanded even further to include changes in greenhouse gas emissions by sources and removals by sinks in agricultural soils and land-use changes.[22]

The reports, annually submitted under these provisions, provide a wealth of information on current trends in agricultural emissions, as well as the policies within Annex I countries that are in place to address these emissions in the Annex I countries. Tens of thousands of pages are produced every year. It is impossible to give a full survey of the content of these reports, so this section will briefly summarize the two most important types – national inventory submissions and national communications – and give some examples from their content, focusing on Annex I countries

---

17 Arts 4(1)(a) and (b) UNFCCC.
18 Art 4(1)(d) UNFCCC.
19 Art 4(2)(a) UNFCCC.
20 Art 4(2)(b) UNFCCC.
21 Art 3(4) KP.
22 Art 3(4) KP.

since the most detailed requirements apply to them. The most important reports are national inventory submissions and national communications.

National inventory submissions consist of the National Inventory Report (NIR) and the Common Reporting Format (CRF). NIRs contain detailed descriptive and numerical information while the CRFs contain summary, sectoral and trend tables for all greenhouse gas emissions and removals.[23] Individual chapters within the submission are devoted to agriculture and to LULUCF, and are (partly) submitted separately because the specific methodology for these categories was developed after the Kyoto Protocol was signed. The EU's 2014 National Inventory Report (dealing with emissions from 1990–2012) includes almost 300 pages on agriculture and LULUCF (more than on any other sector, including energy).[24] The report shows that direct soil emissions of nitrous oxide have been reduced, as well as indirect methane emissions from cattle, caused by the decreasing use of fertilizer and manure. Declining cattle numbers in most EU Member States could also play a part, owing to production quotas[25] and legislation aimed at reducing nitrates in soil and water.[26] Australia, for example, reports that emissions from agriculture decreased by 1 percent between 1990 and 2012, but had increased by 3.3 percent between 2011 and 2012.[27] Their 2012 NIR report indicates that these variations were not caused by legislation:

> From 2002 until 2010 livestock populations declined in response to the prolonged drought conditions which occurred over extensive areas of southern and eastern Australia. Since late 2010 better seasonal conditions have seen farmers begin to rebuild their herds. Between 2011 and 2012, emissions from enteric fermentation increased by 2.0%.[28]

---

23 All reports are available online through UNFCCC 'National Inventory Submissions 2014', accessed 10 October 2016 at http://unfccc.int/national_reports/annex_i_ghg_inventories/national_inventories_submissions/items/8108.php.

24 European Environment Agency (EEA), *Annual European Union Greenhouse Gas Inventory 1990–2012 and Inventory Report 2014* (EEA 2014) Technical Report No 09/2014.

25 Under the EU's Common Agriculture Policy, the milk quota system lead to a strong reduction of animal numbers in the dairy sector, and as consequence also to a reduction of GHG emissions from agriculture.

26 *Ibid* at 529. From the various legal instruments that apply to livestock keeping, the Nitrates Directive (Council Directive 91/676/EEC of 12 December 1991 concerning the protection of waters against pollution caused by nitrates from agricultural sources [1991] OJ L375/1) is considered to have had the largest impact on greenhouse gas emissions from agriculture. This directive aims at reducing water pollution caused by nitrates from agricultural sources.

27 Australian Department of the Environment, *National Inventory Report 2012: Volume 1* (Commonwealth of Australia 2014) 257.

28 *Ibid.*

National communication reports (NCs) detail any policies and measures the country has adopted and their mitigation impact.[29] The NCs submitted by Annex I countries in 2014 are in striking contrast to the accompanying NIRs, as far as agricultural sector information is concerned. For each of the Annex I states, the NIRs contain extremely detailed information on all agricultural sources of emissions, including those associated with land use under LULUCF. In the NCs however, policies and measures receive little attention or are almost absent. The Netherlands' report, for example, explicitly states that no sectoral reduction targets will be imposed on methane and nitrous oxide until 2020. Instead the agricultural sector 'is expected to take cost-effective measures that contribute to emission reductions of greenhouse gases on a voluntary basis,' for instance through adapted cattle feed.[30]

The EU's NC primarily focuses on future impacts of a yet to be fully redesigned Common Agriculture Policy (CAP), especially regarding the second pillar of the CAP, aimed at rural development.[31] However the first pillar of the CAP, aimed at market and income support, states that 'no comprehensive quantified estimates of the impact of current and future actions on greenhouse gas emissions are available.'[32]

3.2 Sector Specific Mitigation Commitments

Comparing the reports on agricultural emissions with those on policies to reduce those emissions demonstrates that under the Kyoto Protocol no sector specific commitments were adopted, nor was LULUCF fully included under the general emission reduction commitment of Article 3(1).[33] Annex A to the Kyoto Protocol lists greenhouse gases that are particularly relevant for agriculture (that is, methane and nitrous oxide). Agriculture is also listed under the sectors and source categories to which the protocol applies ('enteric fermentation, manure management, rice cultivation, agricultural soils, prescribed burning of savannas, field burning of agricultural

[29] All reports are available online through UNFCCC, 'Submitted National Communications', accessed 10 October 2016 at http://unfccc.int/national_reports/annex_i_natcom/submitted_natcom/items/7742.php.

[30] Netherlands Ministry of Infrastructure and the Environment, *Sixth Netherlands National Communication under the United Nations Framework Convention on Climate Change* (Min I&M 2013) 72.

[31] European Commission, *Sixth National Communication and First Biennial Report from the European Union under the UN Framework Convention on Climate Change* (European Commission 2014) 375.

[32] *Ibid* at 374.

[33] Wehrheim and Oleson (n 10) 307, 309.

residues and other').[34] Taking mitigation measures that specifically target the agricultural sector, however, is left to the discretion of the Annex I Parties to the Protocol. The Protocol explicitly mentions 'promotion of sustainable forms of agriculture in light of climate change considerations' as an element of national mitigation policies in Annex I countries that need to be 'implemented and/or further elaborated.'[35] However, promoting sustainable forms of agriculture is a rather general aim that does not necessarily lead to a large reduction of greenhouse gas emissions. For non-Annex I states, Article 10 of the Kyoto Protocol 'reaffirms existing commitments' under Article 4(1) of the UNFCCC by stating that they shall '(f)ormulate, implement, publish and regularly update national and, where appropriate, regional programmes containing measures to mitigate climate change . . . (i) Such programmes would, inter alia, concern . . . agriculture . . .'[36]

In most countries, agriculture is not included in binding mitigation schemes. The EU's emissions trading system (ETS), for instance, does not include agricultural emissions.[37] Including farming in an ETS is difficult because of the measuring problem mentioned above. Usage of uniform emission criteria could solve this problem, but also diminishes the incentives for farmers to change their individual behaviour.[38] Exceptions do exist. California's Cap-and-Trade Regulation (2012) includes projects that are eligible to generate carbon credits, like avoiding methane emissions from installation of anaerobic digesters on farms and carbon sequestration in rural forestry.[39] Australia's Carbon Farming Initiative Act (2011) allows farmers to obtain tradable carbon credits for a range of agricultural emissions avoidance projects, such as avoidance of: methane emissions from livestock's digestive tracts; methane or nitrous oxide emissions from decomposition of livestock urine or dung; methane from rice

---

34 Annex A KP.

35 Art 2(1)(a)(iii) KP.

36 Art 10(b)(i) KP.

37 Wehrheim and Oleson (n 10) 313. Emissions from agriculture have been included in the Effort Sharing Decision (Decision 406/2009/EC of the European Parliament and of the Council of 23 April 2009 on the effort of Member States to reduce their greenhouse gas emissions to meet the Community's greenhouse gas emission reduction commitments up to 2020 [2009] OJ L140/136), which means that agricultural emissions, together with emissions from other non-ETS sectors, are subject to an overall reduction target for each Member State. Individual Member States are free to choose how and where they are achieving this overall target. As a consequence, the extent to which agricultural emissions are cut following domestic policies and measures varies significantly among the Member States.

38 Macintosh and Waugh (n 11) 15.

39 Smith *et al* (n 2) 865.

fields or plants; methane or nitrous oxide from the burning of savannahs or grasslands methane or nitrous oxide from the burning of crop stubble in fields, crop residues in fields or sugar cane before harvest and methane or nitrous oxide from soil, as well as through sequestration offsets projects.[40] New Zealand's ETS requires farmers in certain sectors (meat processors, dairy processors, nitrogen fertilizer manufacturers and importers, and live animal exporters) to report biological emissions of methane and nitrous oxide that are produced on-farm. The original plan to require them to surrender allowances under the ETS was dropped in 2012.[41]

The clean development mechanism (CDM) includes projects aimed at reducing agricultural emissions, but only to a limited extent.[42] In 2013 a mere 2.5 percent of the registered projects were related to agriculture.[43] Only projects aimed at methane avoidance, biogas projects and agricultural residues for biomass are accepted under the CDM. As a consequence, the IPCC concluded that finance streams coming from agriculture projects are marginal from a global perspective.[44]

Within the frameworks of the UNFCCC and Kyoto Protocol, not much attention is paid to mitigating greenhouse gas emissions from agriculture. The focus remains on adaptation, and is further discussed below. In 2008 the Secretariat of the UNFCCC published a technical paper on mitigation challenges and opportunities in the agriculture sector, in which it gave an overview of mitigation practices and identified the relative mitigation potential of each of these practices, as well as methodological challenges and possible technical barriers to their implementation.[45] The report was discussed at a workshop of the UNFCCC's Ad-Hoc Working Group on Long-Term Cooperative Action in 2009,[46] but did not lead to many concrete actions. The workshop concluded, however, that nationally appropriate mitigation actions (NAMAs) should be implemented across

---

40 Macintosh and Waugh (n 11) 15; Celeste Black, 'Linking Land Sector Activities to Emissions Trading: Australia's Carbon Farming Initiative' in Larry Kreiser *et al.*, *Critical Issues in Environmental Taxation Volume XI – Carbon Pricing, Growth and the Environment* (Edward Elgar 2012). See further Chapter ___ of this volume.

41 Climate Change Response (Emissions Trading and Other Matters) Amendment Act 2012 (New Zealand).

42 See Donald F Larson, Ariel Dinar and J Aapris Frisbie, *Agriculture and the Clean Development Mechanism – Policy Research Working Paper 5621* (World Bank 2011).

43 Smith *et al.* (n 2) 864.

44 *Ibid.*

45 FCCC/TP/2008/8 (21 November 2008).

46 FCCC/AWGLCA/2009/CRP.2 (7 April 2009).

the agricultural sector, as 70 percent of the economic potential for mitigation exists in developing countries.[47] If financial support is generated, either under the Green Climate Fund, bilaterally or through private investors,[48] Developing Country Parties to the UNFCCC and the Kyoto Protocol will be enabled to scale up mitigation (and adaptation) activities in the agricultural sector.[49] A recent compilation of information on NAMAs by the UNFCCC's Subsidiary Body for Implementation actually shows that developing countries, such as African countries working together as the African Group, strongly focus on mitigation measures in the agricultural sector.[50]

3.3 Technology Transfer for Mitigation

Common but differentiated responsibilities, as laid down in Article 3(1) of the UNFCCC, constitute a basic principle of international climate law. Various requirements in the UNFCCC and the Kyoto Protocol employ this principle, most notably the provisions on transferring mitigation technologies to developing countries. Article 4(1) of the UNFCCC generally requires all Parties to:

> [p]romote and cooperate in the development, application and diffusion, including transfer, of technologies, practices and processes that control, reduce or prevent anthropogenic emissions of greenhouse gases . . . in all relevant sectors, including the . . . agriculture . . . sector[s].[51]

Developed Country Parties, particularly Annex II states, are even required to 'take all practicable steps to promote, facilitate and finance, as appropriate, the transfer of, or access to, environmentally sound technologies and knowhow to other Parties, particularly Developing Country Parties.'[52] Although this provision imposes a legally binding duty upon developed states, implementation is generally lacking. As a consequence, the topic of technology transfer has gained much attention. In 2007 the Bali Action Plan identified technology development and transfer as a priority area.[53]

---

47 *Ibid* at 2.

48 Mihoko Kawamura and Kenji Utsunomiya, *NAMA Finance Study Examples from the UNEP Bilateral Finance Institutions Climate Change Working Group* (UNEP 2012).

49 *Ibid.*

50 FCCC/SBI/2013/INF.12/Rev.3 (19 January 2015) 4–6.

51 Art 4(1)(c) UNFCCC. See also Art 10(c) KP.

52 Art 4(5) UNFCCC.

53 FCCC/CP/2007/6/Add.1 (14 March 2008).

In 2010 a Technology Mechanism was established 'to facilitate the implementation of enhanced action on technology development and transfer, to support action on mitigation and adaptation, in order to achieve the full implementation of the Convention.'[54]

Despite these recent international developments, developing countries still have difficulty obtaining clean technologies from developed countries, mostly because of the less attractive international investment conditions in these countries and the lack of technical capacity and training systems.[55] The protection of intellectual property rights could be another potential barrier to technology transfer.[56] Polarized debates between some developed and developing countries have prevented meaningful discussions about intellectual property rights from taking place under the UNFCCC.[57] However, the WTO's IPR law, as laid down in the Agreement on Trade Related Aspects of Intellectual Property Rights (TRIPS), actually favours greater access to agricultural technologies and practices, while at the same time protecting intellectual property rights.[58] Nevertheless, transferring technology to developing countries – in practice – still is problematic.[59]

3.4 Relationship to Food Security

We live in a world that is increasingly food-insecure. Between now and 2050, the demand for agricultural products will increase sharply. Global food production needs to increase by 40 percent to meet growing demand, mainly because of population growth (the world's population will grow from 7 billion today to 9 billion in 2050), and a 60 percent rise in global caloric intake (due to greater affluence, particularly in countries like China

---

[54] FCCC/CP/2010/7/Add.1 (15 March 2011).

[55] Smith *et al.* (n 2) 1236.

[56] Matthew Rimmer, *Intellectual Property and Climate Change. Inventing Clean Technologies* (Edward Elgar 2011) 39–82; Markus W Gehring, Marie-Claire Cordonier Segger and Jarrod Hepburn, 'Climate Change and International Trade and Investment Law' in Rosemary Rayfuse and Shirley V Scott (eds), *International Law in the Era of Climate Change* (Edward Elgar 2012) 98.

[57] Bruce Campbell *et al.*, *Agriculture and Climate Change: A Scoping Report* (Meridian Institute 2011) 39.

[58] Art 7 of the Agreement on Trade Related Aspects of Intellectual Property Rights (TRIPS Agreement) (Marrakesh Agreement Establishing the World Trade Organization 1869 UNTS 299, 33 ILM 1197 (1994)) Annex IC.

[59] Robert W Adler, 'Climate Change Adaptation and Agricultural and Forestry Law' in Jonathan Verschuuren (ed.), *Research Handbook on Climate Change Adaptation Law* (Edward Elgar 2013) 242–3. Rimmer, however, argues that the flexibilities of the TRIPS Agreement thought to enable access to agricultural technologies and practices are overrated: Rimmer (n 56) 87.

and India).[60] Climate change negatively impacts food production, so much that the rise in production will be difficult to achieve. By 2050, 56 percent of crops in Sub-Saharan Africa and 21 percent of crops in Asia will experience negative effects of climate change, including shifts in water availability, temperature shifts and changes in the occurrence of pests.[61] To make things worse, a rise in agricultural production would lead to further increases in greenhouse gas emissions from agriculture.[62] In roughly the same period of time, however, global greenhouse gas emissions must sharply decrease to meet the UNFCCC's goal of a less than 2°C rise in global temperature.

Firm mitigation policies could, therefore, negatively affect food production. The IPCC noted that, although feedback between greenhouse gas reduction and food security is not completely understood,[63] large scale biomass energy supplies and carbon sequestration in the AFOLU sector both provide important mitigation measures, but also have potential implications for food security.[64] The large scale use of bioenergy threatens food security in Africa because productive lands for sustainable food production are used instead to produce biofuels.[65] The 2007/2008 global food price spikes are believed to have been partially caused by the rise in biofuel production.[66] Conventional agriculture will also face price increases from emission caps or pricing mechanisms placed upon the use of fuels and fertilizers, as agriculture is heavily energy dependent not only in the developed world, but also increasingly in Latin America and Asia.[67]

The above indicates why Article 2 of the UNFCCC is an important provision when discussing climate policies for the agricultural sector. Article 2 states that the objective of the Convention, and any related legal instruments that the Conference of the Parties may adopt to prevent dangerous anthropogenic interference with the climate system, has to be 'achieved within a time-frame sufficient . . . to ensure that food production is not threatened . . .'. As a consequence, mitigation policy development has to include food security objectives. The concept of 'climate smart agriculture'

---

60 Campbell *et al* (n 57) 1.
61 *Ibid* at 2.
62 *Ibid* at 3.
63 Smith *et al.* (n 2) 837.
64 *Ibid* at 816.
65 *Ibid* at 854.
66 International Centre for Trade and Sustainable Development (ICTSD) and International Food and Agricultural Trade Policy Council, *ICTSD-IPC Platform on Climate Change, Agriculture and Trade: Considerations for Policymakers* (ICTSD 2009) 2.
67 *Ibid.*

does just that: it is an approach to developing technical, policy and investment conditions that will achieve sustainable agricultural development for food security under climate change.[68] Climate smart agriculture focuses on three targets at the same time: sustainably increasing agricultural productivity and incomes, adapting and building resilience to climate change, and reducing and/or removing greenhouse gas emissions where possible.[69] So far, most attention for climate smart agriculture has focused on developing countries, but one can expect this concept to emerge as the leading policy approach worldwide.

3.5 Relationship to Trade[70]

The UNFCCC, in Article 3(5), provides that mitigation measures should not constitute a means of arbitrary or unjustifiable discrimination or a disguised restriction on international trade. Similarly, the Kyoto Protocol states that Annex I Parties, in their mitigation policies, should minimize adverse effects on international trade.[71] These provisions are relevant for agriculture, as the food market is a global market that is severely affected by climate change, as just discussed. This is also a market that is heavily influenced by the provision of trade-distorting subsidies in developed countries,[72] which is important for mitigation as current financial support schemes for farmers may influence the level of their greenhouse gas emissions. A singular focus on production will likely lead to increased use of fertilizers, for example, and may constrain climate smart agriculture by not financially rewarding farmers for switching to agricultural practices aimed at climate change mitigation and adaptation.[73]

In several countries, income support policies are slowly including climate change factors. In the initial proposals to reform the EU's Common Agricultural Policy, the European Commission suggested earmarking 30 percent of direct payments for farmers who apply agricultural practices beneficial to the environment (through crop diversification, maintenance of permanent pasture, the preservation of environmental reservoirs and landscapes, and so on). In addition, the Commission proposed giving increased financial support to agri-environment-climate projects

---

68 FAO, *Climate Smart Agriculture Sourcebook* (FAO 2013) ix.
69 *Ibid.*
70 For more detail on international trade law, see Chapter 10 of this volume.
71 Art 2(3) KP.
72 Adler (n 59) 239.
73 *Ibid* 240.

and organic farming under the EU's rural development policy.[74] In the final stages of the negotiations, however, these proposals were watered down considerably.[75] Yet Regulation (EU) 1307/2013 offers a first significant attempt to promote more climate-friendly agriculture, by introducing measures to implement crop diversification, the preservation of permanent grassland and ecological focus areas.[76] Still, much more rigorous and overarching policies are needed to encourage European farmers to adopt climate smart agricultural practices.

Under the WTO, reducing market distortions caused by income support to farmers has been discussed for years now, albeit without significant progress toward the liberalization of trade in agricultural products.[77] Global trade in biofuels has caused a number of trade disputes about subsidies and non-tariff barriers.[78] International trade law also imposes limits on the transfer of technology, as was explained earlier.

Generally, the relationship between the UNFCCC and the WTO is troublesome. Adler describes it as a chicken and egg problem: the UNFCCC and the Kyoto Protocol prohibit climate policies that restrict international trade (using WTO terminology), while within the WTO, officials and negotiators wait for a comprehensive agreement to replace the Kyoto Protocol before addressing the relationship between the climate change regime and trade law.[79] Campbell *et al.* conclude that, on the one hand, climate change mitigation policies and measures related to agricultural trade remain ungoverned by the UNFCCC, while on the other hand, international

---

[74] *Commission Staff Working Document, Principles and recommendations for integrating climate change adaptation considerations under the 2014–2020 rural development programmes*, SWD (2013) 139 Final, 8–9.

[75] David Baldock and Kaley Hart, 'A Greener Cap: Still within Reach?' (2013), accessed 10 October 2016 at http://www.ieep.eu/assets/1183/Trilogue_Paper_-_April_2013_Final.pdf.

[76] Regulation (EU) No 1307/2013 of the European Parliament and of the Council of 17 December 2013 establishing rules for direct payments to farmers under support schemes within the framework of the common agricultural policy and repealing Council Regulation (EC) No 637/2008 and Council Regulation (EC) No 73/2009, [2013] OJ L347/608.

[77] See Campbell *et al* (n 57) 38. See also Chapter 10 of this volume.

[78] Smith *et al.* (n 2) 867. See extensively Gehring, Cordonier Segger and Hepburn (n 56) 84–117.

[79] Adler (n 59) 239. See on the relationship of the Kyoto Protocol and the WTO, Robert Howse and Antonia L Eliason, 'Domestic and International Strategies to Address Climate Change: An Overview of the WTO Legal Issues' in Thomas Cottier, Olga Nartova and Sadeq Z Bigdeli (eds), *International Trade Regulation and the Mitigation of Climate Change. World Trade Forum* (Cambridge University Press 2009).

trade law may support mitigation measures but has not been designed for that purpose.[80] They too think it is important that a new international climate change agreement be reached, so that countries have no excuse to apply unilateral or plurilateral measures that could lead to trade disputes. These measures, by some countries, are considered discriminatory by other countries and, therefore, are not covered by the general exception under Article XX of the WTO's Global Agreement on Tariffs and Trade.[81]

4. AGRICULTURE IN THE UNFCCC AND THE KYOTO PROTOCOL: ADAPTATION

4.1 The Need for Adaptation

The second pillar of the UNFCCC is adaptation. For agriculture, adaptation is essential. In its Fifth Assessment Report, the IPCC finds that for the major crops in tropical and temperate regions (wheat, rice and maize), climate change without adaptation will negatively impact production with local temperature increases of 2°C or more, although individual locations may benefit.[82] Negative yield impacts for all crops past 3°C of local warming without adaptation are to be expected, even with benefits of higher CO_2 and rainfall (both positively affecting plant growth).[83] There is high confidence that irrigation demand will increase significantly in many areas (by more than 40 percent across Europe, USA and parts of Asia).[84] A wide range of adaptation measures is necessary. According to the IPCC, effective crop adaptation is critical in enhancing food security and sustainable livelihoods, especially in developing countries.[85] Crop adaptation includes altering cultivation and sowing times, crop cultivars and species, and marketing arrangements.[86] When focusing on water availability, switching to more appropriate crop varieties (drought resistant, salt resistant, low water demand) improves irrigation efficiency and reduces demand for irrigation water. Reusing wastewater to irrigate crops is also an important adaptation measure.[87]

80 Campbell *et al.* (n 57) 44.
81 *Ibid.*
82 Porter *et al* (n 3) 488.
83 *Ibid* at 505.
84 *Ibid* at 251.
85 *Ibid* at 514.
86 *Ibid.*
87 *Ibid* at 255.

4.2 Adaptation Obligations under UNFCCC and the Kyoto Protocol[88]

4.2.1 Adopt adaptation policy: NAPAs

All Parties to the UNFCCC have to formulate, implement, publish and regularly update national and regional programs containing measures to facilitate adequate adaptation to climate change.[89] Since 2001 the development of such domestic policies has been guided under the UNFCCC through the guidelines for the preparation of National Adaptation Programmes of Action (NAPA), particularly with the aim of assisting the least developed countries in the process.[90] By 2015, 50 of the least developed countries had submitted their NAPAs, containing detailed priority policies to address their urgent and immediate need for climate change adaptation.[91] Each NAPA reports on the urgent and immediate needs for which further delay could increase vulnerability to climate change or lead to increased costs at a later stage. Almost all primarily focus on agriculture. The Malawi NAPA, for instance, states that Malawi relies on rain-fed agriculture and that droughts have resulted in poor crop yields or total crop failure, leading to serious food shortages, hunger and malnutrition. In addition, floods have also severely disrupted food production. The Malawi NAPA proposes improved early warning systems, improved crop varieties, improved livestock breeds, and improved crop and livestock management practices.[92] Nepal's NAPA, to give another example, formulates extensive policies to increase communities' resilience in the field of agriculture,[93] and proposes very concrete measures to enhance resilience in agriculture and

---

88 See, generally, Jonathan Verschuuren, 'Climate Change Adaptation under the United Nations Framework Convention on Climate Change and Related Documents' in Jonathan Verschuuren (ed), *Research Handbook on Climate Change Adaptation Law* (Edward Elgar 2013). Some paragraphs in s 3.2 are based upon texts from this source.

89 Art 4(1)(b) UNFCCC.

90 Decision 28/CP. 7, FCCC/CP/2001/13/Add. 4, 7–13.

91 The secretariat of the UNFCCC makes all of these available through their website, UNFCCC, 'NAPAs Received by the Secretariat', accessed 10 October 2016 at http://unfccc.int/cooperation_support/least_developed_countries_portal/submitted_napas/items/4585.php.

92 Malawi Ministry of Mines, 'National Adaptation Programmes of Action, (Ministry of Mines, Natural Resources and Environment, 2006) ix, accessed 10 October 2016 at http://unfccc.int/cooperation_support/least_developed_countries_portal/submitted_napas/items/4585.php.

93 Government of Nepal, 'National Adaptation Programme of Action (NAPA) to Climate Change' (2010), accessed 10 October 2016 at http://unfccc.int/cooperation_support/least_developed_countries_portal/submitted_napas/items/4585.php.

food security, such as tree planting around farm lands,[94] better rain water collection, changing crops (replacing potatoes, moving to short duration crops), implementing drip irrigation, and so on.[95] Adopting a NAPA is an important prerequisite for developing countries to receive financial aid as further discussed below.

4.2.2 Reporting on adaptation

Some of the reporting requirements mentioned above are equally relevant for adaptation. Under the Kyoto Protocol, Annex I parties have to submit information on their adaptation policies for various sectors (including agriculture) in their national communication; other parties can do so voluntarily.[96] The NCs indeed pay attention to this. The Australian Sixth NC, for example, reports that research is being carried out to test responses of key Australian crops to increases in temperature and carbon dioxide; monitor ways to manage heat stress in livestock; examine potential new shrub-based forages for livestock; and evaluate the relocation of various crops to northern Australia.[97] Their Intergovernmental Agreement on National Drought Program Reform also includes measures to better support farmers and their families as they prepare for droughts and other challenges.[98] According to the NC, the new approach will include: a farm household support payment, promoting Farm Management Deposits and taxation measures, a national approach to farm business training, a coordinated, collaborative approach to the provision of social support services, and tools and technologies to inform farmer decision making.[99]

4.2.3 Cooperation on adaptation

Under the UNFCCC, all parties have the obligation to 'cooperate in preparing for adaptation to the impacts of climate change; develop and elaborate appropriate and integrated plans for . . . water resources and agriculture, and for the protection and rehabilitation of areas, particularly in Africa, affected by drought and desertification, as well as floods'.[100] This provision also refers to the principle of common but differentiated

---

94 *Ibid* at 69.
95 *Ibid* at 68.
96 Arts 10(b)(ii) and 7 KP.
97 Department of Industry, Innovation, Climate Change, Science, Research and Tertiary Education, *Australia's Sixth National Communication on Climate Change* (Commonwealth of Australia 2013) 176.
98 *Ibid* at 177.
99 *Ibid*.
100 Art 4(1)(e) UNFCCC.

responsibilities, and to the specific national and regional development priorities, objectives and circumstances.[101] Consequently, the focus for agricultural adaptation is mostly on developing countries. As described above, the Kyoto Protocol more explicitly states that national and regional programmes containing measures that facilitate adequate adaptation for various sectors are to be formulated, implemented, published and regularly updated.[102]

It was only through a 2011 COP decision that the cooperation process referred to in Article 4(1)(e) UNFCCC began. In this decision, the COP requested the Subsidiary Body for Scientific and Technological Advice (SBSTA) to consider issues relating to agriculture, with the aim of exchanging views.[103] Such an exchange of views has been initiated, but continues to proceed at a rather slow pace.[104] The process does not have a very ambitious goal ('exchanging views'). By 2014 the SBSTA concluded that it would undertake scientific and technical work in the following areas:

> (a) Development of early warning systems and contingency plans in relation to extreme weather events and its effects such as desertification, drought, floods, landslides, storm surge, soil erosion, and saline water intrusion; (b) Assessment of risk and vulnerability of agricultural systems to different climate change scenarios at regional, national and local levels, including but not limited to pests

---

[101] Art 4(1) UNFCCC.

[102] Art 10(b)(i) KP.

[103] FCCC/CP/2011/9/Add.1, Decision 2/CP.17.

[104] SBSTA 36 initiated the exchange of views on issues relating to agriculture (FCCC/SBSTA/2012/L.19, 25 May 2012). SBSTA 37 (December 2012) continued its exchange of views on issues relating to agriculture. SBSTA 38 invited Parties and admitted observer organizations to submit to the secretariat their views on the current state of scientific knowledge on how to enhance the adaptation of agriculture to climate change impacts while promoting rural development, sustainable development and productivity of agricultural systems and food security in all countries, particularly in developing countries (FCCC/SBSTA/2013/L.20, 13 June 2013). SBSTA 39 (November 2013), acknowledged with appreciation the rich exchange of views by Parties during the in-session workshop on the current state of scientific knowledge on how to enhance the adaptation of agriculture to climate change impacts while promoting rural development, sustainable development and productivity of agricultural systems and food security in all countries, particularly in developing countries, taking into account the diversity of the agricultural systems and the differences in scale as well as possible adaptation co-benefits, and agreed to consider at SBSTA 40 the report of this workshop (FCCC/SBSTA/2013/L.35, 16 November 2013). All the SBSTA documents as well as the various submissions on the topic of adaptation in the agricultural sector by the Parties and observer organisations are available at: UNFCCC, 'Subsidiary Body for Scientific and Technological Advice (SBSTA)', accessed 10 October 2016 at http://unfccc.int/bodies/body/6399/php/view/documents.php.

> and diseases; (c) Identification of adaptation measures, taking into account the diversity of the agricultural systems, indigenous knowledge systems and the differences in scale as well as possible co-benefits and sharing experiences in research and development and on the ground activities, including socio-economic, environmental and gender aspects; (d) Identification and assessment of agricultural practices and technologies to enhance productivity in a sustainable manner, food security and resilience, considering the differences in agro-ecological zones and farming systems, such as different grassland and cropland practices and systems.[105]

This process will take a few more years, as the SBSTA invited parties and admitted observer organizations to submit their views on issues relating to the elements referred to in (a) and (b) by 25 March 2015, and (c) and (d) by 9 March 2016, for consideration at SBSTA 44 and 45, which will take place in 2016.[106]

In 2013 the SBSTA organized a workshop on the current state of scientific knowledge on how to enhance the adaptation to climate change while promoting rural development, sustainable development and productivity of agricultural systems in all countries, particularly in developing countries.[107] The workshop found that adaptation of the agriculture sector is an immediate need and a priority.[108] Developing countries expressed their need for financial support, knowledge and technology sharing, and a strengthened institutional capacity.[109] The UNFCCC and the Kyoto Protocol both outline the need for these services, but the fact that these topics were still being discussed at such a basic level in 2013 underlines the slow progress that has been made. The transfer of technology and financial assistance will be explored in more detail below.

4.2.4 Technology transfer for adaptation

Article 4(1)(c) of the UNFCCC (discussed above) on technology transfer only focuses on mitigation technologies. Article 4(5), however, is relevant both for mitigation and adaptation, and is aimed at implementing the

---

105 FCCC/SBSTA/2014/L.14 (Distr.: Limited 14 June 2014) SBSTA 40.

106 *Ibid.*

107 Workshop held on 12 November 2013 in Warsaw, documents available through: UNFCCC, 'Workshop on the Current State of Scientific Knowledge on How to Enhance the Adaptation of Agriculture to Climate Change Impacts while Promoting Rural Development, Sustainable Development and Productivity of Agricultural Systems and Food Security in All Countries, Particularly in Developing Countries' (2013), accessed 10 October 2016 at http://unfccc.int/meetings/warsaw_nov_2013/workshop/7887.php.

108 FCCC/SBSTA/2014/INF.2 (13 March 2014) 8.

109 *Ibid* at 9.

provisions of the Convention, which obviously also includes the provision to adopt and implement an adaptation policy under Article 4(1)(b). Consequently Developed Country Parties, particularly Annex II states, should take all practicable steps to promote, facilitate and finance the transfer of, or access to, adaptation technologies to Developing Country Parties that will enable the implementation of the UNFCCC's adaptation provisions.[110] The Kyoto Protocol also requires the transfer of adaptation technologies, and more broadly includes the transfer of knowledge, practices and processes that are 'pertinent to climate change'.[111] The IPCC noted that successful implementation of adaptation actions depends on access to technology, among other factors.[112] This certainly is true for adaptation in agriculture. Relevant technological adaptation options include: new crop and animal varieties, genetic techniques, traditional technologies and methods, efficient irrigation, rainwater harvesting, conservation agriculture, food storage and preservation facilities, and second-generation biofuels.[113] For developing country farmers, the transfer of these technologies is of vital importance.[114] The UNFCCC's Technology Executive Committee (TEC) indicates that (a) transfer of knowledge and (b) knowhow on agricultural adaptation technologies do not necessarily only have to go from North to South, but that there can also be a South to South transfer.[115] Despite some initiatives to enhance agricultural adaptation technology transfer, such as the establishment of Climate Change Technology Transfer Centres in Europe and Latin America,[116] the same barriers that make the transfer of mitigation technologies difficult, exist for adaptation technology transfer.[117] The increase in patenting adaptation-specific crops, such as drought resistant seeds, is an additional barrier to agricultural adaptation technology transfer. These patents have led to concerns about accessibility of these crops, particularly for developing country farmers.[118]

---

110 Art 4(5).
111 Art 10(c) KP.
112 Porter *et al* (n 3) 844.
113 *Ibid* at 845 (table 14-1).
114
115 *Ibid* at 7.
116 Focusing primarily on scientific cooperation between Europe and Latin America. See Climate Change Technology Transfer Centres in Europe and Latin America (CELA), 'Welcome to CELA', accessed 10 October 2016 at http://www.cela-project.net.
117 See s 2.2 above.
118 Campbell *et al.* (n 57) 39.

4.2.5 Transfer of financial resources for adaptation

The UNFCCC stipulates that Developed Country Parties have to assist developing countries that are particularly vulnerable to the adverse effects of climate change in meeting the costs of adaptation to those adverse effects.[119] The main instrument that was created to implement this provision is the Adaptation Fund, established during COP7 in Marrakesh in 2001,[120] and operating under the Kyoto Protocol. Its aim is to finance concrete adaptation projects in Developing Country Parties to the Kyoto Protocol, as well as some specific adaptation activities that are not limited to developing countries. The latter is a consequence of another decision taken in Marrakesh, which invites the Adaptation Fund to carry out specific activities relevant for the agricultural sector. These include:

- starting to implement adaptation activities promptly where sufficient information is available to warrant such activities, in the areas of water resource management, land management, and agriculture;
- improving the monitoring of diseases and vectors affected by climate change, and related forecasting, and improving disease control and prevention;
- supporting capacity-building for preventive measures, planning, preparedness and management of disasters relating to climate change, including contingency planning, in particular for droughts and floods in areas prone to extreme weather events; and
- strengthening and, if necessary, establishing national and regional information networks for rapid response to extreme weather events.

The list of projects, funded by the Adaptation Fund, includes many agricultural projects. Examples of these are:[121]

- adapting to climate induced threats to food production and food security in the Karnali Region of Nepal (2015);
- enhancing adaptive capacity and increasing resilience of small and marginal farmers in Purulia and Bankura Districts of West Bengal (2014);
- developing climate resilience of farming communities in the drought prone parts of Uzbekistan (2014); and
- climate smart agriculture: enhancing adaptive capacity of the rural communities in Lebanon (2012).

---

119 Art 4(4) UNFCCC.

120 Decision 10/CP.7, FCCC/CP/2001/13/Add. 1, 52.

121

Projects like these often focus on introducing new irrigation technologies and adapted crop varieties, and usually include technology transfer.

Aside from the Adaptation Fund, two other funds operate under the UNFCCC to finance adaptation projects for the agricultural sector in developing countries. The Special Climate Change Fund (SCCF) and the Least Developed Countries Fund (LDCF) were established during COP7 of the UNFCCC in Marrakesh in 2001.[122] The SCCF is a general fund, aimed at financing activities, programs and measures relating to climate change in a number of areas, including adaptation and agriculture. The LDCF, as its name already indicates, is specifically for financing projects in the least developed countries, particularly projects related to their NAPAs. Both the SCCF and the LDCF are operated by the Global Environment Facility, a financial organization comprised of governments, international organizations such as the World Bank and the United Nations Environment Programme, non-governmental organizations (NGOs) and the private sector, to fund environmental projects in developing countries.[123] The SCCF and LDCF councils have joint meetings, and they both rely mainly on voluntary contributions from Annex I states. As with the Adaptation Fund, many of the projects funded focus on adaptation in the agricultural sector in developing countries. The LDCF project 'Integrating Climate Resilience into Agricultural Production for Food Security in Rural Areas', for instance, aims to enhance adaptive capacity and reduce vulnerability to climate change in a country where 75 percent of the population is employed in agriculture. The project addresses Gambia's inability to effectively predict climate change events, assess potential impacts and deliver short-term alerts or long term warnings, which is considered a major barrier limiting the capacity to adapt to the changing climate.[124] Another project aimed at helping farming communities prepare for increasing climate variability is called 'Integrating Climate Resilience into the Agricultural Sector for Food Security in Rural Areas of Mali'. Under this project, climate change adaptation strategies are taken into account by national and local agricultural sector planning, and implemented through 'Farmers

---

[122] Decision 7/CP.7, FCCC/CP/2001/13/Add. 1, 43.

[123] For more information, see Global Environment Facility (GEF), 'Home Page', accessed 10 October 2016 at http://www.thegef.org.

[124] More information on this project is available from GEF, 'The Gambia – Strengthening Climate Change Early Warning Systems', accessed 10 October 2016 at https://www.thegef.org/project/strengthening-climate-services-and-early-warning-systems-gambia-climate-resilient.

Field Schools', which help Malian farmers explore new technologies and management methods.[125]

There is considerable debate on the proliferation of funds and its inherent risk of overlap and inefficiency.[126] That is why under the negotiations towards the Paris COP, integration of the various funds has been high on the agenda. The biggest problem these funds encounter, however, is the lack of available finance. In the 42 NAPAs that were submitted by 2009, urgent adaptation measures had an estimated total cost of around US$2 billion.[127] In 2010, the LDCF received total pledges of $131.5 million, and the SCCF a total of $49.8 million from Annex I states. Current funding is totally inadequate to address even the most urgent adaptation needs of the least developed countries. It is, therefore, not surprising that the least developed countries raise this issue in every COP, and not without success.

At the Copenhagen and Cancun meetings in 2009 and 2010, developed countries made pledges to provide new and additional resources approaching US$30 billion for the period 2010 to 2012, with a balanced allocation between adaptation and mitigation and prioritized funding for adaptation in the most vulnerable developing countries, including small island developing states and Africa.[128] In addition, the developed countries committed to jointly mobilizing $100 billion per year by 2020 to address the needs of developing countries.[129]

The Green Climate Fund (GCF) was established in Cancun in 2010 to channel these additional funds.[130] The GCF was officially designated as an operating entity of the UNFCCC's financial mechanism under Article 11 during COP17 in Durban in 2011.[131] The GCF's main goal is to 'promote the paradigm shift towards low-emission and climate-resilient development pathways by providing support to developing countries to limit or reduce their greenhouse gas emissions and to adapt

---

125 More information on this project is available at GEF, 'Mali – Integrating Climate Resilience into the Agricultural Food Sector for Food Security in Rural Areas of Mali', accessed 10 October 2016 at https://www.thegef.org/project/integrating-climate-resilience-agricultural-production-food-security-rural-areas.

126 Verschuuren (n 88) 27.

127 See LDC Expert Group, *The Least Developed Countries: Support Needed to Fully Implement National Adaptation Programmes of Action (NAPAs)* (UNFCCC 2009) 17.

128 FCCC/CP/2010/7/Add. 1, 16.

129 FCCC/CP/2010/7/Add. 1, 17. This paragraph was taken from Verschuuren (n 88) 27.

130 *Ibid.*

131 Decision 3/CP.17, FCCC/CP/2011/9/Add.1, 55.

to the impacts of climate change'.[132] The fund will do so by 'channeling new, additional, adequate and predictable financial resources to developing countries' and by catalyzing 'climate finance, both public and private, and at the international and national levels'.[133] So far, the status of pledges and contributions made to the GCF are a bit disappointing. By November 2015, a total amount of US$10.2 billion was pledged to the fund, with actual contracts signed covering about half of this amount.[134]

4.2.6 Adaptation Committee

The 2010 Cancun Agreements established the Adaptation Committee to promote the implementation of enhanced action on adaptation in a coherent manner under the Convention, and to function as the overall advisory body to the COP on adaptation.[135] So far the Adaptation Committee mainly functions as a coordinating body, aimed at aligning the adaptation related activities by the various bodies that operate under the UNFCCC and the Kyoto Protocol. Judging by its yearly reports, the Adaptation Committee has not yet specifically focused its attention on agriculture.[136]

5. AGRICULTURE IN THE PARIS AGREEMENT

Agriculture specifically was hardly mentioned in the various versions of the Negotiating Text for the Paris Agreement or in the final text that was adopted at COP21.[137] The only mention was in the provision on mobilizing finance, where states are called upon to support the integration of climate objectives into other policy-relevant areas and activities 'such as agriculture.'[138] In the final Agreement Negotiating Text by the Co-chairs,

---

[132] The Transitional Committee delivered its final report to the COP in Durban, FCCC/CP/2011/6 58.

[133] *Ibid* at 58.

[134] See GCF, 'Pledge Tracker', accessed 10 October 2016 at http://www.greenclimate.fund/contributions/pledge-tracker.

[135] Decision 2/CP.17, FCCC/CP/2011/9/Add. 1, 19.

[136] Reports on 2012, 2013, 2014 and 2015 at UNFCCC, 'Documents – Adaptation Committee', accessed 10 October 2016 at http://unfccc.int/adaptation/groups_committees/adaptation_committee/items/7584.php.

[137] This section is based upon the negotiating texts and the final agreement, all of which are available through Deconstructing Paris, 'Welcome to Deconstructing Paris', accessed 10 October 2016 at http://paristext2015.com/.

[138] FCCC/ADP/2015/1, 40 (version 11 June 2015) (under 101bis).

all references to 'agriculture' had disappeared.[139] As a consequence, the Paris Agreement, as adopted at COP21, does not refer to agriculture at all.

'[M]aintaining food security' was also mentioned in the full text proposals[140] but in the final Agreement Negotiating Text this reference, too, disappeared, only to reappear in the draft COP Decision's preamble.[141] The latter reference did survive the negotiations in Paris at COP21, so that the preamble to the Paris Agreement now states: 'Recognizing the fundamental priority of safeguarding food security and ending hunger, and the particular vulnerabilities of food production systems to the adverse impacts of climate change.'

Food production also regularly emerged as a topic in the full Negotiating Texts as a limiting factor to mitigation actions (similar to Article 2 UNFCCC, see Section II above). In the final version of the Paris Agreement, only one such reference survived. Article 2 provides the main objectives of the Agreement, one of which is: 'Increasing the ability to adapt to the adverse impacts of climate change and foster climate resilience and low greenhouse gas emissions development, in a manner that does not threaten food production.'[142]

Given the contribution of agriculture to climate change and the impact of climate change on agriculture, it is disappointing that so little attention is paid to agriculture in the Paris Agreement, which sets the tone for the coming years.

The European Union opted for a much firmer approach toward agriculture. In the run-up to the Paris Agreement, the European Commission announced that it would encourage 'climate friendly and resilient food production, while optimising the sector's contribution to greenhouse gas mitigation and sequestration'.[143] For example, the EU proposed including cropland and grazing land management in its official policy from 2020 onward, developing the instruments to do so before 2020. The EU even proposed focusing its future climate change instruments on *all* agricultural activities, to include enteric fermentation, manure

---

139 Ad Hoc Working Group on the Durban Platform for Enhance Action, 'Non-paper, Note by the Co-chairs' (5 October 2015), accessed 10 October 2016 at http://unfccc.int/resource/docs/2015/adp2/eng/8infnot.pdf.

140 FCCC/ADP/2015/1, 21 (under 50).

141 Ad Hoc Working Group (n 140) 10.

142 Art 2(1)(b) Paris Agreement.

143 European Commission, *Commission Staff Working Document. Accompanying the Document Communication from the Commission to the European Parliament and the Council 'The Paris Protocol – A Blueprint for Tackling Global Climate Change Beyond 2020'* (SWD 2015) 17 final, 18.

management, rice cultivation, agricultural soils, prescribed burning of savannahs, field burning of agricultural residues, liming, urea application, other carbon-containing fertilisers, cropland management and grazing land management.[144] With such a wide consideration of agriculture, the EU was actually suggesting that agriculture should be fully included in the Paris Agreement, both as a source of greenhouse gas emissions and as a means of CO_2 absorption and sequestration. This approach would have the agricultural sector undergo a drastic transition from conventional farming to farming that employs climate smart agricultural practices.

Though the Paris Agreement does not mention agriculture much, the document is still important for the sector. Article 4 states that, to keep the increase in global average temperature well below 2°C, a balance needs to be achieved between anthropogenic emissions by sources and removals by sinks of greenhouse gasses in the second half of the 21st century.[145] It is obvious that this automatically implies that drastic mitigation actions are needed to reduce emissions from agriculture and land use. Many of the provisions on adaptation and finance aim at giving increased support to developing countries to meet their adaptation needs, both through greater emphasis on providing financial resources and through the transfer of technology and capacity building.[146] Given the impact of climate change on agriculture and the dependence of developing countries on this sector, it is beyond doubt that implementation of these new provisions will largely focus on agriculture. Under the Agreement, parties are required to engage in adaptation planning processes and building the resilience of socio-economic systems,[147] which could lead to the creation of a National Adaptation Plan. Further integrating the various funds under the Financial Mechanism of the Convention, and strengthening the role of the Adaptation Committee are also actions the parties could take.[148]

---

[144] European Commission, *Energy Union Package. Communication from the Commission to the European Parliament and the Council 'The Paris Protocol – A Blueprint for Tackling Global Climate Change Beyond 2020'* (COM 2015) 81 final, 16.

[145] Arts 4(1) and 2(1)(a) Paris Agreement. Note that the draft texts proposed much stricter end goals, such as zero emissions or full decarbonization by 2050, FCCC/ADP/2015/1, 9-10 (under 17.2).

[146] See for example Arts 7, 9, 10 and 11 Paris Agreement.

[147] Art 7(9) Paris Agreement.

[148] See the decision that accompanies the Paris Agreement, FCCC/CP/2015/L.9.

6. CONCLUSION

Climate change has a profound impact on agriculture and on food security. At the same time, agriculture contributes to climate change to a considerable extent. Fortunately the agricultural sector holds significant climate change mitigation potential through reductions of greenhouse gas emissions and enhancement of sequestration:

> Agriculture offers a wealth of opportunities to deliver simultaneously on improving agricultural resilience to climate change, increasing food production, and lowering emissions. Many of these opportunities use practices, technologies, and systems that are already available and affordable, but need to be tailored to specific contexts and may require incentives from climate finance to ensure adoption. Some interventions also benefit wider environmental services, farming incomes, and agriculture-based economies.[149]

A single policy aimed at achieving greenhouse gas emission reductions, adaptation to climate change and an increase in productivity is, therefore, very much needed. 'Climate smart agriculture' policies have been proposed, but so far remain underdeveloped.

The UNFCCC unfortunately does not provide a powerful stimulus to adopt and implement climate smart agriculture policies, and there is little attention to reducing emissions from agriculture. The current framework focuses on adapting rural areas in developing countries to climate change, particularly through financial instruments. Yet even in that area progress is painfully slow. Much more concrete action is needed to facilitate the transfer of adaptation knowledge and technologies, as well as funds to finance adaptation measures in developing countries. For developed countries, the UNFCCC does not make much of an attempt to address climate change and food security issues. This is a pity, as the developed country agriculture sector plays a major role in addressing the increasing global demand for food. Fortunately developed countries, do not have to wait for the UNFCCC process. The EU, for example, recently announced its intention to implement an ambitious policy for climate friendly and resilient food production that will optimize the agricultural sector's contribution to greenhouse gas mitigation and sequestration. Following this example is vitally important across the globe. It is to be hoped that these initiatives will be adopted by the international community under the UNFCCC process.

---

149 Campbell *et al.* (n 57) 92.

3. Resilience and transformation of agro-ecosystems in a changing climate

Lance H Gunderson

1. INTRODUCTION

Climate has always played a role in the production and procurement of food by human populations. As suggested by many reconstructions, the past 10 to 15 thousand years has been a period of relatively stable climate.[1] Over this period of time, agriculture and agricultural practices have evolved.[2] While climate varies from region to region, it remains an important factor in the distribution of ecosystems that provide the basic needs of human life.[3] As the global population continues to grow, now exceeding seven billion, the need to provide food for a growing population increases. Feeding this population has consistently depended upon the harvest of wild stocks of plants and animals. As a result of increasing harvest rates, overharvesting has contributed to decreasing biodiversity.[4] As wild stocks and hunting and gathering practices have diminished, humans have become increasingly reliant on agriculture for provisioning food.[5]

One way to assess and understand how agriculture may adapt and respond to a changing climate is to use a systems framework. A systems perspective allows for examining how relationships among many elements or components help explain patterns and properties of complex phenomena. An ecosystem is defined as a set of interacting living components including plants, animals and microbes, and non-living components such as soils, air and water, that interact within a specific area or geography over a given time frame.[6] Moreover, an agro-ecosystem is a type of ecosystem that is largely controlled and directed by humans to 'produce, distribute

1 Intergovernmental Panel on Climate Change (IPCC), *Climate Change 2013: The Physical Science Basis*. (Cambridge University Press 2013) 1535.

2 IPCC, above note 1.

3 J Rockstrom *et al.*, 'Ecosystems and Human Well-Being: Synthesis' (2005) 461 *Nature* 472.

4 *Ibid.*

5 W Steffen *et al.*, *The Anthropocene: From Global Change to Planetary Stewardship* (AMBIO 2011) 40.

6 FS Chapin *et al.*, *Principles of Ecosystem Stewardship* (Springer 2009).

and consume food, fuel and fiber'.[7] In addition to the production function at local or small scale, agro-ecosystems include infrastructure, social and political institutions, and economic systems that expand the functions to meet the scale of the planet.[8]

By the end of the 20th century, scientists were beginning to investigate and document changes in earth's climate systems.[9] The long-term patterns of temperature and rainfall that determine climatic categories are undergoing rapid changes. While the planet has undergone large-scale climatic shifts in the past, the current changes are different in that they have been ascribed to the collective activities of humans.[10] Moreover, the relationship between changing climate and agriculture has become the objective of intensive study.[11]

1.1 Agriculture in the Anthropocene

Because of the collective actions of humans over centuries, including the pursuit of agricultural activities, we now live on a human dominated planet.[12] Indeed, many scientists have proposed that we now live in a new geological age referred to as the Anthropocene,[13] which is indicative of the magnitude and manifold ways in which humans have modified our planet, or biosphere. Human activities at the scale of the biosphere include land transformation, modifications of biogeochemical cycles and loss of biodiversity among others.[14] Indeed, climate change is but one of many ways in which the collective actions of humans are affecting global processes.[15]

---

7 JF Cabell and M Oelofse, 'An Indicator Framework for Assessing Agroecosystem Resilience' (2012) 17 *Ecology and Society* 3.

8 *Ibid.*

9 IPCC, above note 1.

10 *Ibid.*

11 EM Bennett and P Balvanera, 'The Future of Production Systems in a Globalized World' (2007) 5 *Frontiers in Ecology and the Environment* 191; PJ Ericksen *et al*, 'What Is the Vulnerability of a Food System to Global Environmental Change?' (2008). 13(2) *Ecology and Society*; TM Jackson, *et al*, 'Building a Climate Resilient Farm: A Risk Based Approach for Understanding Water, Energy and Emissions in Irrigated Agriculture' (2008) 104 *Agricultural Systems* 729; MJ Salinger *et al.*, 'Reducing Vulnerability of Agriculture and Forestry to Climate Variability and Change: Workshop Summary and Recommendations' (2005) 70 *Climatic Change* 341–62.

12 Steffen *et al.*, above note 5.

13 PJ Crutzen and EF Stoermer, 'The Anthropocene' (2000) 41 *Global Change Newsletter* 17, 18.

14 Steffen *et al.*, above note 5.

15 Rockstrom *et al.*, above note 3; Steffen *et al.*, above note 5.

Many of these global scale changes are the result of the increasing capacity and ingenuity of humans to modify ecosystems and ecological processes to produce or procure food.

The production of food has driven many of the environmental changes that can now be measured on a global scale. Documented changes over the past decade[16] include transformation of land for agriculture[17] and modifications of biogeochemical cycles. Nitrogen, phosphorous, carbon and water cycles are now largely controlled by humans, in large part to foster agriculture.[18]

Land transformation for agricultural purposes has resulted in large-scale environmental changes.[19] While agricultural use is a major driver of these changes, other factors influence land use and land use change. Planted crops are found on about 12–15 percent of the earth's surface.[20] The rate of land conversions to agriculture has increased over the past decade, and such conversions have increased the production of food, fuel, and fiber – this production can be described as provisioning of ecosystem goods.[21] Agricultural production of ecosystem goods comes at the expense of other ecosystem goods and services.[22] Another such tradeoff is that the increased food production has led to the loss of biodiversity – another ecosystem good.[23]

The amount of solar energy that is captured via photosynthesis is another indication of the impact of agricultural pursuits on ecological processes at the global level. Based on estimates from scientific studies, as much as 30 percent of the net primary production of the planet is now being used to support human populations.[24] While the green revolution of the 20th century increased the net productivity in most agricultural areas, such gains are due in part to the application of fertilizer and water, as well as technology and modified crop genetics. Moreover, the use of water for

---

16 P Vitousek *et al.*, 'Human Domination of Earth's Ecosystems' (1997) 277 *Science* 494–9; Rockstrom *et al.*, above note 3.

17 BL Turner *et al.*, *The Earth as Transformed By Human Action* (Cambridge University Press 1993) 713.

18 Vitousek *et al.*, above note 16.

19 B Turner *et al.*, *The Emergence of Land Change Science for Global Environmental Change and Sustainability* (Proceedings of the National Academy of Sciences 2007) 20666, 20671.

20 Turner *et al.*, above note 17.

21 R Biggs *et al.*, *Toward Principles for Enhancing the Resilience of Ecosystem Services* (Annual Review of Environment and Resources 2012) 37.

22 Rockstrom *et al.*, above note 3.

23 *Ibid.*

24 Vitousek *et al.*, above note 16.

agricultural production is one of the largest consumers of freshwater on the planet.[25]

Both nitrogen and phosphorous are used extensively as fertilizers to improve crop production. The availability of cheap fossil fuels has led to major modifications in the cycling of nitrogen[26] and phosphorous.[27] Many farmers use legumes (the plant family that includes peas) such as soybean or alfalfa in crop rotations. Many legumes have symbiotic relationships with microbes that live in their roots and act to fix nitrogen, transforming the stable compounds of nitrogen from the atmosphere into forms of nitrogen usable by plants. Currently, humans are adding about twice the amount of nitrogen to terrestrial ecosystems (including agro-ecosystems) than the amount that was fixed by all non-human processes.[28] Concentrations of nitrous oxide, a greenhouse gas, are increasing in the atmosphere due to combustion of fossil fuels. In addition to contributions to climate change, nitrogen compounds in the air also generate acid rain, and nitrogen compounds in runoff contribute to eutrophication.[29] Phosphorus runoff from agricultural lands can contribute to eutrophication in freshwater aquatic systems, wetlands and coastal ecosystems.[30] In addition to the nutrient induced changes in water quality, agricultural activities also alter the way in which freshwater is used.

Many of the freshwater pathways in the hydrological cycle are altered for agricultural activities.[31] Alterations on the local and regional scale in support of crop and livestock production, along with forestry, are the largest freshwater consumers.[32] Many of these activities can result in a loss of agricultural productivity and dramatic changes in aquatic ecosystems.[33] Soil moisture changes associated with irrigation can lead to salinization of soils and resulting losses in agricultural productivity.[34] River diversions

---

25 M Falkenmark and J Rockstrom, 'The New Blue and Green Water Paradigm: Breaking New Ground for Water Resources Planning and Management' (2006) 132(2) *Journal of Water Resources Planning and Management* 129–32.

26 Vitousek *et al.*, above note 16.

27 SR Carpenter, *Eutrophication of Aquatic Ecosystems: Bistability and Soil Phosphorus* (Proceedings of the National Academy of Sciences 2005) 102.

28 Vitousek *et al.*, above note 16; Rockstrom *et al.*, above note 3.

29 Vitousek *et al.*, above note 16; Rockstrom *et al.*, above note 3.

30 Carpenter, above note 27.

31 Falkenmark and Rockstrom, above note 25.

32 J Rockstrom *et al.*, *Water Resilience For Human Prosperity* (Cambridge University Press 2014) 18.

33 L Gordon *et al.*, 'Water in Agriculture for Food Production and Other Ecosystem Services' (2010) 97 *Agricultural Water Management* 512–19.

34 Gordon *et al.*, above note 33.

and impoundments are common causes of shifts in aquatic ecosystems. Regional scale links between water diversions and climate are emerging in areas such as the Sahel[35] or the Amazon.[36] In addition to alterations of nutrient cycles and freshwater flows for agricultural purposes, GHG emissions from agricultural activities are directly contributing to climate change.[37]

Increases in the concentration of greenhouse gases in the atmosphere over the past 50 years have been observed worldwide.[38] Gases such as carbon dioxide, methane and nitrous oxide are all considered greenhouse gases because of the capacity of those gases to capture and recycle heat from long wave radiation. While much of the increase in carbon dioxide is attributed to fossil fuel combustion and industrial development, land clearing, deforestation and soil tilling also contribute a net increase of carbon to the atmosphere.[39] These increases can be reduced through integrative, conservation based activities that minimize soil disturbance, and careful management of crop types.[40]

The preceding paragraphs of this introduction have attempted a brief overview of how global ecological processes have been modified in pursuit of agricultural activities. Agriculture is just one of the human endeavors that has modified the volumes and rates that nitrogen, phosphorous, carbon and water move around the planet. Moreover, this growing body of evidence indicates that these changes are already influencing climate at local, regional and global scales.[41]

The remainder of this chapter is organized into three sections. The first section presents definitions and examples of resilience in agro-ecosystems. Resilience is a model of how systems change over time in abrupt and surprising ways due to both external and internal dynamics. The second section discusses another model of change over time, referred to as transformational change in agro-ecosystems. The context for transformational

---

35 M Scheffer *et al.*, 'Catastrophic Shifts in Ecosystems' (2001) 413 *Nature* 591–96.

36 MD Oyama and CA Nobre, 'A New Climate-Vegetation Equilibrium State For Tropical South America' (2003) 30 Geophys. Res. Letters 2199.

37 P Falloon and R Betts, 'Climate Impacts on European Agriculture and Water Management in the Context of Adaptation and Mitigation – The Importance of an Integrated Approach' (2010) 408 *Science of the Total Environment* 5667–87.

38 IPCC, above note 1.

39 *Ibid.*

40 PR Hobbs and B Govaerts, 'How Conservation Agriculture Can Contribute to Buffering Climate Change' (2010) 1 *Climate Change and Crop Production* 177–99.

41 IPCC, above note 1.

change is one that involves the emergence of new types of complex systems; agricultural, ecological or social. Transformational change involves cross-scale change.[42] The chapter concludes with a discussion of the resilience, adaptation and transformability of agro-ecosystems in a changing climate.

2. ECOLOGICAL RESILIENCE OF AGRO-ECOSYSTEMS

In 1973, CS Holling[43] introduced the concept of 'resilience' in his descriptions of different types of changes observed in ecosystems over time. Holling presented at least three categories of ecological change. Other models of ecosystem change have existed for some time, such as linear or exponential growth, cyclical patterns seen in ecological processes such as daily tides, seasonal temperatures and annual biomass production. Holling's models, however, were novel in that they attempted to describe other, more unpredictable patterns. One model involved the notion of stability and persistence, which was described as the 'ability of systems to absorb changes of state variables, driving variables and parameters, and still persist'.[44] Another model, which he referred to as 'resilience', was 'the size of a stability domain or the amount of disturbance a system could take before it shifted into alternative configuration'.[45] The third conceptual model described the abrupt and sudden changes that characterize many ecological phenomena, such as pest outbreaks and fire.[46]

In the years following Holling's work, ecologists and other scientists have viewed resilience as a system property, with multiple meanings. At the heart of the multiple definitions is the presence (or not) of alternative configurations or 'regimes'. Walker and Salt[47] define a 'regime' as a particular configuration of structures and functions with a unique identity. Some ecologists define resilience as essentially a 'return time', or time of recovery to a particular regime following a disturbance. Ludwig *et al.*[48]

---

[42] LH Gunderson and CS Holling, *Panarchy: Understanding Transformations in Systems of Humans and Nature* (Island Press 2002).

[43] CS Holling, 'Resilience and Stability of Ecological Systems' (1973) 4 *Annual Review of Ecology and Systematics* 1, 1–23.

[44] *Ibid.*

[45] *Ibid.*

[46] *Ibid.*

[47] B Walker and D Salt, *Resilience Thinking: Sustaining Ecosystems and People in a Changing World* (Island Press 2006).

[48] D Ludwig *et al.*, 'Sustainability, Stability, and Resilience' (1997) *Conservation Ecology* 1:7.

concluded that this perspective is applicable to either a linear system or to the behavior of a non-linear system in the vicinity of a local equilibrium. Holling[49] distinguished two types of resilience: engineering and ecological. Engineering resilience is defined as the rate or speed of recovery of a system following a shock.[50] Ecological resilience, on the other hand, assumes multiple regimes and is defined as the magnitude of a disturbance that triggers a shift between alternative regimes.[51] In this sense, a regime shift occurs when the controlling variables in a system (including feedbacks), results in a qualitatively different set of structures and dynamics of these systems.

Walker and colleagues thus define resilience as, 'the capacity to absorb disturbance and reorganize while undergoing change so as to still retain essentially the same function, structure, identity, and feedbacks'.[52] Other ecologists define resilience as 'the ability of the system to maintain its identity in the face of internal change and external shocks and disturbances'.[53] In summary, resilience can describe the capacity of a system to recover after a disturbance (such as a storm or a drought), or it can be used to describe the processes and structures that mediate a shift or transition among different configurations or regimes.

2.1 Regime Shifts and Agricultural Systems

Ecosystems can undergo dramatic changes in structures and or processes that have been described as a shift in system state or regime. A growing number of such regime shifts have been described and reviewed in the literature.[54] The types of regime shifts suggested by ecological resilience theory can be characterized in two general settings with respect to agro-ecosystems. In essence, regime shifts are a result of shifting controls in a system.[55] These can be bottom-up, such as changes in nutrients or

---

49 CS Holling, 'Engineering Resilience vs. Ecological Resilience' in PC Schulze (ed.), *Engineering Within Ecological Constraints* (National Academy Press 1996).

50 *Ibid.*

51 *Ibid.*

52 B Walker *et al.*, 'Resilience Management in Social-Ecological Systems: A Working Hypothesis for a Participatory Approach' (2002) 6 *Conservation Ecology* 14.

53 G Cumming, 'An Exploratory Framework for the Empirical Measurement of Resilience' (2005) 8 Ecosystems 975–87.

54 C Folke *et al.*, 'Regime Shifts, Resilience, and Biodiversity in Ecosystem Management' (2004) 35 *Annual Review of Ecology Evolution and Systematics* 557–81.

55 Holling, above note 43.

water that influence primary production,[56] or top-down, where changes in populations in food webs can result in regime shifts at the scale of the ecosystem.[57] Complex food webs provide resilience to a wide range of disturbances in many marine ecosystems.[58] Coral reefs that have been overfished to the point of removing key species have led to changes or 'flips' in structure.[59] An example of such a structural change includes the shift from a coral dominated reef to an algae dominated reef following a disturbance such as a cyclone,[60] where overharvesting of certain fish species has led to an overabundance of algae. These critical (and missing) fish species are ones that graze (and control) algal abundance.[61] Other examples abound of aquatic food web changes that lead to a regime shift.[62] Such dynamics have been observed in agro-ecosystems, both within the ecosystem as defined by the local field or production unit and the export of nutrients to surrounding ecosystems.[63] Ecosystem flips as a result of changes in nutrient flows, due to fertilization of agro-ecosystems, are present in many freshwater and coastal aquatic systems.

2.1.1 Nutrient induced shifts

Increased flows of nitrogen and phosphorus in the biosphere have also led to regional scale ecological regime shifts. In many of these cases, the changes in nutrient flows can be attributed to runoff of excess fertilizers from crop agriculture or concentrations of livestock. Of course, other non-agricultural human activities can modify these flows. Freshwater lakes subject to increased nitrogen and phosphorus inputs can lead to regime shifts, from clear water lakes with benthic vegetation to eutrophic lakes with recurring algal blooms.[64] Freshwater wetlands can also undergo such

---

[56] L Gunderson and L Pritchard (eds), *Resilience and the Behavior of Large-Scale Systems* (Island Press 2002) 256.

[57] JA Estes and D Duggins, 'Sea Otters and Kelp Forests in Alaska: Generality and Variation in a Community Ecological Paradigm' (1995) 65 *Ecological Monographs* 75–100.

[58] Folke *et al.*, above note 54.

[59] TP Hughes *et al.*, 'Shifting Base-Lines, Declining Coral Cover, and the Erosion of Reef Resilience' (2011) 30 *Coral Reefs* 653–60.

[60] TP Hughes, 'Catastrophes, Phase Shifts, and Large-Scale Degradation of a Caribbean Coral Reef' (1994) 265 *Science* 1547–51.

[61] TR McClanahan, 'Response of the Coral Reef Benthos and Herbivory to Fishery Closure Management and the 1998 ENSO Disturbance' 155 (2008) *Oecologia* 169–77.

[62] RS Steneck *et al.*, 'Kelp Forest Ecosystems: Biodiversity, Stability, Resilience and Future' (2002) 29 *Environmental Conservation* 436–59.

[63] Folke *et al.*, above note 54.

[64] Folke *et al.*, above note 54.

state changes, such as the observed changes in freshwater marsh plants, algae and the biotic community in the Everglades due to runoff from the Everglades Agricultural Area.[65] The appearance and persistence of hypoxic zones in coastal areas such as the Baltic Sea, Chesapeake Bay or Gulf of Mexico are due in part to nutrient runoff from agricultural activities.[66]

Regime shifts in many aquatic systems (from marine to freshwater) can also be attributed to interactions between enhanced nutrients and shifts in trophic structures. These include phase shifts in coral reef ecosystems in the Caribbean[67] and the Great Barrier Reef.[68] Most of these documented regime shifts are attributed to the export of nutrients from agricultural lands to broader areas, and the ecological regime shift occurs outside the agricultural production zone. Other external regime shifts, such as broad scale pest outbreaks also affect local crops as described in the next section.

2.1.2 Alternating state changes: pest outbreaks

The examples presented in the previous section were provided to demonstrate the complexities of regime change in ecosystems. Ecological resilience is the system property that mediates a shift between alternative states or regimes. In some coral reefs, the shift from one regime to another can be reversed,[69] while the reversibility (between regimes) in other systems is uncertain.[70] The next section presents brief overviews of alternating regime shifts in two different agro-ecosystems. Both examples describe outbreaks of agricultural pests; one is the desert locust of northern Africa, the other a white fly that carries a crop pathogen.

2.1.2.1 Desert locust system For millennia, locusts have plagued humans. These plagues occur when the ecological and population dynamics of the locust interact across scales of space and time.[71] A small set of

---

65 Gunderson and Pritchard, above note 56.

66 DJ Conley *et al.*, 'Ecosystem Thresholds with Hypoxia' (2009) 629 *Hydrobiologia* 21–9.

67 TR McClanahan *et al.*, 'Testing for Top-Down Control: Can Post-Disturbance Fisheries Closures Reverse Algal Dominance?' (2011) 21 *Aquatic Conservation-Marine and Freshwater Ecosystems* 658–75.

68 TP Hughes *et al.*, 'Phase Shifts, Herbivory, and the Resilience of Coral Reefs to Climate Change' (2007) 17 *Current Biology* 360–65.

69 *Ibid.*

70 Gunderson and Pritchard, above note 56.

71 M Lecoq, 'Desert Locust Threat to Agricultural Development and Food Security and FAO/International Role in its Control' (2003) 21 *Arab Journal of Plant Protection* 188–93.

ecological factors, food availability and soil moisture in specific regions trigger a shift in the species phenotype from a solitary type locust to a swarming or gregarious type.[72] Originally thought to be two species, biologists now recognize that this phenotypic switch is triggered by an abundance of moisture and food.[73] Once the swarming type is present, population densities can quickly increase to outbreak or plague levels.[74] Although all of the factors that moderate such temporal dynamics are not known, the net result is a dramatic population oscillation, which appeared to cycle on the order of 15 years or so, prior to 1960.[75] Outbreaks can last a dozen years or more, but eventually subside. The spatial patterns of desert locust outbreaks are broad. Beginning in a few isolated source regions in the eastern and southern regions of the Sahara, subsequent populations of the locusts spread across North Africa in complex patterns that eventually covered millions of square kilometers.[76]

Such long term and widespread ecological dynamics pose enormous management problems.[77] Over time, the management has evolved to a twofold approach: monitoring and control. Monitoring of locust populations, meteorological conditions and vegetation conditions at specific localities over two continents is a daunting challenge.[78] The second part of the approach has been the development of an international collaborative control effort to stop the locust spread. From the beginning of the 20th century, an international strategy has been set to prevent the outbreaks. This strategy is applied through collaboration between actors at the local, national, regional and international level under the supervision of the Food and Agriculture Organization of the United Nations (FAO) that acts on the mandate of the member states. Recent experience indicates that problems with funding, planning, implementation, institutional commitment and memory, among others, have led to wide failures and the need for new risk management approaches.[79]

Cascading biological phenomena such as the desert locust outbreaks

---

72 M Lecoq, 'Desert Locust Management: From Ecology to Anthropology' (2005) 14 *Journal of Orthoptera Research* 179–86.

73 *Ibid.*

74 *Ibid.*

75 JI Magor *et al.*, 'Control and Desert Locust Plagues' (2008) 27 *Crop Protection* 1527–33.

76 Lecoq (2005), above note 72.

77 M Lecoq, 'Recent Progress in Desert and Migratory Locust Management in Africa. Are Preventive Actions Possible?' (2001) 10 *Journal of Orthoptera Research* 277–91.

78 Lecoq, above note 71.

79 Lecoq, above note 70.

can be described as alternating phases, or states, of a system.[80] One phase is the low population density (characterized by the solitary insect type). Food and water resources, and perhaps predators, would all act to limit the population size. Pulses of rainfall would increase the food base, and temperature increases would spur regeneration times, triggering a transition to the other state. This alternative phase is one where the gregarious type locust is dominant, and rapidly spreads across space and time with little or no control. Once the energy sources fueling the population increase are exhausted, the erupting population collapses back to the low-density phase.

On the economic and social side, one can also identify different states. During the outbreaks the actors are aiming at controlling the spread by raising funds to apply insecticides, which involves coordinating across countries. During the solitary locust phase the actors are concentrated on monitoring. Hence, the socio-ecological system has two different regimes that correspond to the two locust population phases: a solitary locust/social monitoring phase or a locust outbreak/social control phase. The problem lies in the transition between the first and second socio-ecological states. The longer this transition takes to complete, the more expensive the cost of the crisis becomes.[81]

2.1.2.2 White flies, red tomatoes and a leaf-yellowing virus Globally, Europe ranks second behind China in tomato production and consumption.[82] Driven by a large economic demand year-round for fresh tomatoes, tomato production in Southern France is possible through the use of greenhouses or hothouses. Greenhouse tomato production is profitable because infrastructure and operating costs can be offset by regional sales. Yet, the greenhouse production is not without other risks, such as the spread of disease. Such an outbreak occurred during 2003–2005 in southeastern France with the spread of the Tomato Yellow Leaf Curl Virus (TYLC).[83]

The reaction to the virus outbreak motivated the creation of a set of scientific activities and studies that attempted to summarize what was known about the virus. The work identified the major vector of the virus: a white fly of the genus *Bemesia*, which has caused crop damage in the Mediterranean area since the mid-1970s. Much of the scientific work

---

[80] Ludwig *et al.* (1997), above note 48.

[81] Lecoq (2005), above note 70.

[82] JR Arno, *Evaluation of Tools to Manage White Flies in Europe* (2008) 1 ENDURE Tomato Case Study.

[83] *Ibid.*

focused on identifying the species involved, and identifying the key factors in the appearance and spread of TYLC.[84]

As with the previous examples, at least two regimes can be identified in the White Fly/Tomato/TYLC system. One is a state when the TYLC population is low and contained, the other when the TYLC population is large and spreading. The transition between these two states is called an outbreak. The states are similar to the desert locust in that one state involves a dormant phase and the other is a spreading or outbreak phase.

Such outbreaks are the result of the interactions among a small set of variables that control the population of the White Fly. Outbreaks occur when abundant food within a greenhouse spawns rapid growth of the white fly population. Since the white fly can move within and among greenhouses, the virus spreads as flies feed on the leaves of plants. The white fly populations can fluctuate due to temperature changes that influence reproductive rates, food sources and the abundance of parasitoids and predators.[85]

This section has attempted to review resilience theory that has developed over the past four decades from the study of coupled human and ecological systems. Resilience mediates the transition among two or more qualitatively different system states or regimes. The transition between states is not a result of one factor, but the result of many interacting factors. The way in which these variables interact can increase or decrease the resilience of a system and make it more or less vulnerable to a regime shift. The removal of herbivores from coral reefs (due to overfishing) allows for algae to grow and dominate the reef, while excess nutrients fuel the algal growth. Similarly, the cultivation of tomato plants in greenhouses provides abundant and accessible food for white fly population growth (and the TYLC virus that they carry), while at the same time removing predators (such as birds) of the white fly, which control the fly population. In essence, resilience (or state transition) is determined by shifting feedbacks and controls within a system.[86] Another model of change in ecological systems with humans as primary agents is developed in the next section, described as 'transformative change'.

---

[84] D Gerling, 'Status of *Bemesia tabaci* in Mediterranean Countries: Opportunities for Control' (1996) 6 *Biological Control* 11–22.

[85] *Ibid.*

[86] Folke *et al.*, above note 54.

3. TRANSFORMATIVE CHANGE IN AGRO-ECOSYSTEMS

Transformative change has been used to describe systems in which new and different human and ecological components appear.[87] The human components include institutions, defined by management structures, rules and norms, which guide actions in the system. These institutions could include markets in agro-ecosystems, social and legal structures such as land ownership, or other collective action rules.[88] The ecological components of agro-ecosystems include soils, nutrients, water, plants, genetic diversity of crops and pests, among others. Transformative change occurs when new structures and processes emerge over time in systems. Resilience theory, as described above, was instrumental in the development of a theory of transformational change theory, called panarchy theory.[89] Panarchy theory is based in part on concepts described as an adaptive cycle.[90]

The concept of an adaptive cycle came from examination and review of historical trajectories of dozens of managed systems of people and nature.[91] In essence, an adaptive cycle consists of sequential periods of (1) growth, (2) maturity, (3) release and (4) reorganization. The first period occurs as a system begins to develop. In an agro-ecosystem, an example is a crop planted in a field. Early on, the seeds sprout and the young plants grow at rates determined by conditions of available resources. Over time, biomass or system structure accumulate and more relationships develop as the system becomes more diverse and complex. As structures and resources accumulate to become an identifiable system (such as a corn field), they reach the next stage of development: maturity. Over time the system becomes increasingly connected, less flexible and more vulnerable to external disturbances. As a system matures, it also becomes increasingly vulnerable to external shocks or perturbations. That is, it becomes less resilient as disturbances generate a sudden release of accumulated capital or structure. Disease outbreaks, such as the tomato-yellowing *Bemisia* described above, occur following the maturation phase. Such disturbance can lead to a sudden release period of biomass and structure, and has

---

[87] Gunderson and Holling, above note 42.

[88] E Ostrom, 'A General Framework for Analyzing Sustainability of Social-Ecological Systems' (2009) 325 *Science* 419–22.

[89] Gunderson and Holling, above note 42.

[90] CS Holling, 'The Resilience of Terrestrial Ecosystems: Local Surprise and Global Change' in WC Clark and RE Munn (eds), *Sustainable Development of the Biosphere* (Cambridge University Press 1986) 292–317.

[91] *Ibid.*

been called creative destruction or the omega (end) phase.[92] The collapse or release during this third phase is quickly followed by a reorganization (alpha) phase where a new system emerges, leading to the growth phase of a new cycle. The new growth phase may be very similar to the previous growth phase, or it may be quite different. It is during the omega and alpha phases that regime shifts can occur.[93]

The dynamics conceptualized in the adaptive cycle are for systems at a particular scale range. That is, the dynamics of growth, conservation, destruction and renewal can be observed for specific ranges of structures or focal scale, such as a crop in a field. Linking these scale specific dynamics across scales forms the basis of Panarchy theory.[94]

A panarchy consists of adaptive cycle dynamics at a specific scale range, and coupling across scales. The adaptive cycle provides indications of timings for episodic, cross-scale linkages to occur. Potentially, there are multiple connections between phases at one level and phases at another level. However, in a panarchy, there are dynamics that start at one scale range and, because of positive feedbacks, continue to grow to increasingly larger scales. These are referred to as 'revolting' processes because the changes are dramatic and spread from small scales to larger ones.

Up-scale linkages or connections have been named 'revolt',[95] suggesting that small events can cascade up to larger scales. Most of the time, the larger, broader variables control the smaller and faster processes. However, when the small-scale level in the panarchy enters an omega phase of creative destruction and experiences a collapse, that collapse can cascade up to the next level by triggering a crisis, particularly if that level is at the mature phase when resilience is low. Such processes allow diseases to spread across crops in a field or among crops within a region.

Down-scale interactions in panarchy are captured by the word 'remember'.[96] This type of cross-scale interaction is important at times of change and renewal. Once a catastrophe is triggered at a particular level, the opportunities and constraints for the renewal of the cycle are strongly organized by the status of the larger level. This is exemplified by processes and resources accumulating at a larger level slowing the leakage of nutrients that have been mobilized and released into the soil. The options for renewal draw upon the seed bank, physical structures and surviving species that form biotic legacies that have accumulated during the growth

---

92 *Ibid.*
93 Gunderson and Holling, above note 42.
94 *Ibid.*
95 *Ibid.*
96 *Ibid.*

phases. It is as if this connection draws upon the accumulated wisdom and experiences of maturity, hence the choice of the word 'remember'.

In summary, the cross-scale model of panarchy has been used to describe changes in structures and functions of ecosystems over time.[97] It contrasts long periods of development in which the system develops around specific social and ecological constraints and purposes with short-term disturbances and crises. Those disturbances can be internally generated (such as disease outbreaks) or externally generated (such as a weather event). The panarchy model can form the theoretical basis of understanding how agro-ecosystems change over time. An example of this application is described below using agricultural development in a wetland subtropical environment in the US.

3.1 History of Transformations in Everglades Agriculture

The Everglades region in South Florida provides a good example of an area to explore panarchy theory and transformational changes in agro-ecosystems. Agriculture has been ongoing in the Everglades since the beginning of the 20th century.[98] Located in a subtropical climate, the area was developed because of its year-round growing season and relatively deep accumulations of organic soil. However, the agriculture was enabled by a regional scale water management system that was developed to control water levels.[99] Water management and agriculture have been intimately connected, coevolving in ways where exogenous events have intersected with internal vulnerability to transform system states or regimes.[100]

Water resources in South Florida were developed and managed not only to meet agricultural production, but also to provide water for urban residents and nature conservation. The infrastructure was built and operated to remove unwanted floodwater so that lands could be farmed, to supply water during dry periods, and to maintain the quality of receiving waters. That system of water control has enabled dramatic development of urban and economic development along the southeastern coast of Florida.

---

[97] L Gunderson *et al.*, *Barriers and Bridges to the Renewal of Ecosystems and Institutions* (Columbia University Press 1995).

[98] GH Snyder and JM Davidson, 'Everglades Agriculture: Past, Present and Future' in SMD and JC Ogden (eds). *Everglades: The Ecosystem and it's Restoration* (St. Lucie Press 1994) 85–115.

[99] SS Light *et al.*, 'The Everglades: Evolution of Management in a Turbulent Ecosystem' in LH Gunderson, *et al.* (eds), *Barriers and Bridges to the Renewal of Ecosysyems and Institutions* (Columbia University Press 1995) 103–68.

[100] Gunderson *et al.*, above note 97.

Currently, about eight million people reside in the watershed, and they depend upon this large system for water supply and flood control. This specifically applies to a viable agricultural community, and water that is passed to the environment in this area to support biodiversity and other conservation purposes.[101]

At least four eras of agriculture can be identified in the 20th century. Such eras are defined by three facets of a coupled social ecological system: (1) physical infrastructure to control water, (2) a shift in the types of agricultural crops grown in the area and (3) the presence of new institutions and rules that govern the movement of water to and from surrounding areas.[102] These eras can be described as Early Drainage, Partitioning the Everglades, the Rise of Big Sugar and the Era of Water Pollution.[103]

3.1.1 Early Drainage era

After Florida became a state in 1845, the federal government held claim to the waters and wetlands in the southern peninsula, including the Everglades. As this act was being developed, Congress commissioned Buckingham Smith[104] to survey the 'Ever Glades' region of South Florida. Smith reviewed historical documents, surveys, and maps of the region and made a ground survey of the area. His report was pivotal in establishing agriculture in the Everglades.

The Buckingham Smith survey of South Florida focused on issues of natural history and hydrology. He recounted earlier descriptions of the plants, animals and the tropical climate that supported a unique variety of flora and fauna. His descriptions of the elevations and water levels around the Everglades led to his recommendation that 'draining the Ever Glades of four or five feet of water' would allow for agriculture. He states 'the favorable success of the undertaking cannot be questioned'. He also estimated that drainage would cost $500,000. Because of the tropical climate, he said such drainage would lead to the cultivation of 'coffee, sugar and tropical fruits'.[105] His conclusions were prophetic, especially with respect to sugar production from the Everglades. Following passage of the Swamp Lands Act and Smith's report, the State then created an

---

[101] Light *et al.*, above note 99.

[102] Snyder and Davidson (1994) above note 98.

[103] Light *et al.*, above note 99.

[104] Smith, B. Report of Buckingham Smith, Esq., on His Reconnaissance of the Everglades (1848). US Senate Documents, Committee Report 242, 30th US Congress, 1st Session, Washington DC.

[105] *Ibid.*

Internal Improvement Fund to sell land and begin plans to drain the lands.[106]

Early agriculture was located in the northern Everglades, south of Lake Okeechobee, where the soils were deepest and had the best characteristics for crop growth. Farming was facilitated by a combination of drainage actions. A canal was dug in the 1880s that created a flow way from Lake Okeechobee to the Caloosahatchee River to the west, which lowered the lake water levels by a few feet. To protect the nascent agriculture from lake overflow, earthen levees were built around the southern shore of the lake. Following a flood of the area in 1903, Napoleon Bonaparte Broward was elected Governor of Florida based, in part, on a promise to drain the Everglades and make the land available for family farming. Within a decade of selling small plots, the Everglades Land Company realized that larger, corporate agriculture would be more suitable for the area than small family farms.[107]

By the 1920s, a dozen or more towns were established around the southern end of Lake Okeechobee. Many of the residents farmed. Markets to sell the crops were facilitated by railroad and water access to the rest of the US. The frost-free winters (due in part to the moderating effect of Lake Okeechobee) fostered a growing agriculture that produced winter vegetables and sugarcane. The resilience of the agriculture was severely tested during hurricanes of 1926 and 1928. The 1928 Okeechobee hurricane caused the waters of the lake to rise above the earthen levees and flood hundreds of square miles, at depths up to 20 feet.[108] The storm and resulting flooding led to a loss of about 2,500 lives. Following this event, the federal government refocused on the Everglades and funded the construction of the Herbert Hoover Dike around the lake. The dike was completed by 1938 and was built at a height of over 30 feet to afford flood protection to the agriculture lands. Within a decade, flooding would again test the resilience of agriculture in the Everglades.

3.1.2 Partitioning the Everglades

Extensive flooding occurred again in 1947 as a result of an unusually heavy rainy season and multiple tropical storms. Most of South Florida was flooded for months. The State of Florida appealed again to the federal government for assistance, and the government responded by passing the

---

106 Snyder and Davidson (1994) above note 98

107 Light *et al.*, above note 99.

108 http://en.wikipedia.org/wiki/1928_Okeechobee_hurricane, accessed 1 October 2016.

Flood Control Act in June 1948.[109] The act authorized the US Army Corps of Engineers to develop a plan known as the Central and Southern Florida Project for Flood Control and Other Purposes. The cost of the proposed project (estimated at $70 million in 1949)[110] was justified to some degree by future benefits that would accrue through agricultural production. In 1949, the Florida Legislature created the Central and Southern Florida Flood Control District (FCD) to provide local sponsorship for the federal project. The FCD was an appointed body with large representation from the agricultural community.[111]

The flood control plan called for a large investment in levees, canals and pumps to protect urban and agricultural areas from flooding. The flood control plan partitioned the former connected wetland into specific land uses. As such, the plan demarcated the lands now known as the Everglades Agricultural Area (EAA).[112] The plan included water conservation areas to the south of the EAA. These conservation areas created a land buffer between the agricultural activities and newly established Everglades National Park in the southern Everglades and were designed to receive floodwaters from the EAA. Large pumps were installed as part of the project, which would allow water managers to pump floodwaters from the EAA back into Lake Okeechobee. Construction of the flood control system took more than a decade to complete, at which time the Everglades Agricultural area underwent another transformation.

3.1.3 Rise of Big Sugar

As suggested by Buckingham Smith in 1848,[113] sugarcane has been grown in the area for over 150 years. Because of the abundant rainfall and infrequent frosts, sugarcane grows well in subtropical Florida. In the early 20th century, sugarcane was planted as an experimental crop. Production was limited by deficiency of nutrients in the soil as well as

---

109 Flood Control Act of 1948, Pub. L. No. 80-858, 62 Stat. 1171.

110 Comprehensive Report on Central and Southern Florida for Flood Control and other Purposes; letter from the Secretary Of The Army. 6 May 1948. Referred to the Committee on Public Works United States Government Printing Office Washington: 1949, accessed 14 November 2016 at https://www.archives.gov/research/guide-fed-records/groups/077.html.

111 *See* R. Hamann, 'Florida's Water Management Framework' in JT Scholz and B Stiftel (eds), *Adaptive Governance and Water Conflict: New Institutions For Collaborative Planning* (Resources for the Future 2005) 15, 17.

112 NM Blake, *Land into Water-Water into Land: A History of Water Management in Florida* (University Presses of Florida 1998)

113 Smith, above note 104.

periodic flooding of the soil.[114] Hence, water control was critical to this form of agriculture.

Sugar production in the Everglades, as with other areas in the US, is heavily influenced by policies of the federal government. This began in the New Deal era, when a quota system was established to manage production of sugar.[115] Because the Everglades did not have a history of sugar production, the quotas for the EAA were quite small. Through the 1940s and early 1950s, only about 5 percent of the area supported sugar production. While sugar imported from Cuba was also subject to controls via this quota system, that quota was large. Three-quarters of the international quota of sugar was granted to Cuba, as well as lower tariffs as a result of trade agreements. Indeed as much as one-third of US sugar imports came from Cuba. That all changed in 1959.

Fidel Castro led a successful revolution in Cuba and assumed power in 1959. Castro immediately seized billions of dollars' worth of US assets on the island, especially the most arable land, with the intent on redistributing to the citizens of Cuba. That process, plus Castro's affiliation with the communist party, led President Eisenhower to suspend quotas on Cuban agricultural imports and institute an embargo of products from Cuba. In compensation for the loss of sugar imports the US federal government created price supports for domestic production of sugar.[116]

These shifts in federal policy led to a dramatic increase in sugarcane production within the EAA beginning in 1960. Within a year, 20 percent of the land area in the EAA was growing sugarcane,[117] and the acreage rose to almost 60 percent by the 1980s. Four new sugar mills opened up in the EAA in 1962, a number that has increased to seven mills operating today.[118] As a result, Florida became and remains the highest sugarcane producer in the United States.

The EAA became a different social ecological system as a result of these changes in international agricultural policy. These policies led to the growth in corporate interests and control over the agricultural production system as well as the water management systems. By the 1980s the US Sugar Corporation owned and farmed about 25 percent of the EAA and became the largest sugar producer in the state.[119] In addition to creating

---

[114] Snyder and Davidson, above note 98.

[115] G Hollander, *Raising Cane in the 'Glades: The Global Sugar Trade and the Transformation of Florida* (University of Chicago Press 2008).

[116] *Ibid.*

[117] Snyder and Davidson, above note 98.

[118] *Ibid.*

[119] *Ibid.*

local water control boards, agricultural interests also dominated the state flood control and subsequent water management boards. These state agencies operate the water systems, which were critically important to agricultural production both in supplying water during dry periods and removing water during periods of flooding.

The water management system was designed and developed such that water could be moved away from agricultural areas by two different routes. Levees and canals allowed for drainage, which carried water down gradient, south and east toward Everglades marshes and urban areas. Additionally, large capacity pumps enabled the transfer of large quantities of water into the large lake north of the agricultural area. While the system worked well to achieve water control, the waters carried nutrients. Unforeseen ecological effects of these nutrient laden waters led to further tests of the system's resilience and transformation into a new social-ecological system.

3.1.4 Water quality declines

By the early 1980s, the degradation of water quality became the major environmental issue in South Florida.[120] Recurring algae blooms in Lake Okeechobee were attributed to nutrient runoff from cattle farms to the north of the lake and from crops and agriculture to the south.[121] This led to a shift in policy that stopped the pumping of water from the EAA in the lake. When water could no longer be pumped from agricultural fields to the lake, it was diverted south into the original freshwater marshes of the Everglades. Such diversions occur during periods of high rainfall and high water levels. The native wetland communities have evolved under conditions of low nutrients.[122] As a result, areas in which nutrients from agricultural runoff and other sources had accumulated became vulnerable to an ecological regime shift. The vegetation in these areas shifted from sawgrass plants to cattails, as well as algae and other organisms.[123] In 1988, the US federal government filed a lawsuit against the State of Florida alleging that the state had failed to enforce water quality laws that implemented the federal Clean Water Act, damaging federal resources. The resolution

---

[120] Light *et al.*, above note 94.

[121] NG Aumen, 'The History of Human Impacts, Lake Management, and Limnological Research on Lake Okeechobee, Florida' (1995) 45 *Advances In Limnology* 1–16.

[122] L Gunderson and WF Loftus. 'The Everglades' in WH Martin (ed.), *Biodiversity of the Southeastern United States/Lowland Terrestrial Communties* (John Wiley & Sons 1993) 199–255.

[123] LH Gunderson, 'Ecological Resilience – Theory to Practice' (2000) 31 *Annual Review of Ecology. and Systematics* 421–39.

of the lawsuit led to efforts to limit discharges of phosphorous laden water from the EAA.[124] Subsequently, better management practices by farmers limited nutrient discharges, through limiting fertilizer application. In addition, large areas of constructed wetlands are now being used to remove nutrients from agricultural runoff.

This brief historic overview of the Everglades Agricultural Area was presented in the context of a panarchy model of change/transformation in an agro-ecosystem. Each of the eras represents essentially a new system configuration. These eras represent different agro-ecological systems, defined by the type and abundance of crops that were produced. Each era was also defined by different social and institutional settings as new agencies, rules and laws were created and applied. The transition among these eras was a result of a crisis or instability followed by a reorganization and development of a new agro-ecosystem. Some of the triggering events that created a crisis were related to climatic fluctuations, such as the dramatic amount of rainfall that fell in 1947–48. Others were due to social crises, such as the change in government in Cuba which led to sugarcane becoming a dominant crop in the EAA. These adaptive cycles led to small-scale changes in terms of type of crop produced at a field level, and regional scale changes such as the emergence of sugar processing mills. History also demonstrates how resilience of the Everglades wetland system was exceeded due to nutrient releases, as well as transformational changes. The question of how the EAA and other regional scale agro-ecosystems will respond to climate change is the subject of the next and final section.

4. CLIMATE, RESILIENCE AND AGRO-ECOSYSTEMS

The final section of this chapter will attempt to bring together the ideas and concepts presented in earlier sections on agro-ecosystem changes with notions of climate and climate change.

4.1 Climate and Climate Change

The biosphere is a complex system consisting of physical components such as the atmosphere, oceans and water bodies (liquid and frozen), soils

---

[124] D John, *Civic Environmentalism: Alternatives to Regulation in States and Communities* (CQ Press 1994).

and geology (the lithosphere) and living components, all of which are powered by energy from the sun.[125] Climate, or climate systems, represent a subset of the biosphere complexity and focus study on linkages between the atmospheric and surficial (primarily terrestrial) ecosystems. Weather is defined as short-term conditions of atmospheric state: patterns of temperature, moisture content, pressure, precipitation and wind. Climate is a representation of long-term (decades to centuries) patterns of precipitation and temperature, and the linkages with surface systems (ocean and terrestrial).[126]

Different schemes have been developed to classify climate for specific regions. The Köppen Climate Classification System is based on average estimates (by month) of temperature and humidity to classify different types of climate.[127] In this scheme, climate type is based on calculations from rainfall and temperature data collected over periods of 30 years. Six major categories of climate are used at the broadest level of classification: tropical moist climate, dry climate, moist temperate with mild winters, moist temperate with cold winters, polar climates and highland climates. Further divisions are made based upon the annual patterns of temperature (including freezing periods) and rainfall.[128] The Köppen system corresponds well with biomes or regions with similar plants and animals.[129] Another method is the Thornthwaite Climate Classification System, which is based on moisture regimes in the soil, an important factor in plant growth.[130] In this scheme, categories are defined along a spectrum from humid to arid. Both of these schemes attempt to link atmospheric dynamics and biota, especially the geographic distributions of plant species and assemblages.

Climate change is complex and multi-faceted. The scientific evidence of climate change is based on a number of variables. Some are physical indicators, such as temperature, wind, rainfall, evaporation, runoff, ice cover and freezing cycles.[131] Other biological variables, such as primary

---

125 EP Odum and GW Barrett, *Fundamentals of Ecology* 5th edn (Cengage Learning 2004) 18–19.

126 IPCC, above note 1, see also: http://www.nasa.gov/mission_pages/noaa-n/climate/climate_weather.html, accessed 11 October 2016.

127 M Pidwirny (2011). Köppen, accessed 14 November 2016 at http://editors.eol.org/eoearth/wiki/Köppen_climate_classification_system.

128 *Ibid.*

129 Odum and Barrett, above note 114.

130 CW Thornthwaite, 'An Approach Toward a Rational Classification of Climate' (1948) 38 *Geographical Review* 55–94.

131 IPCC, above note 1.

production or yield in crops, timing of flowering and fruiting cycles, and changes in migratory patterns indicate that climate change is occurring.[132]

The particular ways in which the climate might change are difficult to predict, and this is due in part to the uncertainties of future scenarios and the complexity of the issue. Another complicating factor is translating across scales; the models used to develop future scenarios are at greater spatial scales than most agricultural activity.[133] A robust prediction of climate change is that many facets of climate, including temperature and rainfall, will be less predictable and more variable. Extreme events such as floods will likely occur more often and record temperatures (both high and low) will be more frequently broken. Another expectation is that changing climate will exceed the ecological resilience of many agro-ecosystems, social systems and ecosystems, which will lead to catastrophic shifts.[134] Because of these predictions, resilience has become part of the discussion of how to adapt to or confront changing climate.[135]

4.2 Resilience, Adaptation and Climate Change

Adaptation to climate change in the US has focused largely on either human dominated landscapes (such as cities, suburban or agricultural areas) or ecosystems.[136] Communities, cities, states and regions are planning and acting in response to hazards, whether the source of the peril is a natural disaster or human generated one. In these cases, priority is placed on security and protection of human life, property and infrastructure before and after disturbance events. While traditional responses to natural disasters and climate change have assessed changes in risk, security and

---

132 IPCC, above note 1.

133 TM Jackson *et al.*, 'Building a Climate Resilient Farm: A Risk Based Approach for Understanding Water, Energy and Emissions in Irrigated Agriculture' (2011) 104 *Agricultural Systems* 729–74.

134 EL Tompkins and WN Adger, 'Does Adaptive Management of Natural Resources Enhance Resilience to Climate Change?' (2004) 9 *Ecology and Society* 10, accessed 11 October 2016 at http://www.ecologyandsociety.org/vol9/iss2/art10; SH Schneider, 'What is "Dangerous" Climate Change?' (2001) 411 *Nature* 17–19; WN Adger, 'Social Capital, Collective Action, and Adaptation to Climate Change' (2003) 79 *Economic Geography* 387–404.

135 JM West *et al.*, 'US Natural Resources and Climate Change: Concepts and Approaches for Management Adaptation' (2009) 44 *Environmental Management* 1001–21.

136 West *et al.*, above note 123.

exposure, other concepts such as vulnerability and resilience have been used to frame and guide adaptive responses.[137]

The ability of agro-ecosystems to adapt has been defined as adaptive capacity.[138] Such capacity can be attributed to the ability to deal with change in terms of transforming structures, frameworks and reform.[139] This is similar to the transformation among eras in the history of the Everglades agricultural area as described above. Adaptive capacity is often latent, in the sense that it is difficult to measure and only appears after a shock or stress to the system.[140] It also includes the capacity for creating new configurations to the system and directing future system trajectories of agro-ecosystems.[141]

Some scholars and practitioners have recommended enhancing the resilience of communities, ecosystems, regions and nations to a rising probability of disasters from impending climate change. The US National Academies, in a recent report, defined resilience as 'the ability to prepare and plan for, absorb, recover from and more successfully adapt to adverse events'.[142] Implicit in this approach is the idea that adaptation should involve increasing resilience. Resilience in this context is enhanced by (a) more effectively predicting the frequency and severity of these events, (b) increasing the capacity to absorb the shock of powerful events, and (c) optimizing time of recovery. Implicit in this definition is the idea that the systems (agro-ecosystems, ecosystems, regions or a nation) exist at or near an equilibrium state. In this view, resilience is argued to be a normative property of the system (systems ought to be more resilient). While this argument is germane, it narrows the discussion about creativity, adaptation and transformation of agro-ecosystems with respect to changing climate. Moreover, this definition deals with a general resilience,[143] not a specified resilience which links a

---

137 US National Academy of Sciences, Committee on Increasing National Resilience to Hazards and Disasters, *Disaster Resilience: A National Imperative*. (National Academies Press 2010) 4–6.

138 NL Engle, 'Adaptive Capacity and its Assessment' (2011) 21 *Global Environmental Change – Human and Policy Dimensions* 647–56.

139 R Berman *et al.*, 'The Role of Institutions in the Transformation of Coping Capacity to Sustainable Adaptive Capacity' (2012) 2 *Environmental Development* 86–100.

140 Engle, above note 127.

141 B Bennett *et al.*, 'Toward a More Resilient Agriculture' (2014) 5 *Solutions* 65–75.

142 National Academy of Sciences, above note 126.

143 B Walker and D Salt, *Resilience Practice: Building Capacity to Absorb Disturbance and Maintain Function* (Island Press 2012).

type system with a particular disturbance, such as the resilience of rice crops to a drought. These two facets of resilience (general and specific) are discussed in the next two sections.

4.3 General Resilience in Agro-Ecosystems

General resilience is the capacity of a system to absorb an unexpected shock or disturbance.[144] Such shocks may be a new disease that appears in a crop or livestock, an extreme drought or flood that had not been experienced, or an economic shock such as a rapid increase in energy prices. Climate change will likely increase the occurrence of such events. As climate scientists indicate, changing climate is like putting weather on steroids: normal weather events will occur, but they will be stronger and more frequent.[145]

In essence, confronting such unforeseen or unexpected shocks and events from a resilience perspective entails many unknowns (and perhaps unknowability). There is indication that various forms of capital provide capacity to deal with these types of situations. While agro-ecosystems do focus on natural stocks of capital, such as soil amount, soil fertility, ground water, and crop genetics, the question of how these categories influence general resilience requires much more work in understanding complex linkages in the system. Other forms of capital, such as social capital in the form of public trust, can increase the capacity of a system to respond to unforeseen shocks. Innovation, knowledge and ability to create new ideas can all contribute to this type of resilience. National or regional availability of insurance and banking or other institutions are thought to provide general resilience.[146]

There is some evidence that fostering functional response diversity can increase agricultural resilience. Biodiversity has been shown to provide general resilience to native rangelands and their productivity.[147] Greater biodiversity can provide buffers to environmental shocks such as heatwaves or droughts. More diverse agro-ecosystems are more resilient to pest outbreaks, as suggested earlier in the locust and white-fly examples.

---

[144] Walker and Salt, above note 132.

[145] Noah Besser, *Steroids, Baseball, and Climate Change--What do Home Runs and Weather Extremes Have in Common?*, accessed 11 October 2016 at https://www2.ucar.edu/atmosnews/attribution/steroids-baseball-climate-change.

[146] D O'Connell *et al.*, *The Resilience, Adaptation and Transformation Assessment Framework: From Theory to Application* (CSIRO 2015) 53.

[147] B Walker *et al.*, *Adaptability and Transformability in Social-Ecological Systems* (2004) 9 *Ecology and Society* 5.

Regional diversity in some types of farms can provide resilience to climate and economic variations.[148] Diversity can influence the resilience of ecosystem services in agro-ecosystems.[149]

4.4 Specific Resilience in Agro-Ecosystems

Specific resilience is often captured by the phrase 'resilience of what to what'.[150] For example, O'Connell and colleagues assess the resilience of climate change on irrigated rice systems production in Thailand.[151] They propose that external changes such as rising summer temperatures, more frequent extreme heat events and extreme rainfall events are all factors that will influence production. Temperature may decrease production if the current plant varieties are near a tolerance threshold. Increased temperature and carbon dioxide can also increase rice production. Floods and droughts can result in crop damage, thereby decreasing production.[152]

The complexity of agro-ecosystems makes prediction of changes in resilience as a result of external drivers difficult, if not impossible. Indeed, many authors suggest that resilience is observed after the fact, and a system undergoes a regime shift.[153] As a result, many suggest that it is more important to focus on assessing resilience as a system property rather than trying to measure and predict resilience at any point in time. That is why many authors[154] suggest that a focus on using indicators to assess systems is key to managing resilience.

5. SUMMARY

This chapter has attempted to weave together ideas of how agricultural systems might respond to, cope with and adapt to a changing climate. Climate change is already occurring, as indicated by a growing body of evidence, and will continue to occur in the future.[155] The robust predictions of those changes

---

[148] T Hills *et al.*, *A Monitoring Instrument for Resilience* (CGIAR Research Program on Climate Change, Agriculture and Food Security (CCAFS) 2015). Available 11 October 2016 online at www.ccafs.cgiar.org

[149] R Biggs, 'Toward Principles for Enhancing the Resilience of Ecosystem Services' (2012) 37 *Annual Review of Environment and Resources* 421–48.

[150] Walker *et al.*, above note 136.

[151] O'Connell *et al.*, above note 135.

[152] *Ibid.*

[153] Hills *et al.*, above note 137.

[154] O'Connell *et al.*, above note 135; Hills *et al.*, above note 137.

[155] IPCC, above note 1.

suggest that more extreme events will occur and that more fluctuations in temperature and rainfall patterns will take place. The rates at which sea levels will rise also appear to be increasing, accelerating an ongoing trend and resulting in the increase in flooding and inundation of many coastal zones.

Resilience, adaptation and transformation as models of change in agro-ecosystems were explored. These models come from a growing body of understanding how complex systems of people and nature change over time. Resilience is an inherent property of systems that describes how complex systems respond to disturbances. Moreover, it suggests that these systems can undergo qualitative shifts in structure and function into alternative system states or regimes, and resilience is the property that mediates such transitions. Resilience is not a normative property of systems in the sense that systems ought to be resilient. Rather, these are neutral descriptions of complex dynamics, and people have the capacity in many cases to determine the desirability of a particular system state.

Much more is known about what decreases resilience in a system than what contributes to resilience in a system. In many cases, resilience is only understood after a regime shift occurs. Among the factors that contribute to declining resilience include a decrease in functional diversity, erosion of various forms of capital (such as soil nutrition, or trust in social systems) and limitations on a system's capacity for renewal and regeneration.

Managing for resilience and transformation requires acknowledgement that systems operate and interact across many nested scales. Large-scale systems such as the atmosphere can influence regional patterns of climate. Climate and weather at regional scales can influence changes in agro-ecosystems at the field level, the farm level, and farming community levels. Large-scale economic processes such as globalization and world-wide delivery systems can link local production with markets half a world away. Small-scale processes such as microbial respiration in tilled or drained organic soils contribute to the accumulation of carbon dioxide in the atmosphere. The spread of new and old diseases in crops is another process that starts locally and can spread on a global scale.

Agro-ecosystems are complex adaptive systems. As such, dynamics of such systems are highly uncertain and, for all intents and purposes, unpredictable. Dealing with such uncertainty will require programs that focus on understanding and learning more than development of better policies, laws or programs. A growing scholarship is emerging in this area to help decision makers. Large uncertainties in outcomes and process, however, will likely continue to impede novel and innovative trials and experiments. Yet, the cost of not doing so will result in a less desirable future, rather than one that may be more equitable, just and sustainable.

4. Climate change, complexity, agriculture and challenged governance

Paul Martin[1]

1. INTRODUCTION

Human induced changes to the Earth's atmosphere result in more solar energy being retained than would otherwise be the case. This drives intertwined changes to climate, the hydrological cycle, the land and oceans which affect all living systems, including humanity in many ways – only some of which have been studied. Human actions, partly shaped by governance institutions including environmental laws, both respond to and contribute to these processes. Change will be systemic and pervasive, and much of it not predicted.

Agriculture and rural people are on the frontline of many climate induced changes, through climate impacts upon enterprises and general welfare, and because of the role of agriculture as a potentially adaptive mechanism.[2] Agriculture will also be affected by policies designed to benefit society as a whole, requiring reduction of agricultural emissions, expanding conservation areas, on-farm conservation demands and carbon sequestration.[3] Climate change can also have benign or positive effects, seen with opportunities such as carbon trading or funded conservation.

---

[1] This chapter has been supported by the Australian Research Council (Project No LP110100659) and funding partners including the Australian Cotton Research and Development Corporation; Australian Government Department of Sustainability, Environment, Water, Population and Communities and Australian Organic; and the Invasive Animals CRC research program 4 'Facilitating Effective Community Action'.

[2] United Nations System Standing Committee on Nutrition, 'Climate Change and Nutrition Security: Message to the UNFCCC Negotiators' (2010), accessed 11 October 2016 at http://www.unscn.org/files/Statements/Bdef_NutCC_2311_final.pdf; W Steffan, *Thirsty Country: Climate Change and Drought in Australia* (Climate Council of Australia 2015); H Paul *et al.*, *Agriculture and Climate Change: Real Problems, False Solutions* (EcoNexus 2009); A Hogan *et al.*, *Decisions Made by Farmers that Relate to Climate Change – RIRDC Publication No 10/208* (RIRDC 2011).

[3] Michal Nachmany *et al.*, *The GLOBE Climate Legislation Study: A Review of Climate Change Legislation in 66 Countries* 4th edn (London School of Economics

However, such policies create pressures on rural communities, which are often relatively disadvantaged.[4] Climate related issues also interact with other changes, including pressures on agricultural profitability due to commodity market dynamics and the power of other participants in the agriculture industry;[5] threats to the availability of farm inputs including fertile land, and water and fertilizer;[6] land use fragmentation that complicates resource management; institutional complexity arising from political and consumer expectations and industrialization; and internationalization.[7] Events such as natural disasters, pest and disease attacks, and climate-linked migration compound these pressures. As other chapters in this book indicate, the law has an important role in responding to these and other climate induced challenges.

The human challenges of climate change will force adaptation within the law. This goes beyond the mere creation and interpretation of rules, as the changing nature of the problems that lawyers need to address demands significant innovation in environmental and other laws. As with other aspects of climate change, prediction (even though intrinsically unreliable) is necessary if environmental law is to meet the challenges to come.

Climate change and its consequences are complicated topics. Each sub-topic is complex and hard to understand, and has specific literature.[8] All topics are linked and the number of issues is enormous. Limitations of space and (this author's) intellect require limits to the investigation. In this

---

and Political Science and Grantham Research Institute on Climate Change and the Environment 2014).

[4] For a detailed consideration of the social system aspects of rural natural resource governance, see Paul Martin and Jacqueline Williams, 'Next Generation Rural Natural Resource Governance: A Careful Diagnosis' in Volker Mauerhofer (ed.), *Legal Aspects of Sustainable Development: Horizontal and Sectorial Policy Issues* (Springer 2016).

[5] Bill Vorley, 'The Chains of Agriculture: Sustainability and the Restructuring of Agri-food Markets' (*International Institute for Environment and Development*, 2001), accessed 11 October 2016 at http://pubs.iied.org/pdfs/11009IIED.pdf.

[6] The 'peak Phosphorus' challenge. Dana Cordell and Stuart White, 'Peak Phosphorus: Clarifying the Key Issues of a Vigorous Debate about Long-term Phosphorus Security' (2011) 3(10) *Sustainability* 2027.

[7] Norton Rose Fulbright, 'Food and Agribusiness Survey: Risks and Opportunities in a Changing Global Market' (2014), accessed 11 October 2016 at http://www.nortonrosefulbright.com/files/food-and-agribusiness-survey-2014-118237.PDF.

[8] For example food security (United Nations System Standing Committee on Nutrition (n 2)) or biosecurity (Allan Hawke, *The Australian Environment Act: Report of the Independent Review of the Environment Protection and Biodiversity Conservation Act 1999* (Australian Department of the Environment 2009)).

chapter I have thus limited consideration of rural social consequences and excluded natural disasters. This chapter adopts a novel approach which extends the systems logic of climate science to environmental law.

2. CLIMATE SYSTEMS: THE DANCE OF THE CLOCKWORK DOLLS

Four systems are the building blocks of the natural environment: soil, atmosphere, water and energy systems. Each is complex. The interplay between these systems is the foundation of life: plants, animals, humans and ultimately (at a non-material level) society, economy and man's abstract creations such as culture.[9] Dynamic flows of energy, materials and signals shape the outcomes of these systems. Transactions between individuals and between people and natural systems link man and nature in many ways. Pollution has resulted in more of the sun's energy being held within the earth's atmosphere and changes to the amount of energy fuelling all of our natural systems. Scientific principles suggest that adding substantially more energy into the natural system will result in changes everywhere in that system and that this must flow onto the social system. How and where will these changes occur?; how much will society experience?; and how will man respond? Complex interconnections between open system elements drive the complicated non-linear outcomes, where small changes in one part of some systems can result in unexpected changes elsewhere.[10]

We can visualize the elements and relationships that exist between climate change and agriculture as a set of four Russian 'Babushka' dolls, each large doll containing smaller dolls. Within each smaller doll is a clockwork mechanism that causes them to move in many directions. These nested dolls and the clockwork mechanisms act upon each other, causing unexpected patterns of movement. The input driving the system is the energy from winding the mechanisms. The patterns created and the end positions of the dolls reflect the structure of the mechanism and the dynamic relationship between the dolls. The resulting patterns of movement and the end state positions of the dolls will be difficult to predict. Intervention, such as changing the placement of the dolls or winding the mechanisms to different tensions, will affect the operation of the system in

---

[9] Based upon open system theories of interconnectivity and equifinality. For principles see Ludwig von Bertalanffy, *General System Theory: Foundations, Development, Applications* (George Braziller 1968).

[10] James Gleick, *Chaos: Making a New Science* (Viking Books 1987).

complicated ways. Changes to the amount of energy imported by winding the keys, or by winding each to a different tension, will further affect how the device operates. So too will the original doll placement. Even a puff of wind or bumping the table on which the device stands will have consequences. The dynamic interaction means that there is a vast range of possible outcomes. Even if tension was applied to each spring equally, or the dolls placed equidistant from each other, the results would not be predictable. Climate change is like that; there are so many variables influencing the system that unpredicted outcomes will occur.

Systems thinking is a way of understanding complicated phenomena and the dynamic interactions between elements of a system.[11] This approach delivers insights of the kind popularized by 'chaos theory',[12] and is mainstream in economics (the economic system), climate change (climate models) and studies of nature (for example, the hydrological cycle, population dynamics or various natural systems). Physics informs us that the building blocks of natural systems are energy and matter and that matter is interchangeable with energy.[13] Thus an increase in the energy in any of the four basic systems (because of carbon loading in the atmosphere) will generate change that pervades all natural and social systems. Systems theory suggests that the changes will be non-linear. Predictability will remain a serious problem in shaping policy (though narrowly defined issues where the causal relationships are restricted may appear be more tractable).

Environmental law generally aims to achieve 'real world' biophysical and justice outcomes. Empirical methods are required to understand social and environmental challenges, the operation of legal institutions in practice, and the results of use of the law.[14] One possible method is systems thinking. Ludwig von Bertalanffy is credited with introducing systems thinking into the social sciences,[15] where it finds many uses within economics, health systems, process design and various other applications. Versions of systems thinking range from the descriptive (for example a system

---

[11] Human elements of social systems are typically referred to as 'actors'.

[12] Gleick (n 10).

[13] Famously described in Einstein's formula $E=mc^2$.

[14] Paul Martin and Donna Craig, 'Accelerating the Evolution of Environmental Law through Continuous Learning from Applied Experience in Environmental Jurisprudence – Improving Governance Effectiveness' in Paul Martin and Amanda Kennedy (eds), *Implementing Environmental Law* (Edward Elgar 2015).

[15] Von Bertalanffy (n 9). See also James Grier Miller, 'Living Systems, the Basic Concepts' (*Panarchy*, 1978), accessed 11 October 2016 at http://www.panarchy.org/miller/livingsystems.html.

map in process or computer system design), to sophisticated dynamic or mathematic models which underpin climate or hydrological forecasts, or economic or demographic analysis. Simulations, even games like SimCity,[16] are based upon system models. Systems analysis requires simplification, with the analyst determining the boundaries, the elements, and the important relationships that they see forming the system. No system conceptualization can be considered to be 'true', but they can be useful. Creativity and judgment are required to create useful system models.

Considering only soils and water, increased energy in the system will change the structure and moisture retention of soils, and alter rainfall patterns spatially and volumetrically. In addition, there will be fewer predictable effects (in terms of timing, location and degree) of storms and other natural disasters. The intersection of these may generate challenges including increased erosion, loss of agricultural productivity, desertification, salinization, flooding and reductions in potable or productive water supply.[17] These will not be distributed equally. Areas of particular vulnerability include coastal zones and low-lying islands, and areas that are potentially susceptible to desertification and soil erosion (for example, dry high mountain areas). There will be unexpected water and soil outcomes due to complex interactions.

3. SYSTEM RELATIONSHIPS: AGRICULTURE, GOVERNANCE AND CLIMATE CHANGE

Some systems modelling approaches emphasize the importance of 'stocks', for present purposes equivalent to natural, social or other capital.[18] Six categories of capital have been identified for corporate sustainability reporting: money, goods, intellectual capital, human capital, social and relationship capital, and natural resources.[19] Stocks are fed or depleted by flows. These may be physical (for example, money, or goods), or ephemeral (for example, signals such as communication, information or data). In a

---

[16] Electronic Arts Inc, 'What is SimCity' (*SimCity*, 2014), accessed 11 October 2016 at http://www.simcity.com/en_US/game/info/what-is-simcity.

[17] Paul (n 2).

[18] For a simplified explanation see System Dynamics Society, 'Stocks and Flows', accessed 11 October 2016 at http://www.systemdynamics.org/DL-IntroSysDyn/stock.htm.

[19] Technical Collaboration Group (TCG), 'Capitals Background Paper for <IR>' (IIRC 2013), accessed 11 October 2016 at http://integratedreporting.org/wp-content/uploads/2013/03/IR-Background-Paper-Capitals.pdf.

natural environment relevant physical flows might include nutrients, water, dust, genes, plants or animals, or whatever flow that the analyst identifies as significant. In system terms a farm is a repository of stocks of animals, plants, financial and human capital that draws upon stocks from nature such as water, soil and other ecosystem services of agriculture and exports products such as food and fiber and by-products such as pollutants. The flow of energy is, of course, the basis of all physical systems, transformed in many ways including as matter.

This observation underlines that systems in nature and society are interconnected. Socio-ecological and socio-economic issues are central to the demands made upon resource governance. Climate change impacts on agriculture will affect agricultural economies, and this will have flow-on effects on rural communities and the consumers of agricultural products, and thus on society. There are other less obvious links between these systems including impacts on human health and welfare, effects of changing government policies in matters such as carbon sequestration, the allocation of economic and other supports, or the delivery of rural services.[20]

Signals (for example, data, observations, communication, intelligence and so on) drive behavior. The farmer observes moisture in the soil, the water in the dam, the appearance of his plants and animals, weather forecasts, and information about markets. Animals respond to indicators of the availability of food and shelter, or potential threats. Other relevant signals are internal to creatures and trigger their behavior, such as detected changes in temperature or moisture leading to fruiting, or fertility. Because many natural stimuli are affected by climate change, we should expect many, sometimes subtle, responses by natural systems and by people.

Nature and society import and export material and ephemeral things: they are said to be 'open'. People import energy as food, water or power and export it as communication or work, products or waste. Increases in the energy within Earth's atmosphere will affect all natural systems, and man, in many ways. The climate change effects that have been predicted will be only part of what will occur. Climate change will affect farming directly through changes in weather and water, and indirectly, for example through effects on agricultural markets or public policies. Other changes will follow, such as new plant and animal varieties, strategy changes, changes to agricultural investments, and other consequences radiating out

---

[20] Nicholas Stern, 'Stern Review: The Economics of Climate Change' (HM Treasury 2006), accessed 11 October 2016 at http://mudancasclimaticas.cptec.inpe.br/~rmclima/pdfs/destaques/sternreview_report_complete.pdf.

from the initial changes. Many changes in society will have corresponding legal ramifications.

The coupling of natural/social systems happens through transactions between people, and between individuals and the environment. Driving these transactions are decisions, influenced by various institutions including the law. With industrialization and urbanization, and the recreational and aesthetic uses of nature, the diversity of human/nature transactions has increased. Thus, the agricultural landscape that once would have been characterised by one type of agriculture (for example, broadacre grazing or cropping, or mixed farming) may now involve a complex mosaic of traditional uses, lifestyle developments, mining or gas extraction, eco-tourism or nature conservation. Each of these land uses are likely to involve technologies that did not exist even a decade or less ago, and advanced land use involves higher interactions with soils, water and the atmosphere. New technologies like sophisticated herbicides, pesticides and fertilisers, tilling equipment, land management practices and the like create an increasingly diversified interaction between man and the environment. With increasing population and technology, human impact upon other living beings and upon the four fundamental systems has increased. The following diagram (Figure 4.1) illustrates the direct influences of human activities on the stock of biodiversity in a coastal area with mixed agriculture and other land uses.

If the objective of governance was, in the example provided, protection of biodiversity and optimization of human value, then many transactions would need to be governed simultaneously. An effective strategy is likely to require a 'cocktail' of interventions (for example, regulation, market incentives and enablement through investment or human capacity building) targeting diverse behavior.[21] This suggests that natural resource governance should be conceptualized in systems as well as instrument terms. In this light, governance might be redefined as the network of institutional interventions (including laws) that shape the patterns of human transactions that impact natural and human systems. Its purpose is to optimize the valued flows from nature while preserving the stocks of natural and social capital. Defining governance in these system terms emphasizes the interaction between diverse elements over and above the role of the specific activities and instruments.

---

[21] This is the approach increasingly proposed by contemporary regulatory theory. OECD, *Instrument Mixes for Environmental Policy* (OECD iLibrary 2007); Peter Grabosky and Neil Gunningham, 'The Agriculture Industry' in Peter Grabosky and Neil Gunningham (eds), *Smart Regulation: Designing Environmental Policy* (Clarendon Press 1998); Paul Martin and Neil Gunningham, 'Leading Reform of Natural Resource Management Law: Core Principles' (2011) 28(3) EPLJ 137.

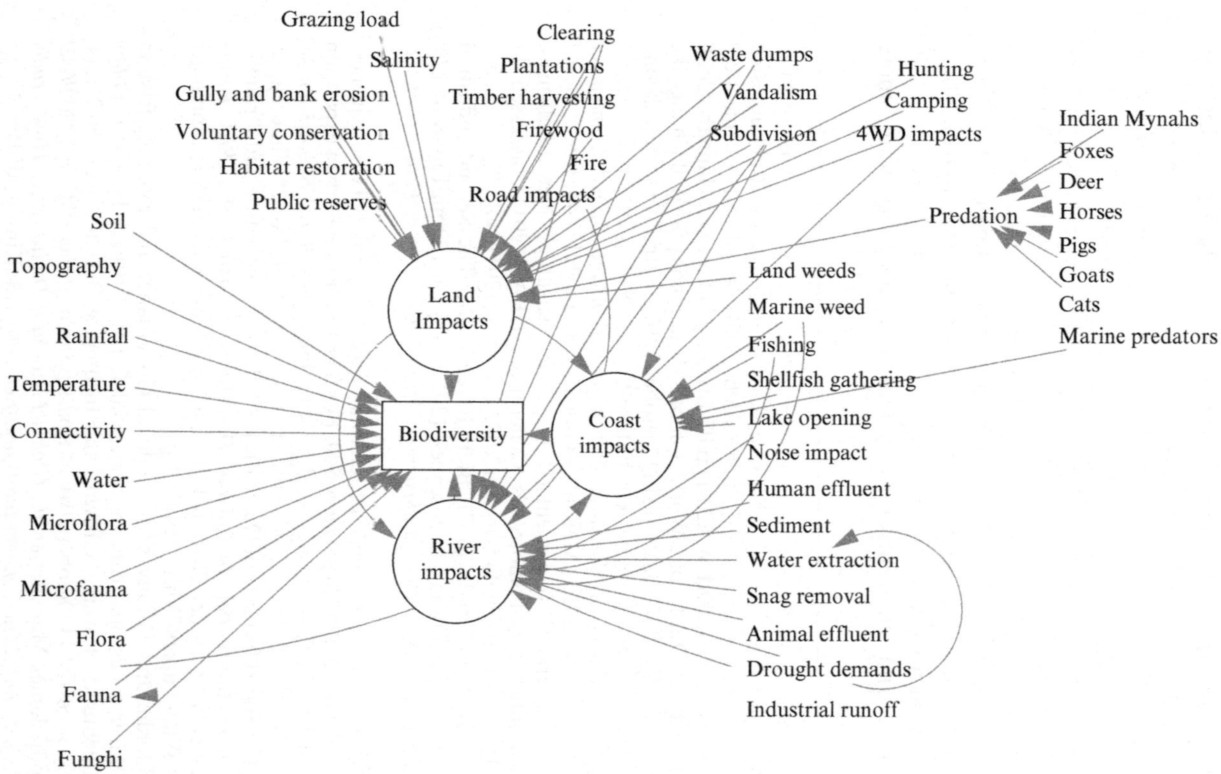

Source: The author.

Figure 4.1 Man and nature interactions and coastal biodiversity[22]

[22] Developed through participative modeling workshops conducted by the author on behalf of the Southern Catchment Management Board, NSW, Australia. For the method see Paul Martin and Miriam Verbeek, *Sustainability Strategy* (Federation Press 2006) 231–8.

Climate induced changes interact with other influences on the socio-ecological system.[23] These include population pressure, resource and food demand,[24] agricultural economics, land use fragmentation, technology and industry change.[25] For example, urban expansion interacts with the declining availability of fertile soils, limiting agricultural land while increasing the fragmentation of land uses.[26] Pressures on water supplies will exacerbate the ensuing conflict. Free trade and economic development policies shift patterns of resource ownership, which exacerbates the proliferation of social justice problems agricultural societies face, even without the effects of climate change.[27] In combination with climate effects, these policies increase turbulence in the rural socio-economic system which then spills over into non-rural society.[28]

Agricultural and environmental systems are also becoming more

---

[23] For example EEA, 'The European Environment – State and Outlook 2010: Assessment of Global Megatrends' (*European Environment Agency*, 2011), accessed 11 October 2016 at http://espas.eu/orbis/sites/default/files/generated/document/en/Global%20megatrends%20NEW%20(1).pdf; Stefan Hajkowicz, Hannah Cook and Anna Littleboy, 'Our Future World: Global Megatrends that Will Change the Way We Live' (CSIRO 2012), accessed 11 October 2016 at http://www.russell.com/au/_campaigns/post-nzis-2014/_pdf/OurFutureWorld_CSIRO_2012.pdf.

[24] Richard Dobbs *et al*, *Resource Revolution: Tracking Global Commodity Markets, Trends Survey 2013* (McKinsey Global Institute 2013) 44; International Fund for Agricultural Development (IFAD), 'New Realities, New Challenges: New Opportunities for Tomorrow's Generation' (IFAD 2011), accessed 11 October 2016 at http://www.ifad.org/rpr2011/report/e/rpr2011.pdf.

[25] Norton Rose Fulbright (n 7); Brett Bryan *et al*, 'The Second Industrial Transformation of Australian Landscapes' (2013) 5(3) Curr Opin Env Sust 278; Brendan Cullen *et al*, *New Rural Industries – Making the Most of Future Climates* (RIRDC 2010).

[26] For broader land use change see Jonathan Foley *et al*, 'Global Consequences of Land Use' (2005) 309(5734) *Science* 570. For a perspective on urban expansion see FM Kelleher, JJ Chant and NL Johnson, *Impact of Rural Subdivision on Agriculture* (RIRDC 1998); for desertification see Pak Sum Low, 'White Paper: Economic and Social Impacts of Desertification, Land Degradation and Drought' (*United Nations Convention to Combat Desertification and Global Risk Forum* 2013), accessed 11 October 2016 at http://2sc.unccd.int/fileadmin/unccd/upload/documents/WhitePapers/White_Paper_1.pdf.

[27] Ian Goldin and Tiffany Vogel, 'Global Governance and Systemic Risk in the 21st Century: Lessons from the Financial Crisis' (2010) 1(1) *Global Policy* 4; Evan Hillebrand, 'Poverty, Growth, and Inequality Over the Next 50 Years' (FAO 2009), accessed 11 October 2016 at ftp://ftp.fao.org/docrep/fao/012/ak968e/ak968e00.pdf; James Manyika *et al.*, *Global Flows in a Digital Age: How Trade, Finance, People, and Data Connect the World Economy* (McKinsey 2014).

[28] Richard Dobbs *et al.*, *Resource Revolution: Meeting the World's Energy, Materials, Food and Water Needs* (McKinsey 2011).

complex.[29] The many interactions and their outcomes reflect 'non-linear' dynamics and are 'equifinal', meaning that change in any part of the system can have effects anywhere else within the system. The apocryphal tale of a butterfly wing fluttering in the Amazon that triggers hurricanes across the United States illustrates the characteristics of complex open systems.[30] Many adaptation problems require legal and other governance interventions to accommodate complex problem-solving approaches. Because innovation (including policy innovation) is accompanied by failures, and because the issues will become increasingly complex, the unreliability of legal (and other) interventions is likely to become a greater, rather than a lesser, concern. There is an increasing realisation that regulatory interventions to protect the environment or the interests of disadvantaged communities often fail to deliver the intended outcomes. This issue of unreliability also applies to other interventions, including market instruments.[31]

Figure 4.1 shows that preserving natural resource stocks is not limited to protection or restoration transactions. The institutional arrangements that drive harm are equally or perhaps more significant. Non-government institutions that drive consumption of resources and produce the distributional outcomes of resource governance are no less significant than pro-environment or pro-justice public arrangements. The role of the private market in shaping socio-ecological outcomes is more powerful than that of government, and social institutions (including peer groups, and those which form social norms) strongly influence transacting patterns. Effective approaches to natural resource governance will thus need to simultaneously address: (1) the phenomena driving exploitation as well as the phenomena driving protection of the environment; and, (2) the interacting roles of government, the market and civil society. Finding methods that can enable this type of holistic intervention is a challenge to traditional environmental policy making and research. These have tended to treat the different types of organisations (government, industry and citizenry) as quite distinct, without addressing the potential synergies and conflicts that are

---

[29] Complexity refers to the elements within a system and their relationships. Complication refers to how difficult it is to understand the system.

[30] Gleick (n 10) 9–10.

[31] See particularly Cl 19 of United Nations, 'The Future We Want – Rio+20 Conference on Sustainable Development' (2012), accessed 11 October 2016 at http://www.un.org/disabilities/documents/rio20_outcome_document_complete.pdf; Organization of American States (OAS), 'Press Release: OAS and UNEP Sign Agreement on Environmental Rule of Law and Sustainable Development' (OAS, 2014), accessed 11 October 2016 at http://www.oas.org/en/media_center/press_release.asp?sCodigo=E-506/14. The IUCN has recently initiated a program on improving the environmental and social effectiveness of legal governance.

present as these different types of actors operate independently. Traditional approaches also tend to focus upon one issue at a time (for example, water or air pollution, over-consumption and waste, or biodiversity loss; or production improvement), using one type of instrument at a time (for example, regulation, subsidies or extension support). The methods required to manage integrated systemic action are quite different from those required to handle isolated instruments and narrowly specified challenges.

4. SYSTEMS THINKING ABOUT LEGAL GOVERNANCE

'Governance' is a term with competing definitions, its use largely depending on the issues being discussed. The essence of its meaning is that governance is a system for governing. Governing means controlling, shaping or influencing human behaviour, with the connotation of the exercise of some form of power. The obvious form of governance power is legal authority, but governance roles are carried out by those with other types of economic or social power, such as by the business sector or civil society, depending on the issue. Thus complementary or competing forms of governance of the environment and social justice issues can be at work in a society – for example government institutions may be operating alongside the rules imposed by financiers, and cultural norms. 'Legal' governance is a subset of the larger system and refers to norms that are supported by the law, whether through regulation, civil or common laws, or rules created by agreements that have legal force.

The traditional content of legal scholarship is doctrinal knowledge, based on interpretive research of legal instruments, and generally applied to specific fact situations or groups of fact situations in particular jurisdictions. Relevant legal instruments include constitutions, agreements, conventions, judgments, statutes and regulations. Alongside doctrinal analysis is a concern for efficient, effective and fair institutions to develop, interpret and apply the law.[32] This is the institutional component of legal work. The attention to laws and factual details means that the scholarship and the practice of law are largely reductionist. Specific issues are examined in detail, interpreting or proposing documents that are often jurisdiction

---

[32] 'Institutions' in this context referring to the organizational structures and networks that create, interpret and apply legal rules. This is distinct from the approach commonly adopted in economics where institutions are discussed as being the rules, which in the context of legal scholarship is not the usual use.

specific, or considering institutions that impact upon a specific aspect of the law. This precision is consistent with the main functions of the law, but limits integrative perspectives.

Legal governance of the environment involves dynamic transactions interwoven with the economic, social and cultural dimensions of man's relationship with the natural world. The more holistic systems approach has not yet greatly informed legal scholarship, but this is changing. A notable influence, particularly in European jurisprudence, has been socio/legal system scholars such as Niklas Luhmann and Jürgen Habermas.[33]

Environmental lawyers interact with systems-based approaches to investigate environmental and social issues, or when they contest policy conclusions derived using models.[34] Economic analysis of the law or environmental issues is also systemic, particularly when models are used. Many environmental lawyers have come to grips (at least at an informed layman's level) with system analyses of climate change, hydrology or biodiversity. A growing number of law researchers have embraced system concepts such as 'resilience', 'adaptive management', and 'complexity'.[35] The influence of system concepts is apparent in other chapters in this book. Systems theory can make a distinct contribution to holistic analysis in environmental law.[36]

---

[33] David Seidl, 'Luhmann's Theory of Autopoeitic Social Systems' (2004) 1, accessed 11 October 2016 at http://www.zfog.bwl.uni-muenchen.de/files/mitarbeiter/paper2004_2.pdf (considers the communicative and autopoetic elements that link both theorists).

[34] Elizabeth Fisher, Pasky Pascual and Wendy Wagner, 'Understanding Environmental Models in Their Legal and Regulatory Context' (2010) 22(2) JEL 251; Paul Martin and Jacqueline Williams, 'Science Hubris and Insufficient Legal Safeguards?' (2014) 31(4) EPLJ 311.

[35] Ooona A Hathaway, 'Path Dependence in the Law: The Course and Pattern of Legal Change in a Common Law System' (2003), accessed 11 October 2016 at http://digitalcommons.law.yale.edu/lepp_papers/270/; Craig Arnold and Lance Gunderson, 'Adaptive Law and Resilience' (2013) 43 Env't L Rep 10426; Dave Huitema *et al.*, 'Adaptive Water Governance: Assessing the Institutional Prescriptions of Adaptive (Co-) Management from a Governance Perspective and Defining a Research Agenda' (2009) 14(1) Ecol Soc 26; Poh-Ling Tan, KH Bowmer and C Baldwin, 'Continued Challenges in the Policy and Legal Framework for Collaborative Water Planning' (2012) 474 *Journal of Hydrology* 84; Elizabeth Kirk and Alison Reeves, 'Regulatory Agencies and Regulatory Change: Breaking Out of the Routine' (2011) 13(3) Env L Rev 155; Martin and Gunningham (n 22) 137; John Dernbach, 'Navigating the US Transition to Sustainability: Matching National Governance Challenges with Appropriate Legal Tools' (2008) 44 Tulsa L Rev 8; John Ruhl, 'Complexity Theory as a Paradigm for the Dynamic Law-and-Society System: A Wake-up Call for Legal Reductionism and the Modern Administrative State' (1996) 45(5) Duke LJ 849.

[36] Ruhl (n 35) 849.

Non-legal instruments interact with the law, within the system of governance.[37] Market interventions rely upon economic incentives and resource allocation mechanisms. These use private agreements and trading instruments that partly rest on legal foundations: property rights, contract law, public and private corporate governance rules and administrative laws. Persuasive and educational strategies rely upon communication and less upon legal support, though this may not always be the case.[38] While the regulatory, economic and social approaches reflect different theories and use different instruments and strategies, their effects are interwoven. Increasingly, legal governance is in practice conducted using a mix of interventions, rather than mutually exclusive choices of legal, market or social instruments.[39] More often than not this is a result of uncoordinated decisions made by different agencies serving different purposes, rather than a coherent strategic choice.

A reductionist approach minimizes attention to the potential conflicts or synergies between legal (and other) arrangements, and the lack of coherence in public policy that is spread across agencies with relatively narrow mandates contributes to this. Environmental law scholarship tends to emphasize national or international regulation, but the segregation from property, administrative, taxation, corporate and constitutional law is arguably counterproductive. The need to think holistically about interventions is reinforced when we consider what drives harmful exploitation of nature. Competing with initiatives to protect and restore the environment (or to equitably distribute costs and benefits) are public and private economic initiatives that increase human impact upon the earth, and allocate resources to the economically powerful.[40] Subsidies for resource use are no less important to governance outcomes than environmental protection laws, and pro-consumption forces like advertising and popular consumer culture drive harm that the law tries

---

[37] Robert Stavins, 'Environmental Policy Instruments: Experience with Market-Based Environmental Policy Instruments' (2001) 1 *Handbook of Environmental Economics* 355.

[38] For example where regulatory compliance, liability risk or other motivations are relevant to the information being communicated.

[39] OECD (n 22); Peter Grabosky and Neil Gunningham (eds), *Smart Regulation: Designing Environmental Policy* (Clarendon Press 1998); Martin and Gunningham (n 22) 137.

[40] Sometimes even as a perverse outcome of pro-environment interventions, e.g., see Paul Martin, 'Embedding Social Justice in the Design of Environmental Regulation' in Yves Le Bouthillier *et al.* (eds), *Poverty Alleviation and Environmental Law* (Edward Elgar 2012) 179.

to counter.[41] Counter-conservation forces also use legal tools, notably property rights, taxation and land use planning. Legal governance of the environment thus involves many aspects of public and private law which intersect, but often in the pursuit of opposing objectives. Legal and political outcomes are also shaped by asymmetries of power and resources. Politically, individual environmental and social justice disputes add up to an ongoing contest over who has legitimacy and power within the governance system, as well as being about the particular contested issue.[42]

5. THE ROLE OF INSTITUTIONS

Structures shape the patterns of flows in a physical system. The hardness and position of rocks shape the flow of water; mountains or tall buildings affect the flow of air; and in social systems rules and organizations shape the flow of signals and resources, helping to define the end state of that system.[43] The institutional structures relevant to climate change and law include, among others, environmental regulations and conventions, and the organizations responsible for creating, interpreting and enforcing rules.[44] Political interest groups and political parties shape the 'architecture' of the law,[45] and with regards to climate change, law and politics are intertwined.

---

[41] Douglas Koplow, *Biofuels – At What Cost Government Support for Ethanol and Biodiesel in the United States* (International Institute for Sustainable Development 2006); Géraldine Kutas, Carina Lindberg, and Ronald Steenblik, *Biofuels – At What Cost? Government Support for Ethanol and Biodiesel in the European Union* (International Institute for Sustainable Development, 2007).

[42] Robert Stavins, 'Introduction to the Political Economy of Environmental Regulations' (*Resources for the Future*, 2004), accessed 11 October 2016 http://www.rff.org/RFF/documents/RFF-DP-04-12.pdf.

[43] These social structures are 'institutions'. Different disciplines define the term differently but all have this system connotation. 'Institutional arrangements' are networks of institutions.

[44] Stephen Dovers, 'Institutions for Sustainability' (*TELA and Australian Conservation Foundation*, 2011), accessed 11 October 2016 at https://digitalcollections.anu.edu.au/bitstream/1885/40972/3/een0101.pdf; Paul Martin and John Becker, 'A Tale of Two Systems: Conflict, Law and the Development of Water Allocation in Two Common Law Jurisdictions: Are Social and Institutional Factors Critical?' (2011) *International Journal of Rural Law and Policy* (special edn) 1.

[45] The fundamental structures and relationships. See Rebecca Henderson and Kim Clark, 'Architectural Innovation: The Reconfiguration of Existing Product Technologies and the Failure of Established Firms' (1990) 1(35) Admin Sci Quart 9; Martin and Gunningham (n 22) 137.

Opponents of stronger legal governance of climate change issues, particularly industries that depend economically upon resource exploitation or consumerism, shape public policy responses to climate change.[46] How to ensure integrity, or at least reduce the most egregious dishonesty, in political processes is an important legal question given the heated politics of these issues.

Many public and private institutional arrangements are involved in agriculture, protection of the environment and climate adaptation. This generates complexity.[47] Complexity in institutional structures makes governance more complicated both for those governing and those being governed. Many rules and institutional arrangements affect the management of resources and the lives of those involved. Institutional complexity, often misleadingly blamed solely on the amount of law, is a source of tension and economic inefficiency.[48] An important new development is the increase of private governance, which interacts in an uncoordinated way with public governance. This adds complexity and transaction costs, but may provide much-needed options for more effective responses to climate change issues that are not restricted by constraints upon public governance such as economic and legal incapacity.[49]

The robust debate over environmental and social governance between the traditional actors (governments) and non-government organizations (including production or marketing chains) is an important development in climate change related governance.[50] Private agricultural governance exists

---

[46] John Ruhl, 'The Political Economy of Climate Change Winners' (2012) 97 Minn L Rev 11; Lauren Rickards, John Wiseman and Yoshi Kashima, 'Barriers to Effective Climate Change Mitigation: The Case of Senior Government and Business Decision Makers' (2014) 5(6) WIREs Clim Change 753; Axel Michaelowa, 'Climate Policy and Interest Groups – A Public Choice Analysis' (1998) 33(6) *Intereconomics* 251.

[47] For example K Prager *et al.*, 'Soil Degradation, Farming Practices, Institutions and Policy Responses: An Analytical Framework' (2011) 22(1) Land Degrad Dev 32; Jouni Paavola, 'Institutions and Environmental Governance: A Reconceptualization' (2007) 63(1) Ecol Econ 93.

[48] Paul Martin *et al.*, *Developing a Good Regulatory Practice Model for Environmental Regulations Impacting on Farmers* (Australian Farm Institute 2007).

[49] Thea Williams, *Environmental Management in Agriculture and the Rural Industries: Voluntary Approaches to Sustainability and Globalisation Imperatives* (Rural Industries Research and Development Corporation, 2009); CERES, 'Digging into Sustainable Agriculture', accessed 11 October 2016 at http://www.ceres.org/roadmap-assessment/roadmap-in-action/explore-by-topic/performance-supply-chain/digging-into-sustainable-agriculture.

[50] Derek Armitage, Rob De Loë and Ryan Plummer, 'Environmental Governance and Its Implications for Conservation Practice' (2012) 5(4)

in various forms including many industry codes and standards, reporting schemes, environmental stewardship and supply 'chain of responsibility' initiatives.[51] A prime example of private governance is the growth in popularity of 'ecolabels': labels on products that attest to their environmental attributes. There are more than 450 such regulatory schemes in 197 countries, covering 25 industry sectors.[52] For the individual farmer this can result in compliance and administrative costs that complicate agricultural transactions. A strategic challenge for agricultural governance is integration of public and private initiatives into a coherent legal governance system that incorporates rules and incentives through private contracts such as self-regulation and co-regulatory approaches. As with the issue of political debate, the lack of public mechanisms such as industry codes to assure the honesty and integrity of private schemes is a systematic deficiency in environmental governance.

Figure 4.1 illustrates the biophysical and social connections that shape biodiversity outcomes. Figure 4.2 below develops the system perspective further. It illustrates the links between the natural environment and human decisions and actions that reshape that natural system the decision makers and some of the institutions that shape the human aspects of transactions.[53] It pinpoints agriculture as one of the uses of water resources, and its role in shaping the biophysical context within which all water transactions occur. Climate change has a direct or indirect effect on virtually all parts of this system. Reliable governance will probably require interventions at different system 'places'. Governance interventions pursuing different objectives for different transactions will either work synergistically or antagonistically in helping or hindering adaptation to climate change. This suggests the need for a governance system focus, as well as a legal instruments focus, to more reliably improve social and environmental outcomes. Increasing natural and social system complexity is adding pressure to resolve institutional incoherence, but piecemeal responses to particular issues will contribute to increasing this problem unless a conscious choice is made to redesign legal governance architectures.

Conservation Letters 245; Thomas Dietz, Elinor Ostrom and Paul Stern, 'The Struggle to Govern the Commons' (2003) 302(5652) *Science* 1907.

51 For agriculture-specific examples, see Philippa Rowland, *National Inventory of Environmental Management Systems in Australian Agriculture* (RIRDC 2009).

52 Anon, 'Ecolabel Index', accessed 11 October 2016 at http://www.ecolabelindex.com.

53 Diagram modified from Martin and Verbeek (n 21) 237. Based on community participative modeling.

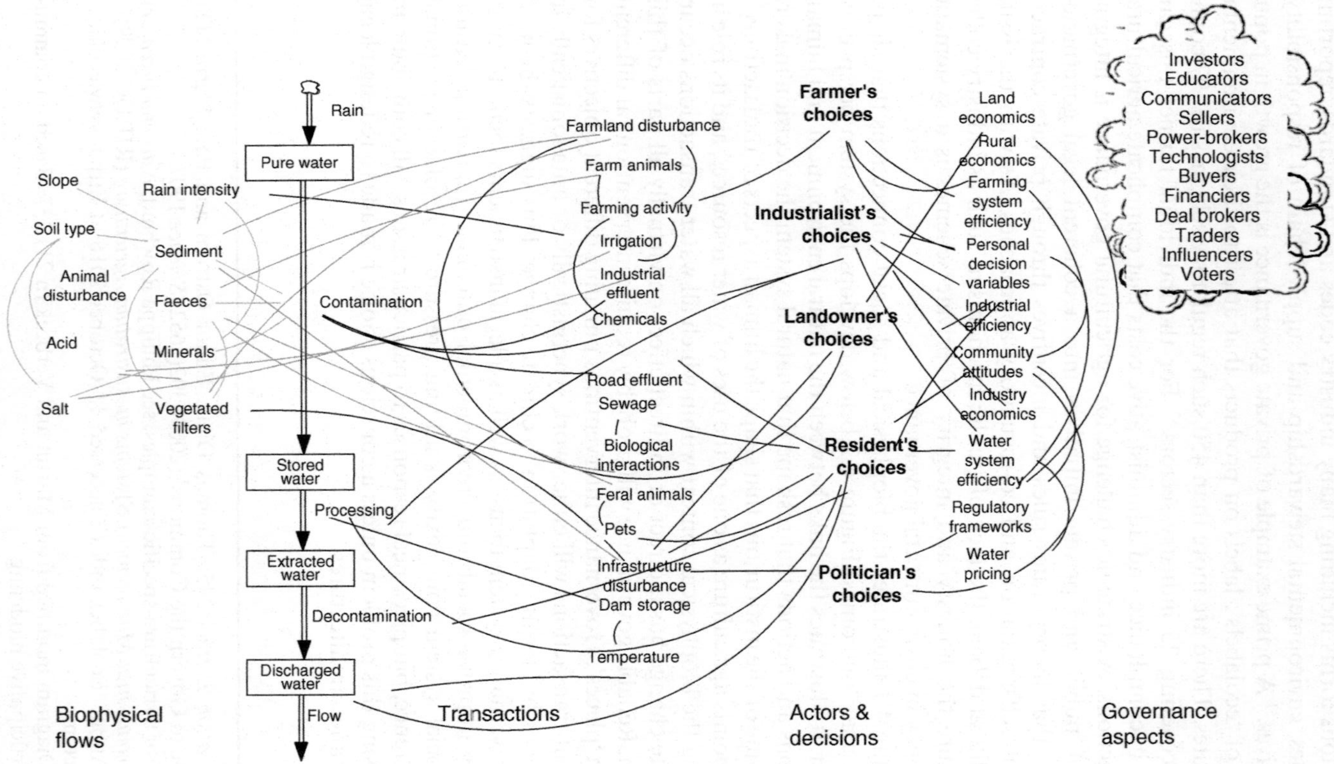

Figure 4.2 Institutions, transactions and water flow

6. WHAT TYPES OF RURAL PROBLEMS MUST BE SOLVED?

How a governance problem is understood determines the strategies used to address it. The dynamic we have discussed suggests that resource governance will become far more complicated, and this will call for innovation within the law. A discussion of the types of problems in agricultural, environmental and social governance to which the law must effectively respond provides further insights into what will be required of the law.

Each issue is characterized by an archetype.[54] A researcher analyzing living creatures may use a population growth or decline archetype (involving an initial population, reproduction, death and dispersal) to predict population growth (or decline) and distribution. In investigating a social system, archetypes of how economic capital accumulates or how conflict evolves might be used. Very complicated problems can arise when different archetypes intersect, increasing the difficulty of analysis.

The traditional archetypical agro-environmental problem is the aggregation of impacts from individual farms. Examples include over-consumption (for example, over-extraction of an aquifer or river, destroying a wild species by overhunting), or contamination (for example, irresponsible disposal of farm wastes resulting in river contamination).[55] The total harm is the sum of the individual impacts. Traditional governance responses to this have been regulation, economic incentives, extension and capacity creation that may include subsidies that target individual behavior. Intellectually these are relatively simple problems compared to others I will now consider, though in practice, management is often challenging and results are variable.

A more difficult version of this basic archetype arises if control or remediation requires that people work together. This might be because the strategies are too hard for individuals to implement even if they are willing, or when the resources needed to solve the problem are more than can be mustered by pooling resources. A typical example is cooperating to sustainably manage a common stock (for example, a fishery), or to build

---

[54] An archetype documents the main elements found in all systems of that type – it is a pattern. A problem archetype encapsulates the main features of re-occurring problems though each will present unique features.

[55] For Australian discussions of agricultural over-extraction or pollution of water see: Peter Nichols, Mary Maher & Associates and Susanne Cooper & Associates, *Australian River Management and Restoration* (Land and Water Australia 2000); Neil Gunningham and Darren Sinclair, 'Curbing Non-Point Pollution: Lessons for the Swan-Canning' (2004) 21(3) EPLJ 181.

infrastructures to protect the resource (for example, a local water source). The tragedy of the commons archetype often arises, and institutional changes such as property rights or a robust community agreement may be needed.[56] With this type of problem, a simple cause translates into the need for a far less simple 'collective action' solution.[57] The simpler the challenge, and the stronger the relationship between the people concerned, the more likely it is that communities can solve the problem without government intervention. Where there are weak interpersonal links, or where the issues are complicated, the need for external intervention is likely to be greater. In this next section I will expand upon the more difficult types of challenges to legal environmental governance.

7. DIFFICULT ENVIRONMENTAL GOVERNANCE CHALLENGES

Intractable problems are increasingly common in rural natural resource governance and often involve complex socio-ecological risks. Their distinctive features have not been well documented in law and policy scholarship, but have significant implications. Climate change coupled with other trends will make them more challenging and more frequent. Significant innovation in environmental law and associated governance will be needed to respond.

The first version of a more difficult type of governance problem involves complex socio-ecological risks. An historical example is the effect of widespread use of DDT upon the population of the iconic species, the Bald Eagle. DDT was an essential agricultural and public health chemical.[58] The use of DDT to control insects led to chemical concentration in the diet of this top predator. This created unexpected negative impacts upon eagle eggs and hence their population. Understanding the causal relationships, and negotiating a resolution once the issue was identified, was very difficult. Notwithstanding a lot of public pressure, it took ten years from the publication of the book *Silent Spring*[59] until the ban on DDT in the USA.

---

56 Garrett Hardin, *The Tragedy of the Commons* (American Association for the Advancement of Science 1968) 162.

57 Elinor Ostrom, 'Analyzing Collective Action' (2010) 41(s1) *Agricultural Economics* 155. For examples in agriculture see Jules Pretty and Hugh Ward, 'Social Capital and the Environment' (2001) 29(2) *World Development* 209.

58 Douglas Allchin, 'Rachel Carson and Silent Spring' in Joel B Hagen *et al.* *Doing Biology* (HarperCollins 1996) 185–96.

59 Rachel Carson, *Silent Spring* (Houghton Mifflin 1962).

The issue is still politically and scientifically contested, though recovery in the eagle population supports the original diagnosis.

A current example of an agricultural issue with complicated biophysical and social interactions and diverse actors is the production and use of biofuels. Depending on contexts and technologies, biofuels can help address climate problems while contributing to agricultural prosperity. In some settings they have also caused significant harms, new risks to the environment and adverse social impacts. Reported problems include the loss of habitats, economic distortion through subsidization, increased competition for agricultural inputs, and increases in food prices affecting the poor. A similar biofuel enterprise carried on in different jurisdictions can have markedly different impacts. The variables causing different results include choices of feedstocks and process technologies, corporate ownership, land management, fuel end uses, the characteristics of the local and international value chain and governance. The factors that can frustrate solutions include competing interests, powerful corporations, an internationally traded commodity, asymmetrical power and information, and institutional impediments;[60] the consequences of which only become apparent when problems emerge or when viable solutions cannot be found.

The two examples above demonstrate that the process of investigation and solution-seeking for complicated problems is typically politicized. If effective action involves many actors pursuing different interests, and scientific and other uncertainties exist, agreeing upon a solution is a complicated socio-political problem. Affected actors try to ensure that the allocation of benefits and costs of intervention are to their advantage. Complex negotiations are needed to overcome limits to government (for example, private property, economic and civil rights), and to agree how intervention will be resourced. The process may involve legal and political conflict over rights and obligations. Difficult issues of accountability, including identification of causes, impacts and forecast outcomes, are interwoven with legal instrumental and institutional questions. As a result, the transaction costs of initiating any intervention is high. Law and politics are difficult to disentangle with these sorts of problems.[61]

---

[60] Robert Bailis and Jennifer Baka, 'Constructing Sustainable Biofuels: Governance of the Emerging Biofuel Economy' (2011) 101(4) *Ann Assoc Am Geogr* 827; John Pickett *et al.*, *Sustainable Biofuels: Prospects and Challenges* (Royal Society 2008); Paul Martin and Elodie Le Gal, 'Biofuels: Why Should We Be Cautious About Policy?' in Yves Le Bouthillier *et al.* (eds), *The Law and Policy of Biofuels* (Edward Elgar 2016).

[61] Andreas Wimmer, Indra de Soysa and Christian Wagner, *Political Science Tools for Assessing Feasibility and Sustainability of Reforms – Discussion Papers*

With regard to these issues, the implementation of solutions is often messy – particularly when there are long delays between the commencement of intervention (which may involve large investments or foregone economic opportunities) and any results. Complete supervision of the process is often challenging because the issues are complicated and the transaction costs of monitoring are high. This is reflected in the number of stakeholder types and interests involved. This archetype involves political and institutional complexity, as well as causal uncertainty, that is greater than the basic archetype. Solutions must be effective on (at least) three levels: changing behavior patterns of individual farmers or other actors at the front line of the problem; creating instruments that sensibly respond to ambiguity and uncertainty; and negotiating sector-wide arrangements that overcome political impasses. However, the risk of law and other policy failure due to misdiagnosis, political dynamics and resourcing problems is far greater than for simpler problems.[62]

8. ACKNOWLEDGING THE SPATIAL FEATURE OF RURALITY

Geographic dispersal of activities and problems is less a concern in urban and industrial settings than it is in agriculture, where it has fundamental effects. For example, controlling salinity induced by climate changes affecting rivers or coasts typically requires coordinated action by landholders across a large area, land use planning or investment by local governments and the cooperation of regional or state agencies. Coordination of dispersed actors is difficult, particularly if there are disagreements about responsibility or cost-sharing and weak private incentives to cooperate. Economics and legal norms further frustrate effective intervention. Constitutional law and rights of private property, when coupled with political norms, can insulate land managers who may not wish to cooperate. If effective action requires their ongoing investment of funds and labor, legal mechanisms that can force unwilling investment are limited. This problem is greater if

---

on Development Policy No 61 (ZEF 2003); Stavins (n 42); Rickards, Wiseman and Kashima (n 46) 753.

[62] Paul Martin and Jim Shortle, 'Transaction Costs, Risks, and Policy Failure' in Claudia Soares *et al.* (eds), *Critical Issues in Environmental Taxation: International and Comparative Perspectives* (Oxford University Press 2010); Cameron Hepburn, 'Environmental Policy, Government, and the Market' (2010) 26(2) Oxf Rev Econ Pol 117.

there is social or political resistance to enforcement. The transaction costs of effectively supervising dispersed activities in rural areas on an ongoing basis can make 'logical' government action unrealistic. Chapter 6 in this book dealing with climate and invasive species institutions demonstrates these institutional difficulties.

When an issue spans international boundaries, the legal and practical limits to governmental power become even more significant.[63] Sovereignty limits the capacity of states to require transjurisdictional cooperation. In practice it is difficult to identify jurisdictional authority and responsibility, to gather evidence or to monitor transactions. Achieving coordinated action to control carbon emissions illustrates these issues, but the same institutional challenge exists with other climate-enhanced agricultural issues including desertification, biodiversity protection and biosecurity.[64] Internationalism (particularly that involving powerful transnational corporations and trade freedoms) makes internationally coordinated governance an intractable element of modern society, notwithstanding the explosion of international legal instruments. Many multinational corporations have more economic power and topic knowledge than the governments trying to control them. These corporations can rely upon free trade and other conventions to help protect their economic autonomy. This leaves few public governance options beyond voluntary or negotiated arrangements, and the power of government may only be its ability to threaten and persuade.[65]

In summary the key features of the complicated problems that challenge traditional legal governance approaches are:

---

[63] The inability of neighboring states of Indonesia, Malaysia and Singapore, to achieve coordinated action over forest fires in that country, is illustrative of trans-jurisdictional difficulties. Transboundary management of water and biodiversity has similar characteristics, though many mechanisms have been developed. Charlotte De Bruyne and Itay Fischhendler, 'Negotiating Conflict Resolution Mechanisms for Transboundary Water Treaties: A Transaction Cost Approach' (2013) 23(6) Glob Environ Change 1841; Michelle Lim, 'Is Water Different from Biodiversity? Governance Criteria for the Effective Management of Transboundary Resources' (2014) 23(1) RECIEL 96.

[64] For example Ian Hannam, 'International and National Aspects of a Legislative Framework to Manage Soil Carbon Sequestration' (2004) 65(3) Clim Change 365.

[65] Kaisa Sorsa, *Private Regulation in Global Value Chain – A Trade Barrier or an Opportunity for Public-Private Co-operation?* (European Consortium on Political Research 2010); David Victor, Kal Raustiala and Eugene Skolnikoff (eds), *The Implementation and Effectiveness of International Environmental Commitments: Theory and Practice* (MIT Press 1998).

(1) The relationships between causes and effects of many problems are not obvious. This generates uncertainty that requires sophisticated investigation to resolve, facilitating political contestation. The continuing contests over whether climate change is 'real', or whether it is human induced illustrates this feature.
(2) Many actors and elements are involved along the causal pathway of complex problems. Competition between interests and values affects the negotiation of viable solutions. The continuing contests over fossil fuel subsidisation for farming and transport illustrate this problem.
(3) Economic issues, including the costs of preventing harms (such as loss of production, and reduced economic activity), and difficulties with the distributional justices of allocation. For example, the economics of legacy infrastructures for electricity generation and distribution provide a counter-argument against policies that support non-traditional energy generation.
(4) The problem has 'vicious' characteristics. A solution to the problem can generate a chain of other problems, some of which may not have solutions.[66] An example is the perverse impacts that have arisen from a well-intentioned shift towards biofuels.

For some rural and climate issues these difficulties are increased by two spatially related issues: (1) the need for coordinated action across large distances, spanning land tenures. Coordination is frustrated by asymmetrical incentives and landholder capabilities, public and private institutional complexity, and the limits of public powers to require private action; and (2) if the issues span state or national boundaries, coordination difficulties are compounded by national sovereignty and international free trade institutions.

With such complicated rural governance issues intervention is likely to require a strategy that addresses diverse stakeholders and transactions across the system. A mix of legal, market and social initiatives will typically be needed to achieve system-wide outcomes. Creating the framework for this coordination will often involve governments, state agencies, market and other intermediaries through new coordinating institutions. This is

---

[66] In a 'vicious' cycle the solution of a problem typically generates another. For example, protecting a hunted animal population using regulation or a market instrument may harm traditional or impoverished users. Managing this social justice problem by exemptions may promote corruption, or undermine protection. For an application to regulatory effectiveness see: Christine Parker, *Reducing the Risk of Policy Failure: Challenges for Regulatory Compliance* (OECD 2000).

in addition to the task of involving farmers, other citizens, and industry or government employees in improved stewardship. Complex coordinated collective action may simultaneously require regulation, incentives, persuasion, capacity building and economic support to align individual behavior to the requirements of the strategy. The likelihood is that effective interventions will affect property rights and can involve constitutional, administrative, as well as international law.

This suggests that new mechanisms for managing governance itself will be increasingly needed. The present cumbersome approaches to finding solutions to carbon and climate issues are, in this light, clumsy experiments hopefully leading to more effective ways of dealing with the new issues of this very complicated type.[67] There is one further complicating consideration: environmental harms that are self-generating and evolving, which pose different management challenges to the more conventional types of problem.

9. A FINAL COMPLICATION: AUTOPOEISIS

The term 'autopoeisis' designates the ability of some systems to self generate or to internally generate change (for example, evolve).[68] An example is genetic mutations, which can be adaptive and sustained, or maladaptive and die out. Social system examples of this phenomenon include the development of culture (for example, art or music), or institutions. An art critic from the very early 20th century, whose art appreciation is attuned to romantic realism, would probably struggle to understand much modern art. The path from realism was self-generating, each new art form built on and departed from the previously favoured aesthetic. Competition between peers involving novelty and abstract ideas was (and still is) a powerful force for cultural evolution.

The role of autopoeisis in the law has been identified by social system theorists, especially Luhmann, and also has relevance to the evolution of legal norms.[69] The significance of autopoeisis has not previously been focused upon as a governance issue, but is very important in the context of agriculture and climate change.

---

67 Dernbach, (n 35) 8.

68 Humberto Maturana, *Autopoiesis and Cognition: The Realization of the Living* (Springer 1980).

69 Andreas Philippopoulos-Mihalopoulos, *Critical Autopoiesis: The Environment of the Law* (Elven 2011) 11; Richard Posner, 'The Decline of Law as an Autonomous Discipline: 1962–1987' (1987) 100(4) Harv L Rev 761.

With living creatures, internally generated change coupled with the availability of resources from the environment is the basis of evolution. Darwinian natural selection allows only those internally generated variations that prove to be useful in securing resources survive.[70] This process can be illustrated with invasive plants, animals or pathogens, where autopoietic change is key to adapting to new environments. The most powerful colonizers tolerate diverse conditions, breed prolifically or rapidly, or alter their behaviour more readily than less adaptive species. Populations spread and increase 'under their own steam'. This ability may be the result of genetic characteristics, but with animals it can also be the result of intelligence. Autopoeisis can be applied to some non-living processes, for example soil erosion or salinity. With climate change, harmful weeds, diseases and animals that possess this adaptive character will exploit the advantage, and processes such as soil erosion and desertification are likely to accelerate.

In contrast with simple harm accumulation, the agricultural harms from erosion or salinity (or from pest animals, weeds or disease) increase once the relevant processes begin. They can intensify and spread across a large area if unchecked. Widespread distribution is likely where wind or water aid expansion (as occurs with many weeds, and some forms of soil erosion), or where the harmful agent is mobile (particularly with invasive animals). With increased winds and storms, due to increased energy in natural systems, such problems are likely to become worse. Human vectors can spread invasion problems purposefully (for example transferring a flowering weed or a pest animal as a pet), or accidentally (for example within cargo on a truck or a boat). Climate refugees can also be a potent vector for agricultural invasives, especially disease, if they avoid biosecurity checks as they try to enter safer countries.

Self-perpetuating problems in agriculture are particularly difficult to manage. Their natural trajectory is to increase exponentially, unless hostile conditions or active human interventions limit this. The resulting losses of natural biodiversity or production, or harms to human and animal species, are not manageable by passive strategies like ceasing harm-causing action or setting aside land from production (unless in combination with active interventions). It is the control, not the problem, which demands sustained inputs of energy and resources. The economic and physical capacity of landholders to continue investment, the legitimacy of obliging them to do so when they are not the 'polluter,' and the problems of accountability for self-propagating harms are significant challenges to rural law and public

[70] Charles Darwin, *On the Origin of Species by Means of Natural Selection, or the Preservation of Favoured Races in the Struggle for Life* (John Murray 1859).

policy. Simplistic 'polluter pays' principles do little to help in deciding what is the appropriate and practical response.

Property rights issues intersect with land use fragmentation, economics and autopoeisis to create intractable public management difficulties. The operators of different types of enterprise do not have the same incentive to control specific problems. For example, in Australia the control of non-native foxes is a concern for producers of lambs or poultry, but largely irrelevant to most grain or horticulture enterprises. Control of weeds that produce burrs will be a concern for wool producers, but not necessarily cattle farmers. In some cases landholders have a private incentive to maintain self-generating problems. Examples include continuing to use grasses for agriculture which also spread and smother ecosystems, or harboring a population of feral animals for hunting (for example, in Australia, feral pigs or deer). The increasing diversity of agricultural land uses means that land use fragmentation adds to the complications of coordinated control. Asymmetrical interests also exist for the protection of threatened species in different parts of the world, such as roaming elephants, predatory wolves or other valued species that compete with some land uses.

Thus, to the six specific complications of rural natural resource governance we must now add the seventh complication, autopoeisis. Many of these complications are made worse by climate change, but climate change alone does not sufficiently explain the increasing difficulties of rural governance.

10. DRAWING THE STRANDS TOGETHER

The interconnection between natural and social systems in agriculture is deep, and involves numerous transactions. Climate change will contaminate many of these. While I will not provide a detailed socio-economic analysis of the capacity of rural communities to respond to the challenges, some aspects should be noted.[71]

Agriculture-dependent communities are on average under more economic stress than non-dependent communities, though there is a lot of variation concealed by the averages. Statistics indicate that national agricultural dependency and relative poverty are often correlated.[72]

---

[71] For a detailed discussion of the social system aspects of rural natural resource governance, see Martin and Williams (n 4).

[72] Based upon comparison of a more agriculture dependent group consisting of Iceland, Argentina, China, Thailand and Indonesia and a less agriculture

Country statistics differ because of national wealth, enterprise types and public policy, but overall farmers are likely to be less well-resourced compared to national average populations. To illustrate, in 2011 in developing countries 55 percent of the population was rural, and more than 70 percent of these were living in extreme poverty.[73] In many countries farmers are typically less educated (due to cultural, spatial and economic considerations); have relatively less access to professional services and infrastructures; and are the subject of economic pressure (largely due to margins within the value chain accruing to non-rural intermediaries).[74] While farmers as a group have relatively high value land holdings, this capital cannot be realized if the enterprise is to stay viable. Finding feasible ways to provide more effective rural natural resource governance is a challenge for which adequate solutions have yet to be found in many jurisdictions. Climate change will, as this chapter has illustrated, create many challenges, not limited to the obvious ones like declining water availability or increased heat stress on crops and animals. These will affect the income and the costs side of the farm budget, adding to the economic pressures on farmers.

It will always be essential to address local issues in particular regions. This is being pursued with climate change specific legislation, and for issues such as those discussed in this book.[75] These chapters show that there is a long way to go, but change is taking place. The recent explosion of natural resource governance instruments spawned by governments, industry and civil society to address specific issues is both heartening and daunting. The number of legal and other environmental instruments has increased

---

dependent group of Switzerland, UK, USA, Canada, and France, with Australia compared to each set. See Martin and Williams (n 4).

[73] IFAD, 'Rural Poverty Report' (2011), accessed 11 October 2016 at http://www.ifad.org/rpr2011/report/e/overview.pdf.

[74] For international information see United Nations, 'Rural Poverty: Resources for Speakers on Global Issues', accessed 11 October 2016 at http://www.un.org/en/globalissues/briefingpapers/ruralpov/developingworld.shtml. In relation to farm viability in Australia see Australian Bureau of Agricultural and Resource Economics, 'Drivers of Structural Change in Australian Agriculture' (Rural Industries Research and Development 2007), accessed 11 October 2016 at https://rirdc.infoservices.com.au/items/07-057. For an examination of the economic dynamics of climate change policy alternatives on farmers in Australia, see Tingsong Jiang, Kevin Hanslow and David Pearce, *On-Farm Impacts of an Australian Emissions Trading Scheme: An Economic Analysis* (Rural Industries and Resource Economics 2009).

[75] Nachmany (n 3); David Markell and John Ruhl, 'An Empirical Assessment of Climate Change in the Courts: A New Jurisprudence or Business as Usual?' (2012) 64(1) Fla L Rev 15.

rapidly.[76] In addition to the labelling schemes discussed earlier there are reportedly more than 2,000 treaties; around 9,000 legislative and regulatory instruments; and very many court decisions.[77] The OECD Environmental Policy Instruments database shows the use of taxes, fees and charges; tradable permits; deposit-refund systems; environmental protection; or exploitation subsidies.[78] The potential for confusion, wasted resource and lost opportunity is apparent. Many of these instruments and initiatives relate directly or indirectly to agriculture, such as those which govern land use, diffuse pollution, control of invasive species, agricultural methods, and the environmental attributes of farm products.[79] The potential for incoherence becomes more obvious when these pro-environment and pro-social justice interventions are considered alongside pro-consumption initiatives such as fossil fuel or other subsidies that contribute to harm. The challenge of climate change has resulted in a plethora of new legal and other governance instruments. Suboptimal performance from the governance system overall is the most likely result of its ever increasing complexity.[80]

There is a pressing need for strategically planned and executed 'whole-of-system' approaches to rural environmental governance, using a mix of instruments to simultaneously target different actors. This is particularly true for the very difficult problem archetypes outlined earlier.

A second fundamental requirement is better ways to identify and

---

[76] Robert Percival, 'History of Environmental Law (1)' (IUCN Academy of Environmental Law 2014), accessed 11 October 2016 at http://www.iucnael.org/en/documents/1109-history-env-law-1-essential-readings-in-environmental-law/file; Robert Percival 'History of Environmental Law (2)' (IUCN Academy of Environmental Law 2014), accessed 11 October 2016 at http://www.iucnael.org/en/documents/1110-history-env-law-2-readings-in-environmental-law/file; Stavins (n 37).

[77] Topic search on Ecolex, 'The Gateway to Environmental Law' http://www.ecolex.org (accessed 11 October 2016).

[78] OECD, 'Database on Instruments Used for Environmental Policy' (2014), accessed 11 October 2016 at http://www2.oecd.org/ecoinst/queries. See also Darren Swanson *et al.*, 'National Strategies for Sustainable Development: Challenges Approaches and Innovations in Strategic and Co-ordinated Action Based on a 19-Country Analysis' (2004), accessed 11 October 2016 at https://www.iisd.org/pdf/2004/measure_nat_strategies_sd.pdf.

[79] National Farmers' Federation, 'Submission to the House of Representatives Standing Committee on the Environment: Inquiry into Streamlining Environmental Regulation, Green Tape, and One Stop Shops' (2014), accessed 11 October 2016 at http://www.aph.gov.au/Parliamentary_Business/Committees/House/Environment/Green_Tape/Submissions.

[80] See Parker (n 66); Paul Martin and Jacqueline Williams, 'Policy Risk Assessment' (CRC for Irrigation Futures 2010).

manage the risk of failure or perverse outcomes of these many interventions. The difficulties in predicting what issues will need to be managed and how they are intrinsic to climate change policy, which have been outlined above, make the risks of policy failures in managing climate change issues associated with agriculture particularly high. These risks increase with unpredictability and complexity.

I close the chapter with an observation about agricultural systems that should give hope. Even though major change to social systems, including governance systems, is very difficult, it is possible. Equifinality and chaos theory suggest that small adjustments can shift whole systems when 'potentiated' by positive dynamics or by small supportive changes to the context of the system. There are many agricultural stewardship initiatives (both government and private), new 'climate friendly' agricultural practices and customer and supply chain strategies to reduce the climate footprint of agricultural produce; some are already being implemented. This gives cause for hope that the challenges we face can be met. This chapter suggests that all of these initiatives can contribute, but beyond them is the need to create a more integrated approach that maximizes the potential synergies and reduces the wasteful conflicts.

5. Climate change and water law in agriculture

Robert W Adler

1. INTRODUCTION

Under international law, every person has a human right to basic life support goods and services, including food, clothing and shelter.[1] All of these are supplied in part by agricultural economies. The global population is expected to grow to nine billion by 2050,[2] and as per capita food consumption increases due to rising affluence and changing food preferences,[3] the United Nations (UN) estimates that agricultural production must increase by 50 to 70 percent by 2050 to meet demand.[4] A growing population will also increase demand for fiber and other agricultural materials, as well as biofuels that increasingly compete with food for agricultural production. Moreover, about a fifth of the world's population and up to 65 percent of people in developing countries rely directly on agriculture and forestry as their source of livelihood.[5] Agriculture represents between 20 and 60 percent of the GDP in most developing countries, and an estimated 2.5 billion households are involved in agriculture worldwide.[6]

Water is essential to this agricultural productivity. Atmospheric moisture and soil water, along with crop management practices and soil quality,

---

1 Olivier De Schutter, 'Climate Change and Human Rights Law' (2009) 50 *Virginia Journal of International Law* 163.

2 United Nations, Department of Economic and Social Affairs, Population Division, '2011 World Population Prospects' (UN 2010 Rev), accessed 11 October 2016 at http://esa.un.org/unpd/wpp/Documentation/WPP%202010%20publications.htm.

3 Bruce Campbell *et al.*, *Agriculture and Climate Change: A Scoping Report* (Meridian Institute 2011) 1.

4 United Nations Environment Programme, Food and Ecological Security, *Identifying Synergy and Tradeoffs* (UNEP Policy Series no. 4, 2011).

5 Campbell *et al.*, above note 3 at 1.

6 United Nations Conference on Trade and Development, *Assuring Food Security in Developing Countries Under the Challenges of Climate Change: Key Trade and Development Issues of a Fundamental Transformation of Agriculture* (2011) 2.

are the key drivers of agricultural productivity.[7] This is characterized by water in sufficient quantities, at the correct place and time, and at an affordable cost relative to the value of the agricultural goods produced and other costs of production. It also requires that water be available with sufficient reliability or predictability to support different kinds of agricultural production. Indeed, agriculture is the dominant source of global water demand (and will remain so for the foreseeable future), accounting for 71 percent of global water withdrawals in 1995 and more than 80 percent in some lower-income countries.[8]

Different agricultural economies may require different amounts of water and at different times.[9] Poor water quality may affect some crops more than others. Water may be available in an agricultural region generally, but not at the location of the best soils or other resources, thus requiring storage and transport to support a successful agricultural economy. Sufficient water may be available to grow particular crops, but other parties may have competing water rights.

Water law regimes, both domestic and international, help human economies use available water supplies for agriculture efficiently, effectively, or equitably depending on the goals established by the governing law.[10] Increasingly, water law also seeks to preserve legitimate human uses while also protecting aquatic and other water dependent ecosystems.[11] At bottom, water law helps to provide sufficient certainty as to the right to use available water, along with reliable mechanisms for resolving disputes among users (including users across international borders and those who enjoy in-stream uses such as recreation and ecosystem protection) to support the investments required to produce agricultural products.

The Intergovernmental Panel on Climate Change (IPCC) has concluded with increasing confidence that climate change is having significant effects on agriculture due to a combination of factors that include water. Climate change is likely to cause shifts in precipitation patterns and other variables that affect water supply (such as temperature, evaporation rate and soil

---

[7] Jiménez Cisnero *et al.*, *Freshwater Resources, in Climate Change 2014: Impacts, Adaptation, and Vulnerability* (Fifth Assessment Report of the Intergovernmental Panel on Climate Change 2014) 251.

[8] Upali A Amarasinghe and Vladimir Smakhtin, *Global Water Demand Projections: Past, Present and Future* (IWMI Research Report 156 2014) 11, 15.

[9] Amarasinghe and Smakhtin, above note 8 at 15.

[10] Robert W Adler, *et al.*, *Modern Water Law: Private Property, Public Rights, and Environmental Protections* (Foundation Press 2013) 1.

[11] A Dan Tarlock, 'Four Challenges for International Water Law' (2010) 23 *Tulane Environmental Law Journal* 369, 385–96.

moisture level). As a result, there is a pressing need for individual nations and the global community to design and implement strategies to adapt existing water law to the potentially significant impacts of climate change.[12] Indeed, from an economic perspective, the IPCC predicts that the most significant impacts of climate change will be to agriculture, particularly in the developing world.[13] Predicted negative effects are more common than positive ones. The IPCC predicts that climate change poses large risks to global food security and is likely to contribute to increased food prices by the middle of the 21st century.[14] Even now, large populations in sub-Saharan Africa, south Asia and elsewhere are undernourished or food insecure.[15]

This chapter explores the impacts of climate change on the relationship between water and agriculture, and the degree to which domestic and international water law will have to adapt to those impacts. It reviews predictions about the likely impact of climate change on water supplies and the stability and reliability of those supplies. The chapter also evaluates the ability of both international water law and domestic water law (using US water law as the prime example) to adapt to those changes.

2. RISKS TO AGRICULTURAL WATER SUPPLIES AND INFRASTRUCTURE

Models designed to assess the potential effects of climate change predict that global food production will decline by the second half of the 21st century and that food prices will rise globally.[16] Approximately 600 million more people could be at risk of hunger by 2080 due to climate change (although not all of those effects are attributable directly to water issues).[17]

---

12 William R Cline, *Global Warming and Agriculture, Impact Estimates by Country* (Center for Global Development 2007) 1; S Mark Howden *et al.*, 'Adapting Agriculture to Climate Change' (2007) 104 PNAS 19691.

13 Purnamita Dasgupta *et al.*, 'Rural Areas' in *Climate Change 2014: Impacts, Adaptation, and Vulnerability* (Fifth Assessment Report of the Intergovernmental Panel on Climate Change 2014) 631.

14 John R Porter *et al.*, 'Food Security and Food Production Systems' in *Climate Change 2014: Impacts, Adaptation, and Vulnerability* (Fifth Assessment Report of the Intergovernmental Panel on Climate Change 2014) 488–9.

15 Porter *et al.*, above note 14 at 490.

16 Gerald C Nelson *et al.*, *Food Security, Farming, and Climate Change to 2050, Scenarios, Results, Policy Options* (International Food Policy Research Institute 2010) xvii–xviii.

17 Human Development Report 2007/2008, *Fighting Climate Change: Human Solidarity in a Divided World* (UNEP 2007).

The predicted impacts of climate change on agriculture are likely to differ dramatically from region to region. Unfortunately, the largest impacts will likely occur in countries and regions that are disproportionately vulnerable due to undiversified economies, shortages of capital to invest in needed improvements in infrastructure, poor access to technological improvements and insufficient earnings from other industries.[18] The most significant declines in food production are expected in portions of Africa and in south central Asia, and to a lesser degree in Latin America. By 2020, an additional 60 to 90 million hectares in sub-Saharan Africa are projected to become more arid, and yields from rain-fed agriculture could decline by as much as 50 percent.[19] Crops could also decline by up to 30 percent in central and southern Asia by the mid-21st century.[20]

2.1 Water Quantity

Climate change will affect precipitation patterns and water supply around the world.[21] Globally, higher global average temperatures will cause greater evaporation from oceans and other large water bodies, leading to more precipitation. These factors will also increase atmospheric moisture.[22] The distribution of rain and snow will also change, likely resulting in higher precipitation in northern latitudes as well as India and parts of central Asia, and lower precipitation in the mid-latitudes.[23] Precipitation is likely to decrease in central and southern Europe, southwestern North America, Amazonia and southern Africa.[24] Overall, for each degree of global average temperature rise, an additional 7 percent of the world's population will face a predicted decrease in renewable water resources of at least

---

[18] Jodie Keane *et al.*, *Climate Change and Developing Country Agriculture: An Overview of Expected Impacts, Adaptation and Mitigation Challenges, and Funding Requirements* (International Centre for Trade and Sustainable Development 2009).

[19] IPCC, *Climate Change 2014, Impacts, Adaptation and Vulnerability* (IPCC 2014) Ch. 9.

[20] *Ibid* at 22.

[21] *Ibid* at 3.

[22] US Global Change Research Program, *Global Climate Change Impacts in the United States* (Cambridge University Press 2009) 41–2.

[23] William Easterling, Pramod Aggarwal *et al.*, *Food, Fibre and Forest Products* in *Climate Change 2007: Impacts, Adaptation and Vulnerability. Contribution of Working Group II to the Fourth Assessment Report of the Intergovernmental Panel on Climate Change* (Cambridge University Press 2007) 275, 281.

[24] Cisnero *et al.*, above note 7 at 240.

20 percent, with the largest impacts in subtropical regions.[25] Even now, about 80 percent of the world's population faces serious threats to water security due to some combination of water quantity and quality issues.[26]

Although climate change is likely to affect surface water conditions immediately, ground water will be affected as well, though less scientific attention has been devoted to groundwater effects.[27] Overall, it is predicted that 34 to 38 percent of the global population will experience a decrease in renewable groundwater resources by the 2080s.[28] For each degree of mean global temperature rise, another 4 percent of the land in the world is projected to experience a reduction in groundwater resources of over 30 percent.[29]

Rising temperatures will increase the amount of water needed to grow crops, especially where soil moisture declines due to heat, changes in precipitation and other factors.[30] Crop selections and planting locations are also likely to change, increasing demand for water in some regions and decreasing demand in others. The effects of climate change on irrigated agriculture differ significantly from rain-fed agriculture.[31] Agricultural systems supported by irrigation are more resilient to reduced precipitation because of water storage and delivery systems that can supply water at the appropriate time, so long as supplies meet agricultural demand. However, reduced precipitation is likely to increase irrigation water demand by more than 40 percent in parts of Europe, the United States and Asia.[32] Despite significant uncertainty and regional variation, some experts project a global increase in irrigation of between 7 to 21 percent by the 2080s.[33]

Livestock is also sensitive to changing precipitation and related conditions that affect forage. Changes in the abundance, condition and plant mix of forage in natural rangeland could impair livestock health by causing reduced nutrition, decreased herd sizes or increased feed costs. Available models are even less developed to predict the impacts of climate change on grazing systems due to uncertainties concerning the effects of CO_2 fertilization, precipitation changes, pests and other factors.[34] Grazing lands in

---

25 *Ibid.* at 232.
26 *Ibid.* at 249–50.
27 *Ibid.* at 244.
28 *Ibid.* at 251.
29 *Ibid.* at 251.
30 US Global Change Research Program, above note 22 at 72.
31 Dasgupta *et al.*, above note 13 at 633.
32 Cisnero *et al.*, above note 7 at 251.
33 *Ibid.* at 251.
34 Britta Tietjen and Florian Jeltsch, 'Semi-arid Grazing Systems and Climate Change: A Survey of Present Modeling Potential and Future Needs' (2007) 44 *Journal of Applied Ecology* 425; Porter *et al.*, above note 14 at 494.

arid and semi-arid regions, where a large number of pastoralist communities derive their livelihoods, will be particularly vulnerable to changes in temperature as well as the amount and seasonal distribution of precipitation. These communities will also be more susceptible to extreme drought events. Those changes will increase water stress to livestock directly and also reduce available forage and other ecosystem resources, particularly in regions such as sub-Saharan Africa and south Asia.[35]

The ability of agriculture to adapt to these changes is complicated further by the volatility in water supply that is likely to result from the uncertainties associated with climate change. Even without factoring in climate change, experts have difficulty projecting agricultural water demand over time, which depends on a wide range of variables.[36] That challenge will be greater due to the uncertainties associated with climate change. Scientists have concluded that 'stationarity', the ability to predict water resources and to conduct physical infrastructure and capital investment planning within reasonably predictable ranges of uncertainty, is now 'dead' due to climate change.[37] There are three main reasons for this conclusion. First, we do not yet have a clear understanding of the likely magnitude of climate change effects over time. There is variability across global circulation models used to project the impact of greenhouse gasses on global average temperatures over time,[38] and uncertainty about the timing and magnitude of greenhouse gas reductions that are likely due to various mitigation strategies and economic conditions.[39] Second, it is even more challenging to predict the relationship between increasing global average temperatures and other climatic conditions, such as the timing, form and amount of precipitation. For example, even where precipitation changes with rising temperatures, the direct impacts of temperature on agricultural production may overshadow effects caused by shifts in water supply.[40] Third, it is challenging

---

35 Porter *et al.*, above note 14 at 502, 508.

36 Amarasinghe and Smakhtin, above note 8 at 15. The four key drivers of agricultural water use in different countries are net irrigated area, the irrigated area of water intensive crops, the efficiency of water use, and crop yields. Id. 17.

37 PCD Milly *et al.*, 'Stationarity is Dead: Whither Water Management?' (2008) 319 *Science* 573.

38 National Research Council, *Advancing the Science of Climate Change* (National Academies Press 2010) 225–7; IPCC, *Special Report on Emissions Scenarios* (IPPC 2001) Ch. 5.

39 United Nations Environment Programme, *The Emissions Gap Report, Are the Copenhagen Accord Pledges Sufficient to Limit Global Warming to 2°C or 1.5° C? A Preliminary Assessment* (UNEP 2010).

40 Porter *et al.*, above note 14 at 497.

to downscale climate projections from global to regional, national or local conditions, although efforts are underway to do so.[41]

2.2 Drought and Aridity

The most significant impacts of shifts in global precipitation will result from increased drought, aridity and desertification.[42] The incidence and severity of drought has already been exacerbated in recent years by the growing mismatch between human population and relatively fixed supplies of usable water. Historical drought records are sparse, making long-term trend studies more difficult.[43] Drought frequency has increased in areas such as Western African and Southern Europe since the 1950s.[44] Livestock mortality during recent droughts in the Horn of Africa reached 80 percent.[45]

An issue related to drought but with more long-term impacts is the increasing desertification of many regions of the world. Desertification already affects one-third of the world's surface and threatens up to one billion people.[46] Even under the most optimistic climate change scenarios, drought and desertification are likely to increase in frequency, impact and geographic range because of expected hydrologic changes.[47] Soil moisture droughts lasting four to six months are expected to double in extent and frequency, and those lasting more than six months are expected to triple by the end of the 21st century.[48]

In terms of impacts on agriculture, precipitation volume alone does not dictate the severity of a drought because hydrologic impacts also depend on

---

[41] United Nations Environment Programme, *Climate Change Science Compendium 2009* (Catherine P McMullen and Jason Jabbourt (eds) 2009) 7; United States Geological Survey, Scientific Investigations Report 2011, *Integrated Watershed-Scale Response to Climate Change for Selected Basins Across the United States* (2012) 5077.

[42] Drought refers to a temporary reduction in precipitation and associated water supplies, while aridity refers to longer term decline in water availability.

[43] Cisnero *et al.*, above note 7 at 239.

[44] *Ibid.*

[45] Dasgupta *et al.*, above note 13 at 621.

[46] Elizabeth Burleson, 'Emerging Law Addressing Climate Change and Water' (2010) 5 *Environmental and Energy Law and Policy Journal* 489, 499.

[47] *Climate Change Science Compendium*, above note 41 at 38; Aiguo Dai, 'Drought Under Global Warming: A Review' (2010) 2 *WIREs Climate Change* 45; Guiling Wang, 'Agricultural Drought in a Future Climate: Results from 15 Global Climate Models Participating in the IPCC 4th Assessment' (2005) 25 *Climate Dynamics* 739.

[48] Cisnero *et al.*, above note 7 at 243.

antecedent conditions. If an extremely dry year follows a long wet period in which a lot of water has been stored in the system (soil moisture, surface water levels, reservoir storage), physical impacts to natural or human water resources may be small. In contrast, even a moderately dry year following a long period of incremental reductions in stored water can have significant hydrologic impacts. The effects of drought on crops, livestock, forests and other agricultural resources also depend on other environmental variables, such as temperature and humidity (which affect evapo-transpiration rates and plant water needs) and the nature and timing of precipitation (snow versus rain; or steady, moderate rainfall versus infrequent downpours).

The effects of drought are particularly significant for forests and agricultural operations that depend on rainfall rather than artificial irrigation, because no storage and irrigation infrastructure has been developed to address the shortage in rainwater for crops. For agriculture such as orchards that rely on artificial irrigation, changes in precipitation amount, timing or type (rain versus snow) can affect storage in dams and distribution networks that provide water for irrigation and other uses.

2.3 Flooding and Water Quality Impacts

Floods,[49] intense storms and hurricanes have already become more frequent as global average temperatures rise.[50] Individual extreme events can have severe impacts on agricultural production, sometimes wiping out entire crops or causing considerable damage to trees.[51] They can also pose serious hazards to agricultural workers, buildings and other infrastructure in agricultural communities.

It is very likely that heavy precipitation events will increase in key agricultural regions such as central North America.[52] Researchers have devoted more effort to studying the agricultural impacts of drought than flooding.[53] It is more difficult to project the impacts of extreme weather events on agriculture through long term modeling because extreme events occur episodically and at unpredictable intervals, meaning that the data needed to develop and run models is scarcer and less reliable. Nevertheless,

---

[49] PCD Milly *et al.*, 'Increasing Risk of Great Floods in a Changing Climate' (2002) 415 *Nature* 514.

[50] *Climate Change Science Compendium,* above note 41 at 2–3; Cisnero et al., above note 7 at 236.

[51] US Global Change Research Program, above note 22 at 74–75; Easterling and Aggarwal, above note 23 at 283.

[52] Dasgupta *et al.*, above note 13 at 621.

[53] Porter *et al.*, above note 14 at 497.

documented experience indicates that the effects of even individual extreme weather events can be significant.[54] Although changes in flood frequency will vary regionally, global flood risk is projected to increase, with the largest expected impacts in Asia, Africa and South America.[55]

Changes in climate are likely to affect water quality[56] at both ends of the hydrological spectrum (increases or decreases in precipitation and water supply), and will affect agriculture in particular.[57] Water shortages can concentrate pollutants by decreasing dilution, which in turn amplifies the effects of pollution from various sources.[58] Intense precipitation can increase runoff of pollutants into water supply systems and exacerbate the frequency and severity of overflows from sanitary sewers, combined sewers and pollution from the malfunctioning water supply and treatment facilities.[59] Heavy downpours will cause more erosion and sedimentation of water bodies, and more runoff of contaminants.[60] This can cause increased levels of pathogens entering the water supply, more waterborne disease, and direct contamination of food crops.[61] Irrigation water supplies can be affected more directly through intensified eutrophication and reduced hydraulic retention times in reservoirs and natural lakes alike.[62]

Increased water temperatures can also affect water quality. Increased stratification of lakes and reservoirs, which result in increasingly colder water layers at greater depths, can deplete dissolved oxygen.[63] Rising temperatures can cause algal blooms and higher pollutant concentrations, such as elevated summer phosphorus levels, greater presence of bacteria and fungi, and increased concentrations and toxicity of some pollutants.[64] Sea level rise can cause saltwater intrusion into coastal aquifers, and extend

---

[54] *Ibid* at 491, Table 18-3 (listing impacts of extreme weather events on cropping systems over the past decade).

[55] Cisnero *et al.*, above note 7 at 232.

[56] US Global Change Research Program, above note 22 at 41; IPCC, *Climate Change and Water* (IPCC Technical Paper VI 2008).

[57] Robert W Adler, 'Agriculture and Water Quality: A Climate-Integrated Perspective' (2013) 37 *Vermont Law Review*, 847, 873–7.

[58] US Global Change Research Program, above note 22 at 46; US Environmental Protection Agency, Office of Water, *National Water Program Strategy, Response to Climate Change* (EPA 800-R-08-001 2008) 11.

[59] US Environmental Protection Agency, above note 58 at 12–14.

[60] *Ibid* at 11.

[61] National Research Council, *Adapting to the Impacts of Climate Change* (The National Academies Press 2010) 20, 41, 47.

[62] Cisnero *et al.*, above note 7 at 237.

[63] US Global Change Research Program, above note 22 at 46.

[64] US Environmental Protection Agency, above note 58 at 8, 38.

saltwater influence farther inland in estuaries and coastal rivers.[65] This can render some water supplies unfit for agriculture and other uses.[66]

3. WATER LAW AS A TOOL TO ADDRESS CLIMATE CHANGE IMPACTS ON AGRICULTURE

Significant steps will be necessary to adapt to the hydrologic impacts of climate change on agriculture. Some of those measures will involve shifts in agricultural practices, such as changes in crop choices (including more drought tolerant species) and cropping patterns, use of agronomic methods to enhance and retain soil moisture, adjustment of planting and harvest times, changes to water storage and delivery capacity, irrigation intensity and efficiency, and so on.[67] Shifting water supplies will also generate legal challenges and other conflicts over water allocation, which will have to be resolved through existing, modified or new systems of water rights.

The certainty of those legal regimes will affect ability to invest in water development and distribution systems. The degree to which those rights are marketable will affect the ability to shift available water to or from agriculture according to societal needs and market demand. Legal regimes will need to resolve conflicts between competing uses of water in ways that encourage efficient use and demand-side responses as well as supply-side responses. Although certainty of rights is important to support investment decisions, flexibility will also be needed in some instances to address the increasing uncertainty in climatic and hydrological projections that suggests the need for more adaptive and integrated systems of water law and governance that account for the full range of expected conditions and affected parties and interests.[68]

Water law will also need to consider tradeoffs between water and energy needs, and consequently between related climate change adaptation and mitigation strategies.[69] Some forms of energy generation designed to reduce GHG emissions, such as hydroelectric or nuclear power, require more water to be left in rivers to power turbines or to cool the power plants, respectively. That water will not be as readily available for agricultural uses

---

[65] National Research Council, above note 61 at 45; US Global Change Research Program, above note 22 at 47; US Environmental Protection Agency, above note 58 at 38.

[66] US Environmental Protection Agency, above note 58 at 16–17.

[67] Porter *et al.*, above note 14 at 514–15.

[68] Tarlock, above note 11 at 404–8.

[69] Burleson, above note 46 at 493–4.

unless power generation occurs upstream of irrigation needs. Increased use of agricultural crops for biofuels will also put pressure on food products, thereby increasing food prices. Systems of integrated land and water use planning may be needed to coordinate multiple uses of water.

Except for international watercourses or other trans-boundary applications, domestic law governs water use and allocation as it affects agricultural and other uses. Water law varies considerably around the world. One source estimates that there are 192 different national water law systems around the world,[70] and it is impossible to describe all of them here. Because US water law reflects considerable variability, this chapter first uses US law to highlight the problems inherent in adapting various water law systems to climate change. Then, it moves to a similar analysis of international law, including customary international law, bilateral treaties and multilateral agreements.

3.1 Domestic Water Law: The Example of the United States

In the United States, state law largely governs water use and allocation, although federal statutes such as the Clean Water Act[71] (CWA) and Endangered Species Act[72] (ESA) are important federal constraints on state laws governing water allocation. Federally funded water projects have also shaped water development throughout the United States, particularly in the West.[73]

The system of riparian rights prevalent in the eastern United States, based on common law principles developed in the United States and Great Britain when water resource conflicts were infrequent, generally assumes abundant supplies that can be divided equitably among competing users. Prior appropriation doctrine, which prevails in the arid western United States, is designed to promote stability and to encourage and reward investment in water infrastructure by choosing 'winners' and 'losers' when water supplies run low, according to the principle of 'first in time, first in right'. Due to abundant storage and conveyance systems, even many arid western states have yet to experience significant problems caused by water scarcity, so this system has not yet been tested extensively in times of extreme drought.

---

[70] Joseph W Dellapenna *et al.* (eds), *Evolution of the Law and Politics of Water* (Springer 2009) 10.

[71] 33 USC §§ 1251-1387 (2006).

[72] 16 USC §§ 1531-44 (2006).

[73] Robert W Adler, 'Climate Change and the Hegemony of State Water Law' (2010) 29 *Stanford Environmental Law Journal* 1.

It is unclear whether existing US state water law is sufficiently flexible to manage the water resource challenges caused by climate change. Existing legal regimes offer some opportunity for adaptation by, for example, facilitating market transfers of water rights between users.[74] Further changes and added flexibility may be needed to adapt water law systems that were developed without consideration of climate change or the hotter, drier and less predictable climate that many regions are likely to experience as a result of climate change. Federal or state legislative action may also be needed to harmonize competing state water allocation laws into a coordinated climate change adaptation and mitigation strategy in the context of interstate water allocation and transfers.

3.1.1 Riparian rights and climate change

The riparian rights doctrine, or a statutory or administrative version of that approach, is used widely in the relatively water rich eastern United States. Under the common law version of this doctrine, all landowners abutting surface waters have an equal right to use those waters. Landowners abutting surface waters were traditionally entitled to the natural flow of the stream undiminished in quantity or quality, and those lacking direct contact with a watercourse had no right to its use.[75] As intensity of water use increased and population density along waterways grew, the law evolved to allow the reasonable use of water by users not abutting water resources.[76]

Most states that initially adopted the riparian rights doctrine have transitioned to statutory or regulatory systems involving administrative permits.[77] States commonly issue time limited water use permits based on determinations of what is 'reasonable use', thus allowing reconsideration as conditions shift due to climate change or other factors. This provides more tools to deal with public water rights, accommodate consumptive uses and address the challenges inherent in applying the rule of reasonableness to a growing population and more intense uses of water that reduce water availability for agricultural use. Administrative permits also have the

---

[74] A Dan Tarlock, *Law of Water Rights and Resources* (first published 1988, Thomson Reuters 2011) Ch. 5.

[75] *Clippinger v. Birge*, 547 P2d 871, 877 (Wash Ct App 1976); *Monroe Carp Pond Co v. River Raisin Paper Co*, 215 NW 325, 327 (Mich 1927).

[76] *Tyler v. Wilkinson*, 24 F Cas 472 (CCRI 1827); *Snow v. Parsons*, 28 Vt 459 (1856).

[77] Ark Code Ann §§ 15-22-215, -217 (2003); Fla Stat 373.219 (2010); Ga Code Ann § 12-5-31 (2012).

potential to impose conservation measures, drought planning and other conditions on water use permits to help manage declining or increasingly variable supplies. The transition to regulated riparianism also allows transfers to, and use by, non-riparian landowners. This greatly increases the amount of land that can be developed, including for agricultural purposes in areas that do not abut waterways.[78]

The riparian rights system adequately balances competing interests as long as each user has sufficient water to continue economically viable uses. Conceivably, a riparian system might also successfully allocate water under conditions affected by climate change. However, conflicts may arise if climate change reduces the available water supply, making it difficult or impossible to support all existing uses. Thus, in riparian rights states in which supplies decline relative to demand, agencies administering water rights and courts adjudicating disputes among users may be forced to prioritize uses.[79] If droughts become more frequent and more severe in riparian rights states due to climate change, and competition increases between agricultural uses and rapidly expanding urban areas and growing urban and suburban populations, changes in riparian water law may be needed to support irrigated agriculture.

The riparian rights system poses at least two potential barriers to climate change adaptation. First, in its traditional form, the system links water use to riparian ownership, thus preventing transfer of water to non-riparian agricultural lands. Second, traditional riparian doctrine limits the degree to which water rights can be conveyed separately from title to land. Although both limitations have been modified to varying degrees in most states, further changes may be needed to facilitate water storage, conveyance and irrigation in jurisdictions that have adopted or modified riparian rights law, particularly as agriculture faces increased competition for water with growing cities.

3.1.2 Prior appropriation and climate change

The western United States developed the prior appropriation doctrine to address some of the limitations in riparian doctrine in a region where agriculture and other early economies (such as mining) required water to be stored and diverted for use on non-riparian lands. In this region, certainty of title to water rights was essential to support the investments in

---

[78] *Stratton v. Mt Hermon Boys' School*, 103 NE 87 (Mass 1913).

[79] Joseph W Dellapenna, 'The Law of Water Allocation in the Southeastern States at the Opening of the Twenty-First Century' (2002) 25 *University of Arkansas Little Rock Law Review* 9.

time and labor needed to finance and construct large irrigation systems. As with riparian rights, state regulation has modified common law prior appropriation doctrine to a significant degree. In most western states, administrative agencies now review applications to divert and use water. Once an agency approves the application, the applicant must develop the water right and prove that water was applied to a beneficial use within the time allowed. The agency then issues an approved water right that limits the amount, purpose, location and seasonal timing of water diversion and use.[80]

The most basic tenet of prior appropriation law, designed to provide certainty of use and investment, is the doctrine of 'first in time, first in right'.[81] Each water right has a priority date coinciding with the date upon which the water was first put to a beneficial use.[82] Senior right holders have use priority when demand for water exceeds the available supply, leaving junior users with less water than their normally allotted amount, or none at all.[83] Because many agricultural lands were developed early in the history of many western states, agriculture often holds a favored position in times of shortage with the exception of more recently developed agricultural enterprises.

In theory, the prior appropriation doctrine also is designed to prevent waste in arid regions or in times of drought because of the requirement that protected uses must be 'beneficial', although some commentators have noted that the prohibition against waste has not been enforced vigorously.[84] In addition, appropriators generally may not hold water rights for future or speculative needs. Thus, if a user does not put a water right to beneficial use within the required period, the water right may revert to the state, and other water users may appropriate the water.[85] Likewise, under

---

[80] Utah Code Ann §§ 73-3-5.6(2) and (6) (2012); Wash Rev Code §§ 90.03.250 and .330 (2004 & Supp 2012).

[81] *Montana v. Wyoming*, 131 SCt 1765, 1780 (2011); *Colorado v. New Mexico*, 459 US 176, 179 n 4 (1982). *United States v. County of Denver*, 656 P2d 1, 12 (Colo 1982) (noting prior appropriation generally governs, in one form or another, the acquisition of water rights in the 19 western states).

[82] *Navajo Dev Co Inc v. Sanderson*, 655 P2d 1374 (Colo 1982); *Meine v. Ferris*, 247 P2d 195, 198 (Mont 1952); *Sand Point Water and Light Co v. Panhandle Dev Co*, 83 P 347 (Idaho 1905).

[83] *Sanpete Water Conservancy Dist v. Carbon Water Conservancy Dist*, 226 F3d 1170, 1173 (10th Cir 2000).

[84] Janet Neumann, 'Beneficial Use, Waste, and Forfeiture: The Inefficient Search for Efficiency in Western Water Use' (1998) 28 *Environmental Law* 919.

[85] *Sears v. Berryman*, 623 P2d 455 (Idaho 1981); *Kearney Lake, Land & Reservoir Co v. Lake DeSmet Reservoir Co*, 487 P2d 324 (Wyo 1971).

the 'use it or lose it' tenet of prior appropriation law, a water right holder who fails to utilize her water right may forfeit that right.[86]

Compared with riparian rights, prior appropriation should provide more certainty regarding allocation during periods of water shortage because the prior appropriation system is designed to determine clear 'winners' or 'losers'. However, because allocation is based entirely on seniority, the doctrine can protect less efficient or less economically valuable water uses. This hardening of existing uses is tempered by the fact that water right holders can convey their rights to others so long as other uses are not harmed in the process.[87] In contrast to riparian doctrine, water right holders under a prior appropriation system can convey these rights separately from the land upon which the water is used.[88] Water markets may therefore develop when water is not available to satisfy the needs of all prospective users. Historically, conversion of agricultural water to municipal and industrial use through water markets has facilitated significant urban expansion,[89] but there is no reason why the opposite could not occur to fuel more valuable agricultural uses in certain scenarios.

One main criticism of the prior appropriation doctrine is its failure to foster efficiency,[90] which will become increasingly important as climate change reduces water supplies in arid regions. Water rights are limited to the amount of water that is beneficially used. Therefore, even when a water user improves efficiency, the saved water is generally made available to other water users according to seniority rather than the water right holder who invested in the water saving improvement,[91] thus reducing incentives for rights owners to invest in more efficient water use methods, such as improved irrigation systems.

---

[86] Idaho Code Ann § 42-222(2) (2003 & Supp 2012); Utah Code Ann § 70-1-3 (2012); Wash. Rev Code § 90.14.160 (2004 & Supp 2012).

[87] *Cline v. McDowell*, 284 P2d 1056, 1059 (Colo 1965); *East Bench Irrigation Co v. Desert Irrigation Co*, 271 P2d 449 (Utah 1954).

[88] Utah Code Ann §§ 73-1-10(1)(a) (2012) (certificates, decrees, or diligence claims) and 73-1-10(2) (2011) (shares in a water company).

[89] Patricia Nelson Limerick with Jason L Hanson, *A Ditch in Time: The City, the West and Water* (Fulcrum Publishing 2012) (discussing development of Denver and its water supply); Mark Reisner, *Cadillac Desert: The American West and Its Disappearing Water* (Viking Penguin Inc. 1986) (discussing water development in the American West).

[90] Adler, above note 73 at 226; Steven J Shupe, 'Waste in Western Water Law: A Blueprint for Change' (1982) 61 *Oregon Law Review* 483.

[91] *Salt River Valley Water Users' Assn v. Kovacovich*, 411 P2d 201 (Ariz App 1966); *Water Supply and Storage Co v. Curtis*, 733 P2d 680 (Colo 1987).

3.1.3 Groundwater law and climate change

Groundwater pumping is a significant source of irrigation water in many agricultural regions of the United States, such as large portions of the Great Plains, Arizona and other parts of the Southwest. Even absent climate change, groundwater pumping to support large-scale irrigation has caused significant groundwater depletion and increased pumping costs (in both dollars and energy, with the latter contributing further to climate change).[92]

Groundwater law in the United States is even more fractured than the law governing surface water allocation, with five major but conceptually overlapping doctrines in different states.[93] Many western states apply prior appropriation to groundwater as well as surface water, in some cases with explicit linkages between groundwater and surface water appropriative rights where the water sources are demonstrably or presumptively connected. Where states adopt the prior appropriation doctrine for groundwater, the central question becomes whether groundwater development would interfere with other water rights holders' uses.[94] Despite the prior appropriation doctrine's clear definition of 'winners' and 'losers', many western states shy away from strict priority enforcement and instead apply the 'rule of reasonableness' in managing groundwater withdrawals. This approach considers a range of factors, including the quantity of water available, the average annual recharge in the basin and the existing rights and their priorities in an effort to make the most water available for beneficial use.[95]

Although more tempered than the more absolute form of prior appropriation, the doctrine remains useful in providing the certainty necessary to support major investments in groundwater pumping and irrigation infrastructure, at least for those with sufficiently senior water rights. Similar to its surface water counterpart, however, the doctrine can lock inefficient senior uses into place and discourage efficiency improvements unless the party investing is focused on water conservation rather than the needs of the next most senior appropriator. Even more extreme in this

---

[92] John Opie, *Ogallala, Water for a Dry Land* (University of Nebraska 1993) 129–32, 146–49.

[93] Tarlock, above note 74 at § 6.1–6.31; Robert E Beck and Amy K Kelly (eds), *Waters and Water Rights* (Michie 2007) § 18.01–23.07.

[94] *Little Cottonwood Water Co v. Sandy City*, 258 P2d 440, 443 (Utah 1953) ('no one can interfere with the source of supply of [a] stream, regardless of how far it may be from the place of use, and whether it flows on the surface or underground, in such a manner as will diminish the quantity or injuriously affect the quality of the water of these established rights.').

[95] *Wayman v. Murray City Corp*, 458 P2d 861, 865 (Utah 1969).

regard is the absolute ownership doctrine (or the 'rule of capture'), which applies in most parts of Texas.[96] The absolute ownership doctrine allows the overlying landowner to pump groundwater from beneath their land without regard to the effect withdrawals may have on others.[97]

Other groundwater allocation regimes borrow more from riparian doctrine and require that scarce resources be shared equitably among competing uses. Correlative rights, a variant on reasonable use, allows groundwater withdrawals based on a reasonable sharing among competing users, generally defined by the proportion of land overlying the aquifer.[98] Some groundwater drawdown is permissible provided it does not unreasonably interfere with other users' reasonable uses of the water source. Some jurisdictions apply a regulated riparian approach to groundwater management, requiring a time limited permit from the state that reflects a reasonableness determination. This determination is revisited upon permit renewal.[99]

Many states are beginning to apply conjunctive surface and groundwater management through active or passive aquifer storage and recovery programs.[100] Conjunctive use recognizes connections between surface and groundwater resources and can provide some insurance against uncertainty by relying on surface water use when supplies are adequate, storing surplus surface water in aquifers, and using stored groundwater during times of drought or other shortages.[101] One example is the Arizona Water Bank, in which Arizona 'banks' unused portions of its Colorado River apportionment in aquifers for later use in both Arizona and Nevada.[102] Aquifer storage and recovery is also being considered to support water dependent

---

[96] *Sipriano v. Great Spring Waters of America Inc*, 1 SW3d 75, 76 (Tex 1999), *Houston & Texas Central Railway Co v. East* 81 SW 279, 280-81 (Tex 1904).

[97] Douglas L Grant, *Conjunctive Management of Hydrologically Connected Surface Water and Groundwater: The Problem of Sustainable Use* (Proceedings of the Rocky Mountain Mineral Law Foundation Fifty-Fourth Annual Institute 2008) § 14.04(2)(a).

[98] Beck and Kelly, above note 93 at § 21.01.

[99] *Ibid* at § 23.02.

[100] National Research Council, above note 61 at 69; Craig Bell and Jeff Taylor, *Water Laws and Policies for a Sustainable Future: A Western States' Perspective* (Western States Water Council 2008) 83–4, 123–35; Utah State Water Plan, *Conjunctive Management of Surface and Ground Water in Utah* (Utah Division of Water Resources 2005), accessed 11 October 2016 at www.water.utah.gov/OtherReports/CMReport/CMReport1bCC.pdf; R David G Pyne, *Groundwater Recharge and Wells: A Guide to Aquifer Storage and Recovery* (CRC Press Inc 1995).

[101] Utah State Water Plan, above note 100 at 29.

[102] Arizona Water Banking Authority, Annual Report (2009).

ecosystem restoration programs in the Everglades,[103] but such programs could be used to benefit agriculture as well in times of shortage. A key issue here is whether conjunctive use systems designed to address temporary shortages during times of drought, under the assumption that stored water will be replenished when precipitation returns to 'normal', will work effectively during more prolonged periods of shortage caused by climate change. Even so, conjunctive use and management could help regions to transition to more arid conditions, providing the time necessary to invest in efficiency improvements or in crop varieties that consume less water.

3.1.4 Interstate apportionment

Interstate rivers and aquifers in the United States pose unique challenges for water managers, and these challenges will be exacerbated as climate change increases incentives to transfer water from wet regions to dry ones. Some states use more water to address new shortages and to support existing or growing agricultural economies, resulting in shortages for downstream neighbors. Although subject to different legal and political constraints that vary across the United States, each of these options poses issues similar to those discussed below in resolving international water conflicts.

Interstate water disputes can be resolved judicially through the original jurisdiction of the US Supreme Court to hear suits between states.[104] To apportion interstate rivers, the Supreme Court applies rules of equitable apportionment[105] that are similar to principles of fair apportionment (discussed below) under both customary principles of international law[106] and multilateral international agreements that adopt or modify those principles. Equitable apportionment is a flexible, fact-specific doctrine that allows the Supreme Court to consider a variety of factors in determining how to allocate water among competing states. These factors include physical conditions, climate, established uses and their importance, and efficiency of existing uses.[107] Under this doctrine, the Court might consider

---

103 National Research Council, *Aquifer Storage and Recovery in the Comprehensive Everglades Restoration Plan: A Critique of the Pilot Projects and Related Plans for ASR in the Lake Okeechobee and Western Hillsboro Areas* (The National Academies Press 2001).

104 US Const art III § 2.

105 *Kansas v. Colorado* 206 US 46 (1907) (balancing equities by comparing the value of water use by competing states).

106 Tarlock, above note 11 at 375.

107 Tarlock, above note 74 at §10.16; *Colorado v. New Mexico (I)*, 459 US 176 (1982).

the importance of continued agricultural productivity if climate change threatens food supplies or other essential societal resources. The flexibility inherent in equitable apportionment allows adjustments if conditions shift significantly and unpredictably, as may be expected from climate change.

For farmers who require expeditious and reasonably certain answers about how much water will be available for their use, equitable apportionment is not likely to be an effective solution to short-term supply problems. The flexibility afforded by the equitable apportionment doctrine, although beneficial in some ways, reduces the certainty on which states and their agricultural and other users can rely in making decisions concerning water use. Moreover, resolving interstate allocation disputes can take many years and thus may not be practicable in the face of a pressing dispute that may arise due to climate change. For example, it took almost a quarter of a century for the Supreme Court to resolve litigation between Kansas and Colorado regarding the Arkansas River system.[108] Moreover, in most interstate water cases the Supreme Court has been reluctant to issue detailed decrees, leaving the details of management to the competing states. As a result, states have been reticent to bring interstate water disputes to the Supreme Court, just as nations resort only rarely to international litigation to resolve bilateral water conflicts.

Many interstate water disputes, therefore, have been resolved or prevented via interstate compacts,[109] for which Congressional ratification is required to ensure protection of national interests.[110] Upon ratification, an interstate compact operates both as federal law and as a contract between states,[111] and in an international context the compact resembles a bilateral or a multilateral treaty. Agreements between states have the benefit of allowing the parties to negotiate specific resolutions to defined problems, and, if sufficiently detailed and carefully crafted, can increase certainty among competing users.

On the other hand, particularly if compacts are crafted without sufficient anticipation of future conditions, the agreements can be difficult to negotiate, leave issues unresolved, and lead to protracted legal battles over ambiguities or unintended consequences.[112] The Colorado River Compact,

---

108 'KS, CO Reach Arkansas River Agreement' (Jan/Feb 2010) 9(1) *Southwest Hydrology*.

109 The Colorado River Compact, 45 Stat 1057 (1928); The Republican River Compact, 57 Stat 86 (1943); The Snake River Compact, 64 Stat 29 (1950); The Red River Compact, 94 Stat 3305 (1980).

110 US Const art I § 10.

111 Tarlock, above note 74 at § 10.25.

112 *Montana v. Wyoming*, 131 S Ct 1765 (2011) (resolving continued dispute over meaning and application of 60-year old Yellowstone River Compact).

for example, was negotiated when average flows were much higher than they are today, meaning that more water was allocated than will likely be available.[113] Upper Colorado River Basin states have fixed delivery obligations defined in acre-feet of water rather than as a percentage of remaining flows, and thus bear a disproportionate burden of future flow reductions. Lower basin states have grown at a faster pace, utilizing most if not all of their Colorado River allocation and creating an incentive to embrace conservation. The fear that climate change may necessitate compact renegotiation[114] and that unused waters may be reallocated creates incentives for states that have yet to develop their full allocation to do so rapidly. It has also created incentives for the basin states to negotiate more specific collateral agreements, such as the current 'shortage agreement' that defines the ramifications of shortages in the basin through 2026,[115] as alternatives to the uncertainty created by the compact or the even greater uncertainty that may result from litigation.

The uncertainty posed by climate change may make it even more difficult to negotiate interstate (or international) water agreements due to the difficulty of predicting future conditions accurately. This possibility suggests that future interstate compacts be drafted with flexibility to anticipate a wide range of conditions, but with sufficient specificity to address the implications of changing conditions with reasonable certainty. Compacts that allocate water on a proportionate basis, for example, may be more flexible under a changing climate than agreements that allocate fixed volumes of water.[116]

3.1.5 Water markets

The free market is one tool to facilitate reallocation of water resources to uses with higher economic value or need, including agriculture, if food or other agricultural products become increasingly scarce.[117] Water markets may become more important in the face of climate change as users seek

113 Robert W Adler, *Restoring Colorado River Ecosystems, A Troubled Sense of Immensity* (Island Press 2007) 20–2.

114 Robert W Adler, 'Revisiting the Colorado River Compact: Time for a Change?' (2008) 28 *Journal of Land, Resources, & Environmental Law* 19.

115 Colorado River Interim Guidelines for Lower Basin Shortages and the Coordinated Operation of Lake Powell and Lake Mead (December 2007).

116 Erik Ansink and Arjan Ruijs, *Climate Change and the Stability of Water Allocation Agreements*, Fondazione Eni Enrico Mattei, Nota Di Lavoro § 16.2007 (2007), accessed 11 October 2016 at http://www.feem.it/userfiles/attach/Publication/NDL2007/NDL2007-016.pdf.

117 Jonathan H Adler, 'Water Rights, Markets, and Changing Ecological Conditions' (2012) 42 *Environmental Law* 93; Terry L Anderson and Peter J Hill,

to transfer water from areas of surplus to regions of scarcity, and from lower to higher value uses. Calls for more robust water markets are premised on the status of water rights as property that can be conveyed, subject to conditions imposed by government regulations. Such conveyances have traditionally fallen into two general categories: transfers from less to more economically valuable uses (for example, from pasture irrigation to municipal and industrial use), and from low to high-demand regions.

Under existing US law, markets designed to reallocate water in the face of shortages caused by climate change or other factors typically can proceed only if they avoid injury to other users. Some states limit transfers that would move water out of basins that suffer from water shortages or that would experience other adverse impacts as a result of the transfer.[118] Federal environmental protection requirements, such as the Endangered Species Act, also may also limit efforts to move water where needed to maintain critical habitats that rely on that water.[119]

In the future, market forces will undoubtedly drive some reallocation of water resources, most likely from agricultural use to domestic, municipal, and industrial uses. Such transfers would take agricultural lands out of production, which would change the face of rural communities, shift agricultural production to other areas, and potentially affect food prices.[120] Transfers will force society to consider how to structure and regulate markets to facilitate climate change adaptation without impairing other values. Should climate change impede agricultural production significantly, the economic value of agricultural uses might rebound relative to municipal or other needs, and the same forces driving agricultural to urban transfers might reverse the flow of money and water.

Improvements in water use efficiency are also likely to occur as the value of water increases, partly as a result of climate change. In the United States, improvements in efficiency have offset increases in population since approximately 1985,[121] but those trends may mask local shortages and further efficiency gains are possible. California water users, for example,

---

Water Marketing: The Next Generation (Rowman & Littlefield Publishers Inc 1996).

118 *Jacobs Ranch LLC v. Smith*, 148 P3d 842, 853 (Okla 2006).

119 *Carson-Truckee Water Conservancy Dist v. Clark*, 741 F2d 257 (9th Cir 1984).

120 Robin Kundis Craig, 'Adapting Water Federalism to Climate Change Impacts: Energy Policy, Food Security, and the Allocation of Water Resources' (2010) 5 *Environmental & Energy Law & Policy Journal* 183.

121 Joan F Kenney *et al.*, *Estimated Use of Water in the United States in 2005* (US Geological Survey, Circular 1344, 2009).

could save 350,000 acre-feet of water per year through water delivery system leak reduction.[122] Transitioning to less water intensive plants creates both a conservation opportunity and a chance to beneficially utilize water that does not meet potable standards.[123] It is not clear, however, whether continued improvement will suffice to adapt to shortages caused by climate change.

3.1.6 Drought law and climate change

Prevailing US water law regimes do not fully address drought in the main body of rules governing water rights and allocation. In many state water law systems, drought is not addressed at all as a distinct concept. There are some notable recent exceptions in the water law of some US states where various water rights and obligations can be modified in the event of drought,[124] but even under those provisions drought is treated as an exception determined on an ad hoc, emergency basis, such as an emergency drought declaration issued by the state governor. All water law regimes allocate the resource in times of both plenty and shortage. Typically, however, they respond to scarcity in the same manner whether it results from meteorological drought, population growth, inefficient use of water, inadequate infrastructure, or other artificial factors. Thus 'winners', under the prevailing legal regimes, have little incentive to improve water use efficiency in ways that might prevent or mitigate drought impacts.

In prior appropriation states, holders of senior water rights receive their full allocations before junior appropriators receive any water. Any reallocation of water rights to more important or more efficient uses is left largely to the market, which as discussed above, operates inefficiently due to various impediments inherent to prior appropriation law. Senior appropriators have little incentive to conserve until resources become so scarce that they face shortages themselves.

Pure riparian rights systems, in theory, allocate scarce water among competing users according to a variety of equitable factors, thus requiring

---

[122] Water Systems Optimization Inc, Prepared for Southern California Edison, Secondary Research for Water Leak Detection Program and Water System Loss Control Study, Final Report, December 2009, as reported in NRDC, *Pipe Dreams: Water Supply Pipeline Projects in the West* (2012).

[123] National Research Council, *Water Reuse: Potential for Expanding the Nation's Water Supply Through Reuse of Municipal Wastewater* (The National Academies Press 2012).

[124] Janet C Neuman, 'Drought Proofing Water Law' (2003) 7 *University of Denver Law Review* 92; Melinda Kassen, 'Statutory Expansion of State Agencies' Authority to Administer and Develop Water Resources in Response to Colorado's Drought' (2003) 7 *University of Denver Law Review* 47.

everyone to share in the risk of scarcity. Even under common law administration of riparian rights, however, courts have often established *de facto* preferences among users.[125] The uncertainty and instability inherent in such a case-by-case allocation system leaves large water users in riparian rights states with little incentive to engage in more sustainable practices. Under more recent statutory and administrative systems of riparian rights (so-called 'regulated riparianism'), statutory preferences operate much as they might in prior appropriation states, and with the same weak incentives for conservation. The administrative process provides, at least, the opportunity for agencies to impose efficiency requirements as a condition of water use permits.

As climate change affects the timing, reliability and amount of precipitation throughout much of the United States and other regions, water users and suppliers will seek protection against shortage. One obvious adaptation is to increase storage. Surface storage, however, is subject to evaporation, and major impoundments pose difficult environmental challenges.[126] Water users may also turn to managed aquifer recharge projects, which capture potable or treated wastewater flows that would otherwise go unused. These flows are then either injected or infiltrated into groundwater aquifers from which they can be withdrawn later, but efficacy may depend on project cost and the suitability of the geology or hydrology to this technology. The severity of projected water shortages, the uncertainty of supply and the financial and environmental costs of securing more reliable water supplies will determine the type and level of adaptation likely to occur.

3.2 International Water Law

International water law will play a critical role in addressing bilateral and multilateral conflicts that can be expected as water scarcity increases in some regions and as flood risk increases in others. Although international water law will govern only a small percentage of disputes generated by climate change, nearly half of the world's population lives in the 261 river basins that cross international borders,[127] and the stakes of those potential conflicts are high. Competition for international water sources could lead to food shortages, economic disruption, increased refugee traffic, and

125 Dellapenna, above note 79 at 9, 14–16.

126 Tarlock, above note 11 at 378–80.

127 Joseph W Dellapenna, 'International Water Law in a Climate of Disruption' (2008) 17 *Michigan State Journal of International Law* 44, 57.

other impacts that could cause or exacerbate international tensions. Water conflict has been a frequent cause of armed conflict throughout history, and military intervention has been threatened in recent years in conflicts over water, such as recent threats by Egypt in the face of increased tension over use of the Nile River Basin.[128] One source cited 42 instances of water-related conflicts since 2010 alone,[129] and such conflicts are only likely to increase in the face of climate change. Water resources are linked through global hydrological systems, and because water is essential to the food supply and other key aspects of the global economy, water governance must involve global institutions.[130] Just as weaknesses in domestic water law will have to be corrected to address the increased stresses posed by climate change, several problems with international water law will require solutions as well.[131]

3.2.1 Customary international water law

Absent a negotiated agreement, international water disputes are governed by principles of customary international water law that were developed largely in the 19th and 20th centuries. Those principles resemble US riparian rights law in some respects and equitable apportionment law in others, but are modified to address the international context.

At times, some nations claimed absolute sovereignty and control over their waters.[132] Downstream nations asserted an equally absolute right to river flows at the international boundary that were undiminished in either quality or quantity.[133] To resolve these claims, many of which were irreconcilable, customary international water law employs principles that resemble modern riparian rights law in some ways and equitable apportionment in others.[134] Under the concept of limited territorial sovereignty, nations have

---

[128] Alice Shih and Trevor Stutz, 'Sink or Swim: Abrogating the Nile Treaties While Upholding the Rule of Law' (2013) 43 *The Environmental Law Reporter* ELR News & Analysis 10786.

[129] Peter H Gleick and Matthew Heberger, *The World's Water, The Biennial Report on Freshwater Resources* Vol 8 (Island Press 2014) 203–8.

[130] Claudia Pahl-Wostl *et al.*, 'Governance and the Global Water System: The Need to Adopt a Global Perspective on Water Issues, Global Governance' (2008) 14 *Global Governance: A Review of Multilateralism and International Organizations* 405–7; Heather Cooley *et al.*, *Global Water Governance in the Twenty-First Century* (Pacific Institute 2014) 1–18.

[131] Tarlock, above note 11 at 370–1.

[132] Opinions of the Attorney General (1898), 21, 274; Tarlock, above note 11 at 373.

[133] Tarlock, above note 11 at 373–4.

[134] Dellapenna, above note 70 at 11.

rights to waters that flow within their borders but are obligated to manage them in ways that account for the needs of other riparian nations, and to share supplies equitably among competing users. Nations are required to use shared water resources in ways that cause no harm to other users: a principle broad enough to address actions that cause downstream shortages, floods or pollution. Finally, customary international law establishes an expectation that nations resolve water disputes peacefully and amicably rather than through armed conflict, and that they observe reasonable procedural mechanisms to resolve disputes. Examples would be providing neighboring nations prior notice of major water projects, affording an opportunity to assess their impacts fully and mitigating impacts on neighboring nations.[135] These principles of international water law are articulated in two nonbinding but influential documents adopted by the International Law Association (ILA): the 1966 Helsinki Rules on the Use of International Rivers and the subsequent 2004 Berlin Rules on Water Resources, the latter of which also incorporates more recent concepts of environmental sustainability, public participation in water governance and integrated water resources management.

Customary international water law establishes principles and expectations that could partially temper the adverse effects of climate change on agricultural water supply by prompting nations to adjust water uses equitably and peacefully, and to cause no significant harm to downstream neighbors. From an agricultural standpoint, upstream neighbors may have incentives to ensure that neighbor states have adequate water supply to prevent famine, malnutrition and consequential breakdown of regional economies, increased refugee pressure and other drivers of regional instability.

These principles are flexible enough to address unpredictable changes in water resources, but this flexibility has potential problems. Enforcement of customary international law is inherently weak, and allocation is subject to the discretion of individual nations. The alternative is litigation in an international tribunal with jurisdiction over the competing parties, but this is often a lengthy, uncertain and expensive remedy. The parties may disagree about what uses and allocations are 'equitable', and about what constitutes 'harm' sufficient to suggest a violation of customary international water law. The absence of expeditious, clear dispute resolution mechanisms could harm agriculture in particular. The loss of adequate water supplies for just one growing season could cause significant harm to food supplies for row crops, and harm could be much greater if herds are decimated or

---

[135] Tarlock, above note 11 at 377.

if perennial plants (such as fruit trees) are damaged or destroyed. This level of uncertainty suggests that, absent the development of more efficient means of enforcing the principles of 'no harm' and equitable allocation, negotiated agreements may provide more certain solutions to water allocation problems caused by hydrological effects of climate change.

3.2.2 Treaties and conventions

In 1997 the United Nations adopted an international convention to govern the use of international waterways,[136] but it is not yet in force because a requisite number of nations have not ratified the convention.[137] The convention essentially codifies existing principles of customary international water law, and therefore does little to improve the manner in which international law is likely to address water issues caused by climate change. A somewhat more specific multilateral agreement is the Helsinki Watercourses Convention, adopted in 1992 to govern water body disputes in Europe.[138] This convention does more to incorporate environmental values into international water law, but still does not include sufficient specificity to address the more acute issues likely to be caused by climate change.

Treaties that address the use and allocation of individual water bodies are more likely to provide the level of specificity necessary to address climate change related water disputes, and hundreds of international treaties have been negotiated to do so.[139] Those agreements can address the competing uses and hydrological variables of individual water bodies and provide the level of certainty necessary to prevent or resolve the kinds of potential disputes that could arise from climate change. This requires the agreements to be written with sufficient flexibility to address a range of potential future hydrological conditions, but with sufficient specificity to avoid significant disagreements over meaning at the time of a drought or other climate-related crisis.

One example of this tension is the United States-Mexico Treaty

---

[136] UN Convention on the Law of the Non-Navigational Uses of International Watercourse, 21 May 1997, UN Doc No A/51/869.

[137] Dellapenna, above note 70 at 11.

[138] Convention on the Protection and Use of Transboundary Watercourses and International Lakes, 19 March 1992, UN Doc E/ECE/1267, 1936 UN Treaty Series 269–356.

[139] Oregon State University and United Nations Environment Programme and the Food and Agriculture Organization of the United Nations, *Atlas of International Freshwater Agreements* (United Nations Environmental Programme 2002).

governing allocation of the Rio Grande and the Colorado River.[140] The treaty provides for modified water allocations in the event of 'extraordinary drought'. The term 'extraordinary drought' is not defined in the treaty, however, leading to looming disputes as to what triggers certain provisions.[141] The term 'drought' is subject to many different physical, economic and social connotations, which renders it an extremely ambiguous and malleable concept for purposes of legal interpretation.[142] In the United States-Mexico treaty, the adjective 'extraordinary' exacerbates this ambiguity.

Other agreements can address pressure to transfer water from regions of abundance to areas of scarcity. Jurisdictions with more abundant water resources are often reluctant to convey water to comparatively dry regions, fearing that excess water may be needed to fuel continued growth. For example, in 1909, the United States and Canada entered into a treaty prohibiting further diversions from the Great Lakes that would result in changes to natural flows or levels.[143] In 2012, drought caused water levels in the Great Lakes to decline to near record lows.[144] The United States subsequently prohibited out-of-basin water diversions from the Great Lakes watershed absent unanimous agreement by all basin states.[145]

Those kinds of unilateral actions may increase as climate change increases water stress and competition. More constructive solutions have been proposed in recent years to manage shared water sources more effectively. Such management is attainable by sharing water more equitably and collaboratively, sharing the benefits of water by trading products generated with water or by compensating adversely affected nations monetarily.[146] It

---

140 Treaty for the Utilization of Waters of the Colorado, Tijuana and Rio Grande Rivers, 13 February 1944, US-Mex, 59 Stat 1219; Charles J Myers, 'The Colorado River, The Treaty with Mexico' (1967) 19 *Stanford Law Review* 367.

141 Stephen P Mumme, 'Managing Acute Water Scarcity on the US-Mexico Border: Institutional Issues Raised by the 1990s Drought' (1999) 39 *Natural Resources Journal* 149, 149–64.

142 Robert W Adler, 'Balancing Compassion and Risk in Climate Adaptation: US Water, Drought and Agricultural Law' (2012) 64 *Florida Law Review* 201, 209–13.

143 Treaty Between the United States and Great Britain Relating to Boundary Waters Between the United States and Canada, US-Can, 11 January 1909. TS No 548.

144 Sam Nelson, 'Drought Drops Lake Michigan Water to Near Record Low' *Reuters* (Chicago, 30 November 2012).

145 42 USC § 1962d-20(b)(3) (2006); Charles F Glass, Jr, 'Enforcing Great Lakes Water Export Restrictions Under the Water Resources Development Act of 1986' (2003) 103 *Columbia Law Review* 1503.

146 Tarlock, above note 11 at 397–402.

can also be promoted through collaborative, shared governance structures rather than unilateral actions that reflect self-interest alone.[147]

3.2.3 Drought and resilience across nations

Differences in national wealth, economic diversity, infrastructure and other factors affect resilience to droughts, floods and other major hydrological disturbances. A comparison of the impacts of two major droughts illustrates the importance of integrating changes in water law with other aspects of agricultural and larger economies.

The California drought of the late 1980s and early 1990s did not result in the massive economic, social, and political dislocation and upheaval that occurred during the droughts of the Great Depression years. Those differences suggest some lessons for the ways in which water law and policy might be applied to reduce the adverse effects of climate change. First, the water resources infrastructure developed by the federal and state governments and the private sector beginning in the early 20th century buffered the region against the full effects of the drought.[148] Colorado River water stored in Lake Mead and Lake Powell during the rainy years of the early 1980s offset shortages from other sources in the southern parts of the state, and California's own system of more than 150 major reservoirs provided additional reserves during the early drought years. Additional water stored in aquifers offset declining surface water supplies. By the end of the drought, however, reserves of both surface water and groundwater had declined. The drought may have caused more serious impacts if it had continued longer, or if it had not followed a wet period. It is unlikely that California will be able to 'store its way' out of the more severe and protracted declines in precipitation, snowpack storage and runoff expected as result of climate change. The drought of 2015 and 2016 – the continuation of a multiple year pattern of low snowpack and other precipitation – may be an early test of resilience to drought in California and other parts of the American West.

California has sophisticated legal and political institutions governing water and other resources. As a result, the state was able to use water banks and other market tools to reduce demand through efficiency improvements adopted, encouraged or required for municipal and agricultural users. Large Southern California municipalities paid farmers in the Imperial Valley to fallow land and to improve irrigation efficiency to allow more

---

[147] Tarlock, above note 11 at 402–8.

[148] Peter H Gleick and Linda Nash, '*The Societal and Environmental Costs of the Continuing California Drought*' (Pacific Institute for Studies in Development, Environment, and Security 1991).

water to be conveyed to cities.[149] The effects of the drought would have been worse absent these actions.

The impact of the drought was also mitigated by the size, diversity, flexibility and economic strength of the United States as a whole. Even within California, if the market functions properly, the reduced gross income from declining yields can be partially offset by increased crop prices.[150] Given the national or international scope of agricultural markets, the absence of trade barriers and the presence of functional markets within the country, crop declines in one region can be offset by increased production elsewhere.[151] Of course, those shifts in regional production cause distributive changes in farm income. Temporary distributive effects even out over time, however, with each region gaining or losing relative market share during different periods. The ability of farmers within a drought stricken region to withstand temporary shifts in production and income depends on their overall economic condition, including cash on hand and other reserves or the existence of government risk allocation programs (through price supports, crop insurance, or otherwise) to carry them through the crisis. In California the agricultural sector was strong enough to bear the impacts of the drought.

In contrast, droughts in countries that lack economic resources, economic diversity, infrastructure and legal institutions as compared to the United States are not nearly as resilient to drought and other climate related disruptions. The United Nations Environment Programme and the Kenyan Government identified a number of additional ways in which unsustainable practices and economies increase vulnerability to drought.[152] Food relief in Africa creates a risk of dependency, in which people in some regions stop growing food and depend entirely on relief. While California had options to shift scarce water among regions and uses, Kenya faces chronic water shortages all the time, with a growing population and fixed or diminishing water resources bringing it ever closer to the lower limits of acceptable per capita supplies.

---

[149] Morris Israel and Jay R Lund, 'Recent California Water Transfers: Implications for Water Management' (1995) 35 *Natural Resources Journal* 1, 1–28; Kevin M O'Brien and Robert R Gunning, 'Water Marketing in California Revisited: The Legacy of the 1987–92 Drought' (1994) 25 *Pacific Law Journal* 1053–80.

[150] Gleick and Nash, above note 148.

[151] US Army Corps of Engineers, *National Study of Water Management During Drought, The Report to the US Congress* (IWR Report 94-NDS-12, 1995).

[152] United Nations Environment Programme and Government of Kenya, 'Devastating Drought in Kenya: Environmental Impacts and Responses, Nairobi, Kenya' (2000).

The UN and Kenyan Government report proposed ways to shift from purely reactive forms of drought relief to preventive strategies designed to decrease drought vulnerability. Those included: a shift to demand based, comprehensive river basin land use and water planning and management that assumed cyclic and increasingly frequent droughts as part of the norm rather than aberration; efforts to promote traditional, community based water management and allocation; programs to de-silt storage reservoirs and plans to enhance storage, using local labor in return for water provided from those sources; a shift to sustainable agricultural practices designed to promote self-reliant communities; and an end to food relief programs not tied to measures to improve sustainable livelihoods.[153]

The international system needs to develop mechanisms to increase the capacity of smaller, less wealthy and less resilient nations to withstand significant climate related impacts. International water law can only ensure the fair allocation of water resources between nations that share watercourses, but the same drought may affect the entire watershed, leaving few options as each riparian nation seeks to maximize their share of a badly depleted resource. If an international agricultural economy allows equitable sharing of food and other agricultural products during times of extreme drought and famine, it could buffer regional impacts in the same way as occurred in the example of California described above. Either better functioning markets or well-designed international aid programs could have such a beneficial effect.

4. CONCLUSION

Significant adaptations will be needed in the laws and institutions governing water resources to address the expected impacts of climate change on agriculture. In areas with declining water supplies, water law will have to adapt to allocate water to the most important uses, and to make difficult choices between environmental and other in-stream uses relative to human withdrawals for off-stream uses. Reallocation of available water supplies could disrupt some economic activities and potentially require them to relocate. Those impacts are likely to generate legal challenges and political battles. Likewise, legal regimes will have to ensure that upstream landowners and governments do not flood downstream parties when precipitation intensifies.

---

[153] United Nations Environment Programme and Government of Kenya, above note 152.

The infrastructure changes necessary to adapt to climate change are likely to be extensive and expensive. In areas receiving increased precipitation, water supply and treatment infrastructure could be threatened by storms, increased flooding and sea level rise. Levees or other protective barriers might be needed to guard against significant and unpredictable flooding risks. Wastewater treatment capacity may have to be expanded in some areas, and more sophisticated treatment technology may be required in others. Areas affected by drought may need additional storage facilities or water supplies, or investments in water efficiency and reuse. Nations have already been investing in new water resource infrastructure for purposes of economic stimulus as well as climate change adaptation.[154] Investments in projects such as rainwater harvesting also can be used to help augment water resources on a more local scale.[155] Changes in farming practices, such as shifts to drought tolerant crops or crop varieties, diversification of farm production from monocultures to integrated farm operations and similar practices can also help shield agriculture from climate impacts.

It is unclear how water law will evolve in response to climate change, and responses are not likely to be uniform around the globe. There are at least four possible scenarios:

(1) climate change induced water conflicts could continue to be addressed within the confines of existing law, which is likely to result in partial or inadequate solutions;
(2) water law will evolve in response to changing conditions;
(3) governments could adopt a more prominent role in water resource management, for example, to resolve conflicts, supervise water transfers, finance or construct additional infrastructure, or change laws that pose a barrier to adaptation, and
(4) legal developments in other fields, such as land use regulation and management, agricultural subsidies and other structural agricultural policies, could exert indirect pressure on water law and incentivize both legal and societal adaptation to climate change.[156]

The fourth possibility suggests that changes to water law cannot function in isolation from other forces that affect agricultural practices. Climate

---

[154] Burleson, above note 46 at 489.

[155] *Ibid.* at 499–503.

[156] A Dan Tarlock, 'How Well Can Water Law Adapt to the Potential Stresses of Global Climate Change?' (2010) 14 *University of Denver Water Law Review* 1, 6–7.

change will require significant adaptations to water resources infrastructure and institutions, as well as distinct changes to agricultural practices and economies that will affect water law and relate to agricultural water use.

6. Climate change and invasive species law in agriculture: a southern hemisphere view of key legal and institutional issues in Australia

Elodie Le Gal

1. INTRODUCTION

Agricultural industries such as livestock and crop production provide food and raw materials to billions of people. They generate income for households, including 70 percent of the world's poor who live in rural areas.[1] To ensure food security and maintain agricultural productivity while sustaining natural resources, environmental services from agricultural lands need to remain sustainable. As defined by the Millennium Ecosystem Assessment (MEA), these services range from supporting services, such as soil formation and nutrient cycling, provisioning services, such as food production and genetic resources, regulating services, such as climate regulation and disease regulation, to cultural services (for example, sense of place).[2] Over recent decades, agricultural assets have been dramatically impacted by a range of human induced environmental threats, among which are climate change and invasive species.[3]

Climate change and invasive species, defined later in this chapter, are partly anthropogenic issues. Complex interactions exist between the two, a significant number of which are uncertain. While the science of climate change still generates vigorous debates, it is identified in the scientific

---

[1] World Bank, 'Agriculture and Rural Development' (2015), accessed 12 October 2016 at http://data.worldbank.org/topic/agriculture-and-rural-development.

[2] Joseph Alcamo *et al.* (eds), *Ecosystems and Human Well-Being: A Framework for Assessment* (Island Press 2003) 5.

[3] Aziz Elbehri (ed.), *Climate Change and Food Systems: Global Assessments and Implications for Food Security and Trade* (United Nations Food and Agriculture Organization (FAO) 2015); CB Field *et al.* (eds), *Climate Change 2014: Impacts, Adaptation, and Vulnerability: Contribution of Working Group II to the Fifth Assessment Report of the Intergovernmental Panel on Climate Change* (Cambridge University Press 2014) 500.

literature as highly likely to alter the distributions of invasive species in urban[4] and agricultural areas.[5]

Invasive species cause significant environmental, agricultural, social and economic costs that directly affect rural livelihoods. Recent wild pig invasions from Virginia to Texas have resulted in massive crop destruction in the United States.[6] The killing of thousands of sheep by wild dogs and foxes in several Australian states has put the sheep industry at risk, throwing people out of their jobs, aggravating the economic vulnerability of country towns and causing psychological stress.[7] In Europe the foot-and-mouth outbreak of 2001 in the United Kingdom has had severe economic and social impacts on British agriculture.[8]

This chapter explores some key relationships between climate change, invasive species law and agriculture. While invasive species law can contribute to climate change adaptation by protecting agricultural values from harmful biological infestations, significant institutional, governance and methodological challenges are likely to jeopardize its effectiveness. We illustrate this on the Australian continent, where the feral animal and weed threats are estimated to respectively cost \$720 million[9] and \$4 billion[10] per year to the Australian economy and agriculture, despite a national biosecurity system often referred to as one of the best in the world. While the specificities of invasive species management are unique for any

---

[4] KV Tubby and JF Webber, 'Pests and Diseases Threatening Urban Trees' (2010) 83(4) *Forestry* 451, 453.

[5] See for example D Kutywayo *et al.*, 'The Impact of Climate Change on the Potential Distribution of Agricultural Pests: The Case of the Coffee White Stem Borer (*Monochamus leuconotus P.*) in Zimbabwe' (2013) 8(8) PLoS ONE, accessed 12 October 2016 at http://journals.plos.org/plosone/article?id=10.1371/journal.pone.0073432. This is also discussed later.

[6] Ari Phillips, 'Wild Pigs are Taking Over the US, Destroying Crops and Uprooting Levees' (ThinkProgress, 27 November 2013), accessed 12 October 2016 at http://thinkprogress.org/climate/2013/11/27/3002821/wild-pigs-hogs-invasive-desctruction/.

[7] Lyndal-Joy Thompson *et al.*, 'Social Impacts of Wild Dogs – A Review of Literature' (ABARES, 2013) 23–9, accessed 12 October 2016 at http://www.wool.com/globalassets/start/on-farm-research-and-development/sheep-health-welfare-and-productivity/pest-animals/wild-dogs-foxes-and-pigs/wp525_wild_dog_management_in_australia.pdf.

[8] F Junker, J Komorowska and F Van Tongeren, 'Impact of Animal Disease Outbreaks and Alternative Control Practices on Agricultural Markets and Trade: The Case of FMD' (OECD, 2009) 16, accessed 12 October 2016 at http://www.oecd.org/tad/agricultural-policies/45013598.pdf.

[9] Ross McLeod, *Counting the Cost: Impacts of Invasive Animals in Australia* (Cooperative Research Centre for Pest Animal Control 2004).

[10] JA Sinden *et al.*, *The Economic Impacts of Weeds in Australia* (CRC for Australian Weed Management 2004).

country, this case study provides broader insights into the institutional improvements likely to be required to protect agricultural systems throughout the world from biological infestations, aggravated by climate change. Designing and implementing effective legal and institutional policy for invasive species management is key to creating a successful pathway to sustainable agriculture and livelihoods, as well as climate change adaptation.

For the purpose of this chapter, the term 'law' includes prescriptive measures and soft law, or quasi-legal instruments, directed to manage human behaviour that either aggravates or alleviates the threats of climate change and invasive species. Invasive species range from terrestrial to aquatic living organisms, and can include: diseases, fungi, parasites, feral animals, weeds, introduced marine pests, insects and other vertebrates;[11] this chapter will particularly emphasize vertebrate pests and weeds.

The investigation draws upon an in-depth review of the international and national literature on the relationship between climate change, invasive species law and agriculture, as well as empirical research conducted within a trans-disciplinary research program that integrates knowledge, experience and skill into the management of invasive species.[12]

This chapter is structured in six key sections. The first section clarifies the concepts of climate change and invasive species, points out some of the interactions identified in the literature between invasive species, climate change and agriculture and highlights the significance of invasive species law for adapting agriculture to climate change. The second section summarizes the institutional and legal framework for invasive species management in Australia. The three following sections explore weaknesses of the current governance system for invasive species management, including: the missing link between invasive species law and their behavioral effectiveness, implementation issues that affect transaction costs, and the deficiencies of institutional arrangements for systemic and community-partnered approaches to invasive species management. In light of these weaknesses, potential innovations are discussed that could improve the effectiveness

---

[11] Australian Department of the Environment, 'Invasive Species', accessed 12 October 2016 at http://www.environment.gov.au/biodiversity/invasive-species.

[12] Invasive Animals CRC, 'Community Engagement', accessed 12 October 2016 at http://www.invasiveanimals.com/research/phase2/community-engagement/. We have extensively discussed institutional issues surrounding invasive species management in the following report which has not been published: Paul Martin *et al.*, 'Effective Citizen Action on Invasive Species: The Institutional Challenge' (Invasive Animals Cooperative Research Centre 2016), accessed 28 November 2016 at http://www.pestsmart.org.au/effective-citizen-action-on-invasive-species-the-institutional-challenge/.

of invasive species management and reduce barriers to climate change adaptation strategies. The conclusion summarizes the findings, highlights the significance for invasive species practitioners, policymakers and researchers, and addresses the need for further research and action to better reconcile invasive species law with sustainable agriculture, while building an effective low-carbon economy.

2. CLIMATE CHANGE, INVASIVE SPECIES AND AGRICULTURE

2.1 The Climate Change Phenomenon

Defined by the Intergovernmental Panel on Climate Change (IPCC) as 'a statistically significant variation in either the mean state of the climate or in its variability, persisting for an extended period (typically decades or longer)',[13] climate change refers to the gradual warming of the Earth's atmosphere and oceans due to the emissions of greenhouse gas from anthropogenic activities.

Clarifying the concept of climate change impacts, Pittock *et al.* explain that:

> [. . .] the impact of a given climate change on a particular system or activity depends on its vulnerability and adaptability. If the system or activity does not adapt to the change, the impacts will be larger. Adaptation may be automatic, or it may be conscious and planned. Automatic adaptation tends to occur in natural systems, such as migration of fauna or flora to more suitable environmental regions. Planned adaptation tends to occur in human systems. This might be as simple as delaying planting of a crop until there is rainfall, or planting a smaller area if a drought is feared. Or it may be more complex, involving research, planning and investment, for example in developing or selecting a different cultivar or crop variety, developing an irrigation system, or switching from growing wheat to grazing cattle or even developing a tourism industry.

Therefore when we talk about climate change impacts we need to be clear whether we are talking about the direct effects without adaptation, or the effects after some automatic or planned adaptation.[14]

To address climate change impacts, two main strategies are available:

---

[13] JT Houghton *et al.* (eds), *Climate Change 2001: The Scientific Basis: Contribution of Working Group I to the Third Assessment Report of the Intergovernmental Panel on Climate Change* (Cambridge University Press 2001) Appendix I.

[14] Barrie Pittock (ed.), *Climate Change: An Australian Guide to Science and Potential Impacts* (Australian Greenhouse Office 2003) 73.

climate change mitigation (human intervention to reduce the sources or enhance the sinks of greenhouse gases)[15] and adaptation strategies. As noted by the IPCC, 'adaptation needs arise when the anticipated risks or experienced impacts of climate change require action to ensure the safety of populations and the security of assets, including ecosystems and their services (*medium evidence, medium agreement*)'.[16] Adaptation options identified by the IPCC include structural or physical, social, and institutional actions. Economic instruments, laws and regulations, government policies and programs are the institutional adaptation options listed to adapt to climate change.[17] As discussed later, strong invasive species management strategies, including those underpinned by invasive species laws, regulations and policies are needed for adaptation to climate change not only in Australia but also in other parts of the world.

2.2 The Elusive Concept of Invasive Species

Invasive species, also referred to as 'alien species', are species occurring in ecosystems to which they are not indigenous. The International Union for Conservation of Nature (IUCN) has defined invasive species with reference to their potential impact on ecosystems' biological diversity as follows:

> Invasive species means an alien species which becomes established in natural or semi-natural ecosystems or habitat, is an agent of change, and threatens native biological diversity.[18]

Invasive species with commercial or non-commercial value can be imported legally or illegally. Many of the invasive plants and animals present in Australia are the result of legal importations for commercial purposes to improve agricultural productivity. For example, cane toads were introduced as biological control agents to control beetles that were damaging sugar cane crops in Queensland.[19] Examples of illegal importation include the

---

[15] Ottmar Edenhofer *et al.* (eds), *Climate Change 2014: Mitigation of Climate Change: Contribution of Working Group III to the Fifth Assessment Report of the Intergovernmental Panel on Climate Change* (Cambridge University Press 2014) 4.

[16] Field *et al.* (n 3) 836.

[17] *Ibid* at 845.

[18] Clare Shine, Williams Nattley and Gündling Lothar, *A Guide to Designing Legal and Institutional Frameworks on Alien Invasive Species – Environmental Policy and Law Paper No 40* (IUCN and World Conservation Union 2000) 2.

[19] Australian Department of the Environment, 'Cane Toads (Bufo Marinus)', accessed 12 October 2016 at http://www.environment.gov.au/biodiversity/invasive-species/feral-animals-australia/cane-toads.

introduction of wildlife and exotic animals protected by the Convention on International Trade in Endangered Species of Wild Fauna and Flora (CITES), which can then become invasive.

Once they are introduced in the natural environment, animals and plants with specific invisibility features may establish, naturalize and invade ecological systems along different possible pathways (the means by which invasive species move). Weeds and insects in particular can be introduced by natural vectors such as wind and birds, and accidentally or deliberately by humans. Pathways for weeds species introduction are many. Sindel *et al.* have identified 17 weed pathways in Australia, for example.[20] Biological infestations often result from a combination of the following elements:

- a complex chain of contributing and interacting processes and events, including ecological and anthropogenic activities;
- human decisions or 'transactions' where a human or organization makes a decision to introduce a new species, or modify a species; and
- autopoetic events (for example, the evolution of species within natural systems) which do not require any human decisions.

Potential legal and regulatory points of intervention may identify those who should be held responsible for managing invasive species under a risk accountability framework. This is illustrated by Figure 6.1, representing the biofuel weed[21] pathway, the actors and institutions potentially involved in the biofuel weed decision-making process, and associated risk management strategies. The various strategies can correspond to different types of market, social and regulatory instruments and institutional categories of response, among which are biosecurity arrangements, disaster responses, agricultural and natural resource management laws, and liability schemes.[22]

In Australia there are more than 50 animals considered invasive and harmful. These include wild dogs, foxes, carps, feral cats and deer. What

---

[20] Brian Sindel *et al.*, 'Pathway Risk Analysis for Weed Spread within Australia – Final Report, Project No UNE61' (*Land and Water Australia*, 2008) 24, accessed 12 October 2016 at http://lwa.gov.au/products/pn22274.

[21] Biofuel crops can be imported into Australia to produce bioethanol (a form of low carbon energy) and become invasive. See Tim Low and Carol Booth, 'The Weedy Truth about Biofuels' (*Invasive Species Council*, 2007), accessed 12 October 2016 at https://invasives.org.au/files/2014/02/isc_biofuels_revised_march08.pdf.

[22] Paul Martin and Elodie Le Gal, 'Concepts for Industry Co-Regulation of Bio-Fuels Weeds' (2010) IUCNAEL e-Journal, Accessed 12 October 2016 at http://www.iucnael.org/en/documents/499-martin-le-gal-iucn-academy-ejournal-article-final/file.

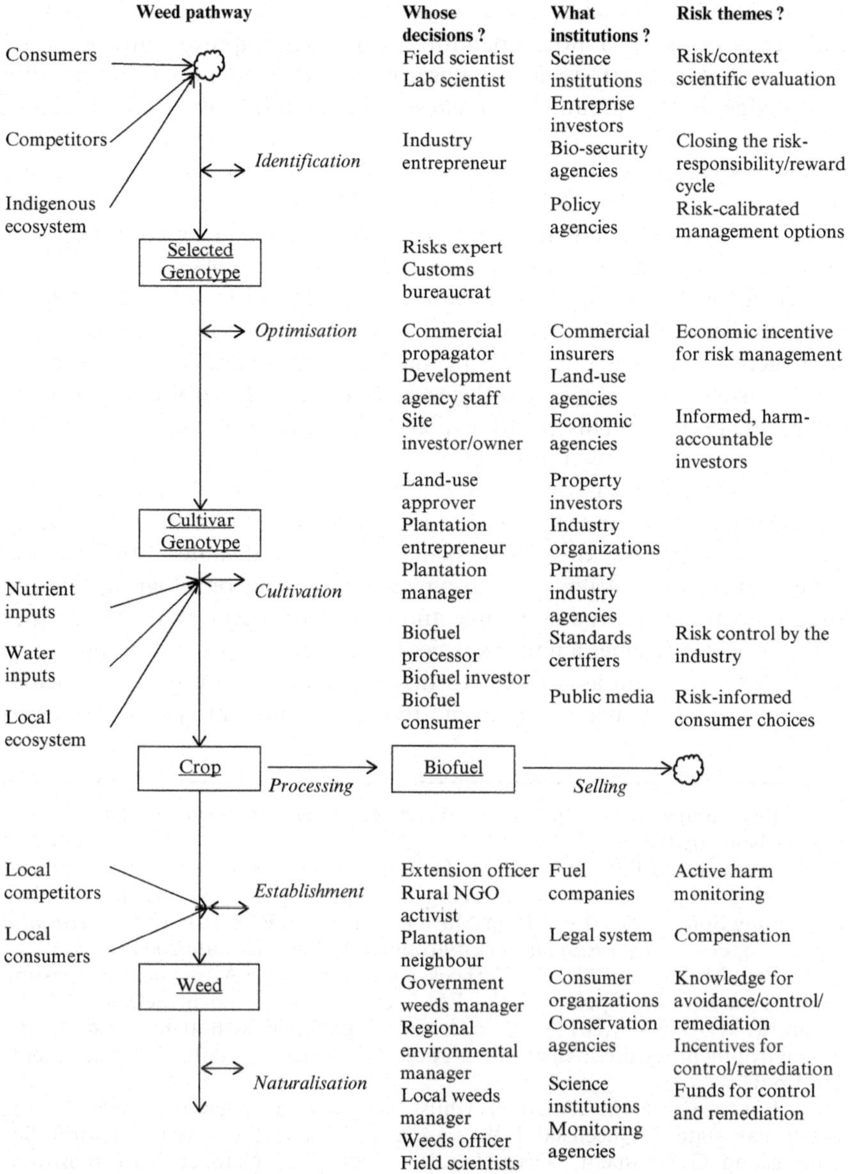

Source: Paul Martin and Elodie Le Gal, 'Concepts for Industry Co-Regulation of Bio-Fuels Weeds' (2010) 1 IUCNAEL e-Journal 7.

Figure 6.1 The biofuel weed pathway

is considered an invasive species is context dependent. A species may be judged as invasive or not depending on the geographical, jurisdictional and cultural context, which themselves are dependent upon scientific knowledge and community awareness. The harmful invasiveness feature of a species may or may not materialize depending upon complex interactions between ecological, biological, geographical and time factors, including climate. The definition of a weed has been described as being time and space dependent.[23] This scientific uncertainty as to whether a specific species may become invasive or not helps to explain what has been described by Miller and Gunderson as 'a profound lack of complete or coherent law and policy'[24] in national legal frameworks for invasive species management. This is illustrated in the Australian context where: there is no consistent definition across the seven States and Territories of the term 'invasive species'.[25] This partly explains why the legal status of invasive species can vary between jurisdictions.[26]

Even the term 'invasive species' is a human construct. People conceptualize differently the invasive species risk depending upon their education, profession, expertise, knowledge, beliefs and cultural values.[27] In Australia for example there is often much disagreement over the ontological status of dingoes between aboriginal people and sheep producers. While the former often consider dingoes a native species to be protected for cultural reasons, the latter view them as a threat to their flocks and economic well-being. Animal welfare activists may argue that any living entity has a right to

---

[23] BM Sindel (ed), *Australian Weed Management Systems* (RG & FG Richardson 2000) 4.

[24] Marc L Miller and Lance H Gunderson, 'Biological and Cultural Camouflage: The Challenges of Seeing the Harmful Invasive Species Problem and Doing Something About It' in Marc L Miller and Robert N Fabian *Harmful Invasive Species: Legal Responses* (Environmental Law Institute 2004) 1.

[25] Paul Miriam and Miriam Verbeek, 'A Comparative Assessment of Existing Policies on Invasive Species in the EU Member States and in Selected OECD Countries' (*DG ENV*, 2011) 372, accessed 12 October 2016 at http://ec.europa.eu/environment/nature/invasivealien/docs/IAS_policies_country%20assessments 2011.pdf.

[26] For example, in Australia whilst deer are considered as feral in the Australian state of Queensland, they are a resource in the state of Victoria. See Queensland Government, 'Deer (Feral)', accessed 12 October 2016 at https://www.daf.qld.gov.au/plants/weeds-pest-animals-ants/pest-animals/a-z-listing-of-pest-animals/photo-guide-to-pest-animals/deer-feral. Under the Victorian Wildlife Act 1975, deer are declared as 'wildlife' and recognized as game hunting resources.

[27] Elodie Le Gal, 'The Effects of Institutional Path-Dependence, Political Dynamics and Transaction Costs on the Potential for Smart regulatory Innovation: An Illustration with the Biofuel Weed Risk Case Study' (2012) 15(2) AJNRLP 219.

live and therefore should not be killed. This difference in values and risk perception, which is underpinned by differing ontological views, explains why certain species are considered pests that should be killed using biological, physical, mechanical, chemical or other control methods, while other species are not.

The term 'invasive' carries a connotation driven by ontological views and subjective risk perception. For some people invasive species may be a risk, yet for others it can be a benefit. This is illustrated with second-generation biofuel crops, which have commercial value and can be used to produce bioethanol and contribute to climate change adaptation. Biofuel plantation crops and biofuel plants on marginal agricultural lands can strengthen economic resilience in rural Australia.[28] However, second generation biofuel crops have become invasive in some countries, devastating agricultural and environmental values in many African countries.[29] Feral camels, which are present in Australia's rangeland ecosystems, affect biodiversity assets and values on aboriginal, conservation, pastoral and Crown lands, and cause economic and social costs by damaging infrastructure and cultural sites.[30] They are, however, perceived as an economic or food resource by certain groups.[31] In a report published by the Desert Knowledge Cooperative Research Centre, it was estimated that in 2007 the income generated by the export of live camels overseas, particularly in South-East Asian, Middle Eastern and African countries was around $272,000 (AU) and $363,000 (AU) per annum.[32] Another research report shows that new camel export markets were being explored in countries such as China.[33] These examples demonstrate that an environmental risk can also become a business opportunity.

Deer, which can generate significant economic revenue from recreational hunting activities, are another example where a certain species is considered invasive by some but as a beneficial resource by others. In the state of Queensland, some species of deer have been declared pest animals and are

---

[28] Australian Government Biofuels Taskforce, *Report of the Biofuels Taskforce to the Prime Minister* (Biofuels Taskforce 2005) 114, 115.

[29] Low and Booth (n 21).

[30] Natural Resource Management Ministerial Council – Vertebrate Pests Committee, *National Feral Camel Action Plan: A National Strategy for the Management of Feral Camels in Australia* (Department of Sustainability, Environment, Water, Population and Communities 2010) 5–7.

[31] *Ibid* at 4.

[32] GP Edwards *et al.* (eds), *Managing the Impacts of Feral Camels in Australia: A New Way of Doing Business* (Desert Knowledge CRC 2008) 240.

[33] David Wu, Chiu-Hing Chan and Crystal Deng, *Australian Camel Meat: China Market* (RIRDC 2011).

controlled under the Feral Deer Management Strategy 2013–2018.[34] In the State of Victoria where they have not been declared pests, deer provide a source of revenue through sport hunting and tourism.[35]

These differences in risk perception and ontological beliefs are informed by political, cultural and social values, all which challenge policymakers to decide whether certain species should be controlled, and how interests should be prioritized. Such decisions are likely to raise conflicts between groups with competing interests, and may generate policy implementation failures. Pests and invasive species issues generate strong political debates among citizens with different ethical values and understanding of the issues. These include discussions about the use of waste products from the culling of pest animals[36] and community opposition to the use of inhumane control methods (for example, camel culls,[37] hunting of wild pigs with trained dogs).[38] Political tension can also emerge from disagreements on policy options. For example, bounties, which are financial rewards for motivating people to control invasive species, can be considered either a waste of public money[39] or an effective control method.[40]

The economic impacts of invasive species on eco- and agricultural systems worldwide have been assessed differently in the literature.[41]

---

[34] Queensland Department of Agriculture, Fisheries and Forestry, 'Feral Deer Management Strategy 2013–2018' (2013), accessed 12 October 2016 at https://www.daf.qld.gov.au/plants/weeds-pest-animals-ants/educational-resources-and-careers/publications/feral-deer-management-strategy-2013-18.

[35] Cameron Wilson, 'More Hunting on the Cards' (*ABC Bush Telegraph*, 11 June 2014), accessed 12 October 2016 at http://www.abc.net.au/radionational/programs/bushtelegraph/hunting-report-sparks--conservation2c-welfare-and-tourism-conv/5515852.

[36] Emily Osmond, 'The PETA Vermin Rabbit Fur Debate' (*Independent Australia*, 1 December 2011), accessed 12 October 2016 at https://independentaustralia.net/australia/australia-display/the-peta-vermin-rabbit-fur-debate,3802.

[37] Sara Everingham, 'Camel Cull to Proceed in Central Australia' (*ABC News* 7 December 2009), accessed 12 October 2016 at http://www.abc.net.au/news/2009-12-07/camel-cull-to-proceed-in-central-australia/1171988.

[38] Caro Meldrum-Hanna, 'Pig Dogging Faces Question of Barbarity or Hobby' (*ABC News* 17 July 2012), accessed 12 October 2016 at http://www.abc.net.au/7.30/content/2012/s3547927.htm.

[39] Michael Green, 'Little Fox, Big Problem' (*The Age*, 3 July 2014), accessed 12 October 2016 at http://www.theage.com.au/victoria/little-fox-big-problem-20140703-3bboh.html.

[40] Anon, 'Fox Eradication Schemes Proves Bountiful' (*ABC News* 14 May 2012), accessed 12 October 2016 at http://www.abc.net.au/news/2012-05-14/fox-eradication-scheme-proves-bountiful/4009084.

[41] Lars J Olson, 'The Economics of Terrestrial Invasive Species: A Review of the Literature' (2006) 35(1) J Agr Resour Econ 178; Wendi Gong, Mike Braysher

Pimentel *et al.* estimated that environmental and economic damage from invasive species cost more than $1.4 trillion per year worldwide.[42] Such data often omit the social, health, psychological and distributional impacts, which are hard to quantify.[43] Invasive species also spread infectious diseases to plants, animals and human beings.[44] Moreover the fallout is not experienced by all citizens, but is instead most felt by landowners and those directly involved (for example, sheep producers impacted by wild dogs). Those responsible for introducing new invasive species into the environment generally remain unaccountable under institutional and legal frameworks, and taxpayers often end up bearing the costs of control and recovery after the fact. This raises distributional justice issues as the environmental, economic and social costs and benefits of invasive species management are not equally shared among citizens. The financial and legal pressures already imposed upon landowners jeopardize their capacity to get involved in effective coordinated action across a 'whole of landscape' approach in invasive species control.

2.3 Interactions between Climate Change and Invasive Species

Relationships between invasive species infestations and climate change are complex. The Australian 'Commonwealth Scientific and Industrial Research Organization' (CSIRO), which is the Commonwealth agency for scientific research in Australia, has identified three models that describe the likely responses of invasive species to climate change.[45] These are the 'rapid distribution change model', the 'relative abundance change model' and the 'gradual distribution shift model'. They each explain distribution scenarios under different environmental conditions. The scientific literature points to difficulties in modeling the impacts of climate change

---

and Randall Jones, 'The Economic Impacts of Vertebrate Pests in Australia' (Invasive Animals Cooperative Research Centre 2008).

42 D Pimentel *et al.*, 'Economic and Environmental Threats of Alien Plant, Animal, and Microbe Invasions' (2001) 84(1) Agric Ecosyst Environ 1, 14.

43 McLeod (n 9); Gerard Fitzgerald and Roger Wilkinson, *Assessing the Social Impacts of Invasive Animals in Australia* (Invasive Animals CRC 2009). See also Thompson *et al.* (n 7).

44 Todd A Crowl *et al.*, 'The Spread of Invasive Species and Infectious Disease as Drivers of Ecosystem Change' (2008) 6(4) Front Ecol Environ 238.

45 Michael Dunlop and Peter R Brown, 'Implications of Climate Change for Australia's National Reserve System: A Preliminary Assessment' (*Department of Climate Change*, 2008) 62–3, accessed 12 October 2016 at http://www.environment.gov.au/system/files/resources/917bb661-b626-44bb-bd52-325645ae7c49/files/nrs-report.pdf.

on invasive species. Hellmann *et al.* identified five potential consequences of climate change on the distribution of invasive species.[46] These include altered transport, altered climatic constraints, altered distributions, altered impacts and altered effectiveness of management strategies.[47] Overall it is predicted that climate change and invasive species, which have been identified with habitat loss as the two most serious threats to Australian native species,[48] are likely to increase the vulnerability of agricultural production systems. Climate change is reported to accelerate the frequency or recurrence of disturbance events (for example, cyclones, droughts, fires and floods) and creates favorable conditions for biological invasions such as weed dispersals.[49] In turn, weed infestations may impact agricultural ecosystem functions, such as nutrient cycling processes, as well as soil composition, crop productivity and food security.[50] Invasive species that disturb native vegetation cover (such as rabbits and wild dogs) harm biodiversity, exacerbating the effects of climate change, and can alter the effectiveness of biological agents to control weeds.[51] With the increasing environmental, economic and social pressures that farmers are likely to face under climate change conditions, it is expected that they will have fewer resources and less time to control biological invasions on their properties.

The complex relationship that exists between climate change, invasive species and agriculture suggests that domestic and international legal architectures for invasive species management will likely have to grapple with 'wicked' problems[52] to design and implement good governance models.

---

[46] Jessica J Hellmann *et al.*, 'Five Potential Consequences of Climate Change for Invasive Species' (2008) 22(3) Conserv Biol 534.

[47] *Ibid.*

[48] Carol Booth, 'Invasive Species: One of the Top Three Threats to Australian Biodiversity, Invasive Species Council' (Invasive Species Council 2009), accessed 12 October 2016 at http://invasives.org.au/publications/invasive-species-one-top-three-threats-australian-biodiversity/.

[49] Tim Low, 'Climate Change and Invasive Species – A Review of Interactions' (Biological Diversity Advisory Committee, 2006) 8, accessed 12 October 2016 at http://www.environment.gov.au/system/files/resources/ab5b298b-7374-4bea-a613-2ffa6fac63fe/files/interactions-cc-invasive.pdf.

[50] Lewis H Ziska and Jeffrey S Dukes (eds), *Weed Biology and Climate Change* (Wiley Blackwell 2010) 85.

[51] Michael D Day *et al.*, *Lantana: Current Management Status and Future Prospects* (Australian Centre for International Agricultural Research 2003).

[52] These are characterized as 'difficult to clearly define, have 'many interdependencies and are often multi-causal', 'are often not stable', usually have no 'clear solution', 'are socially complex', 'hardly ever sit conveniently within the responsibility of any one organization', 'involve changing behavior', are characterized

3. AUSTRALIAN INVASIVE SPECIES LAW

With a population of only almost 24 million,[53] Australia is a vast continent of 7.7 million square kilometers comprising six states[54] and two territories.[55] It is a federation with three levels of government. While the management of invasive species issues is primarily a state/territory responsibility, regulation of the national biosecurity system and matters of national significance[56] fall under the scope of the Commonwealth. This division of powers/responsibilities adds to a fragmented system of invasive species law and implementation arrangements.

As in many other post-enlightenment constitutional republics, no mention of the environment was made in the Australian Constitution when it was drafted. However, section 51 of the Constitution gives the Commonwealth powers to legislate in areas related to environmental protection and natural resource management.[57] The Commonwealth's Environmental Protection Biodiversity Act (EPBA) has been adopted to address broader biodiversity issues, including those that relate to invasive species infestations. Under a framework of cooperative federalism, the Commonwealth has initiated national approaches to coordinate control of invasive species with the States and Territories.

Under section 100 of the Australian Constitution, states and territories retain the power to regulate environmental and natural resource

---

by chronic policy failure. Attempts to address wicked problems often lead to unforeseen consequences'. Australian Public Service Commission, 'Tackling Wicked Problems – A Public Policy Perspective' (Australian Government 2007) 3–5, accessed 12 October 2016 at http://www.apsc.gov.au/publications-and-media/archive/publications-archive/tackling-wicked-problems.

[53] See Australian Bureau of Statistics, 'Population Clock', accessed 12 October 2016 at http://www.abs.gov.au/ausstats/abs@.nsf/0/1647509ef7e25faaca2568a900154b63?OpenDocument.

[54] The six states are New South Wales (NSW), Queensland (Qld), Tasmania (Tas), Victoria (Vic) and Western Australia (WA).

[55] The two territories are the Australian Capital Territory (ACT) and Northern Territory (NT).

[56] Those which are listed in the Environment Protection and Biodiversity Conservation Act 1999 (Cth) include for example world heritage and national heritage properties, wetlands of international importance, listed threatened species and ecological communities.

[57] These include for example a series of federal legislative powers such as interstate and overseas trade and commerce power, corporations power, external affairs power, taxation power, defense power, the power with respect to quarantine, the power with respect with fisheries and the race power. See section 51 of the Australian Constitution.

management matters. Local governments, the third tier of government in Australia, are in charge of designing and implementing local laws relevant to invasive species management (for example, maintenance of local roads and parks, land use planning, land and coast care programs). As promulgated under individual state and territorial legislation, their role may vary from one to another. While in Queensland, local governments are responsible for invasive species regulatory enforcement,[58] in New South Wales[59] and Victoria,[60] state governments have responsibility for invasive species management. By ensuring the management of rodents and other dangerous and disturbing animals, local councils also provide health services to the local population. Although these matters are not normally identified under invasive species laws, they are also relevant to managing biological infestations, as they contribute to protecting the local population from the health and social risks that invasive species can create.

All of these various legislative instruments do not address related matters such as the control of hunting or rules imposed under private law, which can create stewardship conditions under the various leasehold arrangements that govern most of northern Australia. Neither do they include local government health, safety and amenity rules, nor the very many statutory and other plans, such as those governing the operation of public lands. In practice, the invasive species challenge cuts across many different areas of public and private law, showing the sheer breadth of the subject to be scrutinized when exploring governance issues in the context of climate change and agriculture.

Martin *et al.* have identified more than 81 legislative instruments relevant to invasive species management, administered by at least 35 different Commonwealth and State government organizations with varied responsibilities.[61] As a result of its three-tiered system, the Australian model for invasive species management is a complex multi-layered and polycentric federal system[62] characterized by overlapping rules and

---

58 Land Protection (Pest and Stock Route Management Act) 2002 (Qld).

59 See Local Land Services Act 2013 (NSW).

60 See Catchment and Land Protection Act 1994 (CaLP Act) (Vic).

61 Martin *et al.* (n 12), Resource D.

62 For a broad overview of the governance framework for environmental and natural resource management, see Parliament of Australia, 'Turning Back the Tide – The Invasive Species Challenge - Report on the Regulation, Control and Management of Invasive Species and the Environment Protection and Biodiversity Conservation Amendment (Invasive Species) Bill 2002' (Senate Environment Communications Information Technology and the Arts References Committee 2004) ch 2, accessed 12 October 2016 at http://www.aph.gov.au/Parliamentary_Business/Committees/Senate/Environment_and_Communications/Completed_

institutions, including Acts and regulations, policies, private contractual and non-contractual arrangements and soft law, codes of practices, and invasives and Natural Resource Management (NRM) plans and strategies. National Policies which are not legally binding, such as the National Strategy for the Management of Vertebrate Pest Animals,[63] the National Strategy for Ecologically Sustainable Development,[64] the Australian Weeds Strategy,[65] Australia's Biodiversity Strategy 2010–2030,[66] the Australian Animal Welfare Strategy[67] and the National Plant Biosecurity Strategy[68] express stakeholders' strategies to address the invasive species challenge and broader level NRM sustainability issues. States and territories also have their own strategies to address invasive species and biodiversity issues.[69] The diversity of organizations and actors involved in the decision-making process, as well as applicable rules and policies, creates a cumbersome institutional architecture. This jeopardizes the ability of citizens to both understand the legal obligations to which they are subject, and

---

inquiries/2004-07/invasivespecies/report/index; S Wild River, *The Role of Local Government in Environmental and Heritage Management* (Department of Environment and Heritage 2006); JA Bellamy, 'Adaptive Governance: The Challenge for Regional Resource Management' in AJ Brown and JA Bellamy (eds), *Federalism and Regionalism in Australia: New Approaches, New Institutions?* (ANU Press 2007).

[63] Department of the Environment and Water Resources, *A National Strategy for the Management of Vertebrate Pest Animals in Australia* (Department of the Environment and Water Resources 2007).

[64] Ecologically Sustainable Development Steering Committee, *The National Strategy for Ecologically Sustainable Development* (Department of the Environment and Water Resources 1992).

[65] Australian Weeds Committee, 'Australian Weeds Strategy', accessed 12 October 2016 at http://www.weeds.org.au/aws.htm.

[66] Australian Department of the Environment, 'Australia's Biodiversity Conservation Strategy 2010–2030', accessed 12 October 2016 at http://www.environment.gov.au/biodiversity/publications/australias-biodiversity-conservation-strategy.

[67] Australian Animal Welfare Strategy, 'Home Page', accessed 12 October 2016 at http://www.australiananimalwelfare.com.au/.

[68] Plant Health Australia, 'National Plant Biosecurity Strategy', accessed 12 October 2016 at http://www.planthealthaustralia.com.au/national-programs/national-plant-biosecurity-strategy/.

[69] It is beyond the scope of this chapter to detail them all. The following examples will illustrate the many policy documents that are relevant for invasive species management. For the Australian Capital Territory: ACT Pest Animal Management Strategy 2012–2022, ACT Natural Resource Management Plan 2008 – Bush Capital Legacy, Iconic City, Iconic Assets. In New South Wales, are relevant: the NSW Invasive Species Plan 2008–2015, the NSW Animal Biosecurity and Welfare Strategic Plan 2013–2015 and the NSW Biosecurity Strategy 2013–2021.

to identify the public departments and private organizations that are in charge of administering these rules. The existence of multi-layered plans and strategies across jurisdictions offer useful strategic directions for invasives' management. However, they often lack institutional support and the implementation arrangements that are needed to support community action on the ground and facilitate stakeholder engagement in collaborative action. The lack of mechanisms to evaluate objectively whether these arrangements contribute to the intended sustainable outcomes also raises questions about their significance for successfully protecting environmental and agricultural values.

Responsibility for natural resource management, including invasive species issues, is highly fragmented and varies widely between jurisdictions.[70] International instruments relevant for invasive species control to which Australia is a party include the 1992 United Nations Convention on Biological Diversity,[71] the 1973 Convention on International Trade in Endangered Species of Wild Species of Wild Fauna and Flora,[72] and the 1997 FAO International Plant Protection Convention.[73] Institutional arrangements for invasive species are closely linked to biological and chemical control, land law and land use planning, emergency management, biodiversity conservation, water pollution, animal welfare, and sanitary and phytosanitary (biosecurity) measures. The national biosecurity system, which has recently been strengthened following the Nairn Review[74] in 1995 and the Beale report[75] in 2008, is reputed for its tough quarantine standards. The Biosecurity Act 2014, passed by the Commonwealth Parliament in March 2014, provides a risk-based approach to biosecurity.

---

[70] Bellamy (n 62) 104.

[71] 1992 United Nations Convention on Biological Diversity (adopted 5 June 1992, entered into force 29 December 1993) 1760 NTS 79.

[72] 1973 Convention on International Trade in Endangered Species of Wild Species of Wild Fauna and Flora (adopted 3 March 1973, entered into force 1 July 1975) 993 UNTS 243.

[73] The International Plant Protection Convention was first adopted in 1951 and entered into force in 1952. Since then, it has been amended twice, most recently in 1997. See 1997 International Plant Protection Convention (opened for signature 17 November 1997, entered into force 2 October 2005) 2367 UNTS 223.

[74] ME Nairn *et al.*, 'Australian Quarantine: A Shared Responsibility' (Department of Primary Industries and Energy, 1996), accessed 12 October 2016 at http://www.agriculture.gov.au/SiteCollectionDocuments/aqis/about/corporate-docs/nairn/nairn_report.pdf.

[75] Roger Beale *et al.*, *One Biosecurity – A Working Partnership – The Independent Review of Australia's Quarantine and Biosecurity Arrangements Report to the Minister for Agriculture Fisheries and Forestry* (Commonwealth of Australia 2008).

It also provides a framework for a 'shared responsibility' approach to biosecurity issues between the Commonwealth government, states and territories, local communities, and industry stakeholders to ensure an integrated biosecurity continuum. That continuum encompasses activities undertaken pre-border (for example, risk assessment, monitoring, surveillance and response), at the border, and post-border. Because of this a certain number of states, such as New South Wales and Victoria, have begun rethinking domestic procedure for invasive species management, but have yet to provide the sufficient institutional structure (including capacity building and resource mechanisms) to support effective biosecurity community engagement[76] including front line and landowner control work. While the concept of shared responsibility appears to be a promising paradigm to share equitably the costs of invasive species management among citizens, landowners remain the ultimate risk bearers of biological infestations.

4. DO INVASIVE SPECIES LAWS WORK?

There are methodological difficulties in assessing the performance of NRM and environmental laws, including invasive species laws. The effectiveness of environmental laws is context-dependent. There is often a lack of empirical evidence that shows whether specific institutional arrangements work or not.[77] Even when there is sufficient evidence to demonstrate the success or failure of environmental management instruments, lawyers do not necessarily have the skills to understand them.[78] As pointed out by the Australian Bureau of Statistics:

> . . . Australia [is] one of 17 mega-diverse countries that collectively hold around 70% of the world's flora and fauna. Australia is home to an estimated 570,000 different species (147,579 described species), giving it more than 5 percent of the world's plants and animals.[79]

---

[76] Heleen Kruger *et al.*, *Engaging in Biosecurity: Gap Analysis* (Bureau of Rural Sciences, Australian Government 2009).

[77] Michael G Faure, 'Instruments for Environmental Governance: What Works?' in Paul Martin *et al.* (eds), *Environmental Governance and Sustainability* (Elward Elgar 2012) 3.

[78] *Ibid.*, 4.

[79] Australian Bureau of Statistics, 'Australia's Biodiversity' (2009), accessed 12 October 2016 at http://www.abs.gov.au/ausstats/abs@.nsf/Previousproducts/1301.0 Feature%20Article12009%E2%80%9310?opendocument&tabname=Summary&prodno=1301.0&issue=2009%9610&num=&view=.

Despite the existence of an arsenal of legal and normative tools described in the previous section, the 2011 State of the Environment Report states that invasive species impacts are 'high' to 'very high' and that 'there are very limited data on which to assess whether efforts to address problems are having an impact'.[80] It also highlights that:

> It is extremely difficult to assess the effectiveness of management in relation to invasive species and pathogens from some reports from most states and territories . . . These reports mostly list plans, strategies and inputs to management, but rarely report on the effectiveness of processes or on outputs and outcomes . . . Some SoE reports state that actions are not achieving desired results, while this conclusion is implicit in other SoE reports since the effects of invasive species are assessed as getting worse. Some SoE reports conclude that there is not enough information to assess trends or the magnitude of effects.[81]

The report also notes that for invasive species and pathogen control:

(1) 'understanding' is substantially ineffective but improving;
(2) 'planning' is substantially ineffective but improving;
(3) 'inputs' into controls are both ineffective and declining;
(4) 'processes' are ineffective but improving; and
(5) 'outcomes' are ineffective but improving.[82]

These comments suggest that the governance system for invasive species control does not deliver the intended outcomes. This also suggests that current institutional arrangements do not trigger sufficient individual and collective behavioral responses to risk identification and reduction. To trigger sustainable behavioral change, a deeper understanding of how people are likely to respond to laws and regulations is needed. Among other considerations is also the question of the extent to which people have the economic, informational and human capacity to adopt best management practices for managing biological infestations.

Legal strategies for addressing natural resource management and environmental issues, such as climate change and invasive species, are informed by assumptions about how people behave. The specialized literature contrasts government regulation (also referred to as 'command-and-control' or 'prescriptive measures') with market based arrangements.[83] While the

---

[80] State of the Environment 2011 Committee, *Australia State of the Environment 2011* (DSEWPaC 2011) 641.

[81] *Ibid.* at 665.

[82] Ibid. at 666–7.

[83] Jody Freeman and Charles D Kolstad, 'Prescriptive Environmental Regulations versus Market-based Incentives' in Jody Freeman and Charles D Kolstad (eds),

former uses traditional tools such as conditional licensing, permit based systems, prohibition and enforcement, penalties and prosecutions, the latter are underpinned by neo-classical economic theories. They involve a wide range of price based (for example, taxes, levies, subsidies), quantity-based (for example, cap-and-trade, offsets mechanisms) and market friction (for example, certification and eco-labelling schemes) mechanisms. Market-based approaches, referred to as 'economic instruments', incorporate financial incentives/disincentives to motivate people to better manage natural resources and environmental assets. Both of these approaches are informed by the 'rational actor' model, developed by Jeremy Bentham in the 18th century. This model relies on the concept of 'rational man' to predict motivations for compliance behavior.[84] It is expected that the use of financial (dis)incentives through the marketplace as well as punishment mechanisms, such as fines, prosecution and loss of licences, will all encourage needed behavior.

Australian invasive species laws extensively rely on command-and-control approaches, illustrated by bans against import, interdiction to hazardous species, removal orders, and obligations upon private landowners to manage invasive species. They scarcely involve environmental market-based instruments, unlike climate change strategies.

There are many intertwined factors[85] explaining people's 'motivational postures',[86] that is, why people comply or fail to comply with legal rules. The rational actor model, which revolves around the key concepts of deterrence, enforcement and disincentives, is only one of many behavioral models. Other compliance models include the social actor model (which emphasizes shared norms and perceptions and personal ethics)[87] or the ignorant actor model, which insists on the role of education to raise citizens' awareness on laws to change people's behavior.[88]

These models emphasize different motivations, which can correspond to

---

Moving to Markets in Environmental Regulation: Lessons from Twenty Years of Experience (Oxford University Press 2007).

[84] Jeremy Bentham, *Introduction to Principles of Morals and Legislation* (first published in 1789, Clarendon Press 1996).

[85] Peter J May, 'Compliance Motivations: Affirmative and Negative Bases' (2004) 38 L & Soc Rev 41.

[86] Valerie A Braithwaite *et al.*, 'Regulatory Styles, Motivational Postures and Nursing Home Compliance,' (1994) 16(4) Law and Policy 363.

[87] M Wenzel, 'Motivation or Rationalization? Causal Relations between Ethics, Norms and Tax Compliance' (2005) 26 J Econ Psychol 491.

[88] Doreen McBarnett, 'When Compliance is not the Solution but the Problem: From Changes in Law to Changes in Attitudes' in Valerie A Braithwaite (ed.), *Taxing Democracy: Understanding Tax Avoidance and Evasion* (Ashgate 2003).

different instrumental and institutional approaches (for example, market-based instruments, educational and informational tools) for engaging communities on invasive species issues.

Compliance and enforcement issues can undermine the behavioral effectiveness of NRM laws, including invasive species laws. Despite the national biosecurity system, invasive species have been illegally imported and landowners routinely ignore their legal obligation to control invasive species, such as wild dogs or weeds despite their legal compliance obligations. Motivations for managing invasive species risk range from the willingness to comply with regulations and economic self-interest, to the desire to belong to a community protecting public values (for example, non-government organization volunteers such as Landcare volunteers). Little behavioral research has been carried out in the invasive species 'space' to understand the motivational drivers for compliance behavior and design more effective behaviourally-responsive regulatory frameworks to manage biological infestations in the context of agriculture and climate change.

5. IMPLEMENTATION ISSUES

The legal architecture for invasive species management provides the broader framework in which invasive species management techniques and their applications operate. However, there is no 'one size fits all solution' in environmental and natural resource management. Laws, regulations and policies provide some of the instrumental tools necessary to combat biological infestations, but their implementation requires the support of contextualized decision-making systems. A strategic approach to invasive species management requires consideration of the context, as well as ecological, institutional, biophysical, social and political dimensions of the problem to be managed, and for capacity constraints to be understood. Strategizing and contextualizing invasive species management issues are essential to focus available resources on achieving effective invasive species control.[89]

Whether located in peri-urban, rural or remote Australia, communities face key institutional and capacity challenges[90] to their efforts against

---

[89] For an illustration in the biofuel weed context, see Paul Martin and Elodie Le Gal, 'Unpacking the Complexities of Biofuel Policy' in Yves le Bouthillier *et al.* (eds), *The Laws and Policies of Biofuels* (Edward Elgar 2016); Rebecca S Epanchin-Niell *et al.*, 'Controlling Invasive Species in Complex Social Landscapes' (2010) 8(4) Front Ecol Environ 210.

[90] This is discussed in Martin *et al.* (n 12).

biological infestations.[91] The complexity and the normative and organizational fragmentation of the governance system for invasive species have already been pointed out.[92] This has also been demonstrated by research using social network analysis to understand which people and organizations are involved in weed management and the funding and information flows.[93] The specific geographical and institutional context in which invasive species programs are conducted also matters because it will determine whether or not certain management strategies are likely to be supported by current invasive species laws. For instance, rural, remote and peri-urban areas each face distinctive governance challenges. The use of certain control methods such as shooting is prohibited in peri-urban areas, locales which are characterized by a higher density population than in rural locations. In remote Australia, such as in certain areas in Northern Territory, the reliance on community engagement mechanisms to address invasive species challenges cannot be considered – (as in rural or peri-urban areas) as a potential effective management strategy due to a low population density and the insufficient number of volunteers who are likely to engage on invasive species issues. Land tenures are likely to be a key barrier to effective coordinated action among landowners in peri-urban areas because they are much more fragmented than in rural areas.[94]

Another key concern for effective community action is knowing precisely what amount of human and economic resources will be available and/or needed in the future to adequately plan strategic interventions to successfully control invasive species and develop adequate adaptation strategies for agriculture. This is a key political question for which outcomes will highly depend upon on whether the invasive species risk is considered an NRM priority issue by those in charge of allocating resources to community groups. So far, Martin *et al.* have highlighted the difficulties in mapping the flows of resources available throughout Australia to support community efforts in invasive species control.[95] This informational failure

---

91 See Paul Martin, Elodie Le Gal and Darryl Low Choy, *Reconsidering the Effects of Complexity and Fragmentation on Resource Governance* in Basant Maheshwari, Vijay P Singh and Bhadranie Thoradeniya (eds), *Balanced Urban Development: Options and Strategies for Liveable Cities* (Springer 2016).

92 See para III.

93 Lyndal-Joy Thompson, Heleen Kruger and Charlene Trestrail, *Who's Involved with Weeds? A Social Network Analysis of Funding and Information Networks for Weed Management* (Rural Industries Research and Development Corporation 2013).

94 For specific institutional issues surrounding invasive species management in peri-urban areas, see Martin, Le Gal and Choy (n 91).

95 Martin *et al* (n 12), Resource B.

is rooted in the lack of a systematic, comprehensive and centralized cost-accounting system that can record information about financial resource allocation for invasive species management. Overall, the current resourcing model for invasive species management mainly relies on governmental authorities, and therefore taxpayers' contributions, to support the current legal and institutional framework for controlling invasive species. This model funds traditional extension services, legal enforcement and compliance approaches, as well as the design of new institutional controls for invasive species. With a legal duty to control invasive species on their lands, landowners also extensively bear the costs of invasive species management, and often are not the parties responsible for introducing or spreading the invasive species. Additional financial constraints, such as those added by an aging population, are likely to arise from the growing environmental risks of climate change. And, because of competing demands between different public interests (health, education, natural resource management),[96] it is highly likely that there will be fewer and fewer economic resources allocated to communities or NRM stakeholders throughout Australia to support the current invasive species governance system and develop effective climate change adaptation strategies to protect agriculture. This suggests that new alternative financial resourcing approaches are needed to complement government action and develop potentially self-sustaining funding models that can address the current economic resourcing issue. In her work, the author has come across innovative environmental business models which explore new avenues for attracting new investments in biodiversity conservation and pest management, while providing training capacity for those involved in pest management. Applied within the context of agriculture, the concept of payments for ecosystem services could be the basis for developing new resource-allocation approaches for invasive species management.[97]

Besides the key resourcing issues (the lack of manpower in certain remote or rural areas in Australia), as well as insufficient training capacity-building mechanisms in best management practices for control methods, community engagement and effective communication also contribute to limit the effectiveness of the current legal framework for invasive species control.[98] It is expected that invasive species laws will not achieve their intended outcomes if those involved directly on the ground (for example,

---

[96] The Treasury, *Intergenerational Report 2015 – Australia in 2055* (Treasury, Commonwealth of Australia 2015).

[97] FAO, *Payments for Ecosystem Services and Food Security* (2011), accessed 12 October 2016 http://www.fao.org/docrep/014/i2100e/i2100e.pdf.

[98] Paul Martin and Miriam Verbeek, *Measuring the Impacts of Managing*

invasive species practitioners, landowners, biosecurity officers) do not have the relevant tools and methods to effectively use control techniques, such as animal trapping, shooting, baiting or poisoning, or to successfully engage communities on invasive species issues.

These capacity and contextual constraints illustrate some of the key challenges that need to be considered for improving current institutional arrangements for invasive species control. While these governance challenges are specific to the Australian context, it is highly likely that similar constraints exist in other countries. It can be expected that these types of institutional issues will adversely impact the effectiveness of invasive species law, and therefore climate change adaptation response strategies for agriculture. These barriers to sustainable agriculture and livelihoods provide the basis for exploring innovative institutional mechanisms in the invasive species 'space' for ensuring sustainable land management practices both on agricultural and non-agricultural areas.

6. THE EFFECTS OF TRANSACTION COSTS ON INVASIVE SPECIES MANAGEMENT

The concept of transaction costs helps to identify implementation risks and the risk of policy failures that invasive species laws face. It is one method that can be used to analyze qualitatively the effectiveness of environmental policies and laws. The economic theory of transaction costs, which is rooted in Coase's work,[99] focuses attention on resource and information costs at various stages of the 'life cycle' of NRM/invasive species laws from design, implementation and enforcement, to compliance. Transactions are 'the means through which information and resource flow'.[100] They often involve 'costs' which impact upon the effectiveness of legal institutions and arrangements.

There are often high transaction costs associated with collecting relevant information to detect invasive species in a specific geographical area or to monitor through inspection whether invasive species laws have been enforced. The transaction costs involved with detection, prosecution and

Invasive Species (Report prepared for the Invasive Animals CRC commissioned by Department of SEWPAC) (unpublished 2013).

[99] Ronald H Coase, 'The Relevance of Transaction Costs in the Economic Analysis of the Law' in Francesco Parisi and Charles K Rowley (eds), *The Origins of Law and Economics: Essays by the Founding Fathers* (Edward Elgar 2005) 199.

[100] Paul Martin and Miriam Verbeek, *Sustainability Strategy* (Annandale 2006) 45.

law enforcement include those incurred by public authorities. The distribution of public and private land ownership, along with the existence of a counterproductive bureaucratic system and lack of a coherent legal architecture, help explain some of the other types of transaction costs that are a cause of policy implementation failures.

Transaction costs may also occur when innovative institutional arrangements for improving the current invasive species legal framework are explored. In a previous publication, key transaction costs associated with the design and the implementation of a new legal strategy for controlling the biofuel weed risk were identified and explored.[101] Based upon an analysis of experts' responses to a theoretical co-regulatory risk management model for biofuel weed risk control three types of transaction costs have already been identified.[102] These included information costs that related to knowledge issues around weed risk perception and risk perception, knowledge about the scientific, economic, political and social context, and informational costs associated with identifying and searching for technical experts in designing legal innovations for weed control. Other types of transaction costs identified ranged from collective decision-making costs (for example, managing political conflicts and coordination issues, allocating clear roles and responsibilities) to collective operational costs (such as monitoring, enforcement and compliance costs, administration costs, distribution of the social and economic spillovers associated with a new legal strategy for weed control).[103] Transaction cost analysis can contribute to assessing qualitatively the performance of invasive species laws in the context of agriculture and climate change. Such a methodological perspective on environmental governance issues can provide some useful insights into the nature and the extent of the potential institutional areas for improvement in invasive species management and control approaches as well as in climate change adaptation strategies for agriculture in the domestic and international context.

7. A LACK OF A SYSTEMIC APPROACH TO THE INVASIVE SPECIES CHALLENGE

A recent submission on environmental biosecurity to the Australian Senate enquiry on invasive species highlighted the need to adopt a

---

101 Le Gal (n 27) 251–3.
102 *Ibid.* at 252
103 *Ibid.*

whole-of-system approach to governance.[104] Invasive species, climate change, and the complexities of environmental, agricultural, and socio-economic systems create 'wicked problems'.[105] Addressing part of the challenge without examining the broader governance system may raise other unexpected problems for public policymakers. Adopting an institutional systems-based approach can expose the wide range of factors that contribute to support or jeopardize the effectiveness of invasive species laws and therefore some aspects of climate change adaptation strategies in agriculture. Such a methodological approach requires analysis of how institutions, which are driven by beliefs, shape human behaviors in relation to invasive species management. At present, the design of invasive species law is concentrated on the instrument creation and administration aspect of regulation. Research suggests that it does not incorporate other aspects of social sciences, such as behavioral sciences, to inform policymakers about the barriers and drivers to the adoption of best management practices for invasive species management, including communication and engagement.[106] This research finding supports the view that '[i]n Australasia, the recognition of human dimensions as an important component of wildlife management is relatively recent'.[107]

With a focus on systems effectiveness and citizen issues, an integrated behavioral approach to invasive species management would identify different points of legal and regulatory points of intervention, and provide different instruments corresponding to different behaviors and motivations, such as market-based and fiscal instruments, behavioral strategies, prescriptive measures and educational tools.

8. FACILITATING COMMUNITY ENGAGEMENT

The vastness of the Australian continent and mobility of the invasive species problem suggests that community engagement is essential for coordinating and managing actions across fragmented private and public land ownership at a 'landscape scale'. Communities often have a better knowledge of the local context, social networks, and resource constraints than experts. Agricultural communities are the ones who are directly impacted

---

104 Australian Public Service Commission (n 52).

105 *Ibid.*

106 Kelly K Miller, 'Human Dimensions of Wildlife Population Management in Australasia: History, Approaches and Directions' (2009) 36(1) *Aust Wildlife Res* 48, 50.

107 *Ibid.*

by the economic costs generated by invasive animals (for example, direct economic losses from crop loss and stock loss, control costs, opportunity costs in managing the risks of invasive animals). Voluntary action is also needed to avoid the introduction of new invasive species, to monitor pest control methods, or to administer educational programs.

If communities are engaged in the early stages of rule creation and solution-finding, they will likely develop a better sense of ownership of the solutions than they would with a top down, universally prescribed approach.[108] Invasive species laws do not operate in a vacuum; they require strong organizational and institutional capacity to be implemented on the ground by end-users and citizens.

The literature on community engagement suggests that citizens should – in close and equal partnership with policymakers and scientific experts – contribute to identifying the issues to be addressed by sharing power, knowledge and expertise.[109] Under a 'shared responsibility' approach to biosecurity issues between the Commonwealth government, states and territories, communities and industry stakeholders, invasive species programs should also be jointly developed and implemented and accountabilities shared. Community engagement models contrast with traditional extension models in which technical experts and scientists are predominantly in charge of driving the research and decision-making process and transferring knowledge to end-users.[110]

Community engagement may take forms ranging from informing, consulting, involving, collaborating to empowering. These are described in the community engagement model developed by the International Association for Public Participation (IAP2).[111] Different community engagement strategies are required for achieving specific goals. Yet the Australian literature about Natural Resource Management and invasive animal management indicates that research on community engagement and collective action is

---

[108] JC Bridger and AE Luloff, 'Building the Sustainable Community: Is Social Capital the Answer?' (2001) 71(4) Sociol Inq 458.

[109] F Powell and M Geoghegan, 'Community Development and the Contested Politics of the Late Modern Agora: Alongside or Against Neoliberalism?' (2009) 44(4) Community Dev J 430; Sarah J Whatmore, 'Mapping Knowledge Controversies: Science, Democracy, and the Redistribution of Expertise' (2009) 33(5) Prog Hum Geog 587.

[110] Lyndal-Joy Thompson, 'Leadership and Community Engagement for Invasive Animals Management in Australia: Status Quo and Future Vision' (Aglaw Centre, IA CRC 2014 unpublished) 6.

[111] International Association for Public Participation (IAP2), 'The IAP2 Public Participation Spectrum', accessed 12 October 2016 at https://www.iap2.org.au/Resources/IAP2-Published-Resources.

epistemologically informed by the expert-driven approach[112] as opposed to a more citizen-driven (engagement) model. Under the former, the role of citizens in producing new knowledge and finding new solutions to challenges is mostly passive. Natural resource management legal frameworks in Australia approach community engagement issues predominantly in a process-oriented way, mostly requiring minimum participatory consultation processes.[113] They tend not to value community perceptions, community capabilities and community knowledge.[114] To overcome this challenge, policymakers could explore how invasive species legal frameworks could integrate communities into the research process so that they could contribute to the formulation of relevant research questions and co-production of knowledge with technical experts. The incorporation of social and behavioral science methods into policy frameworks could also contribute to a better understanding, from a community perspective, of what barriers prevent them from adopting best management practices for invasive species control under climate change conditions in the agricultural context.

In her article 'Vernacular Knowledge and Environmental Law: Cause and Cure for Regulatory Failure', Robyn Bartel discusses the significance of place-based knowledge for preventing regulatory failure.[115] She argues that current natural resource management legal frameworks are mostly informed by bureaucratic knowledge held by state authorities. The lack of incorporation of 'place-based knowledge' which encompasses 'both biophysical and social and cultural phenomena'[116] held by farmers tends to create '[a] social distance between the governed and the government [which] may lead to regulatory resistance and failure'.[117] She supports her argument with a case study of New South Wales native vegetation

---

112 Thompson (n 110) 6.

113 Tanya Howard, 'The "Rules of Engagement": A Socio-legal Framework for Improving Community Engagement in Natural Resource Governance' (2015), accessed 12 October 2016 at http://ssrn.com/abstract=2553768.

114 For a discussion of how the use of an environmental justice framework could contribute to overcome the lack of incorporation of community perceptions into legal frameworks in the context of low carbon technologies, see Elodie Le Gal, 'Using Social Science Perspective on Risk to Implement an Environmental Justice Analysis – Is this the Right Way Forward to Mitigate the Social Risks of Low Carbon Energy Technologies and Help Policy-Makers Achieve Renewable Energy Targets Worldwide?' in Jordi Jaria I Manzano, Nathalie J Chalifour and Louis J Kotzé (eds), *Energy, Governance and Sustainability* (Edward Elgar 2016).

115 Robyn Bartel, 'Vernacular Knowledge and Environmental Law: Cause and Cure for Regulatory Failure, Local Environmental' (2014) 19 IJESD 891.

116 *Ibid.*, 892.

117 *Ibid.*

protection legislation, which has encountered political resistance from the farming community. In the invasive species space, place-based knowledge is essential for effective control. There are many examples of how knowledge of the specific social context may increase engagement and develop social capital. This is particularly true with indigenous traditional owners. The need to respect, and benefit from, local knowledge is also illustrated with the expertise of trappers and 'weed practitioners' whose local knowledge is difficult to replicate, because it is built over decades of observations in the natural environment.

9. TRANSDISCIPLINARITY

The use of social science to understand how people feel, act, think and behave in relation to invasive species laws and invasive species management could help improve its behavioral effectiveness. Transdisciplinarity[118] and community engaged research[119] are increasingly perceived by the research community as promising paradigms that can improve natural resource management, as well as find practical solutions to complex societal problems in collaboration with communities.[120] Uncertainties remain, however, as to what transdisciplinarity is, how to effectively implement it, and whether it delivers superior research outcomes over traditional monodisciplinary research. Little is known about the 'science of team science', the role of legal researchers in the transdisciplinarity/community-engaged research process, or the coproduction of new knowledge with researchers working from other paradigms, practitioners, policymakers and communities. Beyond biophysical sciences, invasive species involve human and institutional issues, which require the intervention of a wide range of disciplines (for example, humanities, economics, history, psychology,

---

[118] For a clarification of the distinction between the concepts of transdisciplinarity, cross-disciplinarity, interdisciplinarity, multi-disciplinarity, see Malin Mobjörk, *Crossing Boundaries: The Framing of Transdisciplinarity* (Örebro University and Mälardalen University 2009) 29.

[119] Paul Martin *et al.*, 'Researcher Lessons from Community Partnership and Trans-disciplinary Research in a Peri-urban Setting: The WISER Experience in Western Sydney' (Cooperative Research Centre for Irrigation Futures, 2010), accessed 12 October 2016 at http://www.irrigationfutures.org.au/imagesDB/news/0510_web.pdf.

[120] The Swiss network for transdisciplinary research provides extensive resources on domestic and international research projects which intend to engage communities. See Td-net, 'Network for Transdisciplinary Research', accessed 12 October 2016 at http://www.transdisciplinarity.ch/en/td-net/Aktuell.html.

management sciences) to understand how human systems interact with natural systems. As highlighted by Kueffer, the focus of invasive species research has been mostly around natural sciences.[121] He also notes that there are different types of knowledge that are needed to understand environmental issues. These are rooted in different disciplines, and include 'systems knowledge' (solving a problem by understanding its causes), 'target knowledge' ('defining management targets and clarifying conflicts of interests and values') and 'transformation knowledge ('developing management tools to achieve these targets').[122] Integrating various sciences is necessary to engage communities' collective knowledge, which can then be empirically tested and applied on the ground.

The legal science of effective invasive species laws goes beyond regulatory design. Legal scholarship in invasive species could benefit from incorporating non-legal science perspectives and research methods to improve invasive species law and reduce legal and institutional barriers to adaptation strategies. Such an approach shares the view of the American scholar Richard Posner who specifically highlighted the need for the legal discipline to be supported by other disciplines to improve its effectiveness.[123] Other scholars have pointed out that environmental law scholarship has been immature.[124] Others have suggested a more empirical environmental law scholarship that would enable systems improvements and continuous learning.[125] Institutional analysis and transdisciplinarity for environmental legal improvements have significant methodological and cultural implications for environmental lawyers, who are often accustomed to traditional doctrinal legal research. Instead of focusing on the technical aspect of legal instrument design, systems thinking requires exploration of the broader institutional environment surrounding the legal framework. From a practical point of view, this suggests that lawyers should become familiar with the scientific culture and research paradigms associated with the various scientific methods of enquiry from the biophysical and social science areas.

Biological infestations from invasive species which threaten agricultural

---

[121] Christoph Kueffer, 'Integrating Natural and Social Sciences for Understanding and Managing Plant Invasions' in S Larrue (ed.), *Biodiversity and Society in the Pacific Islands* (Presses Universitaires de Provence 2013) 71.

[122] *Ibid.*

[123] Richard A Posner, 'The Decline of Law as an Autonomous Discipline' (1987) 100 (4) Harv L Rev 761.

[124] Elizabeth Fisher *et al.*, 'Maturity and Methodology: Starting a Debate about Environmental Law Scholarship' (2009) 21(2) J Env L 213.

[125] Paul Martin and Donna Craig, 'Accelerating the Evolution of Environmental Law through Continuous Learning from Applied Experience' in Paul Martin and Amanda Kennedy (eds), *Implementing Environmental Law* (Edward Elgar 2015).

and environmental values are aggravated by climate change. They also involve complex interactions between the social and environmental system. New methodological approaches from different disciplines are needed to design and implement evidence-based laws and policies that are behaviorally-effective. A transdisciplinary systems-based approach that can facilitate community action on invasive species issues could contribute to strengthening some of the legal tools used for climate change adaptation in agriculture in the domestic and international context.

10. IMPROVING INVASIVE SPECIES LAWS WITH LEGAL AND INSTITUTIONAL INNOVATIONS

Most institutional arrangements for invasive species control rely on regulatory approaches such as landowners' duties to control invasive species on their properties, or the national risk-based pre-border biosecurity system.[126] Such a regulatory approach lacks flexibility to anticipate uncertain events (for example, whether the invasive trait of a specific species materialize or not) and the wide range of risky behaviors that may be involved in relation to invasive species management.

The many complexities, risks and uncertainties associated with climate change and invasive species raise unique institutional, methodological, behavioral and informational challenges that legal frameworks have to take into consideration. These include high levels of uncertainty about the effect of changing climate on invasions, the effects of biological infestations on agricultural and biodiversity values and assets, the impact upon landholders of reduced production due to crop loss and/or loss of stock, as well as the extent and outcomes of the recovery process (if any) of those who have been impacted. In this context, the role of legal frameworks is essential to address increased competition issues for natural resources and allocate natural resource uses fairly and equitably.

In other natural resource management areas, including climate change adaptation, market-based approaches are implemented to better protect private and public environmental values. These involve risk-based techniques similar to those used in the commercial sector to improve risk-related information. Instruments such as risk-pooling and risk portfolios, financial products such as derivatives and futures, insurance contracts, and

---

[126] Australian Department of Agriculture, 'Changes to Australia's Biosecurity System' (8 September 2015), accessed 12 October 2016 at http://www.agriculture.gov.au/biosecurity/australia/biosecurity-reform?wasRedirectedByModule=true.

risk-pricing mechanisms, as well as financial guarantees (including environmental performance bonds) internalize many of the uncertainties associated with environmental risks in order to adapt to or adjust the possible outcomes from potential external environmental hazards. This suggests further research to explore whether these techniques which are supported by specific institutional arrangements could be transposed within the invasive species context.

While there may be high transaction costs associated with the effective implementation of these approaches,[127] it is proposed that more sophisticated risk ideas within the invasive species context could be worthwhile, provided that distributive, equity and social justice issues are also considered. It is possible to incorporate elements of risk sciences to design 'smart' systems of invasive species governance. One such approach has been detailed in previous publications[128] and revolved around institutional systems-based analysis, 'smart' and 'responsive' regulation, and commercial risk management techniques. The designing of institutional arrangements for environmental insurance schemes or risk pooling funds is one example where potential legal and regulatory innovations could be explored as complementary tools to scientific risk management instruments.

11. CONCLUSION

This chapter has explored the intersection of climate change, invasive species law and governance issues in the agricultural context. It has shown that institutional arrangements for invasive species management, which can enhance climate change adaptation, do not operate in a vacuum. While the discussion centred on the specific Australian context, similar complex and intertwined institutional, cultural, political and methodological barriers to the effectiveness of invasive species laws are likely to arise in other institutional settings, and can limit community engagement and jeopardize sustainable agriculture and the livelihoods of the community. These barriers challenge climate change adaptation strategies for agriculture, thus exposing rural communities to economic, social and environmental risks. Invasive species laws need strong institutional support systems to be successfully implemented by end-users, such as landowners, primary producers, biosecurity officers and other community groups. Community beliefs, values, knowledge, perception of environmental issues, social

---

127 Le Gal (n 27).
128 Martin and Le Gal (n 22).

norms and information and resource flows (for example, scientific information, human and financial resources) are also elements to be taken into consideration to ensure effective implementation of invasive species law. It is also necessary for researchers and policymakers to expand their views and focus on a broader governance system. Those involved in regulatory aspects of invasive species management also need to explore non-legal perspectives on environmental and agricultural issues to create evidence-based laws and policies that are behaviorally effective and will address climate change adaptation issues. These various approaches, which are informed by different research methodologies from different disciplines can, as discussed in this chapter, revolve around key concepts such as community engagement, transdisciplinary, systems-based and strategic thinking, as well as transaction cost analysis. Concentrating exclusively on the legal framework and adopting a top-down approach will likely limit innovations that could address invasive species risk and ensure sustainable agriculture and livelihoods. Instead, we should learn how to enhance continuous improvement, and avoid path-dependent institutional and regulatory solutions. This will likely require a major cultural shift in how we think about and conduct legal scholarship on invasive species law and climate change adaptation policy strategies. Better management of the human and institutional dimension of the invasive species challenge is key to improving how human systems interact with environmental systems, including in the area of invasive species. Community engagement research and the incorporation of behavioral, institutional and social science research into scholarship should inform, in a transdisciplinary fashion, policymakers and communities on how to better control invasive species and implement climate change adaptation strategies for agriculture.

With innovations, including institutional and methodological innovations, there is a risk of failure. In the light of the multi-faceted challenges that the legal framework for invasive species control is currently facing, the risk is worth taking, especially given climate change and food security issues and increasing world population. Supporting new methodological and institutional tools as well as changing people's ways of thinking about invasive species is a long-term process. It requires creativity and perseverance to strengthen capacity-building mechanisms. To overcome the many legal and institutional challenges that jeopardize the effectiveness of invasive species law (and therefore climate change adaptation strategies) an ambitious, challenging but intellectually stimulating research agenda is needed.

7. Genetically modified food crops in an era of climate change: an answer in search of a problem

Rebecca M Bratspies

Since domesticating plants at the dawn of the Neolithic Revolution,[1] human societies have developed hand in hand with agriculture. With barley, maize, squash and wheat as our allies, we have spread ourselves out across five of the six continents, penetrating virtually every ecosystem on the planet.[2] In the process, the global human population increased exponentially: from roughly 6 million[3] at the dawn of the agricultural era to the current 7.4 billion and counting.[4] But, all of this agriculture facilitated development took place during the Holocene Epoch, a roughly 12,000 year period of climatic stability that began at the end of the Paleolithic Ice Age.[5] The Holocene Epoch is now arguably giving way to the Anthropocene,[6] an Epoch dominated by human influences on the global environment, particularly the global climate.[7]

---

1 Vere Gordon Childe, *Man Makes Himself* (Watts 1936) 66–71.

2 This chapter focuses on plant agriculture. There are obviously tremendous climate impacts associated with animal husbandry, including deforestation, methane generation and plant agriculture for feed production.

3 Jean-Pierre Bocquet-Appel, 'When the World's Population Took Off: The Springboard of the Neolithic Demographic Tradition' (2011) 333 Science 560, accessed 13 October 2016 at http://www.sciencemag.org/content/333/6042/560.full.

4 United Nations, Department of Economic and Social Affairs, Population Division, *World Population Prospects: The 2012 Revision, Highlights and Advance Tables*, Working Paper No. ESA/P/WP.228 (2013).

5 Modern humans (Homo sapiens) evolved and dispersed over the earth well before the dawn of the Holocene. However the rise of what we call human civilization, took place largely over the last 12,000 years.

6 See Jan Zalasiewicz, Mark William, Will Steffen and Paul Crutzen, 'The New World of the Anthropocene' (2010) 44 *Environmental Science and Technology* 2228; Paul J Crutzen, 'Geology of Mankind' (2002) 415 *Nature* 23.

7 In 2008 the International Commission on Stratigraphy created a working group on the Anthropocene within the Subcommission on Quaternary Stratigraphy. In 2015 the working group proposed that the Anthropocene be dated to 16 July 1945, the date the first nuclear bomb was exploded. Jan

Because of anthropogenic climate change, the Earth is projected to warm at least 2°C[8] by 2050.[9] In that same time period, global population is projected to top 9.6 billion, and resource use is projected to triple.[10] Loss of biodiversity from habitat destruction associated with all three of these phenomena has become so extreme that researchers are calling it the 'sixth great extinction'.[11] Thus within just over a single human generation, we will have made planet earth significantly warmer, more crowded, and less green. Poised at the intersection of these global forces, agriculture faces a perfect storm. Changes in the climate will negatively affect agricultural productivity, even as population growth demands ever more food. As both climate change and population growth accelerate, their combined impact threatens planetary boundaries[12] and global food security.[13] Growers will either have to glean higher yield from existing farmland under new less favorable conditions, or more land will have to be converted to agriculture.[14]

This is not merely a hypothetical, or a matter for future concern. Indeed the global plant-based agricultural system is already 'being disrupted by climate change'.[15] It is not hard to see how. The drought, wildfires,

---

Zalasiewicz *et al.*, 'When did the Anthropocene Begin? A Mid-Twentieth Century Boundary Level is Stratigraphically Optimal' (2015) 383 Quaternary International 204, accessed 13 October 2016 at http://www.sciencedirect.com/science/article/pii/S1040618214009136.

8 Glen P Peters *et al.*, 'The Challenge to Keep Global Warming Below 2° C' (2013) *3 Nature Climate Change* 4.

9 United Nations, Department of Economic and Social Affairs, Population Division, *World Population Prospects: The 2012 Revision* (UN 2013).

10 United Nations Environment Programme, *Decoupling Natural Resource Use and Environmental Impacts from Economic Growth* (UNEP 2011).

11 Gerardo Ceballos *et al.*, 'Accelerated Modern Human-Induced Species Losses: Entering the Sixth Mass Extinction' (2015) 1 *Science Advances* e1400253, accessed 13 October 2016 at http://advances.sciencemag.org/content/1/5/e1400253.full.

12 The nine planetary boundaries are the stratospheric ozone layer, biodiversity, chemical dispersion, climate change, ocean acidification, freshwater consumption and the global hydrological cycle, land system change, nitrogen and phosphorous inputs to the biosphere and oceans, and atmospheric aerosol loading. Johan Rockström *et al.*, 'Planetary Boundaries: Exploring the Safe Operating Space for Humanity' (2009) 14 *Ecology and Society* 32.

13 David Tilman, Christian Balzer, Jason Hill and Belinda L Befort, 'Global Food Demand and the Sustainable Intensification of Agriculture' (2011) 108 PNAS 20260.

14 Richard Dobbs, *Resource Revolution: Meeting the World's Energy Food and Water Needs* (McKinsey Global Institute 2011) 39.

15 Elizabeth Grossman, 'Climate Change to Pose Serious Threats to Food Distribution' (2015) Earth Island Journal, accessed 13 October 2016 at http://

heatwaves and floods associated with climate change increasingly jeopardize harvests and/or undermine yields. Australia,[16] Brazil,[17] California[18] and the central United States[19] are already being affected and the list goes on. Climate change gives food production systems a one-two punch: for a wide range of crops, productivity declines with rising temperatures.[20] At the same time, extreme precipitation events and floods delay planting and destroy standing crops.[21]

More than 20 years ago, the 196 Parties to the United Nations Framework Convention on Climate Change (UNFCCC), committed themselves to stabilizing greenhouse gas concentrations in the atmosphere at a level that 'would prevent dangerous anthropogenic interference with the climate system', while at the same time 'ensur[ing] that food production is not threatened'.[22]

Yet in 2014, the Intergovernmental Panel on Climate Change (IPCC) stated with *high confidence* that climate change is already negatively affecting agriculture around the world.[23] The severity of these impacts is expected to increase during the course of this century.

Facing a potentially catastrophic climate shift, agricultural policymakers

---

www.earthisland.org/journal/index.php/elist/eListRead/climate_change_poses_serious_threats_to_food_distribution/.

16 Will Steffen, Lesley Hughes and Alix Pearce, *The Heat is On: Climate Change, Extreme Heat and Bushfires in Western Australia* (Climate Council of Australia 2015).

17 Raymond Colitt, 'Climate Change Threatens Brazil's Rich Agriculture' *Reuters* (London, 1 October 2009).

18 Rose Hackman, 'California Drought Will Cost Agriculture Sector $1.8bn This Year, Study Finds' *The Guardian* (London, 2 June 2015).

19 John Eligon, 'After Drought, Rains Plaguing Midwest Farms' *The New York Times* (New York, 9 June 2013).

20 Caroline C Ummenhoffer, Hong Xu, Tracy E Twine, Evan H Girvetz, Heather R McCarthy, Netra Chhetri and Kimberly A Nicholas, 'How Climate Change Affects Extremes in Maize and Wheat Yield in Two Cropping Regions' (2015) *Journal of Climate*, accessed 13 October 2016 at http://journals.ametsoc.org/doi/abs/10.1175/JCLI-D-13-00326.1.

21 Assaf Anyamba *et al.*, 'Recent Weather Extremes and Impacts on Agricultural Production and Vector-Borne Disease Outbreak Patterns' (2014) 9 PLoS ONE e92538, accessed 13 October 2016 at http://journals.plos.org/plosone/article?id=10.1371/journal.pone.0092538.

22 See Article 2 of the United Nations Framework Convention on Climate Change (New York, 9 May 1992), (1992) 31 I.L.M. 849, entered into force 21 March 1994.

23 John R Porter *et al.*, 'Food Security and Food Production Systems' in Christopher B Field, *et al.* (eds), *Climate Change 2014: Impacts, Adaptation, and Vulnerability. Part A: Global and Sectoral Aspects. Contribution of Working Group*

often assume that technology will save the day. Official reports are replete with blithe statements about developing new crops to meet drought and heat challenges.[24] The assumption is that new technologies can and will materialize in response to identified needs. Indeed, assumptions about new (often unspecified) technologies are recited almost as a mantra, with technology expected to provide both mitigation and adaptation responses to climate change. Recently, backers of genetically engineered crops have entered the fray with claims that these crops will play a critical role in both climate change mitigation and adaptation.[25]

Supporters of genetic engineering make two sets of claims about how these crops help farmers respond to climate change. In terms of climate adaptation, the claim is that these crops will be needed to improve yields under the increases in drought and heat conditions associated with climate change. In terms of climate mitigation, the claim is that these crops will temper agriculture's carbon footprint. Thus these crops are offered as an answer both to the effects that climate change will have on agricultural production and to the contributions that plant-based agriculture makes to worsening the climate change problem. Both issues are clearly of major concern to policymakers, farmers and consumers. Genetically engineered crops are often identified as a solution based on assumptions that the crops will improve adaptation to climate change while simultaneously reducing agriculture's contribution to the problem.[26]

This chapter will address two specific narratives about genetically modified organisms in an era of changing climate: the putative role for genetically engineered crops in adaptation to climate change and the putative role for genetic engineering as a mitigation strategy to reduce the climate change trajectory. After a brief introduction to the technology itself, this chapter will take up each set of claims in turn, and will examine how genetically engineered crops might, or might not, be part of a climate change solution.

II to the Fifth Assessment Report of the Intergovernmental Panel on Climate Change (Cambridge University Press 2014) 485, 488.

[24] See, e.g., Charles L Walthall, *et al.*, *Climate Change and Agriculture in the United States: Effects and Adaptation* (USDA Technical Bulletin No. 1935 2013).

[25] Thomas Lee, 'Monsanto Official: Climate Change Makes Genetic Crops More Urgent' *San Francisco Chronicle* (San Francisco, 3 March 2015).

[26] See e.g., The Swiss National Centre of Competence in Research, 'Biotechnology and Climate Change: Challenges and Opportunities' (17 February 2012), accessed 13 October 2016 at http://www.nccr-trade.org/events/biotechnology-and-climate-change-challenges-and-opportunities/.

1. WHAT IS GENETIC ENGINEERING?

Humans have spent much of the Holocene Epoch modifying plants to suit their needs. Farmers discovered relatively early on that desirable traits could be maximized and undesirable traits minimized by selectively breeding individual plants.[27] Gregor Mendel's insights into the mechanisms of genetic inheritance[28] facilitated a more systematized form of selective breeding, giving rise to a host of hybrid crops developed to meet specific agronomic or gustatory needs. As agriculture and food processing transformed into industrial production, there were new demands on selective breeding, this time for crops with traits to facilitate long distance shipping and postharvest ripening. These commercial pressures resulted in radically altered crop plants that bear little resemblance to their wild, uncultivated relatives either in taste or in key agronomic characteristics. All this was accomplished through selective breeding rather than through direct genetic manipulation through the techniques of modern genetic engineering.

However selective breeding operates under a significant constraint: it can only enhance or suppress traits already present in a species.[29] Genetic engineering frees plant breeding from this limitation. A process called transformation,[30] allows a gene to be isolated from any organism and transferred to any other organism – across all barriers of species, class, phylum and kingdom.[31] Recombining the genes themselves[32] avoids the

---

[27] Noel Kingsbury, *Hybrid: The History and Science of Plant Breeding* (University of Chicago Press 2009).

[28] Gregor Mendel, *Experiments in Plant Hybridization* (1865), accessed 13 October 2016 at http://www.mendelweb.org/MWpaptoc.html. Working with pea plants, the monk Gregor Mendel conclusively demonstrated what are now fundamental laws of genetic inheritance. See also, 'Cold Springs Harbor Laboratory, DNA from the Beginning' (2002), accessed 13 October 2016 at http://www.dnaftb.org/dnaftb/.

[29] This constraint is actually less restrictive than it might seem on first glance. Variety is ubiquitous in nature. Plus breeders frequently deploy mutagenic techniques to facilitate selective breeding. Many of these techniques would probably shock advocates clinging to a more romantic notion of selective breeding as somehow 'natural' or 'organic'.

[30] See Plant Protection Act (PPA), 7 U.S.C. §§ 7701-7772 (2000).

[31] The first step of the modern biotechnology revolution dates to Watson and Crick's 1953 discovery of the structure of DNA. James Watson and Fredrick Crick, 'A Structure for Deoxyribose Nucleic Acid' (1953) 171 Nature 737. This paper is generally considered to have ushered in the era of molecular genetics. See also Jeffrey Schell, 'Transgenic Plants as Tools to Study the Molecular Organization of Plant Genes' (1987) 237 *Science* 1176–83.

[32] Stanley Cohen, A. Chang, Herbert Boyer and Robert Helling, 'Construction of Biologically Functional Bacterial Plasmids In Vitro' (1973) 70 PNAS 3240–44.

main constraint on selective breeding: the need to start with sexually compatible organisms. No longer do the existing characteristics of a species define what is possible. Instead, this technology can be used to create organisms that do not, and could not, exist without such intervention.[33]

Since the 1990s, so-called life sciences companies have invested heavily in using genetic engineering to develop new crop varietals. While researchers sought to modify a wide variety of agronomic traits, only two achieved commercial success: insect resistance and herbicide tolerance. Insect resistance involves transferring a gene that prompts crops to endogenously produce chemicals toxic to many common insects. Herbicide tolerance enables farmers to spray broad spectrum weed killers on their crops without killing the crop plant, thereby simplifying weed management systems.

Widespread planting of these genetically engineered crops began in 1996 with the introduction of Monsanto's Roundup Ready Soybeans and *Bt* corn.[34] Since then there have been 380 different approvals for genetically engineered crops around the world,[35] with transgenic corn, cotton, soybean and canola varietals making up the lion's share of approvals and plantings.[36] In the United States alone there are 186 different genetically engineered crop lines that have been approved for planting.[37]

The claims for this technology have been extravagant. Genetically engineered crops have been touted as the answer to world hunger,[38] to

---

[33] Norman Carl Ellstrand, *Dangerous Liaisons? When Cultivated Plants Mate with Their Wild Relatives* (John Hopkins University Press 2003) 171–3.

[34] Monsanto, Company History (Monsanto), accessed 13 October 2016 at http://www.monsanto.com/whoweare/pages/monsanto-history.aspx. Although not widely realized, the first genetically engineered plant was not a commodity crop like maize, soybeans or cotton. It was Calgene's Flavr Savr tomato, which hit the markets in 1994. However the genetically engineered tomato flopped commercially and was withdrawn from the market. George Bruening and JM Lyons, 'The Case of the FLAVR SAVR Tomato' (2000) 54 *California Agriculture* 6–7.

[35] International Service for the Acquisition of Agri-Biotech Applications, 'GM Crop Events Approved in Any Country', accessed 13 October 2016 at http://www.isaaa.org/gmapprovaldatabase/advsearch/default.asp?CropID=Any&TraitTypeID=Any&DeveloperID=Any&CountryID=Any&ApprovalTypeID=Any.

[36] *Ibid.*

[37] International Service for the Acquisition of Agri-Biotech Applications, 'GM Crop Events Approved in the United States', accessed 13 October 2016 at http://www.isaaa.org/gmapprovaldatabase/approvedeventsin/default.asp?CountryID=US&Country=United%20States%20of%20America.

[38] 'Do GM Crops Increase Yields' (Monsanto), accessed 13 October 2016 at http://www.monsanto.com/newsviews/Pages/do-gm-crops-increase-yield.aspx; Maggie Urry, 'Genetic Products Row Worsens' *Financial Times* (London 20 June

pesticide overuse[39] and to negative environmental impacts of agriculture more generally.[40] However their actual track record has been far more tepid. Dueling studies claim that crop yields have either held steady or decreased,[41] while pesticide use has either dropped precipitously or skyrocketed.[42] Since the introduction of genetically engineered crops the problem of food insecurity has increased rather than decreased, leaving more people hungry than at any other point in human history.[43] Many policymakers nevertheless insist that biotech crops are the future of global agriculture,[44] and that these crops will ultimately deliver on the promises made from the very beginning. Indeed, leading agricultural decision makers continue to characterize the technology as 'a crucial part of expanding agricultural productivity in the 21st century'.[45] Climate change

---

1997) 4 (quoting former USDA Secretary Dan Glickman for the proposition that '[g]rowing pest-resistant crops would alleviate world hunger, reduce pesticide damage to the environment, and save rain forests from being cleared for food production').

39 Graham Brooks and Peter Barfoot, *GM Crops: global socio-economic and environmental impacts 1996–2010* (PG Economics, May 2012) 80–82, accessed 13 October 2016 at http://www.pgeconomics.co.uk/page/33/global-impact-2012; Monsanto, 'Biotechnology' (Monsanto), accessed 13 October 2016 at http://www.monsantoafrica.com/biotechnology/default.asp.

40 Brooks and Barfoot, above note 39 at 83–8.

41 Doug Curian-Sherman, *Failure to Yield: Evaluating the Performance of Genetically Engineered Crops* (Union of Concerned Scientists, April 2009) 13, accessed 13 October 2016 at http://www.ucsusa.org/assets/documents/food_and_agriculture/failure-to-yield.pdf; BL Ma and KD Subedi, 'Development, Yield, Grain Moisture and Nitrogen Uptake of Bt Corn Hybrids and Their Conventional Near-isolines' (2005) 93 Field Crops Research 199, 209. For a contrary view, see W Klümper and M Qaim, 'A Meta-Analysis of the Impacts of Genetically Modified Crops' (2014) 9 PLoS ONE e111629, accessed 13 October 2016 at http://journals.plos.org/plosone/article?id=10.1371/journal.pone.0111629.

42 Compare Charles Benbrook, 'Impact of Genetically Engineered Crops on Pesticide Use in the US – the First Sixteen Years' (2012) 24 Environmental Sciences Europe 1, accessed 13 October 2016 at http://www.enveurope.com/content/pdf/2190-4715-24-24.pdf (finding that pesticide use increased by approximately 404 million pounds or 7 percent) with Klümper and Qaim above note 41.

43 Rebecca Bratspies, 'Food, Technology and Hunger' 8 *Law Culture & the Humanities* 1, 7, accessed 13 October 2016 at http://lch.sagepub.com/content/early/2012/09/24/1743872112456990.

44 Ismail Serageldin, 'Opening Statement' (World Bank, undated), accessed 10 July 2015 at http://www.worldbank.org/html/cgiar/publications/mtm97/isopen.pdf.

45 Ismail Serageldin CGIAR Chief and World Bank Vice-President, quoted in 'Harvest of Fear: Should We Grow GM Crops?' (PBS, undated), accessed

mitigation and adaptation are areas where very high hopes are being pinned onto the promise of genetically engineered crops.

2. HOW CLIMATE CHANGE THREATENS AGRICULTURE

The scientific evidence that our world is warming is 'unmistakable'.[46] Indeed, the Fifth IPCC Report, issued in the Fall of 2013, characterized global warming as 'unequivocal' and asserted that it was 'extremely likely' that human activities are driving global warming.[47] This declaration marks a step up in the rhetoric from the IPCC's *Fourth Assessment Report*, which asserted with 95 percent confidence that human activity has had a warming effect, and with 90 percent confidence that the observed increase in average global temperatures since the mid-20th century is attributable to anthropogenic greenhouse gas emissions.[48]

In 2013 alone, human activities contributed 36 billion tons of CO_2 to the atmosphere.[49] For perspective, that figure is 61 percent higher than were 1990 carbon emissions, the reference year for the Kyoto Protocol.[50] After a century of anthropogenic carbon emissions, atmospheric CO_2 is at its highest level in 15 to 20 million years.[51] Worse yet, the global carbon

---

13 October 2016 at http://www.pbs.org/wgbh/harvest/exist/no1.html. While he viewed biotechnology as 'a tremendous help in meeting the challenge of feeding an additional three billion human beings, 95 percent of them in the poor developing countries, on the same amount of land and water currently available', he also recognized that for this to occur, the technology must first be 'safely deployed'.

[46] NOAA, 'Past Decade Warmest on Record According to Scientists in 48 Countries' (28 July 2010), accessed 13 October 2016 at http://www.noaanews.noaa.gov/stories2010/20100728_stateoftheclimate.html.

[47] IPCC, *Climate Change 2013: The Physical Science Basis* (2013), accessed 13 October 2016 at http://www.climatechange2013.org.

[48] IPCC, *Fourth Assessment Report: Climate Change 2007* (2007), accessed 13 October 2013 at http://www.ipcc.ch/publications_and_data/ar4/syr/en/spms2.html.

[49] The Global Carbon Project, 'Carbon Budget 2014' (2014), accessed 13 October 2016 at http://www.globalcarbonproject.org/carbonbudget/archive/2014/GCP_budget_2014_lowres_v1.02.pdf.

[50] Kyoto Protocol to the United Nations Framework Convention on Climate Change, UN Doc FCCC/CP/1997/7/Add. 1 (Kyoto, 10 December 1997) 37 ILM 22, entered into force 16 February 2005.

[51] Aradhna K Tripati *et al.*, 'Coupling of CO_2 and Ice Sheet Stability over Major Climate Transitions of the Last 20 Million Years' (2009) 326 *Science* 1394; Renato Spahni *et al.*, 'Atmospheric Methane and Nitrous Oxide of the Late Pleistocene from Antarctic Ice Cores' (2005) 310 *Science* 1317; Urs Siegenthaler

trajectory is heading in the wrong direction, toward further emissions. The climate impacts from all these carbon emissions are real and growing. Global temperature currently hovers about 0.8°C above pre-industrial levels, dangerously near the climatic tipping point which scientific consensus puts at around 1.5 to 2°C above pre-industrial global temperatures.[52] Atmospheric CO_2 levels already exceed the critical 350 ppm level that many regard as the maximum for keeping climate change within acceptable levels[53] (as of this writing, atmospheric CO_2 concentrations are hovering around 400 ppm).[54] As a result, it is 'virtually certain' that by the end of this century we will experience fewer (and warmer) cold days and more frequent (and hotter) warm days.[55]

Two major sets of human activities are driving this climate change. The first involves adding carbon to the atmosphere by burning fossil fuels, thereby transferring carbon from geologic reservoirs to the atmosphere.[56] The second involves destroying carbon sinks, particularly through deforestation. Agriculture contributes to both activities; the fossil fuel consumed in agriculture (including fertilizer and pesticide production, on-farm mechanics, and transportation of commodities) contributes roughly 705 million tons of GHG emissions annually.[57] At

---

et al., 'Stable Carbon Cycle-Climate Relationship during the Late Pleistocene' (2005) 310 *Science* 1313.

[52] Joeri Rogelj *et al.*, 'Energy System Transformations for Limiting End-of-Century Warming to Below 1.5°C' (2015) 5 *Nature Climate Change* 519; UNFCCC, 'Feeling the Heat: Climate Science and the Basis of the Convention', accessed 13 October 2016 at https://unfccc.int/essential_background/the_science/items/6064txt.php.

[53] James Hansen *et al.*, 'Target Atmospheric CO_2: Where Should Humanity Aim?' (2008) 2(1) *Open Atmospheric Science Journal* 271, accessed 13 October 2016 at http://www.giss.nasa.gov/research/briefs/hansen_13/. Readers interested in an accessible discussion of atmospheric carbon concentrations should visit www.350.org.

[54] National Oceanic and Atmospheric Administration, 'Earth Systems Monitoring: Trends in Atmospheric Carbon Dioxides', accessed 13 October 2016 at http://www.esrl.noaa.gov/gmd/ccgg/trends/global.html. As of 11 July 2015 the global CO_2 concentration was 401.56 ppm. Scripps Institute of Oceanography, 'The Keeling Curve: Latest CO_2 Reading', accessed 13 October 2016 at https://scripps.ucsd.edu/programs/keelingcurve/.

[55] IPCC, '2013: Summary for Policymakers' in TF Stocker *et al.*, (eds.) *Climate Change 2013: The Physical Science Basis. Contribution of Working Group I to the Fifth Assessment Report of the Intergovernmental Panel on Climate Change* (Cambridge University Press 2013) 7.

[56] RJ Andres *et al.*, 'A Synthesis of Carbon Dioxide Emissions from Fossil-Fuel Combustion' (2012) 9 *Biogeosciences* 1845.

[57] FAO, *Greenhouse Gas Emissions from Agriculture, Forestry and Other Land Uses* (FAO 2014), accessed 13 October 2016 http://www.fao.org/resources/infographics/infographics-details/en/c/271780/. For purposes of statistics, the IPCC

the same time, agriculture is responsible for 80 percent of deforestation worldwide.[58]

Genetically engineered crops are touted as a way to address climate change on two fronts: to allow global agriculture to adapt to a radically changed climate, while simultaneously mitigating the scope and scale of the climatic challenges by reducing agriculture's contribution to the problem. In terms of adaptation, advocates claim that these crops are a way for agricultural systems to maintain productivity even in the face of the emerging effects of climate change. This chapter groups these adaptation claims under the label 'how climate change affects agriculture'. At the same time, these crops are promoted as a way to increase production on existing farmland – reducing the need to expand lands under cultivation, reducing fossil fuel inputs and promoting soil carbon sequestration. For purposes of this chapter, this set of mitigation concerns are grouped under the label 'how agriculture affects climate change'. Both sets of claims are addressed below.

3. GENETICALLY ENGINEERED CROPS AS ADAPTATION: ASSESSING THE 'HOW CLIMATE CHANGE AFFECTS AGRICULTURE' NARRATIVE

Agriculture is extremely vulnerable to climate change and growing evidence indicates the significant effects it is already having on plant-based agriculture.[59] Overall at least 30 percent of the variability in global average yields for the six most widely grown crops (wheat, rice, maize, soybeans, barley and sorghum) has been linked directly to variations in

---

lumps plant based agriculture together with animal husbandry as well as with forestry and other land uses. See, e.g., '2006 Intergovernmental Panel on Climate Change (IPCC) guidelines', accessed 13 October 2016 at http://www.ipcc-nggip.iges.or.jp/public/2006gl/. To the extent possible, this chapter will isolate the effects associated with plant based agriculture from the wider category of agriculture, forestry and other land uses.

[58] Gabrielle Kissinger, Martin Herold and Veronique De Sy, *Drivers of Deforestation and Forest Degradation: A Synthesis Report for REDD+ Policymakers* (Lexeme Consulting, August 2012) 5.

[59] Jerry Knox, et al., 'Climate Change Impacts on Crop Productivity in Africa and South Asia' (2012) 7 Environmental Research Letters 034032, accessed 10 July 2015 at http://iopscience.iop.org/1748-9326/7/3/034032/pdf/1748-9326_7_3_034032.pdf (finding robust evidence that climate change will depress wheat, maize, sorghum and millet crop yields in Africa and South Asia).

temperature and precipitation.[60] Extremely high temperatures reduce crop yields by depressing plant productivity, while also encouraging weed and pest proliferation.[61] Changes in precipitation make crops more vulnerable, increasing the likelihood of short run crop failures and of long run production declines. The IPCC Fifth report documents, with high confidence, the negative impact daytime temperatures of more than 30°C (characterized as extreme temperatures) have on crop yields.[62] Assuming that growing regions remain constant, we face the prospect of significant yield decline in the next few decades,[63] with the likelihood of much larger yield declines by the end of the century.[64] Over those same time periods, significant population growth is expected to accelerate demand.

Because United States agricultural production constitutes such a large fraction of the global soybean, corn, wheat and cotton markets,[65] examining climate change's impacts on United States agriculture is a helpful microcosm for appreciating its global consequences. The United States currently supplies about 50 percent of global corn exports, 40 percent of global soybean exports and 20 percent of global wheat exports.[66] Any impact on the productivity of agriculture in the United States will therefore be felt widely. As the United States Department of Agriculture (USDA) projected in 2013, those impacts are likely to be quite significant.[67] For example, the USDA opined that increasing atmospheric CO_2 levels would reduce productivity of crops, and that all production systems will be affected.[68] The USDA further noted that because the pattern and potential

---

[60] David B Lobell and Christopher B Field, 'Global Scale Climate–Crop Yield Relationships and the Impacts of Recent Warming' (2007) 2 *Environmental Research Letters* 014002, 3, accessed 13 October 2016 at http://iopscience.iop.org/1748-9326/2/1/014002/pdf/1748-9326_2_1_014002.pdf.

[61] For perspective, the wheat and maize production lost to climate change is roughly equivalent to the total wheat and maize production of Argentina. Lobell and Field, above note 60 at 4.

[62] Porter, above note 23 at 485, 497.

[63] David B Lobell and Claudia Tebaldi, 'Getting Caught with our Plants Down: The Risks of a Global Crop Yield Slowdown from Climate Trends in the Next Two Decades' (2014) 9 *Environmental Research Letters* 74003.

[64] Wolfram Schlenker and Michael J Roberts, 'Nonlinear Temperature Effects Indicate Severe Damages to U.S. Crop Yields under Climate Change' (2009) 106 PNAS 15594. The range of estimates varies widely depending on how optimistic or pessimistic a warming scenario the researchers adopt.

[65] CL Walthall *et al.*, *Climate Change and Agriculture in the United States: Effects and Adaptation* (USDA Technical Bulletin 1935, 2013) 14.

[66] *Ibid* at 16.

[67] *Ibid.*

[68] *Ibid.*

magnitude of precipitation changes are not well understood, any discussion of how climate change will impact agriculture includes significant uncertainty.[69] Indeed, the executive summary of the Technical Bulletin reads like a litany of doom: climate change will 'exacerbat[e] current biotic stresses on agricultural plants'[70] and will 'have overall detrimental effects on most crops'.[71] Specifically the Bulletin notes that elevated CO_2 concentrations disproportionately stimulate growth of weed species, increasing the risk of crop loss from weed pressure,[72] and that higher temperatures will likely reduce yield.[73]

Rising temperatures are a major concern. Many important staple crops like field corn, wheat and oats experience depressed yields when summer temperatures increase.[74] When temperatures exceed 32°C, many crops experience heat stress.[75] In addition, the frequency with which day/night temperatures exceed a high temperature threshold can affect flowering, fruit set and/or seed production of many crops.[76] Under all climate scenarios the Northeastern United States will experience a significant increase in days exceeding 32°C.[77] At the higher emissions scenarios, most of July is expected to exceed 32°C across virtually the entire Northeast United States by the end of the century.[78] The impact on high value specialty crops like potatoes, apples and cabbage might be especially dire.[79]

In addition to warming, climate change is also expected to dramatically change precipitation patterns. The likelihood of high precipitation events (defined as more than 5 cm in a 48 hour period) will continue to increase.[80]

---

69 *Ibid.*
70 *Ibid.*
71 *Ibid.*
72 *Ibid* at 3.
73 *Ibid.*
74 Porter, above note 23 at 488 (finding that temperature increases of 2°C or more above late-20th-century levels will negatively impact wheat, rice and maize production in temperate zones).
75 *Ibid.* For corn, the critical threshold temperature is 29° C; for soybeans it is 30° C. Because corn and soybeans are the predominant source of feed grains used in cattle, dairy, poultry and hog production, the ramifications will reverberate beyond plant based agriculture.
76 David W Wolfe *et al.*, 'Projected Change in Climate Thresholds in the Northeastern United States: Implications for Crops, Pests, Livestock and Farmers' in Cameron P Wake *et al.* (eds) *Mitigation and Adaptation Strategies for Global Change* (2010) 13, 555–6.
77 *Ibid* at 7.
78 *Ibid.*
79 *Ibid.*
80 *Ibid* at 18.

These events are already resulting in delayed spring planting, field flooding and increased runoff, all of which jeopardize yields. Paradoxically at the same time that high precipitation events are increasing, episodes of drought will become more common, with dry spells projected during times when crops need water.[81]

4. GENETIC ENGINEERING AS THE ANSWER

While the problems climate change poses for agricultural production seem relatively clear, the severity of their impact is less certain. It all hinges on the ability of agricultural producers to adapt to changing growing conditions. In the 'how climate change affects agriculture' narrative, genetically engineered crops are offered as an adaptive strategy: a way to change agricultural activities in order to manage the impacts of climate change by giving farmers tools to capitalize on new opportunities, while minimizing the costs associated with negative climate impacts. We know that adaptation is necessary; the impacts of climate change are already apparent, and future effects are predicted to increase in severity. Indeed, agricultural biotechnology behemoth Monsanto has consistently described these climate change-related agricultural impacts as a potential 'opportunity'.[82]

In assessing the impacts of climate change on agriculture, policymakers generally assume that growers will adapt by switching to more heat tolerant or drought resistant varietals. The projection is that these new varietals will thrive under the new climatic conditions, thereby dampening climate change's impact.[83] But, this kind of adaptation projection assumes that such varietals are available, and that these varietals do not come with features that consumers and growers will find unacceptable. This is where the 'genetic engineering as adaptation' narrative usually comes in. Monsanto explicitly makes the claim that food security in an era of climate change will require agricultural biotechnology.[84] Indeed, the company's 2014 sustain-

---

[81] George W King *et al.*, *Confronting Climate Change in the Great Lakes Region* (Union of Concerned Scientists, April 2003) 4.

[82] Carbon Disclosure Project, Monsanto 2013 Investor CDP Information Request, accessed 13 October 2016 at http://sustainability.monsanto.com/wp-content/uploads/2013/06/cdp-investor-2012.pdf; see also, Kieron Monks, 'Getting Rich from Climate Change? How Business can Thrive in Extreme Conditions' CNN (17 November 2014), accessed 13 October 2016 at http://edition.cnn.com/2014/11/11/business/food-gets-rich/.

[83] Porter, above note 23 at 485–533.

[84] Margaret Zeigler, 'Climate Change Is Another Obstacle to Global Food

ability report proudly claims that its business 'revolves around delivering agricultural innovations . . . to address climate change, resource conservation and, ultimately, how we will collectively feed a global population of more than 9.6 billion by the year 2050'.[85]

Most assessments of climate change's impacts on agriculture are optimistic about the role that new technologies will play.[86] Yet the evidence to date demonstrating that genetically engineered crops will be part of this technological solution is scant. It is not at all clear that these crops can deliver on the claims being made on their behalf. The FAO wryly notes that successful development of drought resistant crops through genetic engineering has 'been anticipated several times . . . but has had very limited impact so far'.[87] Indeed the two decade track record for these crops has been one in which the benefits claimed for such crops repeatedly fail to materialize. Public discourse is beginning to take notice. Even a largely positive opinion piece titled 'Why We Will Need Genetically Modified Foods' had to concede that the tantalizing prospect that genetic engineering could help feed the world's hungry has so far 'been a relatively empty promise'.[88]

Perhaps the clearest example of the chasm between hype about the potential of genetic engineering and its actual ability to deliver on that potential is golden rice. In 2000, *Time Magazine* ran a cover story on golden rice titled 'This Rice Could Save a Million Kids a Year'.[89] A list of widely respected scientists characterized the failure to immediately deploy golden rice as akin to 'a nutritional holocaust'.[90] Genetically engineered to endogenously produce beta-carotene, golden rice was supposed to be a compelling example of how this technology could do more than just gen-

---

Security', Monsanto (undated), accessed 13 October 2016 at http://www.monsanto.com/improvingagriculture/pages/climate-change-an-obstacle.html.

[85] Monsanto, *2014 Sustainability Report* (2014) 9, accessed 13 October 2016 at http://www.monsanto.com/whoweare/pages/corporate-sustainability-report.aspx.

[86] Heleen de Coninck, 'International Technology-Oriented Agreements to Address Climate Change' (2008) 36 Energy Policy 335; W Neil Adger, 'Adaptation to Climate Change in the Developing World' (2003) 3 *Progress in Development Studies* 179.

[87] Hugh Turral, Jacob Burke and Jean-Marc Faurès, *Climate Change, Water, and Food Security* (Food and Agriculture Organization 2011) xxiii.

[88] Davit Rotman, 'Why We Will Need Genetically Modified Foods' *MIT Technology Review* (17 December 2013).

[89] J. Madeline Nash, 'This Rice Could Save a Million Kids a Year' *TIME* (31 July 2000).

[90] Letter to the Editor, *Daily Mail* (London 20 February 2009), accessed 13 October 2016 at http://www.goldenrice.org/PDFs/Daily_Mail_Letter_Feb_2009.pdf.

erate profits for the small group of multinational corporations that dominate the agricultural life science industry.[91] Unlike virtually every other genetically engineered crop, golden rice was being developed by an international nonprofit organization with a mission dedicated to helping poor farmers.[92] As a technical matter, golden rice was more than a good idea; it was an elegant technical solution addressing the vexing problem of vitamin A deficiency, the leading cause of preventable blindness.[93] Life science companies were quick to capitalize on golden rice's noble aspirations, launching slick ad campaigns using the rice in an effort to claim the moral high ground in debates over genetic engineering.

It would be an understatement to claim that golden rice failed to live up to its 'world changing' hype. Pointing out that an at risk child would have to eat 15 pounds of golden rice to meet the recommended daily minimum requirement for vitamin A, Michael Pollen called it 'the world's first purely rhetorical technology' designed 'to win an argument rather than solve a public health problem'.[94] It took more than a decade to bring the beta-carotene content of golden rice to the point that it could potentially provide a significant percentage of a child's daily vitamin A needs.[95] Moreover the solution may have been technically elegant, but it was arguably directed at the wrong problem. In order to convert beta-carotene into vitamin A, the human body needs a diet with sufficient fat and protein. This is a nearly insurmountable barrier for golden rice. The intended beneficiaries of golden rice are malnourished. They lack an adequate diet, with many interrelated nutritional deficiencies. Other nutritional deficiencies, including iron, zinc, protein and fat affect the ability to absorb and

---

[91] Dan Charles, 'In a Grain of Golden Rice, A World of Controversy over GMO Food' *National Public Radio* (7 March 2013), accessed 13 October 2016 at http://www.npr.org/sections/thesalt/2013/03/07/173611461/in-a-grain-of-golden-rice-a-world-of-controversy-over-gmo-foods.

[92] The Golden Rice Project, accessed 13 October 2016 at http://www.goldenrice.org/index.php.

[93] WHO. *Global Prevalence of Vitamin A Deficiency in Populations at Risk 1995–2005: WHO Global Database on Vitamin A Deficiency*. (World Health Organization, 2009).

[94] Michael Pollen, 'The Great Yellow Hype' *New York Times* (New York, 4 March 2001), accessed 13 October 2016 at http://michaelpollan.com/articles-archive/the-way-we-live-now-the-great-yellow-hype/.

[95] Guangwen Tang, Yuming Hu, Shi-an Yin, Yin Wang, Gerard E Dallal, Michael A Grusak and Robert M Russell 'β-Carotene in Golden Rice is as Good as β-Carotene in Oil at Providing Vitamin A to Children' (2012) 96 *American Journal of Clinical Nutrition* 658–64. Unfortunately, this test on healthy school children does not necessarily indicate how malnourished children will handle golden rice.

bio-convert beta-carotene, as do coinfections with intestinal parasites.[96] The problem is more complex than just access to vitamin A or its chemical precursors.

Golden rice is, in many ways, the kind of one size fits all solution that policymakers are increasingly wary of in the context of food security.[97] The problem of vitamin A blindness is deeply embedded in broader issues of food insecurity. More than three decades ago, Amartya Sen convincingly demonstrated that food insecurity is not a production problem.[98] Even in the midst of famine, farmers have produced enough food. Food insecurity is about distribution, about who has access to the food that is produced.[99] For example in the mid-2000s food production exceeded the population's nutritional requirements for Latin America and the Caribbean by over 40 percent. Nevertheless some 45 million people still did not have access to sufficient food, and four million children under the age of five were underweight.[100] Yield alone was not enough.

Like food insecurity, vitamin A blindness is an issue of poverty. It has remained insoluble due to the lack of political will, not the lack of technical know-how[101] or a lack of adequate supplies of vitamin A in the world. Even as the dedicated scientists behind the golden rice project were busy working, there were already low tech, readily available solutions to vitamin A blindness that were simply not being implemented. Had the hundreds of millions of dollars spent developing golden rice been instead used to support the pre-existing, albeit less glamorous, poverty alleviation efforts (or straightforward initiatives to provide vitamin A supplements directly to vulnerable populations) perhaps today's statistics would look very different.

---

96 Marjorie J Haskell, 'The Challenge to Reach Nutritional Adequacy for Vitamin A: β-Carotene Bioavailability and Conversion – Evidence in Humans' (2012) 96 *American Journal of Clinical Nutrition* 1193S; JD Ribaya-Mercado, 'Influence of Dietary Fat on Beta-carotene Absorption and Bioconversion into Vitamin A' (2002) 60 *Nutrition Reviews* 104; OA Eoels, M Trout, and R DuJacquier, 'Carotene Balances on Boys in Ruanda Where Vitamin A Deficiency is Prevalent' (1958) 65 *Journal of Nutrition* 115.

97 Miguel I Gómez *et al.*, *Post-Green Revolution Food Systems and the Triple Burden of Malnutrition* (Food and Agriculture Organization ESA Working Paper 13-02, August 2013).

98 Amartya Sen, *Poverty and Famine: An Essay in Entitlement and Deprivation* (Oxford University Press 1981).

99 *Ibid.*

100 Rodrigo Martinez *et al.*, *Food and Nutrition Insecurity in Latin America and the Caribbean* (United Nations World Food Programme and the Economic Commission for Latin America, 2009) 7.

101 For an analogous argument that food insecurity is about political will formation rather than technology, see Bratspies, above note 43 at 2–3.

None of this diminishes the scale of the problem golden rice was intended to address: in 2014, the IPCC found *robust evidence* that lack of vitamin A continues to affect hundreds of millions of people.[102] However, the saga of golden rice ought to caution for a healthy skepticism toward sweeping claims about genetic engineering's power to provide technological solutions for complex social problems. Aside from the substantive critique of golden rice as a technology, it is also important to consider genetic engineering in the context of the tremendous political and cultural resistance mounted by those opposed to the technology for religious, political or ideological reasons. Those social forces remain a daunting barrier to the deployment of not only golden rice, but all other genetically engineered crops. The existence of these social barriers does not, and should not imply a veto over use of the technology, but does tend to tip the scales away from the case for genetically engineered crops, particularly when alternative technologies are available.

There is of course an argument that climate change is qualitatively different from poverty driven problems like vitamin A blindness. While there is ample evidence to suggest that the poorest among us will face the biggest climate challenges, climate change affects everyone. It jeopardizes all of conventional agriculture by changing the bio-geophysical conditions that underpin current productivity. Moreover the speed with which these conditions are changing seems to make a compelling case that new technologies produced through genetic engineering will be needed to help us adapt. The core of this argument is that we do not have the luxury of time and only genetic engineering can allow us to respond with the necessary rapidity to the agricultural challenges posed by climate change.

Yet actual, on the ground responses suggest otherwise. Many researchers are realizing that conventional breeding not only holds its own, but actually outperforms genetic engineering in terms of producing plants adapted to climate change.[103]

This insight holds true even in the face of massive investments in genetic engineering. Take the case of the Water Efficient Maize for Africa (WEMA)[104] project, one of the initiatives intended to harness genetic engineering both to alleviate poverty and to better equip poor farmers to

[102] Porter, above note 23.

[103] Natasha Gilbert, 'Genetic Engineering Lags behind Conventional Breeding in Efforts to Create Drought-Resistant Maize' 513 *Nature* 292; Doug Gurian-Sherman, *Failure to Yield: Evaluating the Performance of Genetically Engineered Crops* (Union of Concerned Scientists, April 2009).

[104] 'Water Efficient Maize for Africa' (WEMA undated), accessed 13 October 2016 at http://wemanews.aatf-africa.org/; Monsanto, 'Water Efficient Maize for

respond to climate change. A public/private partnership, WEMA is led by the Kenyan based African Agricultural Technology Foundation (AATF).[105] In many ways WEMA is the latest iteration of 'save the world' hype surrounding genetic engineering and it has wealthy, important backers like the Bill and Melinda Gates Foundation, the Howard G Buffett Foundation and USAID. Indeed asserting that 'we believe in the power of technology', the Gates Foundation has invested billions of dollars promoting genetic engineering through WEMA and other projects.[106] In 2014 WEMA announced great preliminary success in developing new drought-resistant seeds. However the seed was produced not through genetic engineering but through conventional breeding.[107] Similarly DuPont recently used conventional breeding to produce a drought resistant corn (*Optimum Aquamax*) that it markets as an alternative to genetically engineered drought resistant crops.[108] Researchers recently introduced a widely adopted flood resistant rice developed not by genetic engineering, but through conventional breeding.[109] These crops drew on the techniques and insights of molecular genetics to facilitate more rapid conventional breeding, thus harnessing modern science without creating genetically engineered seeds.

Thus experience gives lie to the suggestion that only genetic engineering will allow rapid enough development of drought resistant and heat tolerant seeds. While genetically engineered seeds may wind up being part of the climate solution, there is nothing inevitable about such a result. It is, indeed, a choice – one of many alternatives available. Despite the urgent picture painted by the technology's backers, genetic engineering is just one option among a growing array of techniques intended to adapt agricultural production to climate change. Being skeptical of genetically

---

Africa' (Monsanto undated), accessed 13 October 2016 at http://www.monsanto.com/improvingagriculture/pages/water-efficient-maize-for-africa.aspx.

[105] The African Agricultural Technology Foundation can be found at http://www.aatf-africa.org/, accessed 13 October 2016.

[106] Raj Patel, Eric Holt-Gimenez and Annie Shattuck, 'Ending Africa's Hunger' *The Nation* (2 September 2009), accessed 13 October 2016 at http://www.thenation.com/article/ending-africas-hunger.

[107] AATF, 'First Harvest of New Drought-Tolerant Seed Shows Strong Promise of Improved Maize Crops for Smallholder Farmers of Africa' (AATF, 16 May 2014), accessed 13 October 2016 at http://www.aatf-africa.org/media-center/First-Harvest-of-New-Drought-Tolerant-Seed-Shows-Strong-Promise.

[108] 'DuPont's New Drought Tolerant Corn Can Help Improve Yields and Gain Market Share' (TREFIS, 14 March 2013), accessed 13 October 2016 at http://www.trefis.com/stock/dd/articles/173705/duponts-drought-tolerant-corn-can-help-improve-yields-and-gain-market-share/2013-03-14.

[109] Gina Kolata, 'A Proposal to Modify Plants Gives GMO Debate New Life' *New York Times* (New York, 28 May 2015).

engineered crops as a solution to climate change is not the same thing as rejecting any role for technology in solving that problem.

4.1 Mitigation: Assessing the 'How Agriculture Contributes to Climate Change' Narrative

If we are to have a reasonable shot at keeping the rise in global temperature below 2°C, atmospheric greenhouse gases will have to stabilize at about 450 ppm in 2100.[110] The IPCC's most recent report states that greenhouse gas emissions will have to come down by 40 to 70 percent before we will be 'likely' to meet this target.[111] Unfortunately 'business as usual' is 'unlikely' to get us to this goal.[112] The 'business as usual' projections are for over 1,000 ppm in 2100.[113] This emissions gap[114] portends disastrous climate results. Mitigation strategies are urgently needed.

Agriculture has the second largest carbon emitting footprint in the global economy, following the energy sector. In fact, were food a country, it would have the third biggest carbon footprint, behind the United States and China. Agriculture contributes 5.4 billion tons of carbon each year, approximately 10 to 12 percent of annual global GHG emissions,[115] and is responsible for approximately 80 percent of deforestation.[116] While livestock-related emissions contribute the lion's share of this total, synthetic fertilizers are responsible for 13 percent, with rice paddies accounting for an additional 10 percent.[117] Overall roughly 8 percent of global carbon emissions (roughly 3.3 billion tons) are directly attributable to land use changes associated with agriculture, including deforestation.[118] Given that food production will need to increase 50 to 70 percent by 2050 to

---

[110] IPCC, 'Summary for Policymakers' in Christopher D Field *et al.*, (eds.) *Climate Change 2014: Mitigation of Climate Change. Contribution of Working Group III to the Fifth Assessment Report of the Intergovernmental Panel on Climate Change* (Cambridge University Press 2014) 10.

[111] *Ibid.*

[112] *Ibid.*

[113] *Ibid* at 11.

[114] United Nations Emissions Programme, *The Emissions Gap Report 2014: A UNEP Synthesis Report* (UNEP, 2014).

[115] IPCC, 2014 Summary for Policymakers, above note 110.

[116] Gabrielle Kissinger, Martin Herold and Veronique De Sy, *Drivers of Deforestation and Forest Degradation: A Synthesis Report for REDD+ Policymakers* (Lexeme Consulting, August 2012) 5.

[117] United Nations Food and Agriculture Organization Infographic, *Greenhouse Gas Emissions from Agriculture, Forestry and Other Land Uses* (UNFAO, March 2014).

[118] Global Carbon Project, above note 49.

meet the needs of an expanding global population,[119] these numbers raise critical concerns. If nothing is done to stem the carbon footprint of agriculture, growing populations coupled with changes in food consumption patterns (for example, higher demand for milk and meat) will inevitably lead to increased agricultural GHG emissions. At the same time, because agriculture is the second largest emitting sector (after the energy sector), it is a potentially rich source for carbon emissions reductions. Indeed the United Nations Environmental Programme projects that a shift to more sustainable and efficient agriculture could eliminate up to three gigatons of carbon emissions.[120]

This is where promoters of genetically engineered crops come in. There are two basic mitigation claims made about genetically engineered crops. The first mitigation claim is that increased yields will reduce deforestation and other negative climate impacts from farming. If the same amount of farmland can be made to produce more food, the imperative for expanding lands under cultivation diminishes. The second mitigation claim is that genetically engineered crops promote no-till farming, a method that purportedly increases soil carbon sequestration. So the first mitigation claim focuses on avoiding carbon emissions that might otherwise occur, while the second mitigation claim actually asserts that using genetically engineered crops can reduce atmospheric carbon through increased soil sequestration. Each claim will be considered in turn below.

4.2 Do Genetically Engineered Crops Increase Yield?

A major selling point for genetically engineered crops as a tool for both climate adaptation and mitigation has been their purported higher yield.[121] The allure of more yield per acre is obviously appealing to farmers. For policymakers seeking to maximize food production while minimizing carbon emissions, crops that will decrease the pressure to convert more lands to farming are equally attractive. Commercial agriculture is 'the dominant proximate driver of deforestation' in most developing countries.[122]

---

119 United Nations, Food and Agriculture Organization, *How to Feed the World in 2050* (2009), accessed 13 October 2016 at http://www.fao.org/fileadmin/templates/wsfs/docs/expert_paper/How_to_Feed_the_World_in_2050.pdf.

120 United Nations Emissions Programme, above note 114 at 35.

121 W. Klümper and M. Qaim, above note 41.

122 Gabrielle Kissinger, Martin Herold and Veronique De Sy, *Drivers of Deforestation and Forest Degradation: A Synthesis Report for REDD+ Policymakers* (Lexeme Consulting, August 2012) 9.

Crops that boost yield might change this dynamic, mitigating agriculture's contribution to climate change.

Note that this yield-based climate mitigation claim rests on two key preconditions. First switching from conventional seeds to genetically engineered seeds must provide sizable yield gains. Second, in the absence of this yield gain, a greater agricultural land base must be needed to meet demand.[123] Both claims are examined below.

Increased yield claims are a critical component of the climate mitigation case for genetically engineered crops.[124] The idea is that by increasing yield on already cultivated acres, farmers can avoid converting other land to cultivation, thereby avoiding increased carbon emissions. On occasion, these claims have edged into the hyperbolic, with even respected researchers characterizing genetically engineered crops as akin to 'manna from heaven'.[125] For example, in 2010, Berkeley researchers concluded that the yield boost from genetically modified seeds had saved 21 million hectares of land from conversion to agriculture.[126] Aside from the obvious biodiversity advantages of not converting forested or uncultivated lands into farmlands, the researchers asserted that this use of genetically engineered seeds also had significant climate benefits – averting nearly a half gigaton of carbon emissions.[127] In 2014 a different group of researchers claimed that use of genetically engineered crops averted roughly half that figure – just over a quarter gigaton of emissions.[128] The quarter gigaton discrepancy in these estimates is enormous. For perspective, a quarter gigaton is roughly 15 percent of the United States' total annual carbon emissions from motor vehicles.[129] This gulf between estimates about averted carbon emissions should give us pause. It is easy to forget the imprecise nature of this kind

---

123 Klümper and M. Qaim, above note 41.

124 See e.g., Bayer Crop Science, 'Practical Action on Climate Change', accessed 13 October 2016 at http://www.cropscience.bayer.com/en/Commitment/Climate-Change.aspx (describing genetically-engineered crops as 'star players' in the fight against climate change); Syngenta, 'What Syngenta Thinks About . . . Biotechnology', accessed 13 October 2016 at http://www.syngenta.com/global/corporate/en/news-center/Pages/what-syngenta-thinks-about-full.aspx (asserting a connection between genetically-engineered crops and increased yield).

125 Geoffrey Barrows *et al.*, *The Impact of Agricultural Biotechnology on Supply and Land-Use* (CUDARE Working Papers, 14 June 2013), accessed 13 October 2016 at http://escholarship.org/uc/item/3rg0c0fz.

126 *Ibid.*

127 To be precise, the figure is 0.41 Gt. For perspective, this figure would be roughly equivalent to a third of the United States' annual carbon emissions from motor vehicles. *Ibid.*

128 Barfoot and Brookes, above note 39.

129 *Inventory of Greenhouse Gas Emissions and Sinks* (US EPA, April 2015),

of calculation, which necessarily relies on assumption-laden counterfactuals. Moreover averted emissions are not emissions reductions; they are projections about what would have happened but for the variable under consideration, in this case the planting of genetically engineered crops.

Furthermore the yield case for genetically engineered crops is nowhere near as clear cut as their promoters would suggest. Despite consistent industry claims that genetically modified crops were a critical part of the effort to 'solve world hunger', a thorough survey of the available research led the Union for Concerned Scientists to conclude that the most widely adopted genetically modified crops had, to date, produced no increase in intrinsic yield and at most a marginal increase in operational yield.[130] Another recent study of soybean yield found ample evidence that planting genetically engineered seeds did not increase yield.[131] Another recent study concluded that there was, at best, limited evidence that genetically engineered crops either increase yields, or stabilize yields in the face of climate change.[132] These findings sharply contradict industry funded studies, which generally find that genetically engineered crops result in significant yield gains.[133] Given the pedigree of scientists making both sets of claims, it is hard to escape the conclusion that the evidence about crop yield is open to interpretation. The most that can be said about genetically engineered crops either increasing yields or contributing to greater yield stability is that they may, under some circumstances result in some unspecified yield gain. Hardly the kind of claim to make these crops a changing linchpin strategy for climate resilience through either mitigation or adaptation.

---

accessed 13 October 2016 at https://www.epa.gov/ghgemissions/us-greenhouse-gas-inventory-report-1990–2014.

[130] Doug Gurian-Sherman, *Failure to Yield: Evaluating the Performance of Genetically Engineered Crops* (Union of Concerned Scientists, April 2009).

[131] Michelle Marra, Nicholas Piggott and Gerald Carlson, *The Net Benefits, Including Convenience, of Roundup Ready® Soybeans: Results from a National Survey* (National Science Foundation Center for IPM Technical Bulletin 2004-3, September 2004).

[132] Rodomiro Ortiz, Andrew Jarvis, Paul Fox, Pramod K Aggarwal and Bruce M Campbell, *Plant Genetic Engineering, Climate Change and Food Security* (CGIAR Research Program on Climate Change, Agriculture and Food Security Working Paper no. 72, 2014) 20.

[133] Leonard Gianessi, Cressida S Silvers, Sujatha Sankula and Janet E Carpenter, *Plant Biotechology: Current and Potential Impact for Improving Pest Management in the United States* (National Center for Food and Agricultural Policy, June 2002).

4.3 Do Genetically Engineered Crops Promote Soil Carbon Sequestration?

The second mitigation claim about genetically engineered crops is that use of these crops can improve soil carbon sequestration. Indeed a primary climate advantage touted by producers of genetically engineered crops is that these crops promote conservation tillage.[134] This method of cultivating crops involves minimal soil disturbance because it depends on herbicide applications and crop rotation to control weeds. It contrasts sharply with conventional techniques that routinely plough up (or till) fields as a weed control method. There is no question that conservation tillage agriculture can deliver significant agronomic benefits, including improved soil quality, increased rainwater retention and often an increased yield.[135] In terms of climate mitigation, however, the claim is that the undisturbed soil sequesters carbon that would otherwise be released into the atmosphere by the tilling process. This claim rests on two premises: (1) that genetically engineered crops promote conservation tillage, and (2) that conservation tillage sequesters carbon. Both claims are open to debate.

Genetically engineered crops are certainly not a prerequisite for conservation tillage. Around the world, conservation tillage has been and continues to be practiced by farmers who do not plant genetically engineered crops. In the United States for example, the greatest growth of no-till agriculture predates the introduction of genetically engineered crops,[136] and has largely stagnated in the decades since the engineered crops have become available, even as widespread adoption of genetically engineered crops became the

---

[134] Much of the literature refers to 'no-till' cultivation. This chapter uses the more inclusive term 'conservation tillage' because under real farming conditions, the line between no-till and reduced till cultivation is frequently blurred, particularly for the many smallholders who cannot implement no-till. Mark W Rosegrant, *Food Security in a World of Natural Resource Scarcity: The role of agricultural technologies* (International Food Policy Research Institute, 2014) 7.

[135] *Ibid* at 58–60.

[136] According to the United States Department of Agriculture, adoption of no-till methods for soybeans grew from 25% of the soybean acreage in 1990 to 48% in 1995, the five-year period just before the introduction of herbicide tolerant soybeans. J Fernandez-Cornejo and WD McBride, *The Adoption of Bioengineered Crops* (USDA Agricultural Economic Report No. 810, 2002) 27–8, accessed 13 October 2016 at http://www.ers.usda.gov/data-products/adoption-of-genetically-engineered-crops-in-the-us.aspx (attributing the increase to the availability of conventional seeds appropriate for no-till farming, and to financial incentives in the National Farm Bill). Since 1995 rates of no-till farming have remained largely constant despite the introduction of soybeans genetically engineered for herbicide tolerance, crops touted as facilitating no-till farming methods. *Ibid.*

norm.[137] Moreover while farmers engaged in conservation tillage were more likely to adopt genetically engineered seeds, there is no evidence of the converse; that adopting these seeds was likely to promote conservation tillage in farmers who were not already practicing this farming method.[138] Thus to the extent that conservation tillage does in fact contribute to climate mitigation, it is not at all clear that this contribution accrues solely or even primarily because of genetically engineered crops.[139]

Moreover unlike its undisputed advantages in reducing soil erosion and promoting water retention,[140] the claim that conservation tillage sequesters carbon is increasingly suspect. Conservation tillage agriculture does lead to an increase in a higher concentration of carbon near the soil surface.[141] This has often been interpreted as an absolute accumulation of carbon. On the strength of this interpretation, conservation tillage has been promoted as a potent carbon sequestration technique, a powerful form of climate change mitigation. Indeed conservation tillage was prominently featured in the UNEP's annual *Emissions Gap Report* as having a 'significant potential' to sequester carbon.[142]

---

137 USDA statistics break out 'no-till' versus 'other conservation tillage' and 'conventional' acreage. In 2014 USDA data showed that 44% of soybeans were grown under no-till cultivation. National Agricultural Statistics Service, *2012 ARMS – Soybean Industry Highlights* (USDA 2012), accessed 13 October 2016 at https://www.nass.usda.gov/Surveys/Guide_to_NASS_Surveys/Ag_Resource_Management/ARMS_Soybeans_Factsheet/. At the same time, 97% of the soybeans planted in the United States were genetically modified. *Ibid.* Were there a connection between use of these crops and 'no-till' farming, one would expect the rate of no-till to be much higher. If 'other conservation tillage' is included with no-till, the rate of adoption jumps to roughly 70% of the soybeans under cultivation. However the rate of 'other conservation tillage' has remained virtually unchanged since 1995, the year before the introduction of genetically engineered crops. *Ibid.* Thus it would be incorrect to attribute those acres to the advent and adoption of genetically engineered crops.

138 Fernandez-Cornejo and McBride, above note 136.

139 Indeed the International Food Policy Research Institute describes 'no-till' farming as the other end of the spectrum 'from the kind of intensive agriculture associated with genetically-engineered crops.' Mark W Rosegrant, *Food Security in a World of Natural Resource Scarcity: The Role of Agricultural Technologies* (IFRPI, 2014) 4.

140 *Ibid* at 7–9; A Bolliger *et al.*, 'Taking Stock of the Brazilian "Zero-Till Revolution: A Review of Landmark Research and Farmers' Practice" (2014) 91 *Advances in Agronomy* 47–64.

141 Inigo Virto, Pierre Barré, Aurelien Burlot and Clair Chenu, 'Carbon Input Differences as the Main Factor Explaining the Variability in Soil Organic C Storage in No-Tilled Compared to Inversion Tilled Agrosystems' (2012) 108 *Biogeochemistry* 17–26.

142 United Nations Emissions Programme, above note 114 at 24.

However a growing body of experimental evidence shows that conservation tillage alters how carbon is distributed in the soil, rather than affecting the absolute amount sequestered. Earlier research had mistaken this redistribution for additional sequestration. When carbon sequestration across a range of soil depths was taken into account, conservation tillage fields did not actually sequester more carbon than ploughed fields.[143] Studies claiming to find carbon sequestration benefits suffered from a basic methodological flaw: by measuring carbon sequestration only down to about 30 cm[144] these studies missed the carbon redistribution effect, creating an inaccurate picture of carbon sequestration.[145] Moreover, land under conservation tillage is rarely permanently committed. Instead it is ploughed every few years, thus losing much or all of any putative carbon sequestration benefits.[146] Based on these considerations, a recent comprehensive review made it clear that the carbon sequestration benefits of conservation tillage have been 'widely overstated'.[147]

While the soil conservation benefits of conservation tillage are real and significant, the climate mitigation potentials are far more limited.[148] The most that can be said about the role of conservation tillage in climate change mitigation is that it may provide 'a small, but useful additional benefit'.[149] Yet boosters of genetically engineered crops continue to overstate the climate mitigation benefits of conservation tillage, giving a 'falsely

---

[143] David S Powlson, 'Limited Potential of No-Till Agriculture for Climate Change Mitigation' (2014) 4 *Nature Climate Change* 678–83; Bassem Dimassi, 'Long-term Effect of Contrasted Tillage and Crop Management on Soil Carbon Dynamics During 41 Years' (2014) 188 *Agriculture, Ecosystems & Environment* 134–46; ZhongKui Luo, Enli Wang and Osbert J Sun, 'Can No-Tillage Stimulate Carbon Sequestration in Agricultural Soils? A Meta-Analysis of Paired Experiments' (2010) 139 *Agriculture, Ecosystems & Environment* 224–31; DA Angers and NS Eriksen-Hamel, 'Full-Inversion Tillage and Organic Carbon Distribution in Soil Profiles: a Meta-Analysis' (2008) 72 *Soil Science Society of America Journal* 1370–4.

[144] World Bank, Soil Carbon Sequestration Geodatabase, accessed 10 July 2015 at http://www-esd.worldbank.org/SoilCarbonSequestration/#.

[145] JM Baker, TE Ochsner, RT Venterea, and TJ Griffis, 'Tillage and Soil Carbon Sequestration – What do We Really Know?' (2007) 118 *Agriculture, Ecosystems & Environment* 1–5.

[146] Powlson, above note 143.

[147] *Ibid.*

[148] Rolf Sommer and Deborah Bossio, 'Dynamics and Climate Change Mitigation Potential of Soil Organic Carbon Sequestration' (2014) 144 *Journal of Environmental Management* 83–7.

[149] 'The Role of No-Till Agriculture On Climate Change Mitigation may be Overstated' (Rothamsted Research, 31 July 2014), accessed 13 October 2016 at http://www.rothamsted.ac.uk/news/role-no-till-agriculture-climate-change-

optimistic message' about the contribution that conservation tillage (and by proxy genetically engineered crops) can make to climate change mitigation.[150] The climate mitigation potential from conservation tillage simply cannot be a key policy driver for adoption of genetically engineered crops.

4.4 Unintended Consequences

Thus far this chapter has tried to show why decision makers should view climate related claims made about genetically engineered crops skeptically. Close examination of the 'genetic engineering as climate adaptation' narrative shows its many holes. New crops can, and have been developed with drought resistance or heat tolerance traits without using genetic engineering. Thus the most that can be said in terms of adaptation is that these crops are one option among many. Similar flaws beset the 'genetic engineering as climate mitigation' narrative. Adopting genetically engineered crops does not guarantee any reductions in agriculture's massive carbon footprint. Yet promoters of these crops try to cloak them with a sense of climate inevitability, just as prior campaigns tried to portray genetically engineered crops as the inevitable answer to food insecurity.[151] In many ways, these crops truly are an answer in search of a problem.

Aside from the caution to view claims about these crops skeptically, there are also significant, unintended consequences that should be factored into decisions about use of genetic engineering to mitigate or adapt to climate change. A main talking point in favor of relying on genetically engineered crops is the putative two decade history of safe use of these crops.[152] Unfortunately this claim rests largely on the absence of information about negative effects of genetically engineered crops, rather than on direct research and studies. There are precious few in depth pre-market studies (whether human or animal) to support claims about crop safety,[153]

---

mitigation-may-be-over-stated (quoting Prof. David Powlson, lead author of the definitive review study of no-till agriculture's climate impacts).

150 Powlson, above note 143.

151 See e.g., Robert Paarlburg, *Starved for Science: How Biotechnology is Being Kept Out of Africa* (Harvard University Press, 2008).

152 RA Herman and WD Price, 'Unintended Compositional Changes in Genetically Modified (GM) Crops: 20 Years of Research' (2013) 61 *Journal of Agricultural Food Chemistry* 11695; Laura DeFracesco, 'How Safe Does Transgenic Food Need to Be?' (2013) 31 *Nature Biotechnology* 794.

153 Contrary to industry representations about the rigorous food safety oversight to which genetically engineered crops are subject, food safety research is conducted wholly by a proposed crop's proponents. In the United States there is a voluntary opportunity to consult with FDA, but much of the process is shrouded

and the lack of labeling and traceability in many countries[154] makes it virtually impossible to develop this information on a post-market basis. Of course the information gaps are not evidence that that the crops have any negative health effects, but ought to temper public confidence in claims that the crops are demonstrably safe.[155]

Even granting that the assumptions about human health are warranted, there is no question that the widespread adoption of multiple crops engineered to be tolerant to glyphosate, the active ingredient in Monsanto's Roundup, has resulted in the emergence of at least ten species of so called 'superweeds': weed plants resistant to glyphosate.[156] Similarly field evidence of the emergence of evolved pest resistance to *Bt* crops raises serious question about the future of crops genetically engineered to produce *Bt* as a pest management tool.[157] Industry and government claims that these

---

in secrecy. The tests that are done on a voluntary basis are incompletely reported, with no requirement that all research be shared with the agency or the public. The publicly available results are often open to interpretation, both for the data they contain, and for remaining unanswered questions. Rebecca Bratspies, 'Consuming (F)ears of Corn: Biopharming and Public Health' (2004) 30 *American Journal of Law and Medicine* 371–404. Most surprisingly in light of the fact that GM corn, soy and oilseed has been modified to either tolerate herbicides or to endogenously produce pesticides, the patchwork regulatory structure that governs the commercialization of these crops has meant that the possible chronic side effects of the pesticide and herbicide residues themselves have not be subject to study. *Ibid.*

154 Roughly 97% of edible GMOs among cultivated GMOs (soy, corn and oilseed rape or canola, excluding cotton) are grown in South and North American countries where labeling is not required. J Clive, *Global Status of Commercialized Biotech/GM Crops: 2009* (ISAAA Brief No. 41, 2009).

155 Bratspies, Consuming (F)ears of Corn, above note 153.

156 Michael DK Owen, 'Weed species shifts in glyphosate-resistant crops' (2008) 64 *Pest Management Science* 377–87; William Newman and Andrew Pollack, 'Farmers Cope With Roundup Resistant Weeds' *New York Times* (New York, 3 May 2010). Indeed Bayer Cropscience's most frequent pitch for its genetically engineered cotton begins with the phrase 'With weed resistance exploding across America's farmland'. Bayer CropScience, *Stoneville Offers Two New Varieties With GlyTol and LibertyLink Traits* (Bayer CropScience, 9 January 2013), accessed 13 October 2016 at http://www.bayercropscience.us/news/product-news?storyId=0CB58C47-BE79-4A2C-9979-162CE57A055D.

157 Bruce E Tabashnik, Thierry Brévault and Yves Carrière, 'Insect Resistance to *Bt* Crops: Lessons from the First Billion Acres' (2013) 31 *Nature Biotechnology* 510–21. One review reached the carefully worded conclusion that 'the scientific research conducted so far has not detected any significant hazard directly connected with the use of GM crops'. Alessandro Nicolia, Alberto Manzo, Fabio Veronesi and Daniele Rosellini, 'An Overview of the Last 10 Years of Genetically Engineered Crop Safety Research' (2014) 34 *Critical Reviews in Biotechnology* 77–88. The wording needed to be careful because the review documented in some

crops can be a viable part of long-term planning, particularly in light of climate changes already likely to promote pest proliferation, must be read in light of this real and growing resistance problem.

Aside from the emerging resistance challenges that these crops face, genetically engineered crops have been extremely controversial. The United States, Canada and Brazil permit farmers to freely plant these crops with little regulatory oversight. Other countries have been less sanguine about genetically engineered crops. Many countries either ban the crops entirely or permit only a few varietals. The European Union propounds a 'process' form of regulation that imposes a distinct set of approval requirements over all genetically modified crops and food products that contain them.[158]

Given the global nature of food production, it is perhaps unsurprising that one consequence of these different regimes has been repeated problems involving the shipment of unapproved varieties. On more than one occasion, the resulting controversies have roiled global markets. Indeed disputes over the use of this technology have given rise to multiple trade disputes, with the United States and the European Union battling over their differing regulatory regimes in the World Trade Organization.

5. CONCLUSION

It would be foolish to deny that genetic engineering might have something to contribute in the struggle to adapt to or mitigate climate change. However, it is a very different matter to say with any certainty that it will *in fact* make such contributions. For one thing, the discussions at this point are wholly about the *possible* benefits that *might* accrue *if* potential applications of this technology can *successfully* engineer more heat and/or drought resistant varietals. The ability to actually do so remains wholly unproven. To date the most common genetically engineered crops have been highly profitable, both for certain farmers and for so-called

---

detail the research establishing the emergence of weed resistance to glyphosate through gene flow, and the intercession of *Bt* genes into wild plant populations with the attendant risk of accelerating the evolution of insect resistance. By correctly noting that gene flow is a problem not unique to genetically engineered crops, the authors could bracket the significant negative environmental impacts associated with these crops to reach this comforting-sounding conclusion in the face of rising weed resistance and evolving pest resistance.

[158] Commission Regulation 1829/2003, art. 4, [2003] OJ L268/1, 7 (EC); Panel Report, *European Communities – Measures Affecting the Approval and Marketing of Biotech Products*, WT/DS291-93/R (29 September 2006).

life science companies like Monsanto, DuPont and Bayer. However these modifications have done little to promote greater food security, and provide scant evidence that engineering of drought resistance or heat tolerance (traits far more complex than herbicide resistance) will be in the offing. In fact drought resistance and heat tolerance are complex multigene traits where the genetic basis is poorly understood. They are, in short, the kind of challenges for which even advocates of genetic engineering admit that conventional breeding techniques are most useful.[159] As the FAO cautions, climate adaptation is 'likely to be knowledge-rich rather than technology driven'.[160]

Moreover, even assuming that the promoters are right and new crops can be genetically engineered for increased drought resistance and heat tolerance, and also assuming that those crops are widely adopted and provide corresponding yield benefits, history suggests that these crops may still fail to provide their predicted climate benefits. Without measures to improve food security, all of those putative climate benefits will be undermined by an increasingly desperate, exponentially growing population looking to grow food wherever and whenever possible. Increased yield alone cannot solve the problems of global food insecurity because food insecurity is about poverty.[161]

---

159 *New Techniques for Genetic Crop Improvement* (Biotechnology and Biological Science Research Council Position Statement, 2014) 6, accessed 13 October 2016 at http://www.bbsrc.ac.uk/news/policy/2014/141028-pr-position-statement-on-crop-breeding-techniques/.

160 Hugh Turral, Jacob Burke and Jean-Marc Faurès, *Climate Change, Water, and Food Security* (Food and Agriculture Organization, 2011) xxii.

161 Partha Dasgupta, 'Nutritional Status, the Capacity for Work and Poverty Traps' (1997) 77 *Journal of Econometrics* 5–37; World Food Programme, *Hunger and Markets* (Earthscan, 2009) 19–22.

8. Adapting agriculture through land use controls

Keith H Hirokawa

1. INTRODUCTION

Even without the uncertainties associated with climate change, food production involves so complex a system of inputs and outputs, and serves such fundamental human needs, that interruptions and inefficiencies can have terrible consequences for human well-being. Markets, cultures, family and food security, health and security all depend on the secure and reliable production of healthy food. Because climate change interferes with our expectations for climate and ecosystem productivity, it is considered a gargantuan challenge for agriculture, and more generally for civil society.

The chapters collected in this volume provide comprehensive and detailed analyses of resiliency strategies that may be employed at many levels of government to address challenges to agriculture, and the potential of agricultural investments to play a significant role in climate change readiness. This chapter specifically addresses the manner in which land use control strategies are elements of a productive strategy for facilitating resiliency and adaptive capacity. Given the complexity of climate change considerations, in addition to the mutual dependencies of agriculture and local governments, this chapter will address the role of land use regulation in improving resiliency.

Land use control is relevant to a reasoned climate change response. The practice of land use control is essentially an exercise in building (or losing) adaptive capacity in communities. Land use choices, and the regulatory schemes that govern them, determine the location, character and intensity of the community's impact on the environment, vulnerability to change, and resources that can be relied upon. For example, land use regulations influence the locations of different land uses and therefore determine the distances people travel to home, work and play (and therefore determine air, water and land impacts from transportation, as well as infrastructure needs, psychological impacts of travel, and so on). Not surprisingly, land use control is the primary tool that local governments employ to confront new challenges to the social, economic, and environmental fabric of community.

In addition, climate change impacts will first be felt locally, suggesting that local governments and their experiences in land use control are unique assets. Notably, local governments have played a significant role in accelerating both community interest and the availability of community assets to plan for climate change preparedness.[1] Given the realities of climate change and its implications for the viability of communities, many local governments are choosing not to 'wait it out'[2] and are engaging in climate preparedness through regulatory responses that empower communities and improve local adaptive capacity.[3]

This chapter examines some exemplary resiliency strategies that are being employed at the local level as land use regulations.[4] This chapter does not assert that local control over agricultural practices is the best or only answer to the climate challenge. As Professor Erin Ryan notes, '[a]fter all, if local decision-making were always best, there would be no need for a strong federation in the first place'.[5] The land use component of agricultural resiliency is arguably limited because of the role that parochialism plays

---

[1] See Alice Kaswan, 'Climate Adaptation and Land Use Governance: The Vertical Axis' (2014) 39 Colum J Env L 390, 414 (noting that land use planning to address local vulnerabilities is 'spotty and specific implementation measures are the exception rather than the rule.').

[2] Center for Science in the Earth System, Joint Institute for the Study of the Atmosphere and Ocean, University of Washington, and King County, Washington, *Preparing for Climate Change: A Guidebook for Local, Regional and State Governments* (2007) 26.

[3] ICLEI-Local Governments for Sustainability, *Progress and Challenges in Urban Climate Adaptation Planning: Results of a Global Survey* (2012) 3. 'Adaptive capacity' has been defined as 'the ability or potential of a system to respond successfully to climate variability and change, and includes adjustments in both behavior and in resources and technologies'. National Research Council, *Adapting to the Impacts of Climate Change* (The National Academies Press 2010) 160.

[4] This chapter examines the resiliency strategies that local governments across the United States are employing as land use regulations. The focus on the United States is not meant to exclude other approaches or experiences. In the United States, land use is almost entirely a creature of local control. Although some states have developed state-mandated planning schemes, such schemes are aimed at facilitating the planning process, rather than on securing specific planning priorities or pre-empting local prerogative. This focus therefore exposes insights into the local experience and specifically local contribution to climate change readiness.

[5] Erin Ryan, 'The Once and Future Challenges of American Federalism' in Alberto Lopez Basaguren and Leire Escajedo San-Epifanio, *The Ways of Federalism in Western Countries and the Horizons of Territorial Autonomy in Spain* Vol 1 (Springer 2013) 272 (also pointing out that 'for reasons both historical and philosophical, American federalism has proven robust in spite of the alternatives. The critical question is how best to balance the wisdom and interests of the

in defining local environmental and economic concerns. Nevertheless local participation in climate change solutions is considered a necessity, and the types of solutions that local governments produce are important. Land use control is a uniquely local practice that is universally driven by the need to protect local priorities, facilitate local identities, and even to foster local and individual opportunities. Land use control addresses challenges such as climate change because the failure to do so results in the failure of local government to govern.

2. THE LOCAL ISSUE OF CLIMATE CHANGE

Through the traditional land use tools of zoning and planning, local governments have long been in the business of protecting communities from emerging perceived threats – and as local governments have engaged new challenges and developed the details of local environmental concern, we have witnessed the development of local environmental law.[6] Local interaction with local environments focuses on the local concern for human welfare and quality of life, illustrating a type of governance for which there is no federal counterpart. When local governments realize the potential of economic development of preserving ecological benefits through land use regulation, they are engaged in the interaction between community and nature.[7]

---

local and national governments that have remained so robust within our federal system.').

[6] John Nolon, 'In Praise of Parochialism' (2002) 26 Harv Env L Rev 365, 374 ('Communities have long used large-lot zoning as a crude way of protecting open space and its associated natural resources. Upzoning occurred in some suburban areas, aimed principally at lowering development densities to control population growth, maintain residential property values, and contain the cost of servicing development while, incidentally, limiting water use, preventing aquifer contamination, and containing nonpoint source pollution. As the environmental movement evolved and matured in the 1970s and 1980s, the sensitivity of local lawmakers was raised and early signs of the adoption of local environmental law became apparent.'); See also, Keith H Hirokawa, 'Sustaining Ecosystem Services through Local Environmental Law' (2011) 28 *Pace Environmental Law Review* 760.

[7] As the EPA noted in a report entitled *Community Culture and the Environment: A Guide to Understanding a Sense of Place*:

> We live among, and are deeply connected to, the many streams, rivers, lakes, meadows, forests, wetlands, and mountains that compose our natural environment and make it the beautiful and livable place so many of us value. More and more often, human communities realize that the health and

Because local governments are uniquely situated in the environment, the character of their surroundings determines (at least in part) how local communities relate to the environment. Because this association also involves the development and transformation of local environmental resources, local governments have become quite experienced in managing environmental trade-offs. Hence the character of the local environment deeply concerns the well-being of the community, and governance of the community means concern for the local environment.

Given the foregoing it seems a short step to conclude that although governments at all levels will be called upon to address the many challenges of climate change, local governments will face unique needs due to the intense nature of felt impacts at the local level.[8] Indeed local governments regulate the environment because communities depend on their surroundings: local governments define themselves by their environmental opportunities and obstacles in ways that other governmental entities do not.

The role of local governments in environmental regulation is equally important in the climate change era. A recent account of the 'wide-ranging effects' on local governments associated with climate change included '[a]nticipated changes in temperature, precipitation, and natural hazards [that] have the potential to place stress on infrastructure, residential, office, and historic buildings, environmental quality, and both municipal and ecosystem services',[9] in addition to some level of forced agricultural

---

> vibrancy of the natural environment affects the health and vibrancy of the community and vice versa. We value the land, air, and water available to us for material goods, beauty, solace, retreat, recreation, and habitat for all creatures. Throughout the nation, communities are engaging in efforts to protect these treasured natural resources and the quality of life they provide.

Office of Water, US EPA, *Community Culture and the Environment: A Guide to Understanding a Sense of Place* (2002). See also, Community Culture and the Environment at 4 ('Community-based environmental protection recognizes that values held both individually and as a group contribute to the quality of community life. Expression of values through social and cultural practices can create a "sense of community." Many of these values relate directly to the 'place' in which people live, thus creating a strong 'sense of place'.).

[8] New York City Panel on Climate Change, *Climate Risk Information* 5 (2009), accessed 14 October 2016 at http://www.nyc.gov/html/om/pdf/2009/NPCC_CRI.pdf.

[9] ICLEI-Local Governments for Sustainability, *Progress and Challenges in Urban Climate Adaptation Planning: Results of a Global Survey* (2012) 3; National Research Council, *Adapting to the Impacts of Climate Change* (2010) 36–7 ('Examples of other sources of climate [change vulnerability] include population shifts and development choices, such as dense urban development in

and ecosystem migration to locate suitable climate conditions. In the meantime, climatic changes will impact the ability of local governments to provide and maintain public services. Local governments will be particularly challenged by the land use implications of the following interruptions:

- Maintenance and planning for local transportation infrastructure. Increased frequency of extreme weather patterns will change the distribution of various land uses throughout the landscape, transforming transportation needs and methods, while also causing wear on physical infrastructure.
- Availability of local utility infrastructure. Utility systems such as gas, electricity and water may be subject to interruption from extreme weather events, while adapting to changing needs arising from population shifts.
- Provision of water. Climate changes may challenge the adequacy of water supplies and availability of clean water due to drought, storm events and damage to man-made water infrastructure.
- Storm water control, including the removal of wastewater. Extreme storm events and frequent flooding will cause maintenance and performance problems for storm sewer systems, combined with the inability of local wetlands and other deteriorated ecosystem features to perform storm surge control services.
- Waste disposal. Waste disposal infrastructure will be interrupted by flooding and challenges to transportation infrastructure.
- Changing climatic conditions will increase the risk of species extinction and loss of biodiversity.
- Disease and malnutrition. Climate change will cause unexpected changes to water, climate and food production expectations, while changing exposures and vulnerabilities to infectious diseases.[10]

---

drought-prone areas; and places and communities that are especially dependent on climate-sensitive industries such as agriculture, forestry, and tourism. In the built environment, each city's residents and infrastructures will be affected in unique ways.').

[10] See generally, ML Parry, OF Canziani, JP Palutikof, PJ van der Linden and CE Hanson, *Contribution of Working Group II to the Fourth Assessment Report of the Intergovernmental Panel on Climate Change*, 'Summary for Policymakers' (Cambridge University Press 2007) 18; National Research Council, above note 9 at 36–7; ICLEI, above note 9 at 3. For a contextualized example, such as those discussed in Part II–IV, see City of Chula Vista, *City of Chula Vista Climate Adaptation Plan*, accessed 14 October 2016 at http://www.chulavistaca.gov/clean/conservation/Climate/documents/AttA_ClimateActionPlanUpdate_Apr12ProgressReport_FINAL.pdf.

Each of these predicted vulnerabilities can be addressed by developing a systematic land use program for resilient agriculture,[11] both as that concept relates to food production and more generally to food security. As noted below, land use control can be an effective tool in locating sensitive land uses in areas that provide the most protection – and in locating intensive land uses in areas that are durable and less vulnerable to impacts of the built environment.

3. ADDRESSING AGRICULTURE TO IMPROVE RESILIENCY

On a broad scale, resiliency will require securing food access and availability, including reliability in food production methods and minimizing the social, economic and environmental externalities of food production. This section looks more narrowly at agricultural resiliency as a result of adaptive land use control. From the land use perspective, securing agricultural resiliency entails three primary trade-offs involving land use designations, integration of agricultural land uses into the community vision, and community difference.

The first trade-off involves land as a resource and potential (or real) food source. Local governments must consider the amount and location of lands that are used and set aside for food production (and not used for other purposes) so that farming practices and food access meet the needs of the community. The availability of farm lands will help to secure the viability of local foodsheds, and the location of food production will determine the relationship between food security and the ability of the local government to ensure a stable and reliable infrastructure (transportation, water, waste, emergency services and so on). This challenge has an historical backdrop: farmlands have suffered from conversion to other uses, especially farms that are close to public infrastructure and on the fringe of urban and suburban developments. Interestingly although food production has long been valued by civil society, land use controls have traditionally left agricultural lands unregulated and unaddressed. Farmland preservation arguments often find support in the shared purposes of farmlands as

[11] Agricultural resilience has been described as 'a farm ecosystem's capacity to absorb shocks or disturbances, such as those associated with the predicted impacts of climate change, which threaten its continued ability to function'. Mary Jane Angelo and Joanna Reilly Brown, 'Whole-System Agricultural Certification: Using Lessons Learned from LEED to Build a Resilient Agricultural System to Adapt to Climate Change' (2014) 85 U Colo L Rev 689, 716.

open space; yet even as open space, farmlands have often been considered lands that lay in waiting for more dense and intensive development. Once the large acreage farms are sold and developed for other uses, agriculture is displaced and pushed further from denser areas, having greater impacts on public infrastructure and services including decreased access to local foods.

Second, from the land use perspective, farming practices should be integrated as much as possible into the local land use scheme to account for infrastructure needs of agricultural practices and to neutralize conflicts with other locally-important land uses. In many cases, this means that agricultural or other practices must be modified or mitigated to account for the needs of a comprehensive scheme of the community. This challenge recognizes that farming practices are often considered a threat. Farming practices are responsible for significant, sometimes irreversible adverse impacts on ecosystems.[12] It is estimated that the conversion of land to agricultural practices has resulted in the loss of over 100 million acres of wetlands[13] and represents the most significant source of water pollution.[14] Pesticide and fertilizer use results in polluted agricultural runoff and water degradation. Conversion of land to agriculture results in the loss of habitat and displacement of species, interrupting ecosystems both from conversion to farmland and through continuing farming practices.[15] Despite the adverse impacts from farming, agricultural practices have been largely unregulated under current environmental laws.[16]

The third trade-off relates to consistency or uniformity of priorities and value across local borders. Climate change will pose region and community-specific challenges and land use resiliency programs will vary based on the context and severity of the threat in light of local assets and natural capital through which a response can be made. In the context of agricultural resiliency, this means that communities value the benefits of

---

[12] Michael Pollan notes, 'just because food is local doesn't mean it will be organic or sustainable'. Michael Pollan, *The Omnivore's Dilemma* (Penguin Books 2006) 257.

[13] JB Ruhl, 'Farmland Stewardship: Can Ecosystems Stand Any More of It?' (2002) 9 Wash U JL & Pol'y 1, 10.

[14] SR Gliessman, *Agroecology: Ecological Processes in Sustainable Agriculture* (CRC Press 1998).

[15] JB Ruhl, 'Farms, Their Environmental Harms, and Environmental Law' (2000) 27 Ecology LQ 263, 331.

[16] JB Ruhl, 'Farms, Their Environmental Harms, and Environmental Law' 27 Ecology LQ 263, 265 (2000) (As Ruhl notes, the choice is an unwise one: 'the cumulative effects of more than 450 years of crop and livestock farming in America are no longer obscure; if we continue to leave farms unregulated, it is by choice, not by ignorance.').

the built environment in a similar manner to the way they value ecosystem services.

On this point, the study of ecosystem services provides insights into the value of agriculture to local governments.[17] Ecosystem services are the services and products that natural processes provide, such as water and air filtering, habitat provision, soil structure and building materials.[18] From an ecosystem services perspective, agricultural practices can both provide essential services and be highly dependent on the continuing receipt of ecosystem services. For instance, the structure and fertility of local soils may be the critical factor in locating and determining the productivity of agricultural lands.[19] The presence and functionality of watershed features such as forests and wetlands help to deliver clean and metered water for irrigation. Less obvious but nonetheless critical local services include pest regulation, pollination, nutrient cycling and climate control, and can be provided in less visible ways.[20] Importantly, the degree to which such processes benefit local communities (or leave them vulnerable) is dependent on the manner in which the communities are ecologically situated. Communities situated on navigable waterways, in arid locations, at high elevations, or in forest areas will each value local natural capital as resiliency assets in accordance with the benefits that local ecosystems provide. Hence, there is no single, uniform approach to resiliency among communities.

These issues are individually and cumulatively significant for local governments as they plan for climate change. Local responses to climate challenges will be characterized both by limitations in the authority of local governments to act, but also by self-imposed restrictions that concern local priorities, politics and identity. Accordingly, a study of local resiliency strategies is necessarily a study of examples of local resiliency strategies. In what follows this chapter introduces the tools that local governments have

---

[17] Gretchen Daily *et al.*, 'Ecosystem Services: Benefits Supplied to Human Societies by Natural Ecosystems' (1997) 2 *Issues in Ecology* 1, 2; Robert Costanza *et al.*, 'The Value of the World's Ecosystem Services and Natural Capital' (1997) 387 *Nature* 253, 253; Millennium Ecosystem Assessment, *Ecosystems and Human Well-being: Synthesis* (2005) v.

[18] These processes 'provide basic life support for human and animal populations and are the source of spiritual, aesthetic, and other human experiences that are valued in many ways by many people'. EPA Science Advisory Board, *Valuing the Protection of Ecological Systems and Services* (May 2009) 8, accessed 14 October 2016 at https://yosemite.epa.gov/sab/sabproduct.nsf/F3DB1F5C6EF90EE185257 5C500589157/$File/EPA-SAB-09-012-unsigned.pdf.

[19] Vei Zhang et al., 'Ecosystem Services and Dis-services to Agriculture' (2007) 64 Ecological Econ 253, 255.

[20] *Ibid.*

traditionally used to face new challenges, as well as how those tools are being used to foster resiliency and climate change preparedness.

4. TRADITIONAL CHALLENGES AND LAND USE TOOLS: STATIC AND FLEXIBLE

Although resiliency strategies in the agricultural context will vary, any local response will account for (and draw from) the traditional land use questions: property, zoning and planning. At least with regards to the concerns of food production and local food access, local governments use traditional land use control strategies to prevent conversion of agricultural lands, protect farms from urban encroachment, preserve open space and avoid interference with ecosystem services. Local governments have a wide array of tools that are considered more traditional land use control devices, and many communities have been busy implementing innovative plans and strategies to improve resiliency. Central to these tools are the dual and often competing tasks of protecting non-agricultural needs from the impacts of agriculture while protecting the ability of agricultural practices to produce sufficient food supplies to meet our needs.

4.1 Protecting Farms from Displacement

The inevitable growth of urban areas guarantees a collision between farms and residential development. Because the impacts of farms are often felt beyond farm boundaries, non-farming neighbors (often new neighbors) sue to curtail farming operations under nuisance. In response, all states and some local legislatures have acted to protect farming practices from land use development.[21] The Massachusetts case of *Pendoley v. Ferreira*,[22] provides some context for the felt necessity. In this case plaintiffs sought to enjoin a piggery operating nearby, but in an adjacent town. The defendants, the Ferreiras, started their farm in 1949 on 25 acres with 400 pigs and 100 piglets.[23] At that time the impacts of the farming operation on other land uses in the area were not significant.[24] However, over the span of a decade, the farm doubled in size.[25]

---

[21] See Terence J. Centner, 'Creating an "Undeveloped Lands Protection Act" for Farmlands, Forests, and Natural Areas' (2006) 17 Duke Envtl L & Policy F 1.

[22] *Pendoley v. Ferreira* [1963] 345 Mass 309, 310.

[23] *Pendoley*, above note 22.

[24] *Ibid* at 311.

[25] *Ibid* at 144.

The plaintiffs complained of odor and noise. Here, the problem with the Ferreiras' farm was neither negligence nor oversight. Indeed it was agreed that the farm was operated at a high level: 'in the upper 5% to 10% of . . . [comparable] piggeries insofar as quality of operation is concerned'.[26] Rather the incompatible nature of the pig farm to residential living was the basis for relief, or what the court referred to as 'the inherently offensive aspects of any piggery in a residential neighborhood and in the material discomfort which piggeries cause to others'.[27]

Under these circumstances the court allowed the Ferreiras some time for relocation and reestablishment, but also found that 'plaintiffs are entitled to have the offensive operation terminated entirely within a reasonable time'.[28]

The *Pendoley* case illustrates a typical and traditional threat to farming lifestyles and practices: farming lands are often in the way of suburban development. Urban and suburban communities expand into neighboring lands, bringing community needs, infrastructure, lifestyles and values. The noise, dust, odor and even schedule of farming practices may interfere with the expectations that urban immigrants carry with them to the new community.

In response to this threat to agriculture, every state has adopted some form of protection for farming practices. These so-called 'Right-to-Farm' laws often identify specific types of impacts or farming practices that are subject to protection, and they uniformly recognize that the protection afforded prevents competing lifestyles – typically lifestyles introduced to the area after the commencement of farming – from determining the propriety of farms in a particular region. Common Right-to-Farm laws create presumptions that farming practices do not cause nuisances,[29] create conditions that exempt specific agricultural operations from nuisance liability[30] or establish time frames in which nuisance complaints must be raised.[31]

In protecting farmlands and farming practices from competing land uses, local governments have also been creative as well as revealing as they identify the relationship between local identity and farming. For instance the town of Brookhaven, New York declared that farming is an essential tool for protection from intense development, enhancing the local quality

---

26 *Ibid* at 144.
27 *Ibid* at 314.
28 *Ibid* at 315.
29 See, e.g., Mich Comp Laws Ann § 286.473.
30 See, e.g., Kan Stat Ann § 47-1505.
31 See, e.g., Ala Code § 6-5-127 (a).

of life, and is otherwise 'of vital importance to the town's character, environment and economy, as preserving an historically important industry which provides fresh food and horticultural products to local and regional markets'.[32]

The Brookhaven Ordinance notes that 'lands subject to such agricultural use(s) are generally less burdensome upon governmental services than lands subject to other active uses'.[33] Because of the importance of farming to the community character, Brookhaven announced that the typical farming impacts of dust, smoke, noise, vibration and odor are 'presumptively considered to be necessary and essential to bona fide farming, far outweighing any annoyance or nuisance to neighboring properties' occupants'.[34]

Likewise in Malta, New York, the town's legislative body has declared farming to be 'an essential activity' and has adopted an ordinance 'to maintain and preserve the rural tradition and character of the town'.[35] The Malta Ordinance reads: 'farming . . . reinforces the special quality of life enjoyed by citizens, provides the visual benefit of open space and generates economic benefits and social well-being within the community. Therefore the Town of Malta emphasizes to newcomers that this town encourages its agriculture and requests newcomers to be understanding of the necessary day-to-day operations.'[36]

Right-to-Farm laws redefine concepts of property to shield farming practices from private interference. Of course these laws do not guarantee that lands will be set aside or used for farming purposes. Nevertheless they do have a resiliency impact by minimizing the threats to farming practices that may exist in the legal structure of property law. It is noteworthy that some courts have observed constitutional implications in the ways that Right-to-Farm laws address nuisances. In *Bormann v. Board of Supervisors in and for Kossuth County*,[37] the Supreme Court of Iowa found a version of this legislation to constitute a taking. The court reasoned that the protection from nuisance effectively transferred a property right from neighbors to farmers, analogous to an easement on a neighbor's property that preserved the offending activities.[38] Courts in other states, such as Idaho[39]

---

[32] Town of Brookhaven, NY Code 85-405.A (1) (Farmland Bill of Rights).
[33] Town of Brookhaven, NY Code 85-405.A.1
[34] Town of Brookhaven, NY Code 85-405.A.3.
[35] Town of Malta, NY Code 86-1.A, C.
[36] Town of Malta, NY Code 86-1.B.
[37] *Bormann v. Board of Supervisors in and for Kossuth County* [1998] 584 NW2d 309.
[38] *Ibid* at 321.
[39] *Moon v. North Idaho Farmers Ass'n*, [2004] 140 Idaho 536.

and Minnesota[40] have disagreed, finding that nuisance protection does not affect the transfer of property rights.

4.2 Euclidean Zoning and Community Threats

In some instances, public land use controls can act as obstacles against achieving agricultural resiliency. Local zoning laws have historically restricted the choices that landowners can make, often interfering with backyard farming activities such as keeping livestock and other agricultural activities. Local zoning codes have achieved seemingly absurd results in some cases, including the threat of imprisonment in Oak Park, Michigan for planting a vegetable garden.[41]

The Oak Park zoning ordinance includes a detailed landscaping provision for residential properties which allows 'unpaved portions of [a homeowners front yard to] be planted with grass ground cover, shrubbery, or other suitable live plant material'.[42] Oak Park officials determined that vegetable gardens do not qualify as 'grass ground cover, shrubbery, or other suitable live plant material'.[43]

As approved by the Supreme Court in *Village of Euclid v. Ambler Realty Co*,[44] zoning has traditionally served the police power, defined as the public health, safety, welfare and morals. The *Euclid* Court held that private property expectations can be regulated and forced to converge with the needs of the public welfare.[45]

The court analogized the zoning power to nuisance abatement and recognized that when local governments separate incompatible land uses,

---

[40] *Overgaard v. Rock County Board of Commissioners* [2003] 2003 WL 21744235.

[41] Professor Sarah Schindler reports such a story. 'In the quiet Detroit suburb of Oak Park, Michigan, Julie Bass was brazenly breaking the law. An inspection of the crime scene would reveal no assault rifles or illicit drugs, but rather a shovel, raised planting beds, and some seeds. Bass faced ninety-three days of jail time for planting a vegetable garden in her front yard and for refusing to comply with the city's demand to remove it.' Sarah B Schindler, 'Of Backyard Chickens and Front Yard Gardens: the Conflict Between Local Government and Locavores' (2012) 87 Tul L Rev 231, 233. Bass was charged with a code-enforcement violation with the threat of jail time for the misdemeanor; 'Woman Faces Jail Time for Growing Vegetable Garden in Her Own Front Lawn' ABC News (12 July 2011), accessed 14 October 2016 at http://abcnews.go.com/US/vegetable-garden-brings-criminal-charges-oak-park-michigan/story?id=14047214.

[42] Oak Park, Mich, Code §1716(A)(5) (2012).

[43] *Ibid.*

[44] *Village of Euclid v. Ambler Realty Co*, [1926] 272 U.S. 365.

[45] *Euclid*, above note 44 at 388.

they are accommodating visions of community and social, economic and environmental needs. For purposes of climate change readiness, it is relevant that the *Euclid* Court approved this power because allowing local governments to adapt to a 'changing world' could serve adaptive and constitutional goals. The court noted that 'while the meaning of constitutional guaranties never varies, the scope of their application must expand or contract to meet the new and different conditions which are constantly coming within the field of their operation'.[46]

At its core, Euclidean zoning involves the separation of incompatible land uses: land uses that the local community views as interfering with more sensitive and prioritized land uses (such as the single-family home).[47] This equation has resulted in disfavor for farms, which are often thought of as among the 'less-civilized, potentially nuisance-causing uses'.[48] In principle and in practice, zoning often interferes with farming practices by increasing the costs of farming operations through building and site regulation, habitat regulations and storm water controls.

In contrast, local governments have used zoning tools to protect farmlands from urban encroachment, reduce the impacts of agricultural runoff and encourage mixed and multiple-use land uses[49] in neighborhoods that can accommodate such uses. To protect small farms[50] from urbanization, the Kansas legislature has adopted a statute that prohibits interference with farming uses or erection of farming buildings through exercise of the zoning power.[51] Likewise in Suffield, Connecticut,

---

46 *Ibid* at 386–7.

47 Sarah B Schindler, 'Of Backyard Chickens and Front Yard Gardens: The Conflict Between Local Governments and Locavores' (2012) 87 Tul L Rev 231, 253–4 ('This linc of reasoning explains why both gardening and keeping farm animals are prohibited in many localities that hew to traditional Euclidean Zoning.').

48 *Ibid* at 251.

49 See, e.g., Polk County, MN 23.3100 (allowing for siting of wind power generation in agricultural lands).

50 Large farming operations may be considered commercial, rather than agricultural. See, e.g., *Town of Mount Pleasant v. Van Tassell* (1957) 166 N.Y.S.2d 458 (holding that a pig farm did not qualify as agricultural, but rather was commercial due to its intensity).

51 Kan Stat Ann § 19-2960 (d) ('No zoning regulation shall apply to the use of land for agricultural purposes nor for the erection and maintenance of agricultural buildings so long as such agricultural buildings are used for agricultural purposes and no other. Dwellings, garages and other similar accessory buildings shall not be considered as agricultural buildings. All buildings, including agricultural buildings may be regulated as to setback requirements from public roads so as to protect the future use and improvement of such roads.').

subdivision regulations include provisions for agricultural buffer strips. The town's regulation allocates responsibility of avoiding neighborhood conflicts to the developers and residents of property adjacent to an active farm. Developers are required to notify new residents of the practices of neighboring farms, and are also responsible for providing a protective buffer during the development process to minimize the impacts of farming practices. Residents then bear the burden of maintaining the buffer.

The power to adopt and enforce local zoning regulations has enabled local governments to prepare to face new challenges, while leaving local values and identity intact. Hence zoning is not always an obstruction to local food production, especially in communities that have chosen to focus their assets in food production and its associated social, economic and ecological benefits. This attention to local control over resiliency strategies is preserved even in more pervasive, state based controls on farmland conversion. For instance, Oregon's far reaching land use system provides for protection of and accounting for agricultural lands in Goal 3, which provides that '[a]gricultural lands shall be preserved and maintained for farm use, consistent with existing and future needs for agricultural products, forest and open space and with the state's agricultural land use policy'.[52] In the Oregon land use statutes, the legislature declared that the preservation of agricultural lands from urbanization helps to conserve natural resources and the agricultural economy of the state.[53] To protect the availability of farmland, Oregon has identified minimum lot or parcel sizes that apply uniformly to all counties (outside of urban areas): '(a) For land zoned for exclusive farm use and not designated rangeland, at least 80 acres; (b) For land zoned for exclusive farm use and designated rangeland, at least 160 acres; and (c) For land designated forestland, at least 80 acres.'[54] Although there are exceptions to these minimum acreage requirements, they are narrow and are designed to keep farmland preservation intact against direct and piecemeal changes.

4.3 The Promise of Planning

The third land use foundation is land use planning. Land use planning is related to both the property precedents of nuisance and the police power basis for Euclidean zoning. However land use planning diverges in its

---

[52] OAR 660-015-0000 (3).

[53] Oregon Rev Stat Ann § 215.243; *Meyer v Lord* (1978) 586 P2d 367 (Oregon land use system incorporates a clear policy for preserving existing farming lands and operations).

[54] Oregon Rev Stat Ann § 250.780 (1).

goals and scope by involving the process of long term community vision and identity, and as such, offers an opportunity to address climate change preparedness.

Land use planning is often described as a multi-disciplinary attack on the problems that plague unanticipated social change.[55] Equipped with this tool, local governments are empowered to integrate resiliency into the very fabric of the community.[56] Planning allows communities to envision a desirable future, in which the uncertainty of unknown threats 10, 20 or 50 years into the future can be accommodated by community design and control.[57] In this sense land use planning is a resiliency strategy: the process of formulating land use planning requires flexibility, participation and transparency as a method of recognizing and responding to changes that could otherwise threaten local priorities and needs.

Many local governments have already begun the dizzying process of planning for an uncertain future and the formulation of climate change responses. Professor Patricia Salkin has observed that, '[i]ncreasingly, with or without guidance from state enabling acts, local comprehensive plans are attempting to respond to the threats of climate change'.[58] The

---

[55] John R Nolon, 'Comparative Land Use Law: Patterns of Sustainability' (2005) 37 Urb Law 807, 818–19. Professor John Nolon describes land use planning as an experiment that communities must endure to achieve a sense of place:

> City planning is a science and an art concerned primarily with the city's ever-changing pattern. As a pure science, it examines causes (history and etiology) and reciprocal influences of man and environment (urban geography and ecology). As applied science, it synthesizes these findings with those of the economic, sociological, and political sciences as well as the technological branches of statistics, civil and sanitary engineering, architecture, landscape architecture, and other pertinent branches of human knowledge, in an attempt to thoroughly understand conditions and their contexts and trends. As an art, it utilizes these materials, instructs or organizes citizens, molds events, and thwarts or guides trends to bring about the changes in city design which it contemplates.

[56] See generally James A Kushner, 'Smart Growth: Urban Growth Management and Land-Use Regulation Law in America' (2000) 32 Urb Law 211.

[57] Patrick Gallagher, 'The Environmental, Social, and Cultural Impacts of Sprawl' (2000) 15 Nat Res & Env't 219, 219. See also James E Holloway and Donald C Guy, 'Smart Growth and Limits on Government Powers: Effecting Nature, Markets and the Quality of Life Under the Takings and Other Provisions' (2001) 9 Dick J Envtl L & Pol'y 421 (providing an analysis of the implications of changing our way of thinking toward Smart Growth programs).

[58] Patricia E Salkin, 'Sustainability and Land Use Planning: Greening State and Local Land Use Plans and Regulations to Address Climate Change Challenges and Preserve Resources for Future Generations' (2009) 34 Wm & Mary Envtl L & Pol'y Rev 121, 134.

American Planning Association, in a recent report entitled *Planning for Food Access and Community-Based Food Systems*, reports on local planning efforts as follows:

> The comprehensive plan directly influences a community's urban design, land-use, and zoning regulations, which in turn influence the opportunities for food production on public and private land; commercial and community food processing facilities; local distribution networks; food retail diversity, density, and location; and commercial and private compositing operations. Collectively, these factors fundamentally drive the availability of and access to healthy and unhealthy foods in a community. Therefore, tools such as comprehensive planning and zoning regulations—both justified through local governments' police power to address the health, safety, and welfare needs of residents—can be used to improve to community food system and provide children, adolescents, and adults access to healthy, affordable foods. For this reason, many communities throughout the country have been motivated to include food-system-related goals and policies in their local land-use plans as a means for solving the problem of food access.[59]

Land use planning has allowed local governments to secure water for irrigation through the planning process,[60] to ensure adequate transportation to markets, to facilitate local food markets and to capture assets that enable communities to respond adequately to a variety of changing circumstances. The State of New York, for instance, currently encourages and incentivizes local governments to envision and adopt 'Climate Action plans'.[61] Counties, cities and towns across the nation have adopted 'sustainability plans', jurisdictions have adopted 'urban forest plans'[62] and have otherwise incorporated resiliency needs involved in water use, transportation impacts, and even greenhouse gas emissions into their planning process.[63] Many local governments, such as King County, Washington, have recognized in their climate readiness that productive land use planning can help to secure co-benefits of agriculture, including the production

---

[59] Kimberly Hodgson, American Planning Association, *Planning for Food Access and Community-Based Food Systems: A National Scan and Evaluation of Local Comprehensive and Sustainability Plans* (American Planning Association 2012) 21.

[60] Michelle Bryan, 'A Next, Big Step for the West: Using Model Legislation to Create Water-Climate Elements in Local Comprehensive Plans' (2013) 3 Wash J Envtl L & Pol'y 1.

[61] New York State Executive Order No 24 (2009).

[62] See Keith H Hirokawa, 'Sustainability and the Urban Forest: An Ecosystem Services Perspective' (2011) 51 Natural Resources Journal 233.

[63] See, e.g., City of Santa Cruz, 'Climate Adaptation Plan' (2011), accessed 14 October 2016 at http://www.cityofsantacruz.com/home/showdocument?id=23644.

of renewable energy while playing the critical role of local production and provision of food.[64]

When local governments engage in adaptation planning to protect public health, safety and welfare, they manage the economic and social costs of climate change responses. Given the extent of projected global climate change impacts, and the manner in which such impacts can devastate communities (as a matter of scale), local governments illustrate foresight by engaging in the planning process that addresses at least the following elements:

(1) *Assessment of local climate change hazards and vulnerabilities, and inventory of local assets.* Different communities will be disproportionately impacted by changing climates and extreme weather events such as storms and sea level rise in coastal communities, changing temperatures in rural areas and water availability in arid regions. Natural resources (such as species or habitats), built resources (such as roadways or treatment plants) and human resources (such as residential developments, especially with access obstacles) may be more vulnerable climate hazards due to their location.
(2) *Public participation.* Community governance is largely a process of community participation and awareness. Local governments both gather and distribute information needed for coordinated community responses to any new challenge. The role of public participation in resiliency planning is paramount. Greater community involvement can garner better identification of local vulnerabilities and communication of the same, both to and from the residents, resulting in an increased ability of municipal officials to reduce the vulnerability of the community to climate hazards.[65]

---

[64] Center for Science in the Earth System, Joint Institute for the Study of the Atmosphere and Ocean, University of Washington, and King County, Washington, *Preparing for Climate Change: A Guidebook for Local, Regional and State Governments* (2007) 12, accessed 14 October 2016 at http://www.cses.washington.edu/db/pdf/snoveretalgb574.pdf. The County has also adopted a 2012 King County Strategic Climate Action Plan, which sets short-term performance measures and targets, accessed 14 October 2016 at http://your.kingcounty.gov/dnrp/climate/documents/2012_King_County_Strategic_Climate_Action_Plan.pdf.

[65] New York Department of Environmental Conservation, *Climate Smart Resiliency Planning: A Planning Evaluation Tool for New York State Communities* (2014) 30, accessed 14 October 2016 at http://www.dec.ny.gov/docs/administration_pdf/csrptool.pdf ('Public outreach and engagement are crucial to the development of a community that is resilient to climate hazards. Local residents and business owners often have extensive knowledge of past storm, flood and

(3) *Action item selection and adoption of a plan(s).* Although coordination and consistency among the various branches of local governments are essential, this does not mean that the planning process culminates in the adoption of a single document. The community may prefer to separate out climate planning into topics (disaster, food security, water security and so on) or governmental responsibilities (code enforcement, health and safety, education). More importantly, and perhaps even more effectively, local governments can incorporate climate planning actions into existing land-use planning tools, such as comprehensive plans and zoning codes.

(4) *Ongoing information gathering and revision.* Given that climate changes will not be static, local resiliency will depend on ongoing, adaptive planning through a process of community based feedback loops and revision. New vulnerabilities will emerge and prior action items may prove inadequate. As such, implementation of local adaptation actions should be followed by an ongoing assessment of improvements in the community's resilience.

With these elements in mind, land use planning can serve as an effective tool to meet future and uncertain challenges. The strength of land use planning is in the relationship it fosters between local communities and governance: in the process of planning, communities can identify their priorities, preferences, strengths and weaknesses, and can participate in the protection of the community.

5. REPURPOSED LAND USE TOOLS AS CLIMATE CHANGE RESPONSES

As indicated above, local governments are equipped with regulatory tools that can address threats to locally valued resources. As noted, however,

---

erosion effects. By engaging local stakeholders in the identification of hazards, risks and vulnerabilities, municipal officials can build support for risk-averse land-use planning, ordinances and hazard mitigation.

While it is important to involve local stakeholders in the identification of coastal hazards, it is also important to communicate vulnerabilities and risks to the public. Engaging local residents and visitors in storm preparedness is essential to ensuring their safety. In some places, full-time residents are outnumbered by vacationers during months when hazards, e.g., coastal or winter storms, are at their peak. By educating the public on the threat of climate hazards, municipal officials can reduce the exposure and vulnerability of their community.').

traditional land use tools suffer limitations that reflect the purposes they were designed to accomplish. In response to such limitations, many local governments have experimented with more innovative and creative uses of local regulatory authority to address the agricultural and other land use concerns posed by climate change. This section addresses the potential of such innovation to repurpose traditional tools and experiment with various controls that target different impacts of and upon agriculture. In the meantime, and as a complement to these regulatory responses, local governments have experimented with non-traditional tools to encourage residents to grow their own fresh produce and support local farm markets and other local food access mechanisms.[66]

5.1 Governmental Purchase of Agricultural Conservation Easements (PACE)

The majority of states have implemented programs for the purchase of agricultural conservation easements (PACE). PACE programs illustrate the potential resiliency benefits of protecting the amount of available agricultural lands through a local, non-regulatory land use response. PACE programs generally further the protection of agricultural lands through the acquisition of a perpetual or other meaningful development right: typically a conservation easement or conservation covenant, under which a landowner voluntarily assumes a duty to avoid conversion of agricultural lands to another use.[67] In some programs states purchase and hold easements directly, while in others PACE programs facilitate acquisition by land trusts or local governments.[68]

Local governments that employ PACE protections have experimented with many program designs. For instance, Baltimore County, Maryland has authorized the governmental purchase of agricultural easements for purposes of farmland preservation.[69] Under the ordinance the county is directed to identify the most productive farmland parcels that amount to

---

[66] Stephen R Miller, *Boundaries of Nature and the American City, Environmental Law and Contrasting Ideas of Nature: A Constructivist Approach* (Keith H Hirokawa (ed.), Cambridge University Press 2014) 146.

[67] See generally, Jessica Owley, 'Use of Conservation Easements by Local Governments' in Patricia Salkin and Keith Hirokawa (eds) *Greening Local Government* (ABA Publishing 2012) 237.

[68] See, e.g., the Wisconsin Working Lands program and its support for acquisition through its PACE program, accessed 14 October 2016 at https://datcp.wi.gov/Pages/Programs_Services/FarmlandPreservation.aspx.

[69] Baltimore County, Md, County Code art 24, tit 3, §§ 101-205 (2003).

50 acres in size, or a cumulative size of 50 acres when assembled by various parcels. The county decision is based on factors that help the county assess the food production value of the land, balanced against the need for other inconsistent land uses.[70]

However, agricultural lands that are threatened by imminent conversion are not subject to the priority process. The county also may exercise the power of eminent domain. Once acquired, property subject to agricultural easements may be used for limited purposes, with very few exceptions to the preservation goals of the program.

Other notable examples include Boone County, Illinois, which has established a PACE program to protect and maintain the County's rural character. Boone County has declared that farmland preservation furthers the goal of natural resources and environmental quality as well as the protection of historical, cultural and scenic resources.[71] To facilitate acquisition, the Boone County Ordinance has established a land trust commission which is given primary investigatory responsibilities.

The town of Troy, Wisconsin established a Farmland Preservation Commission for purposes of outreach and investigation to support the town's acquisition of development rights on farm land. Although the Troy Ordinance does not authorize the use of eminent domain,[72] rights acquired under this program are held 'in perpetuity' and in trust for the benefit of the townspeople.[73]

5.2 Transferable Development Rights (TDR) Programs

Local governments have recently begun to use transferable development rights (TDR) programs to preserve agricultural lands.[74] In the typical TDR program, the jurisdiction will designate particular lands as sensitive or critical and shoulder those lands with significant land use restrictions. As an alternative to compliance with these regulations, the landowner will be offered the opportunity to sell the rights of development (that are otherwise restricted in the program). Landowners who purchase these development rights may be permitted to increase the density of their development or some other valuable advantage. The benefits of TDR include the low cost opportunities (including little opposition from the

---

70 Baltimore County, Md, County Code art 24, tit 3, § 103(c).

71 Boone County, Ill, Ordinance 06-18 (2006).

72 Troy, Wi, Ordinance § 52-4.D.

73 Troy, Wi, Ordinance § 52-8.

74 Dennis J McEleny, 'Using Transferable Development Rights to Preserve Vanishing Landscapes and Landmarks' (1995) 83 Ill B J 634.

restricted owner and compensation for compliance in these programs), with a high yield of preservation.

As a resiliency tool, TDR programs engage the participation of property owners in the community in deciding how to protect agricultural lands without burdening economic development with stasis. For example, the town of Southampton, New York encourages farming and protects agricultural lands through an Agricultural Overlay District and Agricultural Planned Development District.[75] Under this program, parcels greater than ten acres that are located in the overlay district may encumber their property with an agricultural easement for a ten-year period. During this time, the landowner and town can explore preservation options, including town acquisition or the benefits of the town's TDR program. In a slightly different approach the town of Berthoud, Colorado has adopted regulations for the collection of a fee that supports the acquisition of open space in the Town.[76] Under the ordinance, the fee is triggered by a 're-zoning of land from either a residential, agricultural or transitional zoning district to a district that allows a higher residential density.'[77] Credits are awarded for permanent preservation of open space at the rate of one acre per unit of credit.

5.3 Other Incentive Programs for Farmland Preservation

An additional zoning tool that has been initiated in a few jurisdictions concerns preservation of agricultural lands as open space through the use of available incentives.[78] Like the programs discussed above, incentive programs are aimed at preserving lands that are designated for agricultural use and protecting them from conversion to other land uses. Incentive programs encourage without requiring individual participation and generally do so at little cost to the community. Two examples of effective incentivized land preservation efforts include incentive zoning and tax incentives.

So-called 'incentive zoning' programs stray from traditional uses of the zoning power by offering favorable zoning treatment for voluntary actions. For example, the Town of Beekman, New York adopted an incentive zoning ordinance that secures the voluntary provision of 'community

---

[75] Southampton, NY, Code ch 330, art X (2001).

[76] Town of Berthoud, Colo, Development Code ch 10, § 111 (2008).

[77] Town of Berthoud, Colo, Development Code ch 10, § 111.C. (2008).

[78] This tool was widely employed by local governments as a means to encourage voluntary compliance with green building standards. See Keith H Hirokawa, 'At Home with Nature: Early Reflections on Green Building Laws and the Transformation of the Built Environment' (2009) 39 Envtl L 507.

benefits'[79] from individual property owners. Among other things, a developer in Beekman can work with the town to secure 'permanent' conservation of agricultural lands, a community benefit, in exchange for density increases over permitted residential densities or variances from minimum lot sizes and setbacks, or impervious lot coverage, floor area ratios, building heights or '[a]ny other waivers of provisions of this chapter as deemed necessary and appropriate by the Town Board for the achievement of the community benefit or amenity'.

Other jurisdictions encourage farmland preservation by applying different tax structures and benefits to the owners of farmland. The purposes and effectiveness of differential taxation for preventing farmland conversion have been criticized as 'merely a financial buffer for current farmers [with] limited value for preventing conversion'.[80] Notwithstanding the foregoing, preferential tax treatment for farm land is often justified by evidence that farming demands less commitment from public infrastructure, and differential tax systems reflect an effort to avoid making farmers pay more than their fair share. For example since 1996, the town of Ashford, Connecticut has offered tax abatement for dairy farms, which was subsequently expanded in 2005 to orchards and vineyards, to account for the low cost of providing public infrastructure to these uses.[81]

5.4 Urban Agriculture

Urban agriculture involves a diverse and often competing array of social and economic interests, from organic and sustainable growers who share community values to a more commercial and institutional sense of how and where communities secure food resources. The term is instrumental both in rural areas that feel social and cultural pressure from expanding urban areas, as well as urban communities seeking to embolden food security through direct access to fresh vegetables, fruits and meat products. Urban farming contributes to community empowerment and a sense of place, which helps to reduce crime and maintain robust, green surroundings in urban areas.[82] Perhaps more importantly, the urban

---

79 Beekman, NY, Town Code Ch 155, § 55 (2001).

80 Jeanne S White, 'Beating Plowshares into Townhomes: The Loss of Farmland and Strategies for Slowing its Conversion to Nonagricultural Uses' (1998) 28 Envtl L 113, 115 (explaining that deferred taxation statutes envision repayment after conversion of agricultural lands to nonfarm uses, but have other negative implications).

81 Town of Ashford, Conn, Town Ordinances §§ 16.1–16.4 (2007).

82 See generally, Schindler, above note 47.

agriculture movement has forced zoning administrators to recognize the need to integrate, instead of exclude, agricultural values in neighborhoods that previously seemed incompatible with agricultural use. The resiliency impact of urban agriculture is generally positive, both as a conceptual matter and one of implementation. Urban agriculture has resulted in community education, food access, urban infrastructure improvements (to accommodate the new uses) and community participation in the food production process.

As a subset of urban agriculture, home and community gardens can build resiliency by increasing access to agricultural products. In *Urbanland*, the magazine of the Urban Land Institute, it was reported that in 2010 more than 38 percent of households in the United States maintained a vegetable garden, amounting to approximately 41 million people producing their own vegetables and engaging in community literacy of fresh food production.[83] This figure does not include livestock rearing and maintenance, which is also on the rise.

Backyard and community gardens are facilitated in a number of ways in the land use process.[84] The city of Milwaukie, Oregon facilitates community gardening by supporting and offering opportunities throughout the city, including a demonstration garden in front of city hall to illustrate 'how much food can be grown in a typical 4'x12' community garden plot'.[85] The Milwaukie Municipal Code exempts farming practices from development review,[86] and the City's Comprehensive Plan notes that the

---

[83] Mia Lehrer and Maya Dunne, 'Urban Agriculture: Practices to Improve Cities' Urban Land (18 Jan 2011), accessed 14 October 2016 at http://urbanland.uli.org/sustainability/urban-agriculture-practices-to-improve-cities/.

[84] Universities are significant participants in community and home gardening programs. For instance the University of Wisconsin Extension in Milwaukee County offers garden plots for rent, as well as technical assistance for those desiring to grow their own food, with rental fees for individual plots ranging from $20 to $252 annually. See Garden Rental, accessed 14 October 2016 at http://milwaukee.uwex.edu/agriculture/garden-rental/milwaukeegardens/. The program advertises as follows: 'Gardens and green space are important in the urban environment. Residents at health care facilities, apartments and condominiums may not have the opportunity or space to garden. Unused urban open space can provide the needed land to build gardens. Gardens provide space to grow nutritious food and gardeners can save money on produce. Gardening is a great family recreational activity. Participants share traditions with each other and their children', accessed 14 October 2016 at http://milwaukee.uwex.edu/agriculture/garden-rental/.

[85] See 'Community Gardens', accessed 14 October 2016 at http://www.milwaukieoregon.gov/generalpage/community-gardens.

[86] Milwaukie, Or, Mun Code 19.201.

urbanization of the region (and resulting loss of agriculturally-designated lands) should occur in an 'orderly fashion' without undue pressure on open space needs.[87]

Although home grown produce has had some success in the political arena, home grown livestock has faced difficult hurdles.[88] Common complaints have involved noise, lack of sanitary conditions (spread of disease) and odor. Recently some communities have loosed the strings on livestock at the home. For instance in 2010, Atlanta, Georgia adopted an ordinance that allows (and regulates) keeping of livestock within city limits.[89] The ordinance identifies 'small animals' to include dogs, rabbits, guinea pigs, chickens, turkeys 'and the like'. The ordinance specifies the maximum number of animals by type and also identifies conditions of treatment and maintenance to avoid impacts on neighboring properties. Austin, Texas has adopted an ordinance that allows residents to keep certain livestock and honeybees within city limits. The ordinance identifies specific requirements for the number, treatment and maintenance of animals.[90] This shift in the acceptability of raising livestock in residential areas is rhetorically significant. As noted above, the Euclidean zoning system has often been criticized for being too resistant to evolution in the ways we define the home and the preferences we prioritize to the zoning scheme, including the treatment afforded to agricultural uses. This new direction suggests that local laws are accommodating a more flexible relationship between community needs and residential expectations.

Urban agriculture represents a far divergence from traditional zoning methods and goals. Traditionally, farming has been treated as an ill fit with the values that sustain the home. Nevertheless community thinking on gardens has evolved, and as Sarah Schindler notes, '[b]y allowing residents to use their private property to produce a portion of their food, and to sell their surplus, local governments can incrementally reduce reliance on industrially produced food and in the aggregate avoid related harms'.[91] Through zoning and planning local governments can help to address an observed 'disconnect between consumers and producers' in farm

---

87 Milwaukie, Or, Comprehensive Plan, Environment and Natural Resources Chapter, Agricultural and Forest Lands Element, ch 3.

88 Schindler, above note 47 at 244–5 (explaining the history behind backyard chicken and other livestock bans).

89 Atlanta, Ga, Municipal Code §§18-7 to 18-41 (2010).

90 Austin, Tex, Municipal Code §§3-1-1 to 3-6-25 (2010).

91 Schindler, above note 47 at 262.

markets.[92] As illustrated by the example of urban agriculture, land use laws can be helpful in relocating or otherwise facilitating closer geographical connections between agricultural uses and the markets that they serve. Zoning controls can allow and encourage food production at the home and throughout the community. Euclidean zoning has evolved and adapted and in the meantime, local governments have been learning the ways the zoning power can be used to build resiliency by effectively locating benefits and burdens of particular land uses as a method of distributing community assets.

5.5 Farmers' Markets

Farmers' markets have been a successful tool that local governments have used to facilitate direct markets between farmers and consumers. As suggested by the City of San Francisco, farmers' markets benefit the community by 'promoting better nutrition and healthier living at a time when obesity and other ill effects of poor nutrition and lack of exercise plague our community'.[93] In the meantime, farmers' markets provide an educational benefit, present a social opportunity, and encourage residents to recreate in ways that support their physical and mental well-being. The resiliency benefits of farmers' markets include food access and direct participation by local residents with the agricultural economy of the region. Local governments that encourage and facilitate such markets do so with the specific goal of strengthening the region's foodshed system.[94]

Although farmers' markets are designed around simple principles, cities have viewed this opportunity in light of local resources. For instance, San Francisco has authorized the use of city parks for farmers' markets, subject to use restrictions pertaining to operation of the parks. The community of Dahlonega, Georgia has likewise adopted an ordinance establishing and regulating a farmers' market to operate at the Warwick Street city park.[95] The ordinance restricts farmers' market participation to locally grown produce.[96]

---

[92] Susan A Schneider, 'A Reconsideration of Agricultural Law: A Call for the Law of Food, Farming, and Sustainability' (2010) 34 Wm & Mary Envtl L & Pol'y Rev 935, 954.

[93] San Francisco, Cal, Ordinance 29-07 (6 February 2007).

[94] San Francisco, Cal, Ordinance 29-07 (6 Feb 2007).

[95] Dahlonega, Ga, Ordinance 2001-7 (2 Jul 2001).

[96] Dahlonega, Ga, Ordinance 2001-7 (2 July 2001) (The ordinance states, 'The goods and products sold at the Farmers Market must be locally grown or

In contrast Wichita, Kansas has amended its zoning code to allow farmers' markets as a permitted use.[97] This ordinance allows farmers' markets on private property to be managed and run by private operators. Because the Wichita program considers farmers' markets as accessory uses, the markets themselves cannot dominate the site and also cannot use any of the required areas of off-street parking for the primary use. The ordinance limits farmers' markets to particular days, by signage and by the types of products that can be sold.

5.6 Food Policy Councils

Local governments have experimented with 'food policy councils'. As community organizers, food policy councils can perform investigatory functions by assessing the regional food system, inventorying assets and local obstacles and proposing strategies to incentivize sustainable food production practices.[98] The resiliency impacts of food councils have largely been improved agricultural planning at the local level through the addition of community participation in food access, evolution of community vision with the integration of agriculture and even the introduction of agricultural expertise into local political decision making.

Oakland's experience is exemplary. In 2006, the Oakland City Council authorized the Mayor's Office of Sustainability to develop an Oakland Food Policy. The Resolution sought a plan to assure 30 percent local area

---

locally produced goods or products of high quality. Each participating Vendor, upon making application for a Vendor's license shall sign a statement affirming that all goods or products which such Vendor intends to sell are locally grown or locally produced goods as defined in Section 2 of this Ordinance.' In turn the terms 'locally grown' or 'locally produced goods' are used to mean, 'such goods or products as shall be grown in Lumpkin County, or in any county contiguous to Lumpkin County, which goods are sold at the Farmers Market by those persons or any member of the family of those persons who are substantially responsible for growing or producing said goods or products. In no event shall any goods which have been previously purchased from another source be permitted to be sold at the Dahlonega-Lumpkin County Farmer Market.').

[97] Wichita, Kan, Ordinance 47-025 (18 May 2006).

[98] See, e.g., The City of Austin, Texas has established a 'sustainable food policy board' to serve in an advisory capacity 'concerning the need to improve the availability of safe, nutritious, locally, and sustainably-grown food at reasonable prices for all residents, particularly those in need, by coordinating the relevant activities of city government, as well as non-profit organizations, and food and farming businesses'. City of Austin, Tex, Mun Code 2-1-170.B.

food production, which would be facilitated by an initial food systems assessment and inventory. The resulting assessment informed the city that development of the regional foodshed was a missed opportunity or, from another perspective, could be a ripe opportunity to secure social, economic and environmental benefits.[99] In December, the Office of Sustainability proposed a $50,000 allocation to create a Food Policy Council for Oakland (Oakland FPC), which was subsequently approved by the City Council.[100]

The Oakland FPC took on a variety of far reaching initiatives, including a campaign to expand zoning accommodation for urban gardens and locating an edible greenway mapping tool for designating food space across the city.[101] Amendments to the city's land use regulations were proposed in 2011 that would have allowed growing crops as a home occupation.[102] Expanded, far reaching amendments were proposed in 2014 to allow citizens throughout the city to grow food crops, with restrictions to accommodate the nature of residential density and open space limitations.[103] This work culminated in a 2014 vote by the Oakland City Council to approve

---

[99] City of Oakland, *Mayor's Office of Sustainability, Informational Report on Oakland Food System Assessment* (11 July 2006) (on file with author) ('While California's farming regions are one of the most productive in the world, the majority of the food consumed in the state is imported from out of state and just under half of the state's raw agricultural production is exported. Our findings show that Oaklanders' food expenditures represent a small fraction of the value of the region's agriculture goods. Oakland therefore represents a ripe market for this regional foodshed, and the region is capable of supplying Oakland with most of its food needs. Sourcing food from the regional foodshed could have many positive social, economic, and environmental outcomes for both regional farming and for the City.').

[100] Holly Coleman, *A Report and Resolution Authorizing the City Administrator to Allocate $50,000 from the Williams Energy Settlement Within the City Facilities Energy Conservation Fund (4450) to Provide Startup Funding for the Establishment of a Food Policy Council for Oakland* Oakland City Council Resolution No. 80332 (approved 16 December 2006) (on file with the author).

[101] See 'Priority Conservation Areas For Food Security', accessed 14 October 2016 at http://oaklandfood.org/our-work/policy-initiatives/urban-agriculture/edible-greenways/.

[102] Oakland City Planning Commission Staff Report, June 15, 2011. http://www2.oaklandnet.com/oakca1/groups/ceda/documents/agenda/oak029705.pdf.

[103] See 'Urban Agriculture Citywide Update', accessed 14 October 2016 at http://ec2-54-235-79-104.compute-1.amazonaws.com/Government/o/PBN/OurOrganization/PlanningZoning/OAK029859.htm.

zoning changes that entitled residents to grow their own food: a victory that was largely due to the persistence of the Oakland FPC.[104]

5.7 Water Quality: Water Use and Pesticide Restrictions

In general local governments do not approach the subject of water quality in any manner that resembles the federal water pollution control program. Local governments lack the sophistication or interest in answering the types of questions that are addressed in the Clean Water Act.[105] Nevertheless water quality is a serious concern of local governments since agricultural runoff is a significant source of water pollution. As such it is not a surprise that some local governments have ventured into the arena of water quality protection.

For instance many states are trying to control the release and use of pesticides, and local governments have implemented such policies by seeking to prevent chemical fertilizers and pesticides from finding their ways into local water supplies.[106] The Wisconsin legislature has adopted an annual licensing system for the use of certain pesticides.[107] The Massachusetts legislature provides authority to restrict pesticide use if 'the terms or conditions [of the certification, license, or permit] are being violated or are inadequate to avoid unreasonable adverse impacts on the environment'.[108] Maine provides administrative oversight on the location and manner of pesticide use based on environmental impacts.[109]

A few local governments have dabbled in pesticide restrictions. In 2008 the town of Castine, Maine imposed regulations on the use of fertilizers in non-residential operations.[110] Any proposed non-residential use of pesticides and fertilizers may be approved contingent upon the development of an 'Integrated Pest Management and Nutrient Management Plan'.[111] Pest management plans must incorporate site plans that indicate the areas of chemical use, hydrological conditions at the site and methods of public

---

[104] See 'Urban Ag Campaign Update: The Real Win', accessed 14 October 2016 at http://oaklandfood.org/2014/11/07/urban-ag-campaign-update-the-real-win/.

[105] See A Dan Tarlock, 'Land Use Regulation': The Weak Link in Environmental Protection (2007) 82 Wash L Rev 651; Keith H Hirokawa, 'Sustaining Ecosystem Services through Local Environmental Law' (2011) 28 Pace Env L Rev 760.

[106] Schindler, above note 47 at 256.

[107] Wisc Stat Ann § 94.685 (2014).

[108] Mass Gen Laws Ann ch. 132B, §10 (2015).

[109] Me Rev Stat Ann title 7 § 610 (2015).

[110] Zoning Ordinance of the Town of Castine, Me, Article 4, § 4.7.4 (2008).

[111] Zoning Ordinance of the Town of Castine, Me, Article 4, §4.7.4.12 (2008).

notice prior to pesticide or fertilizer use.[112] A year later, Castine amended its code to require a source water impact analysis and to add some oversight for residential pesticide and fertilizer use.[113]

Water quality regulations by local governments bear little resemblance to the traditional Euclidean zoning tools of separating incompatible land uses. Nevertheless, the resiliency impact is similar. Local governments that regulate water quality impacts do so to minimize land use disruptions across the community, whether the impacts are felt at home, in school, as infrastructure integrity or at the farm. Improvement of pesticide practices and other water quality challenges result in fewer vulnerabilities in land use systems.

5.8 Climate Action Plans

As discussed above, many local governments have crafted land use planning documents to specifically address climate adaptation and mitigation.[114] Some states have adopted enabling programs that encourage and incentivize 'climate action plans' and their implementation in communities. In these plans local governments identify specific policy strategies and recommendations to reduce greenhouse gas emissions,[115] inventory local assets and vulnerabilities and formulate strategies that allow communities to withstand significant climatic changes. Some climate action planning activities have been limited to governmental carbon footprints while others have linked water, transportation, housing and food production in comprehensive land-use planning.

Knoxville, Tennessee adopted a sustainability plan through its 2007 creation of an Energy & Sustainability Initiative. Knoxville sought to improve quality of life for its residents by 'greening' the city. The City's pride is evidenced in their list of accomplishments:

> City operations are leaner, greener, and saving money while reducing impact on the environment. We have more and better infrastructure for folks who bike, walk, or use public transit to get around our city. As a community, we've dramatically increased the Knoxville area's renewable energy capacity and invested in energy efficiency projects for our homes and businesses. We've

---

112 Zoning Ordinance of the Town of Castine, Me, Article 4, §4.7.4.12 (2008).

113 Town of Castine Source Water Protection Ordinance, Castine, Me (2009).

114 See, e.g., local, regional and state governmental initiatives reported by the Environmental Protection Agency in the 'Showcase Communities' program: https://www.epa.gov/statelocalclimate/climate-showcase-communities-program.

115 See, e.g., Metropolitan Washington Council of Governments, *National Capital Region: Best Practices and Policies to Reduce Greenhouse Gases* (2008).

> increased the amount of waste that is recycled or mulched instead of sent to a landfill. Our local economy is poised for growth in clean and advanced energy markets that offer good jobs for our local workforce.[116]

The city's plan is underscored by relevant challenges to community and governance needs. The plan identifies action items to improve community engagement, energy use, infrastructure needs and development, transportation, growth challenges, waste disposal and food security.[117] To combat the development of 'food deserts' and other city challenges, the plan identifies five action items. First the city has embarked on developing its urban forests as an ecologically significant resource. Second the city will educate its residents on appropriate home gardening practices, including species and technical assistance. Third the city plans to host staff from AmeriCorps VISTA for guidance on facilitating development of community gardens. Fourth the plan directs the city to develop either a zoning mechanism or distribution of property opportunities through leases of otherwise unused public property to accommodate home and community gardening deficiencies. Finally the plan envisions adoption of landscaping guidelines to ensure safe use of pesticides and herbicides.[118]

Although many communities remain reluctant to commit resources to climate preparedness, participating communities are better equipped to address agricultural resilience and food security. Local governments that have engaged in the planning process have already asked the questions that determine the state of affairs during a climate change response. They are in a better position to know which community and agricultural assets are essential, token or even useful as trade-offs.

---

[116] City of Knoxville, TN, *2014 City of Knoxville Energy & Sustainability Work Plan* (2014) 3, accessed 14 October 2016 at http://www.knoxvilletn.gov/UserFiles/Servers/Server_109478/File/Sustainability/2014workplan.pdf.

[117] *Ibid.* ('Knoxville, like many cities, has neighborhoods in our urban core that have been designated as "food deserts" – places with few local grocers, where food prices are often high, and healthy food choices are narrow. Many of those neighborhoods also suffer blight from abandoned or neglected properties and face other symptoms of disinvestment. To address food deserts, the City of Knoxville has been working with community partners to connect land, farming jobs, processing facilities, food transit, sale, and composting to create a complete, sustainable urban food cycle that eliminates areas of food insecurity.').

[118] City of Knoxville, TN, *2014 City of Knoxville Energy & Sustainability Work Plan* (2014), note 116 above, 13.

6. CONCLUSION

In its recent Climate Change Adaptation Plan,[119] the EPA identified local governments and communities as key partners in responding to the threats of climate change. For instance in identifying water change as a critical vulnerability, the EPA's Climate Change Adaptation Plan focuses on understanding local decision makers as central to adaptation goals.[120] It is no longer novel to insist on local participation in formulating the climate change response, as 'local governments may be the most important players',[121] at least because 'many adaptation strategies will have to be intensely local in implementation'.[122]

Land use controls, and the local governments that wield them, are critical considerations in a climate change response. At no other level of governance are considerations of home and community identity more pervasive. And at no other level of governance do citizens feel more empowered. And yet reliance on land use controls for resiliency purposes is a complicated task. The pervasive and perennial problems that confront local governments as they turn their attention to the impacts of climate change involve financial ability to implement a vision of resiliency,

---

[119] EPA's planning initiative was in response to Executive Order 13653 – Preparing the United States for the Impacts of Climate Change, 78 Fed Reg No 215 (6 November 2013), accessed 14 October 2016 at http://www.gpo.gov/fdsys/pkg/FR-2013-11-06/pdf/2013-26785.pdf. The resulting Climate Change Adaptation Plan is supported by 17 Climate Change Adaptation Implementation Plans that address specific resource-based and region-based climate response needs.

[120] US EPA *Climate Change Adaptation Plan 24* (June 2014) http://epa.gov/climatechange/Downloads/EPA-climate-change-adaptation-plan.pdf. ('While there is relatively high confidence in our ability to project temperature increases due to climate change, projected changes in precipitation and its effects on hydrology at the local scale are less certain. Therefore, a key challenge will be how to help local decision makers understand potential local impacts, and how to make long-term plans under a new range of uncertainty about future hydrologic conditions. Water resource managers will also need to consider the local impacts of climate change as they grapple with other challenges – including population growth, land use changes, economic constraints, and a variety of stressors to the quality and quantity of our nations waters.').

[121] Patricia E Salkin, 'Sustainability and Land Use Planning: Greening State and Local Land Use Plans and Regulations to Address Climate Change Challenges and Preserve Resources for Future Generations' (2009) 34 Wm & Mary Envtl L & Pol'y Rev 121, 147.

[122] Robin Kundis Craig, 'Stationarity is Dead – Long Live Transformation: Five Principles for Climate Change Adaptation Law' (2010) 34 Harv Envtl L Rev 9, 20–1.

authority to act within a federalist system and community commitment on an issue that retains enough political controversy to call into question the police power justification for action.

This chapter identifies the traditional land use tools wielded by local governments and suggests some areas where land use control can be redirected or repurposed to serve the goals of resiliency. In coordination with commitments from and cooperation with other levels of governance, land use controls can address many of the vulnerabilities that agriculture imposes and faces in the climate change era.

9. Much ado about methane: intensive animal agriculture and greenhouse gas emissions

Michelle Nowlin and Emily Spiegel

1. INTRODUCTION

When people think about the drivers of global climate change, few identify agriculture as a leading source of greenhouse gases (GHGs). Much of the discussion is focused on culprits such as coal fired power plants and the nation's fleet of personal vehicles.[1] While these traditional subjects of the Environmental Protection Agency's (EPA's) regulatory attention under the Clean Air Act[2] are significant sources of GHG emissions and thus necessary to control, it would be folly to ignore agriculture's role, both as a driver of global climate change and as a key partner in the quest for solutions. It is also a critical segment of the world's economy, one that is most vulnerable in a changing climate. How we grow our food is ever more critical to human survival.[3]

According to the United Nations Food and Agriculture Organization (FAO), '. . . current country pledges to reduce [greenhouse gas] emissions will deliver no more than one third of what is needed by 2020 to avoid a 2° C rise in global temperature'.[4] This projected shortfall underscores the importance of examining every sector to quantify its emissions and identify mechanisms to reduce them. Agriculture is a significant source of emissions of two of the most potent GHGs: nitrous oxide (N_2O) and methane (CH_4), accounting for approximately 70 percent and 25 percent

---

1 See *Massachusetts v. EPA*, 549 US 497 (2007).

2 42 USC § 7401-7626.

3 Ridge Shinn, 'Comments from Sustainable Livestock Farmers on the 2015 Dietary Guidelines Advisory Committee Recommendations' (8 May 2015), accessed 14 October 2016 at https://www.organicconsumers.org/news/comments-sustainable-livestock-farmers-2015-dietary-guidelines-advisory-committee.

4 PJ Gerber, H Steinfeld, B Henderson, A Mottet, C Opio, J Dijkman, A Falcucci and G Tempio, *Tackling Climate Change Through Livestock – A Global Assessment of Emission and Mitigation Opportunities* (Food and Agriculture Organization of the United Nations 2013) 2.

of emissions, respectively, across all economic sectors in the United States.[5] Of agriculture's total GHG contribution, approximately 30 percent is attributed to the livestock sector.[6]

Despite the significant role agriculture plays in shaping the climate, it has thus far evaded regulation in the United States. This is unsurprising, as the industry enjoys considerable freedom from the nation's regulatory attentions, especially in the area of environmental protection.[7] However, livestock production in particular has been linked with numerous adverse impacts to natural resources and public health,[8] and many studies document production and waste management measures that are effective in reducing noxious odors and water pollution:[9] measures that could also substantially reduce the industry's GHG footprint. As we seek ways to lower GHG emissions and forestall the effects of global climate change, we must remove the protections long afforded the agricultural industry and adapt existing regulatory tools to address its contributions.

---

[5] Renee Johnson, *Climate Change: The Role of the US Agriculture Sector and Congressional Action* (Congressional Research Service July 2009) 4. It is important to note that these numbers likely underestimate agriculture's actual contribution: Estimates for GHG emissions from the agricultural sector exclude on-farm fuel usage, which is generally attributed to the energy sector, and food processing and distribution, which are attributed to the industrial and transportation sectors, respectively. Id. at 5; JW Erisman *et al.*, 'Agricultural Air Quality in Europe and the Future Perspectives' (2008) 42 Atmospheric Environment 14, 3209–17, accessed 14 October 2016 at http://ac.els-cdn.com/S1352231007003329/1-s2.0-S1352231007003329-main.pdf?_tid=4059fde2-da3d-11e4-8328-00000aab0f26&acdnat=1428091913_0dc32714e725243ef641f3f5b3c3b72e. More recent numbers suggest even greater contributions, with methane emissions from enteric fermentation and manure management accounting for 22.5% and 8.4%, respectively, of total US emissions from anthropogenic activities in 2014, and agricultural soil management activities accounting for 78.9% of US emissions of nitrous oxide in 2014. US EPA, Inventory of US GHG Emissions and Sinks: 1990-2014 (April 2016). https://www.epa.gov/sites/production/files/2016-04/documents/us-ghg-inventory-2016-main-text.pdf.

[6] Johnson, above note 5 at 5.

[7] See, e.g., Michelle Nowlin, 'Sustainable Production of Swine: Putting Lipstick on a Pig?' (Summer 2013) 37 Vt. L. Rev. 4, 1079; JB Ruhl, 'Farms, Their Environmental Harms, and Environmental Law' (May 2000) Ecol. L. Q.; Ad Hoc Committee on Air Emissions from Animal Feeding Operations *et al.*, *Air Emissions from Animal Feeding Operations: Current Knowledge, Future Needs* (National Academies Press 2003) 130.

[8] *Putting Meat on the Table: Industrial Farm Animal Production in America* (Pew Commission on Industrial Farm Animal Production in America 2008) 5.

[9] See, e.g., Nowlin, above note 7, at 1127–8 (documenting several technologies for odor and water pollution reduction).

This chapter will therefore examine the structure of the US livestock industry, research done to quantify its GHG emissions and explore mitigation measures and regulatory tools that are emerging to address and reduce its contributions. This chapter concentrates on industrial-scale dairy, beef cattle and swine operations, which current research identifies as more significant sources of GHG emissions than poultry operations (due to a range of inherent and deliberate differences in production, from digestive chemistry and rates of feed conversion to animal housing and manure management practices). In discussing mitigation opportunities, this chapter concentrates on manure management; although scientific studies estimate methane emissions from enteric digestion (that is, 'cow burps') exceed emissions from manure management, experts working in the field consider manure management strategies to be a more economically efficient source of reductions.

2. BACKGROUND

2.1 Brief History of American Livestock Systems

It is instructive to understand how the livestock industry evolved into the structure examined in this chapter. Domesticated hogs were introduced to North America in what is now Florida by Hernando de Soto in 1539.[10] Well into the 20th century, many farmers kept a few hogs as efficient converters of table scraps, harvest residues and acorns into meats that could be sold and stored.[11] These free ranging animals dropped their manure in the woods and on fallow pastures, and farmers slaughtered them in the fall, bringing in some extra cash and meat for the winter months.[12] As the country became more populated, farmers began to raise pigs on fenced pasture and, by the 1970s, confined and concentrated production was the

---

[10] Mick Vann, 'A History of Pigs in America' The Austin Chronicle (10 April 2009), accessed 14 October 2016 at http://www.austinchronicle.com/food/2009-04-10/764573/ (discussing the origins and history of pig farming in the US).

[11] See generally Mark Essig, *Lesser Beasts: A Snout-to-Tail History of the Humble Pig*, Chapter 8 (2005) (quoting Adam Smith's *The Wealth of Nations* and China's Chairman Mao in characterizing pigs as inexpensive to raise and beneficial to the soil); Daniel W Gade, *The Cambridge World History of Food,* Hogs II.G.13 (Kenneth F Kiple and Kreimhild C Ornelas (eds), 2000) (discussing the history of pigs in America).

[12] Gade, above note 11.

industry norm.[13] By the 1990s, '82 percent of swine in the United States were reared in some kind of confinement'.[14]

The shift in production practices is similar for dairy and beef cattle. The western hemisphere's first domesticated cattle arrived on Christopher Columbus's ships in 1493.[15] Early on, cattle were fed and used primarily on a local scale and, like hogs, relied on foraged nutrition.[16] A number of technological advancements and economic factors brought changes to traditional farming practices. One impetus came from the advent of refrigerated shipping and the attempt to sell to the British export market. In 1875, the first shipment of refrigerated beef (ten carcasses) was sent from the United States to England.[17] By the end of the next year, the leading exporter was sending England three million pounds of fresh beef per month.[18] The profitability of foreign trade led to changes in feeding practices. British tastes at that time preferred fatty, marbled beef, which could be better achieved through corn feeding than foraging on western US grazing lands.[19] When the United States Department of Agriculture (USDA) adopted a grading system that used fatty, British-style beef as the value standard, it reinforced the idea in American markets that corn fed beef was of higher quality and worth a higher price.[20] Thus, despite the abundance of native grasses for forage, American farmers shifted to a corn-based diet that made it necessary to bring the cattle into a confined area for feeding.

It wasn't until the 20th century, with the advent of practices such as pasteurization, corn feeding and commercialized cattle feedlots, that the

---

[13] See generally Essig, above note 11.

[14] Shawn Archibeque, Karen Haugen-Kozyra, Kristen Johnson, Ermias Kebread and Wendy Powers-Schilling, *Near Term Options for Reducing GHG Emissions from Livestock Systems in the US* (Nicholas Institute 2012) 18.

[15] McTavish *et al.*, 'New World Cattle Show Ancestry from Multiple Independent Domestication Events' (2013) 110 *Proceedings of the National Academy of Sciences of the United States of America* 15, 1398–1406, accessed 14 October 2016 at http://www.pnas.org/content/110/15/E1398.full.pdf.

[16] 'The American Dairy Industry' (USDA, 2015), accessed 14 October 2016 at http://www.nal.usda.gov/exhibits/speccoll/exhibits/show/the-american-dairy-industry/early-history; MacDonald and McBride, 'The Transformation of US Livestock Agriculture: Scale, Efficiency, and Risks' (USDA Economic Research Service, 2009), accessed 14 October 2016 at http://www.ers.usda.gov/webdocs/publications/eib43/13805_eib43c_1_.pdf; Corah, 'ASAS Centennial Paper: Development of a Corn-Based Beef Industry' (2008) 86 *American Society of Animal Science* 12,3635-9.

[17] Jeremy Rifkin, *Beyond Beef: The Rise and Fall of the Cattle Culture* (Plume, 1992) 88.

[18] *Ibid* at 88.

[19] *Ibid* at 93.

[20] *Ibid* at 96.

US dairy and beef industries grew to their current production capacity.[21] As of 1 January 2015, there were 29.7 million beef cattle and 9.3 million dairy cows in the United States.[22] Today's dairy and beef farms are fewer in number but larger in herd size as farms have become increasingly industrialized.[23] In addition, in 2007 approximately 82 percent of US dairy cows in the largest dairy states were part of 'conventional operations', characterized by confinement and lack of grazing access.[24] In parallel, most beef cattle are sent to feedlots where they are confined in pens before being shipped to slaughter.[25]

Technology and market forces thus encouraged more concentrated and centralized production of animals, with hog and cattle production concentrated in the Midwest, proximate to where the grain was grown. In the last several decades, falling transportation costs, changes in land values and consolidation of the corporations that dominate the livestock industry have further influenced production trends; although the Midwest supports the majority of livestock production in the United States, production is now concentrated in different geographic regions, depending on livestock species.

Along with these shifts in location, the size and number of livestock producers have changed significantly: farms are fewer in number but larger in size.[26] With concentrated production of animals comes concentrated production of manure, increased emissions of GHGs and increased challenges for managing the manure in ways that minimize negative impacts on neighboring properties and natural resources. The following section places these issues in context.

---

[21] 'Background of Dairy Production in the US' (EPA, 2012), accessed 14 October 2016 at http://www.ers.usda.gov/topics/animal-products/dairy/background.aspx; MacDonald and McBride, above note 16; Corah, above note 16.

[22] 'US Cattle Inventory Up 1 Percent from 2014' (USDA NASS, 2015), accessed 14 October 2016 at http://usda.mannlib.cornell.edu/usda/nass/Catt/2010s/2015/Catt-01-30-2015.pdf.

[23] 'Dairy: Overview' (USDA ERS, 2012), accessed 14 October 2016 at http://www.ers.usda.gov/topics/animal-products/dairy.aspx; MacDonald and McBride, above note 16.

[24] 'Dairy 2007, Part I: Reference of Dairy Cattle Health and Management Practices in the United States' (USDA, 2007), accessed 14 October 2016 at http://www.aphis.usda.gov/animal_health/nahms/dairy/downloads/dairy07/Dairy07_dr_PartI.pdf.

[25] MacDonald and McBride, above note 16.

[26] See MacDonald and Simon, *Cattle, Soyanization, and Climate Change: Brazil's Agricultural Revolution* (Brighter Green 2011) 1.

2.2 What is a CAFO?

This chapter focuses attention on the intensive and concentrated production of animals for food. The facilities that specialize in this method of production are sometimes referred to as intensive livestock operations, factory farms, industrial livestock facilities or concentrated animal feeding operations (CAFOs). CAFO is a term of art used in the federal Clean Water Act to describe a category of 'point sources', or discrete sources of water pollution that are subject to permitting requirements if they discharge pollutants (another statutory term that includes animal waste) to waters of the United States.[27] For ease of reference, the term 'CAFO' will be used herein, without regard to actual permitting status, to describe the class of livestock operations that raise the majority of the nation's food animals and contribute the majority of the industry's greenhouse gas emissions.

In the United States, most meat and food animals are raised in CAFOs.[28] According to regulations enacted by the EPA, a CAFO is a facility in which large numbers of livestock are raised in confined settings. The regulations specify that to be considered a CAFO, the animals must be confined for a cumulative total of 45 days during any 12-month period in an enclosure in which no grass or vegetation is sustained during the normal growing season.[29] In addition, the facility must house the equivalent of 1,000 animal units per year to qualify as a CAFO.[30] The regulations translate the animal unit requirement to the specific numbers of each animal species typically raised for food, such as swine, turkeys or dairy cattle. The

---

[27] 33 USC § 1362.

[28] Ad Hoc Committee on Air Emissions from Animal Feeding Operations, *et al.*, *Air Emissions from Animal Feeding Operations: Current Knowledge, Future Needs* (National Academies Press 2003) 28–9; Many types of food animals are raised in CAFOs in the United States, including hogs, poultry, sheep, cattle and cows. In other countries, other types of animals – including dogs and horses – are raised in CAFOs. See Hal Herzog, *Some We Love, Some We Hate, Some We Eat: Why It's So Hard To Think Straight About Animals* (Harper Perennial, 2011) 186 (discussing dog meat).

[29] 40 CFR § 122.23(b)(1).

[30] The EPA's regulations use the term 'animal unit' with reference to the specific number of different species of animals. See 40 CFR §122.23(b)(4)(i)–(xiii) (for 'animal unit' references). The actual number of animals varies depending on species (e.g., 2,500 hogs each weighing more than 55 pounds). For hogs, the number also varies depending on the animals' stage of growth. For example, for farrowing operations (nurseries), the regulatory limit for categorization as a 'large' CAFO is 10,000 hogs each weighing less than 55 pounds. *Id.* § 122.23(b)(4)(v). Facilities that employ similar production practices but raise fewer than 1,000 animal units are referred to as Animal Feeding Operations (AFOs).

legal definition refers only to the size of the operation and does not include the practices employed to manage, treat or dispose of the waste produced by the animals.

In more colloquial terms, a CAFO is a farm that produces livestock in a factory like setting. The 'barns' in which the animals are confined are long, low metal buildings that resemble warehouses. The inside configuration of the buildings and the methods of waste removal vary considerably depending on the type of animal produced.

In hog CAFOs, barns house 800–1,200 pigs each, depending on life stage.[31] Feeding and watering troughs are located throughout the buildings, ensuring the animals have a steady supply of food to accelerate their growth to full market weight. The hogs typically stand on floors constructed of concrete or other durable material, with wide slats through which the animals' excreta falls into a deep underground pit that is periodically flushed out with water.[32] In some collection systems, the manure is stored up to several months between cleanings; in 'flush' systems, pumped water is used to remove collected manure several times per day.[33] The collected waste is transferred via gravity or flush water into 'lagoons', open air retention ponds several acres in size that store millions of gallons of waste. The waste slurry stored in the lagoons contains high levels of nutrients, such as nitrogen, phosphorus, potassium, calcium, magnesium and boron, along with metals such as copper and zinc.[34] The nutrient rich liquid layer is siphoned off to irrigate agricultural fields surrounding the facility;[35] solids collect at the bottom of the lagoon and must be removed periodically and properly applied to land disposal sites.[36]

---

[31] Ad Hoc Commission, above note 7 at 37.

[32] US EPA, Agriculture 101: Common Manure Handling Systems, accessed 14 October 2016 at http://infohouse.p2ric.org/ref/02/01244/www.epa.gov/agriculture/ag101/dairymanure.html.

[33] Id.

[34] See, e.g., Ardeshir Adeli *et al.*, 'Effect of Long Term Swine Effluent Application on Selected Soil Properties' (2008) 173(3) *Soil Science* 223–35 (discussing the effect of swine effluent on soil quality); JP Zublena *et al.*, 'Soil Facts: Swine Manure as a Fertilizer Source' NC Extension Service, accessed 14 October 2016 at https://www.bae.ncsu.edu/extension/ext-publications/waste/animal/ag-439-4-swine-fertilizer.pdf last updated Dec 1997 (discussing the nutrient content of manure).

[35] See 'National Resource Conservation Service Nutrient Management' No. 590 (2012), accessed 14 October 2016 at http://www.nrcs.usda.gov/wps/portal/nrcs/detailfull/national/technical/?cid=nrcs143_026849 (discussing standards for nutrient management).

[36] *Wastewater Technology Fact Sheet: Anaerobic Lagoons* (EPA 2002) 2, accessed 14 October 2016 at http://www.bvsde.paho.org/bvsacd/leeds/lagoons.pdf.

Confinement practices in dairy CAFOs vary depending on the size and type of operation. Lactating cows are milked twice per day in a 'milking parlor' separate from the housing area.[37] When not being milked, cows may be housed in freestall barns, dry feedlots or on pasture. Manure from the cows is collected and stored in a variety of ways depending on the size of the operation. To reduce labor costs, large dairies frequently use automatic flushing systems similar to those used for hogs.[38] On drylots, the manure is collected and stored in uncovered open-air piles, where it is exposed to weather. Dry manure piles are sources of nitrous oxide, ammonia, hydrogen sulfide and particulate matter, whereas the manure ponds are sources of methane, ammonia, hydrogen sulfide and volatile organic compounds (VOCs).[39]

Other food animals, including sheep, goats and horses, are raised in a similar fashion. However, due to US market preferences, these facilities are much smaller and fewer in number. Consequently, this chapter focuses on the GHG contributions of the dairy, cattle and swine sectors.

2.3 Global Production Trends

Worldwide meat consumption is on the rise, with annual per capita consumption expected to hit 45 kg by 2030, up from an average of about 24 kg in the 1960s.[40] Per capita meat consumption in industrialized countries is disproportionately high (increasing approximately from 60 kg to 100 kg annually), but developing countries' consumption is growing rapidly (approximately from 10 kg to 37 kg over the same time period).[41] In China alone, meat consumption has quadrupled over the past three decades and will likely continue to rise.[42]

Steadily climbing worldwide demand for meat exerts constant pressure on livestock production to keep pace. The Intergovernmental Panel on Climate Change's (IPCC) *Fifth Assessment Report* documents

---

37 In smaller operations, milking may take place in the stalls.

38 *Development of Emissions Estimating Methodologies for Lagoons and Basins at Swine and Dairy Animal Feeding Operations* (Draft) (EPA Feb 2012) 2-2, accessed 14 October 2016 at https://www.epa.gov/afos-air/development-emissions-estimating-methodologies-lagoons-and-basins-swine-and-dairy-animal.

39 *Ibid* at 2-3.

40 World Health Organization, 'Availability and Changes in Consumption of Animal Products', accessed 14 October 2016 at www.who.int/nutrition/topics/3_foodconsumption/en/index4.html.

41 *Id.*

42 Mia MacDonald and Sangamithra Iyer, *Skillful Means: The Challenges of China's Encounter With Factory Farming* 1 (Brighter Green 2011).

huge worldwide growth in animal production between 1970 and 2010, and estimates that the global population increased by approximately one-third for cattle, buffalo, sheep and goats; two-fold for pigs; and almost four-fold for poultry, from 6 to 22 billion birds.[43] Both consumption and production are growing fastest in developing countries. Meat production in developing countries as a whole surpassed that in developed countries in the mid-1990s, and developing countries' share of global meat production is expected to reach about two-thirds by 2030.[44]

In many countries, small scale animal husbandry has traditionally played a social and economic role beyond the provision of commodity meat. As the president of the UN International Fund for Agricultural Development notes, '[f]or families with no access to financial services, owning livestock is also a form of asset accumulation and management'.[45] The livestock sector is particularly important for women, who comprise about 70 percent of its workers worldwide.[46] Livestock is a major source of income for poor women; more than ten times as many women own livestock as land.[47]

Despite the social importance of small scale husbandry, most livestock production in developing and middle income countries is shifting away from that model and toward the industrial model that already dominates the global North.[48] In China, for example, animal agriculture is following the US trend toward vertical integration, with large corporations owning not just the livestock production facilities, but also slaughterhouses and feed

---

[43] P Smith *et al.*, '2014: Agriculture, Forestry and Other Land Use (AFOLU)' 821, Figure 11.3 in O. Edenhofer *et al.* (eds), *Climate Change 2014: Mitigation of Climate Change, Contribution of Working Group III to the Fifth Assessment Report of the Intergovernmental Panel on Climate Change* (Cambridge University Press 2014).

[44] H Steinfeld, P Gerber, T Wassenaar, V Castel, M Rosales, and C de Haan, *Livestock's Long Shadow: Environmental Issues and Options* (Food and Agriculture Organization of the United Nations 2006) 34.

[45] Kanayo Nwanze, 'Livestock and Livelihoods, Remarks by IFAD President at ILRI's 40th anniversary' (6 November 2014).

[46] *Id.*

[47] *Id.*

[48] Partly as a result of the worldwide shift to industrial production, and partly because most of the sector's growth is projected to take place in less emission-intensive commodities (e.g., chicken instead of beef), global average emission intensity of livestock production is likely to decrease slightly even without concerted mitigation efforts. Gerber *et al.*, above note 4 at 100. However, the decrease in emissions intensity is unlikely to fully offset the growth in emissions from expanded overall production. *Id.*

companies.[49] According to a study conducted in 2005, China had 53,210 factory style farms producing pigs, dairy cows, beef cattle and sheep.[50] Hog farms, especially, are consolidating: from 1991 to 2009, the number of hog farms decreased by 70 percent while average farm size grew from under 1,000 head to over 8,000.[51] Similar trends are occurring in India, where poultry production is becoming more like that in the United States. 'About 90 percent of the more than two billion "broiler" (meat) chickens produced in India each year are raised in industrial style facilities', with the remaining 10 percent produced in more traditional, small backyard flocks.[52] Although dairy and beef production is not yet heavily industrialized, the number of animals produced is staggering, with approximately 200 million cows and 105 million buffalo.[53] In Brazil, production of poultry, pork, veal and eggs is also centered on the industrial model of confinement production that dominates the US livestock industry.[54] While cattle feedlots remain rare (most cattle are free ranging), 'much of the pasture they graze on has been created in areas of great biological diversity, specifically in the Amazon rainforest and the Cerrado—the Brazilian savannah—both of which are enormously important to the global climate'.[55]

The transition to intensified agriculture also changes the dominant emissions sources within the sector. FAO notes that '[i]n low productivity regions, enteric fermentation is the main emission source. In industrialized regions, feed production and processing, and manure together are as important a source of emissions as enteric fermentation'.[56] In addition to direct emissions, increasing production increases emissions from land conversion. In the United States, more crops must be grown to supply

---

[49] MacDonald and Iyer, above note 42 at 5, 8. In 2014, the Chinese company Shuanghui International Holdings Ltd. bought the American company Smithfield Foods Inc., the world's largest hog farming and pig processing company in a $4.7 billion acquisition that ensures the Chinese market's access to vertically integrated swine production systems in the US: Prudence Ho and Yvonne Lee, 'Smithfield Foods Owner Plans $5 Billion IPO' *The Wall Street Journal* (Hong Kong, 6 November 2013).

[50] MacDonald and Iyer, above note 42 at 5.

[51] Sam Brasch, 'How China Became the World's Largest Pork Producer' (Modern Farmer, 11 March 2014).

[52] Mia MacDonald and Sangamithra Iyer, *Veg or Non-Veg? India at the Crossroads* (Brighter Green 2012), 5, 14–15.

[53] *Ibid* at 5.

[54] MacDonald and Simon, above note 26, 21.

[55] *Ibid* at 1. Agriculture more generally is 'the main cause of conversion of natural ecosystems and loss of biodiversity'. *World Development Report 2008: Agriculture for Development* (World Bank 2008) 199.

[56] See Gerber *et al.*, above note 4 at 27.

industrial feedlots. In nations with less intensive systems, land often must be cleared for grazing, including land in rainforests or native cerrados that might otherwise serve as a carbon sink. Shifting production patterns mean that the relative importance of various emissions sources will vary by location and over time. Developing countries will likely present the greatest future potential for mitigating GHG emissions from the livestock sector because they raise the most livestock. But advances in technologies, management practices and legal tools to mitigate emissions in countries with established industrial farming systems will contribute to the worldwide response.

3. GHGS AND INTENSIVE LIVESTOCK PRODUCTION

Although CAFOs are notorious for their noxious and debilitating odors and dust,[57] relatively few studies have quantified their pollutant emissions, and fewer still have measured their GHG emissions,[58] although that is quickly changing. Existing studies point to significant emissions of methane (CH_4) and nitrous oxide (N_2O), both potent greenhouse gases with 20 and 300 times the global warming potential of carbon dioxide (CO_2), respectively.[59] In the United States, these emissions have been increasing since the 1990s, due mainly to changes in the livestock industry's structure and methods of production, which strongly influence the quantity and mix of emissions of greenhouse gases and other

---

[57] Susan S Schiffman *et al.*, 'The Effect of Environmental Odors Emanating From Commercial Swine Operations on the Mood of Nearby Residents' (1995) 37 Brain Res. Bull. 369; Gwangpyoko Ko, *et al.*, 'Investigation of Bioaerosols Released from Swine Farms Using Conventional and Alternative Waste Treatment and Management Technologies' (2008) 42 Envt'l Sci. Tech. 8849, 8852; Ad Hoc Commission, above note 7 at 55.

[58] See, e.g., *Development of Emissions Estimating Methodologies for Lagoons and Basins at Swine and Dairy Animal Feeding Operations* above note 38; see also also BJ DeAngelo, FC de la Chesnaye, RH Beach, A Sommer and BC Murray, 'Methane and Nitrous Oxide Mitigation in Agriculture' (2006) 27 The Energy Journal, 90.

[59] 'USDA Fact Sheet: Biogas Opportunities Roadmap: Voluntary Actions to Reduce Methane Emissions, Increase Energy Independence and Grow the Economy' (USDA 2014), accessed 14 October 2016 at http://www.usda.gov/wps/portal/usda/usdahome?contentidonly=true&contentid=2014/08/0166.xml; *Greenhouse Gas Emissions: Nitrous Oxide Emissions*;_*Greenhouse Gas Emissions: Methane Emissions* (EPA 2012), both accessed 14 October 2016 at https://www.epa.gov/ghgemissions/overview-greenhouse-gases.

pollutants.[60] Emissions have also increased in other parts of the world, including the Netherlands, India and China, as livestock production intensified during the past three decades.[61] Emission type and intensity vary across different species of livestock and their feed and manure management regimes. The leading sources of emissions from dairy, beef cattle, and swine production in the United States are enteric fermentation, liquid manure storage and treatment systems, and land application of manure.[62]

3.1 Enteric Fermentation

Methane producing microorganisms called methanogens live in the gastrointestinal tracts of animals, especially in the rumens of cud chewing animals.[63] They aid the animals' digestion in a process called enteric fermentation, which produces methane as a byproduct[64] that the animal typically belches out. Ruminant animals (that is, beef and dairy cattle, buffalo and sheep) are the major source of enteric emissions, but some non-ruminant animals, including pigs, also emit methane.[65] The beef industry is responsible for an estimated 72 percent of methane emissions

---

[60] See *Greenhouse Gas Mitigation Opportunities for Livestock Management in the United States* (Nicholas Institute for Environmental Policy Solutions 2012) 2; Johnson, above note 5 at 2; *US Agriculture and Forestry Greenhouse Gas Inventory: 1990–2008* (Climate Change Program Office, Office of the Chief Economist, US Department of Agriculture June 2011) Technical Bulletin No.1930. 159pp.

[61] JW Erisman *et al.*, 'Reduced Nitrogen in Ecology and the Environment' (2007) 150 *Environmental Pollution* 1, 144; Erisman *et al.*, above note 5, at 3209–10.

[62] Archibeque *et al.*, above note 14 at 24. See also U.S. EPA, 1990–2014 430-R-16-002 2-17, Figure 2-10, *Inventory of U.S. Greenhouse Gas Emissions and Sinks* (2016) accessed 5 January 2017 at https://www.epa.gov/sites/production/files/2016-04/documents/us-ghg-inventory-2016-main-text.pdf.

[63] *Quantifying Greenhouse Gas Fluxes in Agriculture and Forestry: Methods for Entity-Scale Inventory* (USDA Climate Change Program Office 2014) 2-19, accessed 14 October 2016 at http://www.usda.gov/oce/climate_change/Quantifying_GHG/USDATB1939_07072014.pdf.

[64] *Id.*

[65] *2006 IPCC Guidelines for National Greenhouse Gas Inventories* Vol 4 'AFOLU' (IPCC 2006) Annex 1A; see also Francis Thicke, *A New Vision for Iowa Food and Agriculture* (Mulberry Knoll, 2010) 42 (stating that in contrast to dairy and beef cattle operations, very little of the methane generated by hog operations is produced by the hogs' digestion of feed; nearly 90 percent of methane emissions from hog production is caused by the storage of manure in lagoons) (citing US EPA, 1990–2010 430-R-12-001 6-1, Inventory of US Greenhouse Gas Emissions and Sinks (2012), accessed 14 October 2016 at https://www.epa.gov/ghgemissions/us-greenhouse-gas-inventory-report-1990-2014).

from enteric fermentation[66] and is thought to have the greatest impact on climate change within the livestock sector.[67] Enteric emissions alone account for 25 percent of all anthropogenic sources of methane in the United States[68] and for 70 percent of annual methane emissions from all US agricultural production.[69]

3.2 Manure Storage and Management

Manure management is a significant challenge for livestock operations, especially CAFOs, which concentrate thousands of animals (and their manure) in one location.[70] Cattle and swine produce significant quantities of manure. For example:

> the average feedlot steer produces over 47 pounds of manure every twenty-four hours. On a standard feedlot containing 10,000 head, nearly 500,000 pounds of manure are produced daily. To place the problem in perspective, the organic

---

[66] John N Moore and Kale Van Bruggen, 'Agriculture's Fate Under Climate Change: Economic and Environmental Imperatives for Action' (2011) 86 Chi.-Kent L. Rev. 87, 94. Note that this does not mean that beef cattle produce enteric emissions at higher rates per head; the difference is attributable to national herd size. The US dairy herd numbers approximately 9 million milk cows (*Farm Milk Production* (USDA Economic Research Service 2012), accessed 14 October 2016 at http://www.ers.usda.gov/topics/animal-products/dairy/background.aspx (providing an estimate of the dairy herd in 2007)), whereas the nation's cattle inventory numbers 89.9 million (*Beef Industry Statistics* (National Cattlemen's Beef Association 2012), accessed 14 October 2016 at http://www.beefusa.org/beefindustrystatistics.aspx (providing an estimate of the cattle population in 2015). IPCC guidelines estimate enteric emissions per head in North America to be more than double for dairy cattle what they are for other cattle (128 and 53 kg CH_4/head/year, respectively). 'IPCC Emissions from Livestock and Manure Management' (2006) Table 10.11.

[67] D Donahue, 'Elephant in the Room: Livestock's Role in Climate and Environmental Change' (2008) 17 Mich. St. J. Int'l L. 95, 111.

[68] John N Moore and Kale Van Bruggen, 'Agriculture's Fate Under Climate Change: Economic and Environmental Imperatives for Action' (2011) 86 Chi.-Kent L. Rev. 87, 94; *Inventory of Greenhouse Gas Emissions and Sinks: 1990-2008* (US EPA April 2010).

[69] Archibeque *et al.*, above note 14 at 4.

[70] See *Putting Meat on the Table: Industrial Farm Animal Production in America* (Pew Commission on Industrial Farm Animal Production in America 2008) 6 (noting the trend toward larger, concentrated operations; the routine use of formulated feeds with low-level antibiotics; and that the 'waste management practices of [industrial food animal production] facilities can have substantial adverse [e]ffects on air, water, and soils').

> waste generated by a typical 10,000-head feedlot is equivalent to the human waste generated in a city of 110,000 people.[71]

The largest hog farms, housing up to 800,000 hogs, can produce over 1.6 million tons of waste annually, a figure 50 percent larger than the waste from a human city with 1.5 million people.[72]

Globally, methane and nitrous oxide emissions from livestock manure comprise the fourth largest source of greenhouse gases in the agricultural sector.[73] The majority of GHG emissions from manure storage and management at CAFOs consist of methane and nitrous oxide. Emissions of each gas depend primarily on the type of manure and the waste management system employed.[74] Most nitrous oxide is produced under aerobic conditions, such as in 'stirred' liquid manure systems or actively aerated compost operations; both techniques are designed to denitrify (release nitrogen from) the manure.[75] Methane is produced from the decomposition of manure under anaerobic conditions, such as liquid storage in a lagoon. When manure is stored in a lagoon, it begins to decompose; because there is insufficient oxygen for complete decomposition, the

---

[71] Rifkin, above note 17 at 221.

[72] 'Concentrated Animal Feeding Operations: EPA Needs More Information and a Clearly Defined Strategy to Protect Air and Water Quality from Pollutants of Concern' (GAO 2008) 5, accessed 14 October 2016 at http://www.gao.gov/new.items/d08944.pdf.

[73] The three largest sources of GHG emissions in agriculture are soil N_2O emissions from fertilizers, enteric emissions, and methane emissions from rice paddies. BJ DeAngelo, FC de la Chesnaye, RH Beach, A Sommer and BC Murray, 'Methane and Nitrous Oxide Mitigation in Agriculture' (2006) 27 *The Energy Journal* 89, 92.

[74] For example, one study measured GHG emissions from different manure handling systems used by dairies and feedlots. During a three-month period over the summer, scientists measured the emissions of CH_4, N_2O, and CO_2 from the manure slurry, manure stockpile, and compost piles. During this time, 'total GHG emissions for the beef manure were 51 g CO2e/kg dry matter (DM) for the compost, 76 g CO2e/kg DM for the stockpile, and 230 g CO2e/kg DM for the slurry. A similar response was observed for the dairy manure emissions: 207 g CO2e/kg DM for the compost, 301 g CO2e/kg DM for the stockpile, and 397 g CO2e/kg DM for the slurry.' Archibeque *et al.*, above note 14 at 17. The study's authors also calculated the potential GHG reductions if all the manure was composted rather than stockpiled or stored as slurry. Based on these calculations, they estimated that composting would achieve a 0.7 Tg CO2e/yr reduction. 'If all of the manure was handled as slurry and the CH_4 was collected and used for energy, a 1.08 Tg CO2e/yr reduction could be achieved.' *Id.*

[75] *2006 IPCC Guidelines for National Greenhouse Gas Inventories* Vol 4 'AFOLU' (IPCC 2006) Annex 1A; see also de Angelo *et al.*, above note 74 at 92; Archibeque *et al.*, above note 14 at 17.

system becomes anaerobic and produces more than 300 volatile organic compounds, including methane.[76] Lagoons are also responsible for *indirect* emissions of nitrous oxide. Lagoons release nitrogen into the atmosphere as ammonia, which later may be transformed into nitrous oxide.[77] The increased use of liquid manure management systems in dairy cattle and swine operations has contributed to an approximately 56 percent increase in those sectors' GHG emissions since 1990.[78] Ambient air temperatures also influence the quantity of GHG emissions from manure management systems, with higher temperatures resulting in greater emissions of methane and ammonia.[79] Storage time is also influential, as is relative humidity and pH.[80] These variables make it challenging to measure total emissions precisely.[81] Nevertheless, sufficient information is available to understand the impact different manure management systems have on emissions, to estimate those emissions and to identify important mitigation measures.

3.3 Land Application

Livestock manure has long been used to fertilize crops and forage. This process takes place naturally on grazing landscapes, and through direct application to crops and sprayfields in confinement-based livestock operations. However, over-application of manure is common in some areas and can result not only in water pollution through the run off of nutrient rich effluent, but also in increased GHG

---

[76] Thicke, above note 65, at 45; Roland W Melse and Maikel Timmerman, 'Sustainable Intensive Livestock Production Demands Manure and Exhaust Air Treatment Technologies' (2009), 100 Bioresources Tech. 5506, 5506; Ad Hoc Committee, above note 7 at 21.

[77] Gerber *et al.*, above note 4 at 20; Erisman *et al.*, 'Reduced Nitrogen in Ecology and the Environment' above note 61, at 141, 143.

[78] Nicholas Institute 2012 Policy Brief, 4.

[79] Archibeque *et al.*, above note 14 at 17, 29.

[80] *Ibid* at 18; see also Ad Hoc Committee on Air Emissions from Animal Feeding Operations, *et al.*, *Air Emissions from Animal Feeding Operations: Current Knowledge, Future Needs* (National Academies Press 2003) 111 (noting that relative humidity and manure pH affect manure emission rates for ammonia, a precursor to some GHGs).

[81] See, e.g., RR Sharpe and LA Harper, *'Methane Emissions from an Anaerobic Swine Lagoon'* (1999) 33 Atmospheric Env't 3627, 3632 (noting that research conducted by EPA and others, including their own, produces widely varying estimates); Erisman *et al.*, 'Reduced Nitrogen in Ecology and the Environment' above note 61, at 144 (noting that farm-level research shows a gap between nitrogen input and output, which can reach up to 40%).

emissions.[82] Soil microorganisms can use methane as a carbon source, but when concentrations overwhelm their metabolic capacity, the excess methane is released.[83] Nitrous oxide is emitted when high levels of organic carbon in manure cause soil microbes to increase their respiration rates, which depletes soil oxygen and creates the anaerobic conditions under which denitrification (and the release of N_2O) takes place.[84] Nitrous oxide is also produced when manure is deposited on the ground in a feedlot or on pasture and rangelands by grazing animals.[85]

As with manure storage, emissions from land application of manure are influenced by a number of variables including ambient air temperature, relative humidity and soil type, as well as crop type, timing of manure application relative to the crop's growing cycle, and tillage practices.[86] For example, studies estimate that applying swine manure in the fall results in methane emissions twice as high as those from springtime application.[87]

3.4 Effects of Climate Change on Livestock Production

Livestock production is not only a contributor to climate change, but also a sector deeply affected by it. Climate change influences livestock production through five main mechanisms: feed grain production, pasture and forage availability, heat tolerance of the animals themselves, pathogens and pests, and waste management systems.

One way animal agriculture will feel the effects of climate change is through its impacts on feed crop production. Climate change affects crop production through a complex set of interactions that make the ultimate effects on yield difficult to predict. Temperature, light, carbon dioxide and precipitation levels interact to determine plant germination, growth rate and grain qualities.[88] Nevertheless, some generalized

---

[82] Erisman *et al.*, 'Reduced Nitrogen in Ecology and the Environment' above note 61, at 141, 143, 146; Erisman *et al.*, above note 5, at 3211.

[83] F Montes *et al.*, 'Mitigation of Methane and Nitrous Oxide Emissions from Animal Operations: II. A Review of Manure Management Mitigation Options' (2013) 91 J. Anim. Sci. 5070, 5084.

[84] *Ibid* at 5084.

[85] Archibeque, above note 14 at 16 (internal citations omitted).

[86] See Johnson, above note 5 at 6.

[87] Rudek, above note 43 at 78.

[88] See, e.g., J Hatfield *et al.*, 'Ch. 6: Agriculture' in *Climate Change Impacts in the United States: The Third National Climate Assessment*' 150, 154 (see http://nca2014.globalchange.gov/report/sectors/agriculture_(US Global Change Research Program)) (noting interactions among carbon dioxide, temperature, solar radiation, and precipitation).

conclusions can be drawn. For example, CO_2 increases are associated with lowered nitrogen and protein content in some important feed crops, such as soybean and alfalfa.[89] The reduced nutritional content lowers the crops' ability to support livestock populations.[90] This may result in pressure to produce more feed grains to support existing levels of production.

Like feed grains, pastures will respond to climate change according to a complex set of interactions. In addition to the direct effects of temperature and CO_2 concentrations, pasture composition is influenced by factors such as plant competition, plant-animal interaction and water availability.[91] In much of North America, climate change will lengthen the growing season for cattle forage but reduce forage quality.[92] As with feed grains, higher concentrations of atmospheric carbon dioxide tend to reduce plant nitrogen and protein concentrations, requiring more acreage to supply animals with equivalent nutritional value.[93] Many weeds and invasive species have genetic traits that result in large growth responses to high CO_2 levels, so increased CO_2 may also lead to weedier, less productive pastures.[94]

An additional factor affecting livestock response to climate change is the heat tolerance of livestock animals themselves. Heat stress can affect animals' heart rate, respiration, blood chemistry, hormones, metabolism and digestion, as well as their behavior.[95] It thereby affects animal mortality, productivity and reproduction. Some of these changes are beginning to manifest. The 21st century has already seen increased livestock mortality from extreme heat events, including heat waves that killed over 5,000 head

---

89 See, *ibid* at 157.

90 *Ibid.*

91 JR Porter *et al.*, *Climate Change 2014: Impacts, Adaptation, and Vulnerability. Part A: Global and Sectoral Aspects. Contribution of Working Group II to the Fifth Assessment Report of the Intergovernmental Panel on Climate Change* '2014: Food Security and Food Production Systems' (Cambridge University Press 2014) 502. Soil pH also affects plant growth and pasture composition. See University of Minnesota Extension Service, *Breeding Herd Nutrition Home Study Course* 'Lesson Six: Pasture Management' (2015) 2, accessed 5 January 2017 at http://web.archive.org/web/20151002165058/http://www.extension.umn.edu/agriculture/beef/components/homestudy/nlesson6.pdf.

92 Porter, above note 91 at 508.

93 See Thomas R Karl, Jerry M Melillo, and Thomas C Peterson, *Global Climate Change Impacts in the United States* (Cambridge University Press 2009) 77.

94 *Ibid* at 78.

95 Nigel Key, Stacy Sneeringer and David Marquardt, *Climate Change, Heat Stress, and US Dairy Production* (USDA 2014) 3.

of cattle per event in some US states.[96] In 2011, heat related losses in US livestock production exceeded $1 billion.[97]

The heat stress from several consecutive hot nights can reduce production of meat, milk and eggs.[98] Heat stress increases the amount of energy an animal spends maintaining itself, thereby lowering the energy available for food production.[99] High temperatures also generally decrease animal feeding rates and, subsequently, growth rates.[100] Elevated temperatures in the spring and summer, when cattle are breeding, reduce their conception rates.[101] Swine and chickens are similarly impaired in high temperatures: swine have lower conception rates, and chickens lay smaller eggs with thinner shells.[102]

Exacerbating these concerns is the fact that in the drive for ever higher yielding animals, a lack of resilience to high temperatures is inadvertently bred into commodity livestock. Growers desire the highest yield possible from their animals to maximize profits, but '[a]s productivity increases . . . so metabolic heat production increases and the capacity to tolerate elevated temperatures decreases', meaning that '[o]ver the long term, single trait selection for productivity will tend to result in animals with lower heat tolerance'.[103] The biological processes by which animals turn feed into meat or milk also produce heat, so elevated external temperatures are particularly hard on the most productive animals. Growers can mitigate these effects, but providing additional shelter, water and electricity for cooling is likely to increase production costs.[104]

Adding to the production concerns above, the IPCC predicts with high confidence that climate change will expand geographic ranges for livestock

---

96 Karl *et al.*, above note 93 at 78.

97 See, e.g., Hatfield *et al.*, above note 88 at 157.

98 *Ibid* at 155. *See also* MacDonald and Iyer, above note 52, 4 (Possible temperature increases in India of between 2.3 to 4.8 degrees Celsius by 2050 will add to heat stress in animals used to produce milk and affect reproduction and the amounts of milk each animal provides. Crossbred cows may be most vulnerable to higher temperatures.)

99 Key *et al.*, above note 96 at 4.

100 Porter, above note 92 at 508 (The IPCC asserts 'high confidence' in this claim.)

101 Hatfield *et al.*, above note 88 at 157.

102 Key *et al.*, above note 95 at 3.

103 Porter, above note 91 at 502; See also Hatfield *et al.*, above note 88 at 157 ('Meat animals are managed for a high rate of weight gain (high metabolic rate), which increases their potential risk when exposed to high temperature conditions.')

104 See Key *et al.*, above note 95 at 4–5 (noting that several heat stress mitigation options are available to producers, but many of these methods are costly).

pathogens such as West Nile virus and schistosomiasis.[105] Disease distribution changes in the United States are also expected for afflictions that 'are sensitive to temperature and moisture, such as anthrax, blackleg, and hemorrhagic septicemia', and those that 'lead to increased incidence of ketosis, mastitis, and lameness in dairy cows'.[106] Livestock disease management will likely require a greater investment of resources both at the research level and on farm.

Climate change may also affect farms by exacerbating other environmental problems and increasing costs associated with addressing them. For example, '[i]ncreased storm intensity might be expected to increase the likelihood of upset of large open-air lagoon treatment systems [on swine farms], and increased severity has the potential to inundate the barns'.[107] Concentrated Animal Feeding Operation (CAFO) owners will likely need to increase expenditures on waste management systems to avoid spills from extreme weather events, or face increased liability for waste overflows.

Livestock production and climate change influence each other through a complex set of iterative causes and effects. Increasingly, livestock producers will find that production choices are shaped and constrained by the ways that animals' feed sources, pathogens, bodies and behaviors respond to increasingly warmer production conditions. The myriad effects that livestock production is already experiencing from climate change should be a compelling reason for that sector to commit to reducing its GHG emissions.

4. APPROACHES FOR REDUCING GHGS FROM CAFOS

Given the assortment of GHG emission sources throughout the livestock production process, there are multiple intervention points at which production changes may reduce emission levels. Some of these interventions are already well understood, and there is great potential to reduce emissions through broad scale adoption of current best practices. Keeping output constant, the livestock sector's 'emissions could be reduced by approximately 30 percent . . . if producers in a given system, region and

---

105 Porter, above note 91 at 512.

106 Hatfield *et al.*, above note 88 at 158.

107 J Rudek and V Aneja, 'Climate Vulnerabilities of the Swine Industry' (2013) 2 *Climate Vulnerability* 77, 77.

agro-ecological zone were to apply the practices of the 10 percent of producers with lowest emission intensity'.[108]

Further emissions reductions will require more fundamental changes. The best first steps will vary by region, animal and current production system.[109] Some of the measures to reduce GHG emissions include changing feed stocks, changing manure management practices and transitioning to different production systems.

4.1 Changing Feed Stocks

Animal feed selection influences GHG emissions in at least three ways. First, livestock feed production (including land conversion for that purpose)[110] and transportation have their own emission footprints.[111] Second, feed type and digestibility affect the level of enteric emissions from ruminant species. Both feed production and enteric emissions are significant emissions contributors, accounting for about 45 percent and 40 percent of livestock sector emissions, respectively.[112] Third, feed stock digestibility affects animal weight gain, altering overall production efficiency and emissions per unit of product.

For ruminant species, changing feed content has potential to mitigate enteric emissions by altering animals' gut chemistry.[113] Possible

---

[108] Gerber *et al.*, above note 4 at 45–6.

[109] For example, the FAO identifies reducing mortality rates as the best contributor to mitigation in South America, where a significant portion of the sector's emissions come from animals that never end up as meat. See Gerber *et al.*, above note 4 at 72.

[110] See, e.g., Daniel C Nepstad *et al.*, 'Responding to Climate Change and the Global Land Crisis: REDD+, Market Transformation and Low-Emissions Rural Development' 368 (22 April 2013) Phil. Trans. R. Soc. B; Douglas C Morton *et al.*, 'Cropland Expansion Changes Deforestation Dynamics in the Southern Brazilian Amazon' (26 Sept 2006) 103(39) Proc. Nat'l Acad. Sci. USA 14637–41.

[111] Emissions from producing feed stocks are typically attributed to crop agriculture rather than animal production, so a full discussion of feed stock production is outside the scope of this chapter, but a full consideration of feed stock production emissions should inform efforts to mitigate emissions through changes in animal feed.

[112] Gerber *et al.*, above note 4 at 17. The IPCC emissions categories break down livestock emissions into two main categories – enteric digestion and manure management – leaving feed production to a separate agriculture subsector. Hongmin Dong *et al.*, "Emissions from Livestock and Manure Management," *2006 IPCC Guidelines for National Greenhouse Gas Inventories*, Vol 4 'Agriculture, Forestry, and Other Land Use' (IPCC 2006) 10.7.

[113] For non-ruminant species, changing feed may have some potential to reduce emissions from manure, but there is less evidence in support of this

feed additives to reduce methane production include lipid supplements, pharmaceuticals to inhibit rumen bacterial growth and antimicrobial essential oils.[114] These supplements show varying degrees of success in reducing enteric emissions, but can have undesirable side effects such as reducing production efficiency and contributing to the development of antibiotic-resistant bacteria.[115]

FAO finds that regional differences in emission rates are largely attributable to low feed digestibility and subsequent low slaughter weights and high slaughter ages in some regions.[116] This combination of factors drives up the emissions per unit of product because the animal spends more time emitting throughout its longer life, only to produce less product in the form of consumable meats. Feed enhancements to improve feed conversion efficiency may be attractive emission mitigation options in those regions. However, transitioning from forage to more efficient crop feed stocks may contribute to increased emissions from feed production, so total emission effects may defy straightforward calculation. Additional study and modeling are needed to assess potential trade-offs.

4.2 Changing Production Systems

One factor contributing to overall emission rates in animal agriculture is the composition of the world's livestock population. Emission intensities vary significantly among animal species and among breeds. For example, chickens emit far less per unit of protein than cattle.[117] These differentials create an opportunity for emission reductions through transitioning from high-emitting to lower-emitting livestock. Some evidence suggests that this change is happening already; most growth in the worldwide livestock industry is occurring in low-emitting species.[118] This transition is mainly driven by market forces (lower emitters like chicken are often less expensive to raise than beef or dairy herds) rather than efforts to reduce GHG emissions within the livestock sector. These market forces will likely be

approach. Archibeque *et al.*, above note 14 at 59 (noting that although several feeding strategies held promise for reducing methane emissions from cattle, most feeding strategies for swine were largely ineffectual, with the exception of lowering the protein content in swine feed).

114 Archibeque, *et al.*, above note 14 at 12–14.

115 *Ibid.*

116 Gerber *et al.*, above note 4 at 26; see also at 20 ('Poorly digestible (i.e. fibrous) rations cause higher CH_4 emissions per unit of ingested energy.')

117 Smith, above note 43 at 839.

118 Gerber, above note 4 at 100.

sufficient to reduce emission intensities, but not to reduce total emissions in a sector whose growth outpaces its efficiency gains.[119]

Livestock GHG emission rates vary by breed as well as by species. The dominant breeds raised in concentrated production systems tend to be particularly high-emitting, a trend that continues as more countries move production toward consolidated systems. For example, in India, '[i]ndigenous cattle emit less methane than so called "high-yielding" breeds, but the latter are gaining in popularity among larger Indian dairy operations'.[120]

Selecting breedstock for traits associated with lower GHG emissions, rather than focusing single mindedly on output, could also reduce the sector's climate impact. However, that transition is unlikely to be simple or cheap. 'Changing livestock species . . . is a much more extreme, high-risk, and, in most cases, high-cost option than changing crop varieties.'[121] Employing this strategy would likely require significant further research, extension and incentive or support programs for growers.

Another production system intervention, most commonly considered with respect to cattle, is to change production models. Three archetypal models include the intensive feedlot (CAFO) approach that predominates in the United States, the extensive pasture based systems prevalent in South America, and management intensive grazing, which is a pasture based system of rotational grazing designed to improve pasture quality through carefully managed grazing and rest periods.[122] The relative merits of the various production systems for minimizing GHG emissions depend on myriad factors including enteric emissions from feed, manure management systems and soil carbon sequestration to offset emissions.

Changing from extensive pasture grazing to management intensive grazing is likely to improve soil carbon retention and production efficiency.[123] There is little evidence that such a change reduces enteric

---

119 *Ibid.*

120 MacDonald and Iyer, above note 52, 9.

121 Karl *et al.*, above note 93 at 78.

122 See M Eve *et al.*, *Quantifying Greenhouse Gas Fluxes in Agriculture and Forestry: Methods for Entity-Scale Inventory*. Technical Bulletin Number 1939 (Office of the Chief Economist, US Department of Agriculture, July 2014) 3-32, accessed 14 October 2016 at http://lter.kbs.msu.edu/docs/robertson/ogle+et+al.+2014+chapter+3+usda+ghg+report.pdf for distinctions among terms including 'rotational grazing' and 'management-intensive grazing'; here, 'management intensive grazing' is used broadly to describe grazing practices that move livestock among land units so that some areas of rangeland are grazed while others rest.

123 Archibeque, *et al.*, above note 14 at 20. But see Eve *et al.*, above note 122 at 3-32, suggesting that too few studies have been conducted on pasture soil organic carbon to evaluate how rotational grazing affects its retention.

emissions per animal, but it does increase animal productivity, which reduces emissions per unit of product.[124] Changing from extensive pasture grazing to intensive feedlot systems may reduce enteric emissions, but at significant expense and with further research required to address possible emissions leakage issues (for example, increased emissions from manure management).[125] Shifting production systems can be costly at the farm level, so benefits and tradeoffs in GHG emissions should be more fully researched. Policy interventions may be required if this strategy is pursued.

4.3 Changes to Manure Management

Manure management on CAFOs in the United States consists primarily of aggregating and storing the manure and periodically spreading it on fields to dispose of the nutrients. Despite some composting in dry stacks and some digestion of solids in lagoons, minimal processing occurs. As a consequence, much of the nutrient value in the manure is lost to the environment and transformed into pollutants.[126] Citizens, public health workers, environmental protection advocates and scientists who have documented the industry's impacts on water quality have thus been interested in technological advancements to improve manure management on CAFOs and thereby reduce impacts on communities, natural resources and climate change.

The technology that has received the most attention and support from policymakers in the US is anaerobic digestion.[127] With anaerobic digestion, manure is collected and heated, producing methane gas as a natural byproduct. In this way, it functions similarly to an anaerobic lagoon. But where a lagoon emits methane and ammonia into the ambient air, a digester collects those gases, then cleans and processes them as a biogas that can be used to power generators and provide a renewable source of energy.[128]

Other technologies may also be effective in reducing GHG emissions. Legal action against hog CAFOs in Missouri[129] and North Carolina[130]

---

124 Archibeque, *et al.*, above note 14 at 13.

125 *Ibid* at 20.

126 Matais Vanotti, *Development of Clean Technologies for Management of Wastes from Pig Production and their Environmental Benefits* (Engormix.com, 22 October 2012).

127 See discussion in Section 5.2 below.

128 See Ad Hoc Commission, above note 7 at 47.

129 *State ex rel Nixon v. Premium Standard Farms, Inc.*, Case No. CV99-0745, Consent Judgment 1 (Cir. Ct. MO, 2001).

130 Agreement Between the Attorney General of North Carolina and Smithfield

resulted in settlements that intensified research into improved waste management and accelerated the transfer of new technologies to working farms. These technologies, which were developed to improve air and water quality in affected communities, have the additional benefit of reducing GHG emissions from manure management practices.

In Missouri, an advisory panel approved several technologies: the Crystal Peak Fertilizer (CPF) Process, permeable lagoon covers with Advanced Nitrification/De-nitrification (AND), and a 'sustainable technology system' consisting of barn scrapers and an anaerobic digester.[131] The systems have been successful in reducing odor and dust, generating renewable energy to run the systems and offset the farms' electricity needs, and reducing greenhouse gas emissions.[132] In North Carolina, the state's Attorney General forged an agreement with Smithfield Foods, Inc. to research, develop and implement 'Environmentally Superior Technologies' that met five separate environmental performance criteria.

Several of the technologies included systems for the capture of methane, used to heat greenhouses or generate electricity. Although the Smithfield Agreement did not require measurements of GHGs, the developers of one technology (Super Soils) measured them and concluded that it substantially reduces emissions. Testing documented a 98.9 percent reduction in GHG emissions, which generated sufficient carbon credits to provide an economic return of 91 cents/hog.[133] Several technologies are ripe for adoption, and others are being researched, for reducing CAFO GHG emissions. But with cost and inertia impeding adoption of changes to feed stocks, to production systems and to manure management, legal intervention may be necessary to prompt changes in the industry.

Foods, Inc. (25 July 2000), accessed 14 October 2016 at http://infohouse.p2ric.org/ref/11/10597.pdf.

[131] *State ex rel. Koster v. Premium Standard Farms, Inc.* 02CV217957, Judgment Extending Consent Judgment (1 September 2010); see also Smithfield Foods 2010 Annual Report 13, accessed 14 October 2016 at http://www.smithfieldfoods.com/pdf/past-reports/smi_ar_10.pdf (referencing technology studies and financial considerations from the Missouri settlement with Premium Standard Farms).

[132] *Koster* Judgment, above note 131, at 1.

[133] Matias B. Vanotti *et al.*, 'Greenhouse Gas Emission Reductions and Carbon Credits from Implementation of Aerobic Manure Treatment Systems in Swine Farms' (2006) 5, 7, accessed 14 October 2016 at http://www.researchgate.net/publication/43271027_Greenhouse_Gas_Emission_Reductions_and_Carbon_Credits_from_Implementation_of_Aerobic_Manure_Treatment_Systems_in_Swine_Farms.

5. FEDERAL REGULATION PURSUANT TO THE CLEAN AIR ACT

Legal actions such as those described above illustrate the complexities of the relationship between consumers and the livestock industry. As our consumption of meat increases, so too does our concern about the impact of its production on natural resources and communities. It also reflects the failings of livestock policy: subsidies and exemptions have proliferated in recent decades as policymakers focus on increasing production, without taking into account the environmental impacts.[134] This section discusses the current state of regulatory programs to address GHG emissions from CAFOs.

5.1 Overview of the Clean Air Act

The federal Clean Air Act (CAA)[135] regulates the emission of air pollutants and delegates to the EPA the authority to develop and enforce a program of rules and permits designed to protect public health and welfare.[136] The CAA is a technically complex statute, and a comprehensive overview of the statute is far beyond the scope of this chapter. However, a basic overview is necessary to understand the ways in which emissions of air pollutants from CAFOs are, and may be, regulated and reduced.

The CAA defines 'air pollutant' broadly to include 'any air pollution agent or combination of such agents, including any physical, chemical, biological, radioactive . . . substance or matter which is emitted into or otherwise enters the ambient air. Such term includes any precursors to the formation of any air pollutant . . .'.[137] The EPA cannot possibly regulate all emissions of every substance into the air, however, and the CAA thus limits the EPA's authority over those pollutants the agency has determined 'may reasonably be anticipated to endanger public health or welfare'.[138] Once the EPA makes a finding of endangerment and includes the pollutant on a published list, the pollutant is then subject to the CAA's regulatory

---

134 Erisman *et al.*, above note 5, at 3209, 3213. See also Donahue, above note 67, at 102 (noting a lack of policy instruments to address the impacts of livestock production).

135 42 USC §§ 7401–7671q.

136 42 USC § 7401.

137 See 42 USC § 7602(g).

138 42 USC § 7408(a)(1). The CAA defines welfare to include impacts to soil, water, crops, weather and climate, among other resources. 42 USC § 7602(h).

provisions.[139] Since enactment, the EPA has established National Ambient Air Quality Standards (NAAQS) for six 'criteria' pollutants regulated under the CAA: sulfur dioxide (SO_2), nitrogen dioxide (NO_2), particulate matter of two sizes ($PM_{2.5}$ and PM_{10}), carbon monoxide (CO), ozone (O_3) and lead.[140] The EPA also regulates Volatile Organic Compounds (VOCs) and ammonia under this program, as they are considered to be precursors of ozone and $PM_{2.5}$, respectively.[141]

The EPA also has authority to regulate emissions of toxic air pollutants that are less pervasive than the criteria pollutants but have the potential for more serious impacts on human health. Although this authority was present in the 1970 CAA, it was not until Congress amended the statute in 1990 and substantially redesigned the air toxins program that the EPA began regulating Hazardous Air Pollutants (HAPs) in earnest. In amending the program, Congress listed 189 air pollutants as HAPs[142] and directed the EPA to periodically review and revise this list by rule, adding those pollutants that:

> . . . threaten adverse human health effects (including, but not limited to, substances which are known to be, or may reasonably be anticipated to be, carcinogenic, mutagenic, teratogenic, neurotoxic, which cause reproductive dysfunction, or which are acutely or chronically toxic) or adverse environmental effects whether through ambient concentrations, bioaccumulation, deposition, or otherwise . . .[143]

Congress also directed the EPA to develop technology-based standards for each category of industrial sources that emit HAPs. The technology-based standards must be designed to achieve the maximum degree of emission reductions possible for that source category, and are referred to as Maximum Achievable Control Technology (MACT) standards.[144]

To protect public health and welfare from the effects of air pollution, the CAA directs the EPA to develop, issue and enforce permits for

---

139 42 USC § 7408(a)(1).

140 42 USC § 7409(a), (b), (c); see generally 40 CFR pt 50. These pollutants are referred to as 'criteria pollutants' because EPA regulates them by developing health-based criteria for setting the maximum permissible concentrations in the ambient air for each pollutant. See 42 USC § 7408(a)(2); Envtl. Prot. Agency, *The Plain English Guide to the Clean Air Act*, accessed 14 October 2016 at https://www.epa.gov/clean-air-act-overview/plain-english-guide-clean-air-act.

141 42 USC § 7412(k)(2); EPA, *The Plain English Guide to the Clean Air Act*, above note 140.

142 42 USC § 7412(b)(1).

143 42 USC § 7412(b)(2).

144 42 USC § 7412(d)(2)

major stationary[145] and mobile sources[146] that emit air pollutants. It also gives the agency authority to initiate enforcement action and assess civil penalties against any facility that violates the terms of its permit.[147] States play an important role in implementing the CAA by developing State Implementation Plans (SIPs) to ensure that state air quality meets federal air quality standards.[148] A SIP must include permitting programs for stationary sources of air pollution.[149] In areas that are in compliance with the NAAQS, stationary sources must obtain a permit pursuant to the CAA's Prevention of Significant Degradation (PSD) program that imposes emissions limitations that reflect the Best Available Control Technology.[150] In addition, major sources are required to obtain an operating permit pursuant to Title V[151] of the Act, which consolidates all the facility's obligations, including compliance with emissions limitations for HAPs, inspection, monitoring and reporting requirements.[152]

5.2 Regulation of GHGs under the CAA

In compliance with the Supreme Court's order in *Massachusetts v. EPA*,[153] the EPA compiled and analyzed a large body of scientific evidence that resulted in its issuance of an 'Endangerment Finding'[154] for six greenhouse

[145] A major source is any stationary source with the potential to emit more than 100 tons per year of any air pollutant. 42 USC. § 7602(j). For hazardous air pollutants, a major source is any stationary source with the potential to emit more than 10 tons per year of any HAP, or 25 tons per year of any combination of HAPs. 42 USC § 7412(a)(1).

[146] Title II of the CAA, 42 USC §§7521-7590.

[147] 42 USC §§ 7413, 7477.

[148] See 42 USC § 7407(a) (requiring states submit an implementation plan specifying the manner in which national primary and secondary ambient air quality standards will be achieved and maintained within the state).

[149] 42 USC § 7410(a)(2)(C). A separate section of the statute sets different requirements for areas that violate NAAQS, termed 'non-attainment' areas. CAA Title 1, Part D, 42 USC §§ 7501–7515.

[150] 42 USC § 7475(a)(4). The PSD permits are sometimes referred to as pre-construction permits, because they are required prior to the start of construction of a new source of modification of an existing source. In non-attainment areas, stationary sources are subject to more stringent permitting requirements and must install more effective emission control technology. See 42 USC § 7503.

[151] 42 USC § 7661a.

[152] 42 USC § 7661c.

[153] 549 US 497 (2007).

[154] Endangerment and Cause or Contribute Findings for Greenhouse Gases Under Section 202(a) of the Clean Air Act, 74 Fed. Reg. 66, 496 (15 December 2009).

gases, including methane and nitrous oxide, that the agency determined contribute to climate change and thus endanger human health and welfare. Once it issued the Endangerment Finding, the EPA initiated rulemaking to determine how best to regulate GHG emissions.[155]

Because greenhouse gases are emitted in far greater volumes than other pollutants, the EPA estimated that millions of industrial, residential and commercial sources of GHG emissions would exceed the CAA regulatory emission threshold of 100/250 tons per year (tpy) CO_2e.[156] Accordingly, the EPA decided to limit the initial GHG permitting requirements to the largest sources by 'tailoring' the emissions threshold for GHG emissions to 75,000 or 100,000 tpy CO_2e[157] and phase in coverage of smaller sources over time.[158] The EPA also asserted that because of this initial high threshold, livestock operations would not likely be subject to regulation of GHGs under the CAA.[159]

In 2014, the US Supreme Court determined that the Tailoring Rule[160] exceeded the EPA's statutory authority under the CAA, and also ruled that the EPA lacked the authority to regulate GHG emissions separate from other pollutants.[161] In accordance with that ruling, the EPA may regulate GHGs only from sources that are already subject to the CAA's PSD permitting requirements.[162] Thus, to determine whether CAFOs are subject to regulation of their GHG emissions, we must examine the EPA's regulation of pollutant emissions from CAFOs under other provisions of the CAA.

In the absence of a coordinated federal program to regulate GHGs, prioritize mitigation measures and provide incentives for sources to reduce their emissions, states are creating their own programs. These programs include approaches as varied as carbon trading programs in

---

155 75 Fed. Reg. 17004 (2010) (the Triggering Rule).

156 75 Fed. Reg. 31514, 31534-36 (the Tailoring Rule).

157 75 Fed. Reg. 31514, at 31523.

158 See generally 75 Fed. Reg. 31514. EPA estimated that the largest facilities – power plants, refineries, and cement production facilities – are responsible for approximately 70% of stationary source emissions of GHGs in the United States. EPA Fact Sheet, Final Rule: Prevention of Significant Deterioration and Title V Greenhouse Gas Tailoring Rule, accessed 14 October 2016 at https://www.epa.gov/sites/production/files/2015-12/documents/20100413fs.pdf.

159 Claudia Copeland, 'Congressional Research Service, Air Quality Issues and Animal Agriculture: A Primer' (22 December 2014) 23, accessed 14 October 2016 at http://nationalaglawcenter.org/wp-content/uploads/assets/crs/RL32948.pdf.

160 The Tailoring Rule refers to EPA's decision to restrict GHG permitting requirements to the largest sources. See above note 157.

161 *United Air Resources Group v US EPA* (2014)134 S.Ct. 2427.

162 *Ibid* at XX.

California,[163] Oregon[164] and the Northeast;[165] renewable energy portfolio standards in North Carolina[166] and Pennsylvania;[167] and tax credits and other incentives for adopting GHG-reducing technologies in Oregon,[168] Washington[169] and Wisconsin.[170]

5.3 Regulation of Air Pollutants from CAFOs under the CAA

A number of compounds emitted by livestock facilities (such as particulate matter, VOCs and ammonia) can have serious adverse impacts on human health and welfare[171] and, as such, are pollutants regulated under the CAA. However, the EPA and state regulatory agencies have not regulated emissions from CAFOs, even though they are not exempt from compliance with the CAA's permitting and technology requirements.[172] CAFOs emit pollutants from many different structures (such as barns, lagoons, manure piles and land application sites) and there are few discrete points from which air samples and measurements can be taken.[173] Moreover, factors

---

[163] AB 32 directs the California Air Resources Board (CARB) to develop and implement a statewide program that would reduce the state's greenhouse gas emissions to 1990 levels by 2020, achieving a reduction of approximately 15% under a no-action scenario. See CARB, Assembly Bill 32 Overview (5 Aug 2014), accessed 14 October 2016 at http://www.arb.ca.gov/cc/ab32/ab32.htm. The program is funded by fees assessed to approximately 250 industrial sources of GHGs. *Id.*

[164] OAR 345-024-0680(1).

[165] Regional Greenhouse Gas Initiative Model Rule (December 2013), accessed 14 October 2016 at https://www.rggi.org/docs/ProgramReview/_FinalProgramReviewMaterials/Model_Rule_FINAL.pdf.

[166] NCGS § 62-133.8.

[167] 52 Pa. Code § 75.1

[168] OAR 330-090-0105.

[169] Wash. Rev. Code 82.16.110–82.16.130 (2014); Wash. Admin. Code § 458-20-273 (2015).

[170] Wis. Stat. § 77.54(56)(a) (2014).

[171] See e.g., '6 Comm. on Acute Exposure Guideline Levels, Nat'l Research Council, Acute Exposure Guideline Levels for Selected Airborne Chemicals' (2008) 59–60 (explaining that ammonia can cause severe irritation and burning of the eyes, skin, and respiratory tract, and in higher concentrations, even death); '9 Comm. on Acute Exposure Guideline Levels, Nat'l Research Council, Acute Exposure Guideline Levels for Selected Airborne Chemicals' (2010) 177–8 (describing how hydrogen sulfide can cause death in humans through respiratory failure, most times occurring in confined spaces, such as manure pits, animal processing facilities, and sewers).

[172] See Animal Feeding Operations Consent Agreement and Final Order, 70 Fed. Reg. 4958, 4958 (proposed 31 Jan 2005).

[173] Ad Hoc Commission, above note 7 at 44, 75.

such as meteorological conditions, topography, size of operation, type of animals raised and on-farm management practices affect the amount and rate of emissions of various compounds.[174] Consequently, emissions cannot be measured with precision and instead must be estimated.[175]

In the absence of sound protocols to reliably measure and estimate emissions from CAFOs, in 2005 the EPA entered into an agreement with the livestock industry to conduct a National Air Emissions Monitoring Study (NAEMS)[176] that would provide sufficient data to create those protocols. Specifically, the study would quantify pollutant emissions from dairy, pork, egg and broiler production facilities (open air feedlots were specifically excluded from the study[177]); provide reliable data for developing and validating emissions models for livestock and poultry production and for comparison with government regulatory thresholds; and promote a national consensus on methods and procedures for measuring emissions from livestock operations.[178] Thousands of CAFOs entered into the Air Compliance Consent Agreement (the Agreement) with the EPA: participating operations paid a civil penalty, ranging from $200 to $100,000 depending on size, and agreed to enroll in the EPA-designed air emissions monitoring program.[179]

The Agreement specified that the EPA would use the data to develop emissions-estimating methodologies (EEMs) that would estimate the daily and annual emissions of ammonia, hydrogen sulfide, particulate matter and VOCs from different types of livestock operations, and which could be used to clarify the operations' regulatory obligations pursuant to the CAA.[180] In exchange for their participation, the EPA agreed not to sue

---

[174] *Ibid* at 57, 169.

[175] *Ibid* at 98; Written Testimony of Dr Albert Heber, Professor, Agricultural and Biological Engineering Department at Purdue University, for the US House of Representatives Committee on Science and Technology's Hearing on 'Monitoring, Measurement and Verification of Greenhouse Gas Emissions II: The Role of Federal and Academic Research and Monitoring Programs' (22 April 2009) 10, accessed 14 October 2016 at http://www.hq.nasa.gov/legislative/hearings/2009%20hearings/4-22-09%20Heber.pdf.

[176] Animal Feeding Operations Consent Agreement and Final Order, 70 Fed. Reg. 4958, 4959 (proposed 31 Jan 2005) at 4958.

[177] *Ibid* at 4959.

[178] Claudia Copeland, 'Congressional Research Service, Air Quality Issues and Animal Agriculture: EPA's Air Compliance Agreement' (20 July 2012) 11.

[179] Animal Feeding Operations Consent Agreement, above note 172, 70 Fed. Reg. at 4959. Overall, the facilities and industry trade associations contributed approximately $14.6 million to fund the NAEMS.

[180] The Agreement also provided amnesty from EPA enforcement pursuant to the Comprehensive Environmental Response, Compensation and Liability

participating operations for past civil violations of the CAA, as well as violations of the reporting requirements for ammonia articulated in the Comprehensive Environmental Response, Compensation, and Liability Act (CERCLA) and the Emergency Planning and Community Right-to-Know Act (EPCRA),[181] pending completion of the study and the EPA's promulgation of the relevant emission factors.[182] According to the terms of the Agreement, the EPA's publication of emission factors would trigger a requirement for all participating entities to apply those factors to their own operations, and then comply with their obligations under the statutes.[183]

Although thousands of individual CAFOs agreed to participate, the NAEMS was carried out on only 25 sites and focused on emissions from confinement areas (barn sites) and manure management systems (open air sites, which also included naturally ventilated dairy corrals); land disposal sites were excluded. It took two years to develop the study protocol and quality assurance plan, and to set up the monitoring equipment at participating sites. Between 80 and 300 variables were measured at each site; it was estimated that more than 2.4 billion data points would be generated and evaluated.[184] The EPA published its initial findings in 2011, five years after the Agreement took effect.[185]

Act (CERCLA), 42 USC §§ 9601 et seq., and the Emergency Planning and Community Right-to-Know Act (EPCRA), 42 USC §§ 11001 et seq. Animal Feeding Operations Consent Agreement, above note 172, 70 Fed. Reg. at 4961.

[181] CERCLA and EPCRA require reporting of ammonia and hydrogen sulfide emissions in excess of 100 pounds per day.

[182] Animal Feeding Operations Consent Agreement, 70 Fed. Reg. at 4961. The Agreement did not limit EPA's ability to take action in the event of an imminent and substantial danger to public health or the environment, nor did it limit the ability of state and local governments to enforce their own odor and nuisance laws. *Ibid.*

[183] Community-based and environmental organizations challenged the Consent Agreement as exceeding the EPA's enforcement authority and violating the Administrative Procedure Act's notice and comment rulemaking requirements, but the court rejected the challenge. See *Ass'n of Irritated Residents v. US EPA*, 494 F.3d 1027, 1028 (D.C. Cir. 2007).

[184] Written Testimony of Dr Albert Heber, Professor, Agricultural and Biological Engineering Department at Purdue University, for the US House of Representatives Committee on Science and Technology's Hearing on 'Monitoring, Measurement and Verification of Greenhouse Gas Emissions II: The Role of Federal and Academic Research and Monitoring Programs' (22 April 2009) 9–10, accessed 14 October 2016 at http://www.hq.nasa.gov/legislative/hearings/2009%20hearings/4-22-09%20Heber.pdf.

[185] *Office of Air Quality planning and Standards, Draft Development of Emissions Estimating Methodologies for Lagoons and Basins at Swine and Dairy Animal Feeding Operations* (EPA 2012) vii https://yosemite.epa.gov/sab/sabproduct.

Despite the time and effort that went into the NAEMS and the amount of data collected, it fell far short of its touted goals. The EPA was able to develop only three alternative EEMs for ammonia emitted from swine and dairy housing structures and manure storage units, and to date has not finalized an EEM for ammonia emitted from lagoons.[186] Moreover, due to limited data collection and availability, the agency has not developed EEMs for VOCs or particulate matter and is still in the process of developing EEMs for hydrogen sulfide.[187] At this time, it is uncertain what the EPA's next steps will be, and the agency has yet to set any requirements to restrict, or even monitor, air emissions of any pollutants from CAFOs.[188] The agency's own Science Advisory Board roundly criticized its efforts, asserting that the study's small sample size was insufficient to develop a process-based model to predict emissions across the industry.[189]

Although the NAEMS was not designed to measure emissions of GHGs, researchers did collect data on carbon dioxide emissions at the 15 barn sites that participated in the study, as well as methane emissions at five sites. They also collected data on methane emissions at ten of the open-air monitoring sites, representing four dairy farms and six swine farms. Researchers measured nitrous oxide at one swine operation and one dairy operation.[190] In addition, the dairy industry took advantage of the infrastructure created for the NAEMS and 'funded a project to add all three major GHG to all the dairy sites for the last few months' of the study; it also paid to extend research on three of the barn sites to obtain

---

nsf/0/AE6639DD6B79360E852579A4004E5529/$File/PDF+for+Development+of+Emissions+Estimating+Methodologies+for+Lagoons+and+Basins+at+Swine+and+Dairy+Animal+Feeding+Operation.pdf.

[186] Draft Development of Emissions, above note 185 at 5.7.

[187] Draft Development of Emissions, above note 185 at viii.

[188] See *Environmental Integrity Project, et al. v. US EPA*, Complaint for Declaratory and Injunctive Relief, ¶ 50, 1:15-cv-00139, US Dist. Ct. D.C. (filed 28 January 2015) (challenging EPA's failure to respond to Plaintiffs' Petition, filed in 2011, asking the EPA to take action under the CAA and issue a finding that ammonia gas pollution endangers public health and welfare, designate ammonia as a 'criteria pollutant' under CAA § 108, and establish NAAQS for ammonia under CAA § 109).

[189] Letter from Dr David T Allen, Science Advisory Board, to Bob Perciasepe, US EPA, re: SAB Review of Emissions-Estimating Methodologies for Broiler Animal Feeding Operations and for Lagoons and Basins at Swine and Dairy Animal Feeding Operations (19 April 2013), accessed 14 October 2016 at http://yosemite.epa.gov/sab/sabproduct.nsf/0/08A7FD5F8BD5D2FE85257B52004234FE/$File/EPA-SAB-13-003-unsigned+.pdf.

[190] Written Testimony of Dr Albert Heber, above note 184.

additional GHG emissions data.[191] Not only did the NAEMS result in the generation of data on GHG emissions from livestock operations, it also provided insight into the infrastructure that would be needed to conduct a robust study, the challenges of conducting such a study, and the costs involved. But at present, NAEMS's failure to lead to any regulatory restrictions on air pollutants from CAFOs means that GHGs from CAFOs also remain unregulated under the Clean Air Act.

5.4 Reporting CAFO GHG Emissions under the CAA

Although CAFOs are not subject to GHG regulation, they may still be required to report their emissions. Tracking this information is important because livestock manure management systems alone 'account for 7.5 percent of total anthropogenic [methane] emissions and 4.7 percent of [nitrous oxide] emissions in the US',[192] and estimates of total emissions from manure management have increased steadily since 1990.[193] Methane emissions alone increased by 65 percent during this time, due primarily to substantial increases in production of dairy cattle and swine and increased use of liquid waste management systems.[194]

Due to these reports, the EPA included CAFOs that emit more than 25,000 metric tons of carbon dioxide equivalent per year in its 2009 rulemaking to quantify national GHG emissions from major sources in the United States.[195] The Mandatory Greenhouse Gas Reporting Rule does not purport to regulate the emissions of GHGs, but rather is a reporting mechanism by which the EPA can gather data on large emission sources and determine whether future regulation is necessary. As with the NAEMS, the reporting rule excludes land disposal of manure and applies only to confinement houses and manure management systems, including anaerobic lagoons, deep bedding systems for cattle and swine, poultry houses of all types and composting operations, as well as anaerobic digesters promoted

---

191 *Ibid.*

192 'Mandatory Reporting of Greenhouse Gases' 74 Fed. Reg. 56260, 56339 (Oct. 30, 2009) (codified as 40 CFR pts. 86–7, 89–90, 94, 98, 1033, 1039, 1042, 1045, 1048, 1051, 1054, 1065). In 2007, methane emissions from manure management systems totaled 44 million metric tons CO2e, and N_2O emissions were 14.7 million metric tons CO2e. *Ibid.*

193 *Inventory of US Greenhouse Gas Emissions and Sinks: 1990–2014* (EPA April 2016) 5-2, accessed 14 October 2016 at https://www.epa.gov/ghgemissions/us-greenhouse-gas-inventory-report-1990-2014.

194 *Ibid* at 5-9.

195 US EPA, Final Rule: Mandatory Reporting of Greenhouse Gases, 74 Federal Register 56260- 56519 (30 October 2009).

for their ability to capture methane and reduce GHG emissions.[196] It also excludes enteric fermentation: the largest source of methane on beef, dairy and sheep feedlots. Owners or operators of CAFOs covered by the rule are required to collect emission data, calculate GHG emissions, and follow specified procedures for quality assurance, recordkeeping and reporting.[197]

The rule's threshold emission level to trigger reporting is 25,000 tpy CO_2e.[198] To reduce the burden of measuring actual emissions to determine whether an individual livestock operation needed to report under the rule, the EPA included a table of population thresholds for beef, dairy, swine and poultry operations that are estimated to emit less than 25,000 tpy CO_2e of emissions.[199] For example, CAFOs that raise fewer than 34,100 swine, 29,300 beef cattle or 3,200 dairy cattle would not be required to report.[200] Across the industry, the EPA estimated that the reporting rule threshold would affect only 100–110 of the largest CAFOs in the country, approximately one percent of the total number of operations.[201]

A key objective in promulgating the reporting rule was to allow businesses to track their emissions and compare them with other industry participants to accelerate data sharing and the identification of cost effective ways of reducing emissions. Facilities covered by the rule were required to begin reporting their emissions starting in 2010, with the first reports due in 2011.[202]

However, before this requirement took effect, Congress took action to protect the livestock industry from participating: it inserted a provision into the EPA's budget appropriations prohibiting the agency from expending any funds to implement the GHG Reporting Rule with regard to CAFOs.[203] Congress also inserted a separate amendment prohibiting the agency from promulgating regulations that would subject CAFOs to permitting for GHG emissions under Title V of the Clean Air Act.[204] Congress has continued to include these prohibitions in subsequent

---

196 40 CFR 98.362.

197 See generally 40 CFR Subpart JJ §§ 98.360–98.367 (2015).

198 40 CFR § 98.360(a) (2015).

199 Mandatory Reporting of Greenhouse Gases, 74 Fed. Reg. at 56337JJ; 40 C.F.R. Subpart JJ § 98.360(a)(1), Table JJ-1 (2015).

200 Mandatory Reporting of Greenhouse Gases, 74 Fed. Reg. at 56,485; 40 CFR PT. 98, SUBPT. JJ, TBL. JJ-1.

201 *Guide for the Agriculture and Livestock Sectors* (EPA Sept 2009) 2.

202 *Greenhouse Gas Reporting Program Basic Information* (EPA 5 February 2013), accessed 14 October 2016 at http://www.epa.gov/ghgreporting/basic-info/index.html.

203 PL 111-88.

204 *Ibid.*

appropriations bills for the EPA, including its most recent appropriations for FY 2015.[205] The EPA has advised the livestock industry that the restriction on reporting GHG emissions applies only to the EPA's expenditure of funds and does not alter any requirements in the regulations themselves,[206] but the agency has also issued a blanket statement that the GHG reporting program does not include emissions from agriculture.[207] As of this writing, the EPA's online reporting database does not include CAFOs,[208] so it is not possible to determine whether any CAFOs are reporting their emissions voluntarily.

6. VOLUNTARY AND INCENTIVE BASED GOVERNMENT PROGRAMS

There is currently no regulatory mechanism for reporting or controlling GHG emissions from CAFOs. There are, however, a number of government-run, incentive-based programs that may be used to reduce or mitigate emissions. Improvements in animal feed and manure management can reduce GHG emissions, and such efforts may be supported by 'existing voluntary federal and state agricultural programs that provide cost sharing and technical assistance to farmers, predominantly for other production or environmental purposes'.[209] At the federal level, these programs are administered by the US Department of Agriculture (USDA) and funded through the Farm Bill.

---

205 PL 113-235.

206 See frequently asked question 282: 'Can someone tell me if livestock operations need to report this year or not? I have heard from EPA personnel, Cattlemen industry people and state officials, but continue to get conflicting information.' Greenhouse Gas Reporting Program's Frequently Asked Questions, EPA, accessed 14 October 2016 at http://www.ccdsupport.com/confluence/pages/viewpage.action?pageId=91980293 (indicating that the restriction in subpart JJ of Part 98 applies to EPA's expenditure of funds and does not alter the requirements under part 98).

207 See frequently asked question 406: 'What is the scope of this data' Greenhouse Gas Reporting Program's Frequently Asked Questions, EPA, accessed 14 October 2016 at http://www.ccdsupport.com/confluence/pages/viewpage.action?pageId=141983782.

208 See http://www.epa.gov/ghgreporting/basic-info/index.html (accessed 14 October 2016).

209 Johnson, above note 5 (summary).

6.1 USDA Farm Bill Programs

The USDA has a wide range of responsibilities that include implementing food assistance programs to aid low income families, managing national forests, regulating the safety of meat and poultry products, and conserving natural resources.[210] Many of the agency's programmatic responsibilities are set forth in the Farm Bill, an omnibus law containing policies and funding for food, nutrition and agricultural programs that the USDA administers. The Farm Bill now includes an energy title and provisions dedicated to livestock.[211] The total funding for these two programs – a mere one percent of the Farm Bill – belies the significance of their impact on livestock production and the sector's GHG emissions.

The key USDA program for livestock is the Environmental Quality Incentives Program (EQIP), which dedicates at least 60 percent of its budget to conservation practices for livestock producers.[212] Through EQIP, NRCS may enter into contracts, up to a maximum length of ten years, with producers who agree to implement eligible environmental and conservation practices in exchange for cost share and incentive payments, as well as provide technical assistance to producers for designing and implementing the specified conservation practices.[213]

EQIP was originally created to assist small, independent producers but its funding and mission have grown in the 19 years since its creation. In 2002, Congress expanded the program to allow large scale livestock production and processing companies to participate.[214] Correspondingly, Congress significantly increased the program's funding, most recently

---

210 Ad Hoc Committee, above note 7, at 146. Several USDA divisions provide significant support to livestock producers, such as conducting research on animal genetics and pasture management, administering outreach and training programs, and providing technical and financial assistance to help livestock producers reduce or mitigate the environmental impacts of livestock production. These divisions include the Natural Resources Conservation Service (NRCS), the Farm Service Agency (FSA), the Agricultural Research Service (ARS) and the National Institute for Food and Agriculture (NIFA) (formerly the Cooperative State Research, Education, and Extension Service). *Ibid.*

211 Although the 2008 Farm Bill had a title dedicated to Livestock, this Title was eliminated in the 2014 Farm Bill. Compare PL 110-246 with PL 113-79.

212 HR 2642-82, Sec. 2203 (f). EQIP is funded through the Conservation Title of the Farm Bill.

213 Ad Hoc Commission, above note 7 at 148 (citing 16 USC §§ 3839aa(2)(B)–(C) (2006)).

214 *The 2002 Farm Bill: Provisions and Economic Implications* (USDA Economic Research Service 2002) 87, accessed 6 January 2017 at https://www.ers.usda.gov/webdocs/publications/ap022/13780_ap022_4_.pdf. See also Farm Security and

authorizing $8 billion for fiscal years 2014–2018.[215] Through this funding, the program has the potential to significantly influence waste management practices.

NRCS maintains a comprehensive list of conservation practice standards (CPSs) designed to minimize adverse impacts to natural resources.[216] Many CPSs address waste management challenges presented by the geographically concentrated production of livestock in CAFOs. For example, there are CPSs for managing both routine and disaster related animal mortalities, constructing and operating waste lagoons and composting facilities for manure treatment, and proper closure of waste management structures.[217] Because NRCS authorizes the manure management systems that are eligible for cost share assistance, it has subsidized the costs of building the lagoons and irrigation lines for sprayfields that account for much of the industry's GHG emissions.[218] With the adoption of a CPS for anaerobic digesters,[219] the agency is now in a position to subsidize the transition to waste management systems that will reduce both GHG emissions and the industry's reliance on fossil fuels for electricity and fuel. Initial capital costs for anaerobic digesters range from $500,000 to $1 million,[220] which can be offset by payments through EQIP of up to $450,000 per entity.[221] To reduce emissions further, NRCS could also work with the EPA to develop CPSs for other manure treatment options.

Rural Investment Act of 2002, Pub. L. No.107-171, 116 Stat. 134, 253, Sec. 2301 (May 13, 2002).

215 Megan Stubbs, 'Congressional Research Service, Conservation Provisions in the 2014 Farm Bill' 8 (PL 113-79) (24 April 2014).

216 See 'USDA Nat. Res. Conservation Serv., Nat'l Conservation Practice Standards' (USDA), accessed 14 October 2016 at http://www.nrcs.usda.gov/wps/portal/nrcs/detailfull/national/technical/cp/ncps/?cid=nrcs143_026849.

217 See *ibid.*, 'Nat'l Conservation Practice Standard' Nos. 316, 313, 317 and 360, respectively.

218 *Ibid.* See also Andrew Martin, 'In the Farm Bill, a Creature from the Black Lagoon' *New York Times* (13 Jan 2008), accessed 14 October 2016 at http://www.nytimes.com/2008/01/13/business/13feed.html.

219 See 'USDA Nat. Res. Conservation Serv., Nat'l Conservation Practice Standards' No. 366, above note 216.

220 '[T]he initial capital cost of an anaerobic digester with energy recovery is between $0.5 million and $1 million at a large-sized dairy operation, and annual operating costs are about $36,000. Initial capital costs for a digester at a larger hog operation is about $250,000, with similar operating costs.' Johnson, above note 5, 7.

221 HS 2642-82, 83, Sec. 1240G (describing payments available in fiscal years 2014–2018). An increased payment rate and advanced payment are available for socially disadvantaged and beginning and limited resource farmers, including veterans. USDA, EQIP Fact Sheet (2014), accessed 14 October 2016

Another relevant part of the Farm Bill is the Energy Title, which was added to the Farm Bill in 2002.[222] It encourages farmers to invest in energy efficiency and clean energy production, measures that reduce GHG emissions.[223] In the past two versions of the Farm Bill, Congress has created programs and directed funding to increase the use of manure as a source of energy. In the 2008 Farm Bill,[224] for example, Congress authorized several programs to aid the development of energy facilities that use biomass, defined to include animal waste and animal waste byproducts (such as fats and oils).[225] It also mandated a study that has accelerated interest and, correspondingly, funding for additional research and deployment. The study report, *Manure Use for Fertilizer and for Energy*,[226] concludes that manure-to-energy projects are a 'win-win' solution to both manure management and energy crises, especially in geographic areas where there is excess manure production.[227] By using excess manure to produce energy,

---

at http://www.nrcs.usda.gov/wps/portal/nrcs/detail/national/programs/financial/eqip/?cid=stelprdb1242633.

[222] Farm Security and Rural Investment Act of 2002, Pub L No 107-171.

[223] *Mitigating Global Warming Through the Farm Bill: Measuring the Potential Greenhouse Gas Savings of the Farm Bill's Energy Title Programs* (Midwest Environmental Law and Policy Center 2007) 3, accessed 14 October 2016 at http://farmenergy.org/wp-content/uploads/2009/05/elpc_ghg_fbenergy_paper.pdf ('Through grants and loan guarantees, together with biofuel production incentives, these programs help to fund the construction of wind energy, anaerobic digesters, biofuels, energy efficiency and other projects.').This report documents that through the Renewable Energy Systems and Energy Efficiency Investments Program, USDA has issued 'over 1100 grant and loan guarantee awards for renewable energy and energy efficiency projects worth over $1.5 billion in total investments.' This program, together with the Biomass Research and Development program (Section 9008 of the 2002 Farm Bill), 'help to accelerate biomass use and production which ultimately should save millions of metric tons of avoided CO2 emissions'. *Ibid* at 4.

[224] Food, Conservation, and Energy Act of 2008, PL 110-246, 122 Stat 1651.

[225] Food, Conservation, and Energy Act of 2008, PL 110-246, 122 Stat 1651. Sec. 9001(12)(B)(ii)(lll).

[226] James MacDonald, *et al., Manure Use for Fertilizer and for Energy / Report to Congress* (USDA Economic Research Service June 2009), accessed 14 October 2016 at https://naldc.nal.usda.gov/download/46168/PDF.

[227] See also John N Moore and Kale Van Bruggen, 'Agriculture's Fate Under Climate Change: Economic and Environmental Imperative for Action' (2011) 86 Chi.-Kent L. Rev. 87, 101 ('Effective manure and fertilizer management can reduce methane and nitrous oxide emissions, two greenhouse gases that are even more damaging to the environment than carbon dioxide. Legislation that employs these practices can create economic opportunities for agriculture and avoid the damaging long-term costs of climate change to agricultural production, thus creating a "win-win" future for farmers and the environment.')

farmers realize benefits not only from avoided purchases of electricity or direct sales of electricity, but also from reduced costs associated with transporting and using the combustion residuals for fertilizer. The study concludes that manure-to-energy projects will create societal benefits as well, by replacing newly mined fossil fuels in energy production, replacing synthetic fertilizers[228] and capturing methane emissions. The study notes that such benefits have already led to proposals 'to support manure-to-energy projects through State utility mandates (to purchase electricity from farms and to invest in renewable production sites), through subsidies for capital costs, and through direct subsidies and credits for energy production'.[229] The study concludes that with increased federal support there could be 'substantial growth in energy applications for manure'.[230]

The 2014 Farm Bill (the Agricultural Act)[231] aims to provide that support, authorizing substantial additional funding in the Energy Title to expand and accelerate the production and use of bio-based energy, including biofuels, allocating both mandatory and discretionary funds for research, development and deployment of technologies. The initiatives include up to $575 million in mandatory and discretionary funding to develop and construct biorefineries through 2018,[232] and up to $62 million to encourage biorefineries to replace fossil fuels with energy from renewable biomass (including animal waste).[233] Congress also authorized $23 million per year in funding between 2014 and 2017, along with $20 million in funding for 2018, for the Biomass Research and Development Initiative, a collaborative effort among the USDA, the Department of Energy and the EPA.[234] In this program, Congress has explicitly directed the agencies to coordinate their policies and processes to more effectively promote the production of biofuels and biobased products, attempting to secure commercial biofuel production at prices that are competitive with fossil fuels.[235]

---

[228] MacDonald, above note 226.
[229] *Ibid.*
[230] *Ibid.*
[231] Pub L 113–79.
[232] HR 2642-280-281, Sec. 9003 (7 USC § 8103(g)).
[233] HR 2642-282, Sec. 9004 (7 USC § 8104).
[234] See Agricultural Act of 2014, PL 113-79 § 9008, 128 Stat 624, 931, amending 7 USC § 8108.
[235] See Agricultural Act of 2014, PL 113-79 § 9008, 128 Stat 624, 931, amending 7 USC § 8108(b)(1), (e)(2)(A). See also Randy Schnepf, CRS, Energy Provisions of the 2014 Farm Bill (12 March 2014).

6.2 Other Federal Programs

Several federal programs outside the Farm Bill aim to reduce the livestock sector's contributions to climate change by advancing technologies: primarily anaerobic digesters that capture methane emissions and convert them to energy sources. For example, the USDA and the EPA are working together through the AgSTAR program to promote the use of anaerobic digesters at livestock facilities across the country 'to reduce methane emissions and achieve other environmental benefits through outreach, education, tools and partnerships.'[236] According to program officials, more than 8,000 livestock operations show potential for conversion to anaerobic digesters for waste management. Converting these operations would reduce methane emissions 'by over 34 $MMTCO_2e$ annually, roughly equivalent to emissions from 6.5 million passenger vehicles', while generating 'more than 13 million megawatt-hours of energy per year and displac[ing] about 1,670 megawatts of fossil-fired generation.'[237] As of May 2015, the program estimated there were a total of approximately 261 anaerobic digesters operating on commercial livestock farms in the US, mostly on dairy operations located in the upper Midwest, Northeast and the West Coast.[238] As of May 2016, only 40 hog farms had installed anaerobic digesters.[239] Since 1994, EPA estimates that the AgSTAR program has reduced 19 million metric tons of CO_2e from livestock operations, the equivalent of 2.6 million homes' electricity use for one year, and has generated 5.3 billion kilowatt hours of energy.[240]

In his 2013 Climate Action Plan, President Obama noted that to be

---

[236] Biomass Research and Development Board, 'Federal Activities Report on the Bioeconomy,' at 27 (February 2016), http://www.biomassboard.gov/pdfs/farb_2_18_16.pdf.

https://www.epa.gov/sites/production/files/2014-12/documents/agstar-handbook.pdf.

[237] US EPA, 'AgSTAR Data and Trends,' https://www.epa.gov/agstar/agstar-data-and-trends; see also US EPA, 'Market Opportunities for Biogas Recovery Systems at U.S. Livestock Facilities,' Table 1 (2011), https://www.epa.gov/sites/production/files/2014-12/documents/biogas_recovery_systems_screenres.pdf.

[238] http://www2.epa.gov/agstar/livestock-anaerobic-digester-database. (accessed 14 October 2016).

[239] *Id.*

[240] US EPA, 'Celebrating 20 Years of Anaerobic Digestion', https://www.epa.gov/sites/production/files/2016-05/documents/agstar_20th_infographic.pdf; see also US EPA, 'AgSTAR Accomplishments', https://www.epa.gov/agstar/agstar-accomplishments (noting that in 2014 the AgSTAR program offset 1.20 million metric tons of CO_2e, which is equivalent to the 109,498 homes' energy use for one year).

effective, any plan to confront climate change must address methane. He therefore directed his administration 'to develop a comprehensive, interagency strategy to reduce methane emissions and promote cutting-edge technologies that help farmers, energy companies, and communities convert methane into a renewable energy source and grow America's biogas energy industry'.[241] Pursuant to this directive, the EPA, USDA and DOE collaborated on a study called 'Biogas Opportunities Roadmap: Voluntary Actions to Reduce Methane Emissions and Increase Energy Independence',[242] which outlines voluntary measures these federal agencies can take to lower methane emissions through the use of biogas technology. Biogas typically contains 50–70 percent methane and the study focuses on certain biological materials, including manure and litter produced by beef, dairy, swine and poultry operations, with significant methane emissions. The document affirms that biogas systems 'have the potential to capture methane that would escape into the atmosphere and utilize it to create energy', such as electricity, heat and engine fuel.[243]

The Biogas Opportunities Roadmap identifies four primary goals: promoting the use of biogas through existing programs, fostering investment in biogas systems, strengthening markets for biogas, and improving coordination and communication among agencies. The Roadmap seeks to leverage the government's ability to shape markets by collecting industry financial and technical data on the performance of anaerobic digesters, revising federal procurement guidelines, reviewing interconnection standards for electric utilities and natural gas pipelines, and reviewing net metering policies. The Roadmap identifies EQIP as one of the primary USDA programs for funding and promoting the use of biogas, and directs NRCS to review its CPSs to ensure they reflect the full range of environmental benefits provided by anaerobic digestion of animal waste (such as manure separation, nutrient recovery and methane destruction).

In December 2015, USDA, DOE, and EPA published a joint report detailing the progress achieved since the Biogas Opportunities Roadmap

---

[241] 'Fact Sheet: Biogas Opportunities Roadmap: Voluntary Actions to Reduce Methane Emissions, Increase Energy Independence and Grow the Economy' (USDA 2014), accessed 14 October 2016 at http://www.usda.gov/wps/portal/usda/usdahome?contentidonly=true&contentid=2014/08/0166.xml.

[242] USDA, USEPA and USDOE, 'Biogas Opportunities Roadmap: Voluntary Actions to Reduce Methane Emissions and Increase Energy Independence' (August 2014), accessed 14 October 2016 at https://www3.epa.gov/climatechange/Downloads/Biogas-Roadmap.pdf.

[243] 'Other byproducts of biogas systems include non-energy products such as nutrient rich soil amendments, pelletized and pumpable fertilizers, and even feedstock for plastics and chemicals.' *Ibid* at 6.

was published in August 2014: 'Biogas Opportunities Roadmap Progress Report'.[244] Notable achievements and funding allocations highlighted in the report include USDA's allocation of nearly $1.8 million to subsidize anaerobic digester system costs through EQIP and the Rural Energy for America Program (REAP);[245] EPA's publication of its final rule for Renewable Fuel Standards which specifies that a certain portion of annual renewable fuel volume requirements must come from cellulosic advanced fuels, which can include biogas;[246] and DOE's completion of a biogas resource assessment study, which concluded that potential bio-methane production in the United States could reach 16 million metric tons annually.[247] Policy challenges highlighted in the report include effective nutrient management at dairy CAFOs, securing USDA Organic Certification for fertilizers made from the solid and liquid digestate output from anaerobic digesters, creating a North American Industry Classification System code for biogas, and reducing the amount of NO_x produced by the combustion of biogas.[248]

In a separate initiative, the USDA announced the launch of its Climate Smart Agriculture and Forestry Strategy in April 2015.[249] Through this initiative, the USDA will encourage broader deployment of technologies to reduce methane emissions from cattle, dairy and swine operations. The USDA references a variety of methods, such as increased use of solids separators and composting of waste, but more ambitiously states its plan to support 500 new digesters over the next ten years, as well as expand the use of covers on 10 percent of anaerobic lagoons used in dairy cattle and hog operations.[250] In addition to reducing emissions, the USDA announced plans to help farmers and ranchers increase carbon sequestration in soils through practices such as rotational grazing and improved forage management. By 2025, the USDA plans to support improved grazing management on an additional four million acres, for a total of 20 million acres.[251]

In May 2016, USDA published a report highlighting the success of

---

[244] USDA, DOE and EPA, 'Biogas Opportunities Roadmap Progress Report', (December 2015), accessed 6 January 2017 at https://www.epa.gov/sites/production/files/2015-12/documents/biogas-roadmap-progress-report.pdf.

[245] *Id.* at 3.

[246] *Id.* at 4.

[247] *Id.*

[248] *Id.* at 9–10.

[249] See http://www.usda.gov/wps/portal/usda/usdahome?contentidonly=true&contentid=climate-smart.html (accessed 14 October 2016).

[250] 'Building Blocks for Climate Smart Agriculture & Forestry – Fact Sheet' (USDA 2015), accessed 14 October 2016 at http://www.usda.gov/documents/climate-smart-fact-sheet.pdf.

[251] *Ibid.*

Climate Smart Agriculture and Forestry Strategy programs.[252] This report sets forth a goal of achieving a reduction of 21.1 million metric tons of CO_2e per year by 2025 through the installation and operation of the 500 new anaerobic digesters and through capturing methane by covering lagoons and waste storage ponds at 10 percent of dairy cattle and swine operations.[253] In furtherance of this goal, USDA, through the Renewable Energy for America Program, awarded $12.5 million in grants and loans to subsidize the construction of 17 anaerobic digesters and biogas systems in California, Maine, Massachusetts, Michigan, New York, North Carolina, Ohio, and Washington.[254]

The USDA has also addressed livestock GHG emissions through public-private partnerships. In 2009, the USDA entered into an agreement with the nation's dairy industry to reduce the industry's GHG emissions by accelerating the adoption of waste-to-energy projects and energy efficiency improvements. The collaboration resulted in nearly 180 awards from the USDA to finance the development and construction of anaerobic digesters on dairy CAFOs across the United States.[255] The parties renewed the agreement in 2013, and the industry committed to reducing its greenhouse gas emissions by 25 percent by 2020.[256] In this renewed agreement, the USDA pledged to accelerate and streamline the process for adopting anaerobic digesters, develop guidance for NRCS to offer enhanced payment rates for anaerobic digesters and use EQIP to promote improved sustainability measures and technologies to reduce dairies' GHG emissions. The parties also agreed to create a national repository of open source lifecycle assessment data that would be accessible via the USDA National Agricultural Library.[257]

Although it has not promulgated a technology standard for controlling emissions from livestock operations, these various funding and incen-

---

[252] USDA, 'USDA Building Blocks for Climate Smart Agriculture and Forestry – Implementation Plan and Progress Report,' (May 2016), accessed 6 January 2017 at http://www.usda.gov/documents/building-blocks-implementation-plan-progress-report.pdf.

[253] *Id.* at 20.

[254] *Id.* at 8.

[255] 'USDA and Dairy Producers Renew Agreement to Reduce Greenhouse Gas Emissions and Increase Sustainability of Dairy Production' (USDA 24 April 2013), accessed 14 October 2016 at http://www.usda.gov/wps/portal/usda/usdahome?contentid=2013/04/0076.xml.

[256] *Ibid*, see also http://www.usdairy.com/sustainability/industry-commitment/about (accessed 14 October 2016).

[257] To access the repository, go to http://www.lcacommons.gov/ (accessed 14 October 2016).

tive programs evince the government's preference for anaerobic digesters. According to the EPA: 'biogas recovery systems are effective at confined livestock facilities that handle manure as liquids and slurries, typically swine and dairy farms . . . Anaerobic digesters are particularly effective in reducing methane emissions but also provide other air and water pollution control opportunities.'[258] Equally important from the industry's perspective, anaerobic digesters are easily added on to existing lagoons and slurry ponds, thereby reducing costs for enhanced treatment. However, even with these advantages, initial capital costs for an anaerobic digester are high, ranging from $500,000 to $1 million. Thus, despite government incentives and cost sharing programs, widespread adoption of anaerobic digesters is unlikely. It may require affirmative government regulation to control emissions, or the development of a robust market for carbon offsets, to change this trend in the United States.[259]

6.3 Bioenergy and Biofuel Standards

As discussed above, some emissions from livestock waste can be captured and repurposed as renewable energy. Renewable energy mandates and renewable fuel standards in many jurisdictions now recognize the potential contributions of animal waste and provide incentives for developing and deploying these GHG reducing technologies.

At the federal level, livestock methane is recognized as a renewable energy source under the Energy Policy Act of 2005, and is among the sources of renewable energy authorized to receive incentive payments.[260] The Energy Policy Act also addresses transportation fuel, including animal wastes as sources of cellulosic biomass ethanol and waste derived ethanol, which both count as the equivalent of 2.5 times their volume of other types of renewable fuels under the Clean Air Act's Renewable Fuel Program.[261]

---

[258] A biogas recovery system is an anaerobic digester that captures emissions and combusts them to produce electricity, heat or hot water. USDA-EPA 'Collaboration to Advance the Deployment of Anaerobic Digesters at US Livestock Facilities', p. 3. US EPA, 'US Anaerobic Digester: A Status Report' (October 2010), accessed 6 January 2017 at http://pdf.thepdfportal.org/?id=34077#.

[259] See Donald Hornstein, 'The Future of Food Regulation: The Environmental Role of Agriculture in an Era of Carbon Caps' (2010) 20 *Health Matrix* 145, 164; Joe Reedek and Viney Aneja, 'Climate Vulnerabilities of the Swine Industry' in RA Pielke, Jr (ed.), *Climate Vulnerability: Understanding and Addressing Threats to Essential Resources* (Academic Press 2013) 86.

[260] Energy Policy Act of 2005 §§ 202(b)(2) & 202(e); 42 USC 13317(b) & 42 USC 13317(e)(1).

[261] Energy Policy Act of 2005 § 1501.

The distinction between agricultural wastes and other biofuel stocks may be necessary to encourage development and adoption of waste emission capture technologies, which can be more expensive to implement than planting biofuel crops.

7. CONCLUSION

The US livestock industry has been slow to change its production and management practices to address climate change (or other environmental impacts, for that matter). This may be because advanced technologies can be expensive and difficult to maintain. Yet there are a variety of federal and state programs that provide financial incentives to help livestock producers implement practices and install systems that reduce GHG emissions. Several hundred livestock producers have adopted new methods and technologies through these programs. Even so, the overall adoption rate remains quite low. Thus it is likely that factors other than cost are at work, such as the economic structure of the industry (in which the farmers themselves, not the companies they are contracted with, shoulder the costs of regulatory compliance) and the laissez faire regulatory policy for agricultural operations.[262] To effectively mitigate the livestock sector's contributions to global climate and environmental change, the United States must develop and implement consistent and coherent policies across federal administrative agencies that encourage and, where necessary, force the industry to change its practices.

Of course this involves an 'end-of-pipe' solution, that is, the use of advanced technologies to manage, treat and dispose of the large volumes of waste generated at a typical CAFO. Other options exist, such as changing the production system itself through the use of less intensive, pasture based animal husbandry. Consumers could also choose to consume less meat based proteins, thereby reducing demand and forcing reductions in livestock production.[263] These approaches, while beyond

---

[262] Personal Communication between author and Charles Adair, Program Manager of Duke Carbon Offsets Initiative (5 June 2015); Ad Hoc Committee on Air Emissions above note 7, at 130; JB Ruhl, 'Farms, Their Environmental Harms, and Environmental Law' (May 2000) 2 Ecol. L. Q. 263.

[263] See, e.g., Peter Scarborough *et al.*, 'Dietary Greenhouse Gas Emissions of Meat-Eaters, Fish-Eaters, Vegetarians and Vegans in the UK' (2014) 125 *Climate Change* 179 (comparing average dietary GHG emissions by diet group and finding that diets with fewer meat products had lower average GHG emissions); Samuel Soret *et al.*, 'Climate Change Mitigation and Health Effects of Varied Dietary

the scope of this chapter, are nonetheless necessary to consider as we collectively come to terms with the impacts our agricultural practices have on the planet.

Patterns in Real-Life Settings throughout North America' (2014) 100 supp. 1 Am. J. Clin. Nutr. 490S, 491S (comparing annual dietary GHG emissions from vegetarian, semi-vegetarian, and non-vegetarian diets, and finding GHG emission reductions of approximately 30% for vegetarian diets and 20% for semi-vegetarian diets from a baseline non-vegetarian diet).

10. Adaptation of agricultural trade and investment rules to climate change

*Christian Häberli**

1. INTRODUCTION

Most scientists now accept that structural developments attributed to climate change, such as increasing temperatures, changes in precipitation patterns, carbon dioxide (CO_2) concentrations, water availability and rising sea levels also have an impact on agriculture.[1] The good news is that food production at the global level appears to be able to cope with such structural changes. While no single country or producer is immune, credible estimates based on productivity records show that this world could possibly feed twice the number of its present population, even in a changing climate and without using substantially more land and forest resources.[2] At the same time, while poor developing countries and producers have always been, and still are, the smallest greenhouse gas (GHG) emitters, they are likely to be among the most severely climate affected. This means that their food security is perhaps the single most important equity issue in the whole climate change discussion. For subsistence farmers and poor consumers in those countries, climate change is a major, new and highly complex cause for (additional) food insecurity.

This chapter is a legal scholar's examination of the climate change challenges faced by food producers, national policymakers and regulators of international trade and investment. The chapter asks whether the food value chain is climate change resilient under the present international regulatory framework. It starts with an overview of recent literature on climate

* Christian Häberli, PhD, Fellow at the University of Bern's World Trade Institute (WTI). Research for this chapter was funded by the Swiss National Science Foundation under a grant to the National Centre of Competence in Research on Trade Regulation.

1 Global Commission on the Economy and Climate, *The New Climate Economy Report. Programme of Work Conducted by a Global Partnership of Eight Leading Research Institutes* (World Resources Institute 2014).

2 Hector E Maletta, 'Trends in World Agricultural and Food Trade' (*Centro de Investigación de la Universidad del Pacífico,* 2014), accessed 7 November 2016 at http://papers.ssrn.com/sol3/papers.cfm?abstract_id=2484252.

related issues with a need for regulatory review and change. The research hypothesis is that a legal analysis of the issues presented here has to take into account the specific situation of weak states, small farmers and poor consumers, in respect of trade liberalization and (foreign) investment. This is an under-researched issue of policy space: one that should be considered when discussing how to address climate change and food insecurity at the national and international levels. Currently available literature shows a remarkable lack of analytical research on the legal aspects of climate change-related agricultural production and trade policies. Exceptions are, for instance, studies on the negative impact of the Doha Round's failure to limit subsidies abuse[3] and on the 'land grab' potential of foreign direct investment (FDI).[4] The reason for this lack of legal analysis might be that, except for voluntary principles and guidelines, no 'best agricultural practices' for climate change have been adopted at the international level, let alone codified so as to serve as international, mandatory standards that could inform national policies. But farmers and hungry consumers cannot wait for scientists, governments and international organizations to agree on legally secure solutions when faced with a rising tide of uncertainties. Hence, a regulatory review and some proposals for adjustments may serve as a basis for progress.

Climate change affects poverty, hunger and malnutrition. Mitigation and adaptation efforts need to consider a wide range of issues including: production subsidies, productivity, non-food uses of agricultural land such as biofuels for local production and for exports, biotechnology, standards and labelling, carbon taxation and emission trading, water and fish, risk insurance and export restrictions, food stockpiles, mitigation of various vulnerabilities, gender, local and international migration, and violence. This book addresses most of these issues in more detail. From a regulatory perspective the relevant question would be whether the present policies and tools promoting food production, investment and trade are good enough to cope with the additional challenge of global warming.

The conclusion is that today's regulatory deficits are anything but climate change-resilient. What is required first and foremost are multilaterally agreed upon, climate-smart, best farming and processing practices. These

---

[3] Valdete Berisha-Krasniqi *et al.*, *The Development Promise: Can the Doha Development Agenda Deliver for Least-Developed Countries? IFPRI Note 14* (International Food Policy Research Institute (IFPRI) 2008); Alan O Sykes, 'The Questionable Case for Subsidies Regulation: A Comparative Perspective' (2014) 2(2) JLA 473.

[4] Lorenzo Cotula, 'The New Enclosures? Polanyi, International Investment Law and the Global Land Rush' (2013) 34(9) TWQ 1605.

standards would then need to be enshrined in a credible multilateral environmental agreement, and protected against legal challenges in the World Trade Organization (WTO) similar to internationally agreed upon food safety standards. This exercise quite possibly implies some new and climate-specific trade and investment rules for agriculture, water, and aquaculture. Absent such rules, climate change might even increase the already serious collateral damage done, particularly to net food-importing developing countries (NFIDC), and drive poor smallholders out of business.

This chapter was drafted before the adoption of the Paris Agreement on Climate Change in December 2015. A preliminary analysis of the relevant WTO rules for agriculture in respect of climate change mitigation was presented by the same author in June 2016.[5]

2. CLIMATE CHANGE CHALLENGES TO AGRICULTURE

2.1 What are the Issues?

How will climate change affect agriculture by 2050, in a world with more than 9 billion mouths to feed with more and better food? The International Food Policy Research Institute (IFPRI) in Washington argues that while climate change might benefit some regions, for many farmers it will increase the challenges of maintaining production and increasing productivity, and of managing the new risks.[6] The capacity of the agricultural system as a whole to supply timely, sufficient and adequate quantities of food at constant real prices will be subject to continuous stress tests. For food to remain affordable, the global trade regime must ensure that agricultural trade flows reflect changes in comparative advantages, and balance world food supply and demand. National governments will have to provide additional and adequate support for land and water productivity-enhancing investments by public and private investors. At the same time, their policies must not have a negative impact on other countries.

The World Bank notes that, by and large, the first Millennium

---

[5] Christian Häberli, WTO Rules Can Prevent Climate Change Mitigation for Agriculture. Draft Working Paper dated 24 June 2016, presented at the SIEL's Fifth Biennial Global Conference: International Economic Law in a Diverse World? accessed 6 January 2017 at (http://papers.ssrn.com/sol3/papers.cfm?abstract_id=2800011) (Johannesburg, 7–9 July 2016).

[6] Gerald C Nelson *et al.*, *Food Security, Farming, and Climate Change to 2050: Scenarios, Results, Policy Options* (IFPRI 2010).

Development Goal of halving hunger has been met and that extreme poverty has receded even faster.[7] Today, the world enjoys an all-time record number of cereal harvests, with globally affordable consumer prices and remunerative production prices. Low fuel prices, high FDI and productivity growth contribute to an overall expansion of 'commercial' agricultural production, including non-food end uses (feed, fibre and fuel). And all this takes place without using substantially more of the agricultural land still available on all continents and especially in Africa. This is not to belittle the massive acquisition of agricultural land by FDI (also called 'land grabbing'), often at the expense of poor farmers and even taking place in World Bank-financed projects.[8] At any rate, in such a vast and dynamic field food prices will continue to fluctuate, and the most vulnerable are faced with new and growing existential problems. It is thus essential to have the adequate tools in place to act quickly when new climate change-induced food price crises arise.[9]

This section starts with a layman's discussion of the widely differing climate change data. While the forecast variations may confound policy-makers, a legal analysis must take into account the differences and respond to the generally recognized need for regulatory action. The other issues raised here are the development dimension of climate change and agriculture, the productivity challenge, and the specific questions of biofuels, water and fish. Finally, there is a brief discussion of the resilience issue under the most likely climate change scenarios.

2.2 Differing but not Contradictory Figures, and their Interpretation

In his widely quoted 2007 study, William Cline assumes a 'business as usual' scenario in which GHG emissions continue to increase.[10] He predicts that by 2080 agricultural productivity would be 8 percent lower in

---

[7] The world achieved Goal 1 five years ahead of schedule. See World Bank, 'Millennium Development Goals: Goal 1 – Eradicate Extreme Hunger and Poverty by 2015', accessed 7 November 2016 at http://www.un.org/millenniumgoals/poverty.shtml.

[8] World Bank, 'World Bank Acknowledges Shortcomings in Resettlement Projects, Announces Action Plan to Fix Problems – IBRD Press Release No 2015/332/ECR' (4 March 2015), accessed 7 November 2016 at http://www.worldbank.org/en/news/press-release/2015/03/04/world-bank-shortcomings-resettlement-projects-plan-fix-problems. See Section 2.2.

[9] World Bank, 'International Food Prices Hit Four-Year Low - Press Release No 2015/122/POV' (30 September 2014), accessed 7 November 2016 at http://www.worldbank.org/en/news/press-release/2014/09/30/international-food-prices-four-year-low.

[10] William R Cline, *Global Warming and Agriculture: Impact Estimates by*

developing countries, 8 percent higher in high-income countries, and 3 percent lower globally, even without the so-called (and hotly debated) carbon fertilization effect. Individual developing countries would face even larger declines. India, for example, could see a drop of 30 to 40 percent.

A recent study by the Organisation for Economic Co-operation and Development (OECD) finds that, by 2050, the real prices of rice and wheat could increase by about 25 percent from 2005 prices, even if agricultural productivity continues to increase.[11] The real price of maize could even increase by 50 percent. Climate change could negatively affect the growth rates of yields for most commodities in most countries, and is likely to affect especially the poorest populations, thereby increasing food insecurity in many regions.

Using a global economy-wide model (and drawing upon earlier World Bank research results), Valenzuela and Anderson assess (i) the overall effects on farm product prices of expected yield changes, and (ii) the expected adverse effect of higher temperatures and humidity in the tropics on the productivity of unskilled workers in developing countries.[12] Given the considerable uncertainties surrounding all climate change effects, their calculations are based on a wide range of yield productivity and labour shock simulations, indicating changes in international agricultural prices, national food consumption, self-sufficiency, and net farm income. The exogenous yield shocks, which can be attributed to climate change in 2030, vary by region and sector. Percentage change in land productivity might thus range from −20 percent for rice and wheat in South Africa to +19 percent for rice, wheat, fruits and vegetables, oilseeds and cotton in many temperate countries. Extremes appear possible for coarse grains, for which productivity in South Africa could decrease by 42 percent, but increase by 22 percent in Argentina. In terms of global food security, the tentative findings of Valenzuela and Anderson are less pessimistic than those of earlier studies, without contradicting them. Because of considerable underinvestment in such activities in most developing countries, they emphasize that expanding agricultural research and development (R&D) must become a top priority. Such investments could reduce the need

---

Country (Center for Global Development and Peterson Institute for International Economics 2007).

[11] A Ignaciuk and D Mason-D'Croz, *Modelling Adaptation to Climate Change in Agriculture – OECD Food, Agriculture and Fisheries Papers No 70* (OECD 2014).

[12] Ernesto Valenzuela and Kym Anderson, 'Climate Change and Food Security to 2030: A Global Economy-Wide Perspective' (2011) 11(1) *Economía Agraria y Recursos Naturales* 29.

perceived by governments to use border measures to increase or maintain food self-sufficiency, as well as the inefficient production structures by permanently shielding their producers from fluctuations in international prices. This in turn would increase global food security.

A study of UN Environment Programme experts based on forecasts by the United Nations Framework Convention on Climate Change (UNFCCC) shows increasingly unsustainable global food consumption patterns. At the same time, climate change might diminish productivity and reduce yields in some areas by as much as one-third.[13]

Quantitative forecasts remain tentative and fraught with imponderables. Nonetheless, it appears that warmer regions and their agriculture could be affected more than others (especially temperate zones), some of which would actually benefit from climate change.[14] A major study finds that the length of growing periods could decline by 5 percent or more across a broad area of the global tropics. Where this length falls below 120 days, cropping might become too risky to pursue as a major livelihood strategy. Additional stress factors are temperatures above 30°C, increasingly variable rainfall, reduced rainfall per rain event and increased rainfall intensity and drought risks. Even though production conditions are increasingly precarious and the net food production index might stagnate in all tropical areas, this study defined and identified 'poverty hotspots' characterized by high exposure, high sensitivity and low coping capacity in west, central and east Africa, India, Bangladesh and Southeast Asia.[15]

2.3 Development Challenges and Opportunities

The development dimension of climate change as it impacts agriculture calls for a new focus, and urgent attention. Today, and for a long time to come, the majority of the hundreds of millions of small and landless farmers, and pastoralists, cannot produce enough food even for their families. This issue is compounded in economies where the manufacturing and services sectors fail to offer better alternatives. Hence, it is particularly urgent to find out how climate change affects these producers.

---

[13] William Moomaw *et al.*, *The Critical Role of Global Food Consumption Patterns in Achieving Sustainable Food Systems and Food for All* (UNEP 2012).

[14] Hector Maletta, 'From Hunger to Food Security: A Conceptual History' (Universidad del Pacífico, 7 July 2014), accessed 7 November 2016 at http://papers.ssrn.com/sol3/papers.cfm?abstract_id=2484166.

[15] Polly Ericksen *et al.*, *Mapping Hotspots of Climate Change and Food Insecurity in the Global Tropics – CCAFS Report No 5* (CGIAR Research Program on Climate Change, Agriculture and Food Security (CCAFS). 2011).

Thornton, Herrero and Jones studied the nature and importance of mixed crop and livestock farms which, especially in South Asia and in sub-Saharan Africa (SSA), are still the principal means of food production.[16] They found that shifts in climate and climatic variability were likely to increase the challenges faced by countries largely dependent on agricultural commodities. Consequently, adaptation options for affected households are intensification (or extensification) of production, diversification and new forms of risk management.

While poor consumers are particularly vulnerable to food price increases, poor producers rarely benefit from them. There are many reasons for this apparent paradox, but this is why Hertel and Rosch find that climate action can yield the highest returns for agriculture as the predominant form of economic activity of poor people.[17] Using UNFCCC-generated data from 2007 and 2010 these authors calculate that at a carbon price of US$100 per ton CO_2 equivalent, the combined mitigation potential of agriculture and forestry is greater than that of any other individual sector in the economy. Consequently, agricultural (GDP) growth is 2.2 times as effective at reducing poverty as growth in non-agricultural GDP.

Lybbert and Sumner consider that greater complexity and greater urgency are the twin imperatives of climate change action. This must be kept in mind when discussing policies and institutional reforms aimed at improving climate change adaptation and mitigation in agriculture. Agriculture's share in global GDP is about 4 percent but most scholars put its GHG emissions at over a quarter of the total. Therefore, policies and institutions must support the development and diffusion of new technologies, and provide incentives for further technological breakthroughs.[18]

This proximity of challenges and opportunities logically leads economists and international financial institutions to argue that the best way to mitigate climate change is by additional private and public sector investments in agricultural productivity. The above-mentioned IFPRI study

---

[16] Philip K Thornton, Mario Herrero and Peter G Jones, 'Adaptation to Climate Change in Mixed Crop–Livestock Farming Systems in Developing Countries' in Ariel Dinar and Robert Mendelsohn (eds), *Handbook On Climate Change and Agriculture* (Edward Elgar 2011).

[17] Thomas W Hertel and Stephanie D Rosch, 'Climate Change, Agriculture, and Poverty' (2010) 32(3) AEPP 355.

[18] Travis Lybbert and Daniel Sumner, *Agricultural Technologies for Climate Change Mitigation and Adaptation in Developing Countries: Policy Options for Innovation and Technology Diffusion - ICTSD–IPC Platform on Climate Change, Agriculture and Trade Issue Brief No 6* (International Centre for Trade and Sustainable Development (ICTSD) and International Food and Agricultural Trade Policy Council 2010).

asserts that such investments would do even more to reduce suffering: relative to their baseline outcome in 2050, a 40 percent increase in productivity growth would reduce the number of malnourished children by an additional 37 percent (that is, by 19.1 million children).[19] Their caveat is the uncertainty of climate change prediction, implying that climate-specific investments are still a difficult proposition. However, IFPRI argues that supporting physical and human capital investments should begin immediately as a win-win way of increasing the efficiency of land, water, and nutrient use. This will promote growth, climate resilience and mitigation of agricultural GHGs. According to IFPRI the investments needed to cope with climate change up to 2050 seem feasible, at least under conditions of relatively free international trade. Thereafter, however, the possibility of ever-increasing temperatures becomes potentially much greater.

2.4 Production and Productivity as a Key for all 'Four Fs'

Social scientists and policymakers still disagree on the basics of international food security governance, but it is generally understood that all four elements of food security – availability, access, utilization and supply stability – are affected by climate change. The Commission on Sustainable Agriculture and Climate Change set up in 2011 within the UNFCCC concurs that agriculture must produce more food for more people with more purchasing power. Given the limited availability (and erosion) of land suitable for agriculture, the food security challenge requires not only more production, but also continuous productivity increases (keeping in mind that the 'Four F's', food, feed, fuel and fibre, are competing for those agricultural lands). Only two pertinent aspects can be addressed here.

The resurgence of input subsidy programmes in many African countries is not unproblematic, and this includes present WTO subsidy rules, including the little-examined Article 6.2 of the Agreement on Agriculture. Nin-Pratt *et al.* find that land productivity correlates positively with fertilizer use, but note considerable variability in the efficient use of fertilizer across countries.[20] They also posit that such programmes should be complemented by capital investments increasing labour productivity.

A particular sustainability problem affects subsistence agriculture where it consumes more energy for production and household uses than is locally available from renewable sources such as wood and cow dung. In a regional

---

[19] IFPRI, Nelson *et al.* (n 6).

[20] Alejandro Nin-Pratt, *Agricultural Intensification in Africa: A Regional Analysis – IFPRI Discussion Paper 01433* (IFPRI 2015).

study, Wessels *et al.* conclude that wood and charcoal still supply the majority of sub-Saharan Africa's rural energy needs, but that the long term supply of fuelwood is in jeopardy. This underscores the importance of providing affordable energy for rural economic development.[21] Here again, the difficult subsidy issue warrants further studies. While the environmental harm of certain energy subsidies has been recognized, for agriculture a comprehensive analysis from a climate perspective is still lacking.[22]

2.5 Biofuels

Generally speaking, while biofuels could negatively affect local food security, those produced for local use lessen dependence on fuel imports. A calculation of their effect on GHG emissions must include input use and conversion technology.

In countries like Argentina and Mauritius, the carbon balances for soy biodiesel and bioethanol made from sugarcane and corn appear to be better than that of fossil fuels, because their agricultural production systems use little energy and few agrochemicals.[23]

Hochman *et al.* consider that second generation biofuels can be part of a solution for global warming, and propose reducing trade barriers for sugarcane as a relatively clean first generation biofuel, even though subsidies for flex fuel cars might still be needed.[24] They also argue that policies should target as closely as possible the sources of market failures (for example, emissions with net social and environmental costs should be taxed).

Biofuels also raise issues of land use changes (LUC). Warner *et al.* see direct and indirect effects resulting from the increasing demand for

---

[21] KJ Wessels *et al.*, 'Unsustainable Fuelwood Extraction from South African Savannas' (2013) 8 Environ Res Lett, accessed 7 November 2016 http://iopscience.iop.org/article/10.1088/1748-9326/8/1/014007.

[22] Ilaria Espa and Sonia E Rolland, *Subsidies, Clean Energy, and Climate Change* (ICTSD and World Economic Forum 2015). Also see Amelia Porges and Thomas L Brewer, *Climate Change and a Renewable Energy Scale-Up: Responding to Challenges Posed to the WTO* (ICTSD and World Economic Forum 2013), and ICTSD, *Tackling Perverse Subsidies in Agriculture, Fisheries and Energy – Information Note* (ICTSD 2012).

[23] Marcelo Regúnaga and Agustín Tejeda Rodriguez, *Argentina's Agricultural Trade Policy and Sustainable Development – ICTSD Issue Paper No 55* (ICTSD 2015).

[24] Gal Hochman, Deepak Rajagopal and David Zilberman, 'Biofuels and Climate Change' in Ariel Dinar and Robert Mendelsohn (eds), *Handbook On Climate Change and Agriculture* (Edward Elgar 2011).

crop based biofuels. They developed a system dynamics tool intended to complement existing LUC modelling approaches, and to improve the understanding of global LUC drivers and dynamics under diverse scenarios and varying model assumptions. In all scenarios cropland must expand substantially; there is potential for supply shortfalls and associated upward pressure on prices, and biofuels might also lead to higher land use intensity for other food commodities (for example, cattle raising).[25]

2.6 Water

Water security is a central issue of adaptation to climate change, not only in agriculture. The OECD estimates that by 2050, over 40 percent of the world's population will be living in water-stressed areas and calls for inclusive stakeholder consultations, including at the international level.[26] In their forecast studies for 2030, EU experts estimate that between 1.9 and 2.6 billion people are likely to lack freshwater; more migration linked to the effects of climate change is expected, predominantly via South-South flows, and with potentially destabilizing local effects.[27] The effects of foreseeable climate change on crop water requirements, availability and quality of water, and other agriculture-related factors are uncertain. The potential increases in the frequency and severity of extreme events such as floods and droughts as a result of climate change would have substantial negative impacts on agricultural production. Such changes in extremes are expected in the same study to be the most costly scenarios.

Adaptation to climate change is both private and collective: policies for water supply and demand adaptation in agriculture imply on-farm and watershed management in competition with other water uses (ecosystems) and other users (urban and industrial). New risk management systems are required for droughts and floods, including prevention and insurance. It may be politically costly to reduce water allocation rights, or to increase the price of water to reflect projected decreases of average water availability. However, unless regulatory mechanisms are in place to govern the use of water 'saved' by efficiency measures, farmers often use the additional water

---

[25] Ethan Warner *et al.*, 'Modeling Biofuel Expansion Effects on Land Use Change Dynamics'(2013) 8 Environ Res Lett, accessed 7 November 2016 at http://iopscience.iop.org/article/10.1088/1748-9326/8/1/015003.

[26] Organisation for Economic Co-operation and Development (OECD), *Stakeholder Engagement for Inclusive Water Governance* (OECD 2015).

[27] European Strategy and Policy Analysis System (ESPAS), *Global Trends to 2030: Can the EU Meet the Challenges Ahead?* (ESPAS 2015).

to increase food production by expanding irrigated land or by converting to higher-value, higher-profit commodities.[28]

Irrigation, where water is available, is a generally recommended technology for climate adaptation. Using data from a World Bank survey of more than 9,000 farmers across 11 African countries, a cross-sectional approach estimates how farm net revenues are affected by climate change, compared with current mean temperatures. Revenues fall with warming for dryland crops (temperature elasticity of –1.9) and livestock (–5.4), whereas revenues rise for irrigated crops (elasticity of 0.5); such crops are located in relatively cool parts of Africa, and buffered from the effects of warming by irrigation. Hence, a warmer climate has little net aggregate effect initially, as the gains for irrigated crops offset the losses for dryland crops and livestock. Warming is likely to directly reduce dryland farm income, whereas irrigated farms are generally less sensitive to climate change.[29]

It is worth reiterating that, while water efficiency is critical to eradicating hunger and poverty, technological solutions alone are not sufficient. The Stockholm International Water Institute points out that better irrigation technology rarely saves water, and that interventions to improve efficiency when excess water is used productively by others will not reduce hunger or eradicate poverty. Instead it suggests water accounting exercises to identify the benefits and scale of such opportunities for specific projects.[30]

2.7 Fish

Any discussion on climate change, water availability and food production must include fish. Ocean fisheries and marine aquaculture are an important but often overlooked component of world food security. Changes in catch composition have direct implications for coastal fishing communities, particularly those in tropical developing countries which tend to be socio-economically vulnerable to the effects of climate change. Continued warming in the tropics to a level that exceeds the thermal tolerance of tropical species may further reduce the catch potential in these regions.[31]

---

[28] OECD, *Climate Change, Water and Agriculture: Towards Resilient Systems* (OECD 2014).

[29] Pradeep Kurukulasuriya, 'Will African Agriculture Survive Climate Change?' (2006) 20(3) World Bank Econ Rev 367.

[30] Chris Perry, Dennis Wichelns and Pasquale Steduto, 'The Myth that "Water Efficiency" will Eradicate Hunger and Poverty' (2014) 4 *Stockholm WaterFront* 10.

[31] W Cheung, R Watson and D Pauly, 'Signature of Ocean Warming in Global Fisheries Catch' (16 May 2013) 497 *Nature* 365.

These findings by Cheung *et al.* matter, because one billion people depend on farmed and wild-catch fish as their primary source of protein.

The African Development Bank estimates that Africa loses one million tonnes of fish annually due to illegal fishing: that is the equivalent of 10 million cows. It estimates that bad management of the world's fisheries is threatening 19 million people with undernourishment.[32] In their study, Cheung *et al.* also noted that increasing sea surface temperatures are changing catch composition around the world. Although fishing statistics are particularly uncertain, they suggest that catch composition in tropical regions has already stabilized, implying that these regions have already become too hot for all but the warmest-water species.[33] Taking Senegal as an example, Moomaw and Blankenship see a link between the rise in unsustainable fish consumption in more developed countries, overfishing and marine resource depletion, and increased food accessibility concerns in poor countries.[34]

Trade plays an important role. China, for example, is not only becoming the world's largest rice importer, but its food security also depends increasingly on imported fish and fishmeal. With greater climate variability and expected increases in temperature and ocean acidification in the coming decades, volatility in fish production is expected to worsen. The Intergovernmental Panel on Climate Change projects that the ocean fish catch potential in the Southeast Asian tropics may be reduced by 40 to 60 percent, owing to fish migration. Hence, marine fish and marine aquaculture are likely to become increasingly unreliable sources of food, especially in tropical seas. This also means that resilience-focused governance policies for marine aquaculture are particularly important.[35]

2.8 Vulnerabilities and Resilience

A growing body of evidence confirms the hypothesis that climate change disproportionately affects poor developing country producers.

---

[32] African Development Bank (AfDB), 4/16 *Market Brief* for the period of 3 to 7 June, 2013, and sources quoted therein (on file with author).

[33] Cheung, Watson and Pauly (n 31).

[34] Case Study 3.4 Commercial Fishing and the Depletion of Artisanal Fisheries: Senegal in William Moomaw and Sara Blankenship, *Charting a New Course for the Oceans: A Report on the State of the World's Oceans, Global Fisheries and Fisheries Treaties, and Potential Strategies for Reversing the Decline in Ocean Health and Productivity* (Tufts University 2014).

[35] Robin Kundis Craig, 'Re-Tooling Marine Food Supply Resilience in a Climate Change Era: Some Needed Reforms' (2015) 38 Seattle U L Rev 1189.

A recent study on key farming systems and agro-climates in 11 African countries with data from more than 10,000 farm household surveys lists different reasons for the adaptation constraints and, hence, the vulnerability of farm households. The most climate vulnerable farmers are likely to be specialized crop and livestock producers (mono systems), particularly under dryland conditions in arid and semi-arid regions. This also indicates how difficult an African green revolution will be, considering the current high reliance on dryland systems (more than 95 percent of the land). Substantial public and private infrastructure and support investments are therefore required, including support for the improvement of access to information, finance, critical inputs and output trade.[36]

According to Ifejika Speranza, smallholder farmers are among the worst hit by climate change, owing to their dependency on rain-fed agriculture. This comes at a time when selling or leasing out of agricultural lands in some African countries takes place solely to produce food, feed or fuel, often for export. Such lease agreements highlight the need for an international reference framework. Land tenure and security of tenure play key roles in smallholder farming and are crucial for various adaptation practices. Also useful are early warning systems, crop research, extension services, rural radios, rural finance and microfinance, weather indexed crop insurance, payment for ecosystem services and strategic grain reserves. Monitoring and evaluation will be necessary to ascertain how effective these services are in supporting climate adaptation, especially by the very poor and in very marginal areas.[37]

3. MITIGATION AND ADAPTATION

Climate change is not a new phenomenon, nor is the challenge at national and local levels totally different from other structural changes. American grain farmers, for instance, faced rapidly evolving geo-climatic conditions when they pushed wheat production into environments previously considered unsuitable for cultivation. Olmstead and Rhode have noted that between 1839 and 2007 median annual precipitation for North American wheat production fell by half, and the median annual temperature norm

---

[36] Rashid M Hassan, 'Implications of Climate Change for Agricultural Sector Performance in Africa: Policy Challenges and Research Agenda' (2010) 19 J Afr Econ (AERC Suppl 2) ii77.

[37] Chinwe Ifejika Speranza, *Resilient Adaptation to Climate Change in African Agriculture* (Deutsches Institut für Entwicklungspolitik (DIE) 2010).

sank by 3.7°C. These changes, which occurred mostly before 1929, were addressed mainly by new biological technologies.[38]

The multifaceted challenges presently facing agriculture in a climate change and development context demonstrate the need for mitigation and adaptation on a new level. Environmental impacts may be positive or negative, but exports from areas with insufficient rainfall can be problematic. Assessments *in situ* must include 'virtual water' defined as 'the amount of water that is embedded in food or other products needed for its production', or the 'water footprint' of producers or consumers, as expressed by source, time and location.[39] These developments would then call for regulatory review and action.[40] What is clear is that climate change works differently in different countries and regions. The Center for Global Development based in Washington, D.C. is mapping the impacts of climate change with a comprehensive dataset tracking four dimensions of climate impact: extreme weather, sea level rise, agricultural productivity loss, and overall impact. The resulting region and country rankings show the impact of (i) direct risks (from physical climate impacts alone), and (ii) overall vulnerability (direct risks adjusted for a country's ability to cope with climate impacts). When adjusted for coping ability each map shows a strong correlation between vulnerability and country income levels. It also provides criteria for mitigation and adaptation finance.[41] Climate change adaptation, obviously, also varies for (big and small) producers and consumers (and will be different for rural and urban net food buyers). Here the focus is on poor farmers' resilience, keeping in mind that most subsistence farmers are net food buyers.

The following subsections discuss climate change mitigation and adaptation for (1) different country situations, (2) biotechnology, (3) trade and investment measures, (4) export restrictions, (5) carbon taxation, (6) risk insurance, (7) gender issues, (8) urban agriculture, (9) migration and (10) hunger and violence.

---

[38] Alan L Olmstead and Paul W Rhode, 'Adapting North American Wheat Production to Climatic Challenges, 1839–2009' (2011) 108(2) PNAS 480.

[39] Fitzgerald R Temmerman, *Trade in Water in International Law: Bulk Fresh Water, Irrigation Subsidies and Virtual Water* (2015 (forthcoming)).

[40] JR Beddington *et al.*, 'What Next for Agriculture after Durban?' (20 January 2012) 335(6066) *Science* 289.

[41] David Wheeler, *Quantifying Vulnerability to Climate Change: Implications for Adaptation Assistance – Centre for Global Development (CGD) Working Paper 240* (CGD 2011).

3.1 Different Countries, Different Solutions: China, Vietnam, Ethiopia and South Asia

China's agricultural producers are big GHG emitters. But according to Wang, Huang and Rozelle, China also makes important and efficient mitigation efforts to combat, for instance, low efficiency of fertilizer use.[42] The government has begun to promote technology aimed at calibrating fertilizer dosages according to the characteristics of soil. In addition, conservation tillage (CT) has been considered as a potential way to create carbon sinks. Over the past decade, the adoption of CT and established pilot projects were promoted in more than ten provinces. Finally, extending intermittent irrigation and adopting new seed varieties for paddy fields are also supported and promoted as part of the overall campaign to reduce GHG emissions.

Vietnam is likely to be among the countries hardest hit by climate change. This may even reverse its role as a leader in leveraging agriculture for development. In 2005, approximately 60 percent of the employed labour force was engaged in agriculture. But almost half of the country's agricultural area would face inundation with a two metre rise in sea levels. Using an innovative yield function approach, Yu *et al.* examine how a changing climate may affect rice production and how Vietnamese farmers are likely to adapt to various climatic conditions, taking into account sample selection bias and endogeneity of inputs.[43] They suggest that although climate change can potentially reduce rice production, farmers will respond mainly by adjusting the production portfolio and levels of input use. Nonetheless, they also posit that public investment in rural infrastructure and human capital will have to support farmers in the adaptation process if production levels and farm incomes are to be sustained.

Farm household strategies depend on their on-farm productivity and expected risks and returns on investments, as well as employment alternatives. Based on a simultaneous equations model for Ethiopia, Di Falco *et al.* identified the main drivers behind adaptation and food productivity increases as: access to credit, formal and farmer-to-farmer extension, and information on climate change. Interestingly, the most vulnerable farm

---

[42] Jinxia Wang, Jikun Huang and Scott Rozelle, *Climate Change and China's Agricultural Sector: An Overview of Impacts, Adaptation and Mitigation – Issue Brief No 5* (ICTSD and International Food and Agricultural Trade Policy Council 2010).

[43] Bingxin Yu *et al.*, *How Are Farmers Adapting to Climate Change in Vietnam? Endogeneity and Sample Selection in a Rice Yield Model – IFPRI Discussion Paper 01248* (IFPRI 2013).

households were the least likely to adapt even though they would benefit the most from adaptation.[44] In another study the same author found that increasing rainfall variability increases risk aversion and may decrease the propensity of farmers to invest in new crop varieties and to adopt better water and soil conservation strategies.[45]

South Asian agriculture preponderantly consists of rice and wheat-based production systems. In their 'willingness to pay' (WTP) study for this region, Taneja *et al.* found that 'climate-smart' agriculture (CSA) can mitigate the negative impacts of climate change and improve its efficiency.[46] CSA requires a comprehensive package of measures to achieve the desired objectives, and adoption is largely dependent on farmers' preferences: their capacity and WTP. To assess farmers' choices and their WTP for different climate-smart technologies and other interventions, the authors use scoring and bidding protocols from focus group meetings in two distinct regions of the Eastern and Western Indo-Gangetic Plain (IGP). Through the bidding approach, farmers implicitly expressed their WTP for new technologies that could transform current agricultural practices into relatively low carbon, more productive farming methods. The authors find that laser land levelling (LLL), crop insurance and weather advisory services are the preferred interventions in Eastern IGP, whereas farmers in Western IGP prefer LLL, direct seeding, zero tillage, irrigation scheduling and crop insurance. Actual large scale adoption of the preferred climate-smart technologies and other interventions would also require access to adequate funding as well as capacity building among technology promoters and users.

3.2 Biotechnology

There exists vast and often heated debate on the merits of genetically modified plants in the climate change context. Only two studies are referred to here, without a view on their scientific robustness.

For Lybbert and Sumner, biotechnology plays an important role but is

---

[44] Salvatore Di Falco, Marcella Veronesi and Mahmud Yesuf, 'Does Adaptation to Climate Change Provide Food Security? A Micro-Perspective from Ethiopia' (2011) 93(3) Amer J Agr Econ 829.

[45] Salvatore Di Falco, 'Adaptation to Climate Change in Sub-Saharan Agriculture: Assessing the Evidence and Rethinking the Drivers' (2014) 41(3) Eur Rev Agr Econ 405.

[46] Garima Taneja *et al.*, *Farmers' Preferences for Climate-Smart Agriculture: An Assessment in the Indo-Gangetic Plain – IFPRI Discussion Paper 01337* (IFPRI 2014).

not a quick and easy solution to addressing climate concerns and other agricultural development issues; its efficacy depends on the broader technological, economic, environmental and political circumstances of the region.[47]

Similarly, Quaye *et al.* consider that biotechnology may reverse the disturbing food insecurity trends on the African continent, but they recommend that African countries first establish effective bio-safety regulatory frameworks and commit resources to capacity building and infrastructure for biotechnology development.[48] This includes research on tropical and subtropical staple foods suitable for the needs of small scale farmers in SSA countries.

3.3 Trade and Investment Measures

Climate change is likely to add to the classic (and mostly cyclical) price drivers such as stock-to-use ratios, crude oil and manufacturing prices, the United States dollar exchange rate, interest rates, and income variations. Baffes and Dennis see here an important role of trade and investment, given that they can bring about agricultural adaptation, economic growth and environmental change.[49] The increasing demand for cash crops, especially feedstuffs, is a new development that possibly calls for international action. A still under-studied consequence of additional export opportunities is that they seem to turn land into a movable asset, often with less labour and thus less local food security, but with more social problems.

Here too, the development dimension requires careful attention. Reilly argues that international trade could mitigate the impact of climate change on agriculture.[50] However, he also recognizes that where domestic policies attempt to isolate producers and consumers from climate change or from the trade effect of changes abroad, the overall economic impact can be negative. He points out that agriculture in developing countries is often hampered not only by high input prices and low domestic prices, but also by surplus dumping from subsidized competitors. All these policies

---

[47] Lybbert and Sumner (n 18).

[48] Wilhelmina Quaye *et al.*, 'Climate Change and Food Security: The Role of Biotechnology' (2012) 12(5) AJFAND 6354.

[49] John Baffes and Allen Dennis, *Long-Term Drivers of Food Prices – Public Policy Research Working Paper 6455* (World Bank 2013).

[50] John Reilly, 'The Role of Growth and Trade in Agricultural Adaptation to Environmental Change' in Ariel Dinar and Robert Mendelsohn (eds), *Handbook On Climate Change and Agriculture* (Edward Elgar 2011).

create avoidable economic costs and reduce the attractiveness of domestic and foreign investments. The continuation of such distortions could thus further aggravate climate change effects.

3.4 Export Restrictions

Governments use agricultural export prohibitions and restrictions to combat domestic price volatility and to increase revenue. Most economists agree that export restrictions are bad for global food security, and are an additional disincentive for local farmers. The weak WTO disciplines in the General Agreement on Tariffs and Trade (GATT), Article XI:2, and Agreement on Agriculture, Article 12, fail to solve this sensitive issue (and food aid abuse). Alarmed by the last food crisis, three G8 and G20 Summits had decided to exempt at least the purchases by the World Food Programme from such restrictions, but then failed to harvest even such a 'low hanging fruit' at the WTO.

The well-studied case of Tanzania illustrates the trade and climate problem of outright export bans. The World Bank forecasts increasing poverty in the country due to climate change.[51] While its maize economy is subject both to external market scares and domestic weather shocks, the latter exert a stronger influence. Here, the intermittent imposition of export bans act as a disincentive for the local agrarian economy.[52] Rowhani *et al.* have shown that maize yields in Tanzania are affected by shifts in the growing season, as well as by greater intra-seasonal variability.[53] Droughts in other countries might offer new net income opportunities from maize exports, unless the present export bans are maintained.[54] Despite the relatively modest trade volumes involved, the regional impact of these export restrictions is even greater than on the global scale. This is because African food markets are weakly integrated with world markets. Consequently, regional trade is seen as a special means to mitigate local shocks. Unpredictable and restrictive trade policies such as in Tanzania exacerbate the impact of domestic

---

[51] Syud Amer Ahmed *et al.*, *Climate Volatility and Poverty Vulnerability in Tanzania – World Bank Policy Research Working Paper 5117* (World Bank 2009).

[52] John Baffes, Varun Kshirsagarz and Donald Mitchell, *Domestic and External Drivers of Maize Prices in Tanzania* (World Bank 2015).

[53] Pedram Rowhani *et al.*, 'Climate Variability and Crop Production in Tanzania' (2011) 151(4) Agr Forest Meteorol 449.

[54] Syud Amer Ahmed *et al.*, 'Agriculture and Trade Opportunities for Tanzania: Past Volatility and Future Climate Change' (2012) 16(3) Rev Dev Econ 429.

weather shocks rather than contributing to national and regional food security.[55]

Going beyond the general scholarly condemnation of export restrictions, an interesting World Bank research report proposes a kind of social contract between producers and consumers by comparing two types of insurance schemes, through transfers and through trade policies.[56] Do *et al.* agree that export restrictions can have a negative impact on other markets, because such restrictions can generate a counter-cyclical demand shock that amplifies the effects of the underlying food shortage. They nonetheless caution against a systematic condemnation of trade insulation practices. In their opinion, export restrictions could work like an insurance policy and transfer resources away from the producers to the consumers when food prices are high, and do the reverse when prices are low. Similarly, Ravaillon agrees that regulated trade (through taxes or even export bans addressing the issue of a *decline in food availability*) might then be a defensible policy response for helping vulnerable groups relative to feasible alternatives.[57]

3.5 Carbon Taxes

Carbon dioxide is increasingly used as feedstock for fuels, chemicals and polymers, but carbon capture and utilization by and for agriculture remains unrecognized in international environmental agreements. Carbon pricing helps lowering overall emissions while allowing businesses the flexibility to find their own most efficient solutions. A 2014 World Bank survey showed that governments responsible for 54 percent of global GHG emissions and 52 percent of GDP, as well as thousands of businesses, planned toward or favoured carbon taxes.[58] The implications of carbon taxation on the international competitiveness of the agricultural sector have also been studied for British Columbia (Canada). The potentially negative effects of such climate smart policies call for policy remedies to mitigate the effects of the tax on firm profitability and international competitiveness.[59]

---

[55] Nicholas Minot, *Transmission of World Food Price Changes to Markets in Sub-Saharan Africa* (IFPRI 2011).

[56] Quy-Toan Do, Andrei A Levchenko and Martin Ravaillon, *Trade Insulation as Social Protection – IBRD Policy Research Working Paper 6448* (World Bank 2013).

[57] Martin Ravaillon, 'Famines and Economics' (1997) 35(3) J Econ Lit 1205.

[58] IBRD Feature Story dated 22 September 2014 (on file with author).

[59] Nicholas Rivers and Brandon Schaufele, 'The Effect of Carbon Taxes on Agricultural Trade' (2014) 63(2) Can J Agr Econ 1.

3.6 Insurance

Insurance for agricultural risks has long been discussed in literature. Bertolt Brecht describes Julius Caesar as the world's first inside trader.[60] New for producer risk management in non-subsidizing developing countries are weather and other insurance schemes, and seasonal weather forecasts. How can such schemes help climate adaptation?

According to Garrido *et al.* the growth potential is immense.[61] Without insurance buttressing cropping decisions, extreme events lead to capital disinvestment and the destruction of natural assets. For agriculture, the classic weather insurance schemes against hail are expanding, often with public participation through market regulatory frameworks and financial support (premium subsidies, public reinsurance, tax rebates and direct insurance participation). The authors also point out that insurers and government agencies can provide meteorological data and forecasts, as well as risk and hazard analysis, and loss adjustment techniques. Also required, at least initially, are financial investment incentives, credit access and index insurance. Examples from Russia, India, Argentina, Mexico and Brazil show that the role of insurance as a risk management tool for farmers is increasing everywhere. However, for the poorest regions, current agricultural insurance remains inadequate. One promising exception mentioned by the authors is a scheme by which Ethiopian farmers purchase unsubsidized index insurance as part of an adaptation package. Other types of risks insured are single risks like hail and fire, combined risks and yield insurance.

An important consideration which has been little discussed in a climate change context is the mutually exclusive nature of different risk management schemes. Indeed, producer price guarantees, production quotas and subsidized safety nets such as under the US Farm Bill may defeat market based insurance schemes both nationally and globally, despite arguments to the contrary raised by insurance support defenders.[62]

---

[60] Bertolt Brecht, *Die Geschäfte des Herrn Julius Caesar* (Suhrkamp 1998) 332.

[61] Alberto Garrido *et al.*, 'Insurance as an Adaptation to Climate Variability in Agriculture' in Ariel Dinar and Robert Mendelsohn (eds), *Handbook On Climate Change and Agriculture* (Edward Elgar 2011).

[62] Joseph W Glauber, *Agricultural Insurance and the World Trade Organization – IFPRI Discussion Paper 01473* (IFPRI October 2015).

3.7 Gender Issues

Building human, organizational and institutional capacity for more effective climate change adaptation in developing countries presents both challenges and opportunities. Given the largely undervalued contribution by women, gender action seems to offer some of the greatest opportunities. Somewhat surprisingly, only a few specific case studies seem to be available.[63]

Aberman argues that vulnerability to climate change is exacerbated when individuals' asset bases are limited or insecure.[64] She recognizes the importance of gender dynamics in climate change adaptation and the use of such assets in this process; control over assets is strongly shaped by gender patterns, with women typically controlling fewer low value assets, which then are more likely to be disposed of during shocks. She thus sees a key strategy for adapting to climate change for women and men through producer groups, primarily as a tool to facilitate asset development through common purchase of large farm appliances (physical capital), group loans (financial capital) or capacity development (human capital).

A study on male and female farm managers in West Africa showed that women face additional sources of life cycle risk, particularly health risks associated with fertility and childcare.[65] Female farm managers were found less likely to purchase agricultural insurance and more likely to invest in savings for emergencies, even though those who did purchase more

---

[63] IFPRI has published a number of baseline studies, presented in Nancy Johnson *et al.*, *Gender, Assets, and Agricultural Development: Lessons from Eight Projects – IFPRI Discussion Paper No 01436* (IFPRI 2015). For Asia, see Caitlin Kieran *et al.*, *Examining Gender Inequalities in Land Rights Indicators in Asia – IFPRI Discussion Paper 01429* (IFPRI 2015). For Africa, see Valerie Mueller *et al.*, *Filling the Legal Void? Experimental Evidence from a Community-Based Legal Aid Program for Gender-Equal Land Rights in Tanzania – IFPRI Discussion Paper 1434* (IFPRI 2015). For an interesting World Bank country study see Daniel Ayalew Ali *et al.*, *Investigating the Gender Gap in Agricultural Productivity: Evidence from Uganda - Policy Research Working Paper 7262* (World Bank 2015). On the importance of social context in gender and climate variability studies see Deborah Rubin, *Qualitative Methods for Gender Research in Agricultural Development. IFPRI Discussion Paper 01535*, May 2016.

[64] Noora Aberman, *Climate Change Adaptation Assets and Group-Based Approaches: Gendered Perceptions from Bangladesh, Ethiopia, Mali, and Kenya – IFPRI Discussion Paper 01412* (IFPRI 2015).

[65] Clara Delavallade *et al.*, *Managing Risk with Insurance and Savings: Experimental Evidence for Male and Female Farm Managers in West Africa – IFPRI Discussion Paper 01426* (IFPRI 2015).

insurance obtained higher average yields and were better able to manage food insecurity and shocks.

Climate change projects should therefore pay greater attention to gender as well as to other social, political and cultural issues. Also, strengthening of organizational capacity for local organizations would help trigger mind-set and organizational culture changes.[66]

3.8 Urban Agriculture

For the first time in history a majority of humankind now lives in cities. Spontaneously, poor urban dwellers in particular have taken to producing fruits and vegetables and even small livestock wherever possible. So far, however, most urban planners have taken few steps to encourage these developments, which are naturally welcome from both a climate and local food security perspective. Producing food locally has the additional benefit of mitigating climate change through reduced food miles.

Cities like Beijing and Shanghai, as well as smaller cities in China, are incorporating space in the cityscape for urban farming in appreciation of its added 'green' value. Reported examples are rooftop vegetable farms. In Zhejiang Province (China) one rooftop paddy farm yielded over nine tons per hectare.[67]

Other successful examples are reported by the International Network of Resource Centres on Urban Agriculture and Food Security (RUAF Foundation). This NGO, with member organizations in Africa, Asia, the Middle East, Latin America and Europe, seeks to contribute to the development of sustainable cities by actively promoting resilient and equitable urban food systems.[68]

3.9 Migration

Where sustainable agriculture and/or absorption of the agricultural labour force in other economic sectors and in the same location are not possible,

---

[66] Catherine Ragasa *et al.*, *Organizational and Institutional Issues in Climate Change Adaptation and Risk Management: Insights from Practitioners' Survey in Bangladesh, Ethiopia, Kenya, and Mali – IFPRI Discussion Paper 01279* (IFPRI 2013).

[67] Paul S Teng, *Food Security: Cities as Part of the Solution and not the Problem – Commentaries No 142/2012* (RSIS Centre for Non-Traditional Security Studies (NTS) 2012).

[68] RUAF Foundation 'Home Page', accessed 7 November 2016 at http://www.ruaf.org/.

the potential consequences are migration, accelerated urbanization and violence.

The notion of 'climate refugees' existed before significant global warming occurred. Asia and the Pacific is the global area with the highest absolute number of natural disasters and of populations affected. Hence, this region is highly exposed to climate impacts; it is also home to highly vulnerable population groups, who are disproportionately poor and marginalized. In 2012, the Asian Development Bank (ADB) presented what it called 'the first significant effort to identify policy and other responses to impacts of environment events on human mobility within the Asia and Pacific region'. In the large countries of the region, the number of workers reliant on agriculture remains high, increasing the potential for climate change to impact substantially on livelihoods. The report highlights 'environmental hot spots' at particular risk of flooding, cyclones, typhoons and water stress (including rising sea levels and salinization). Consequently, the ADB expects further increases in rural emigration and cross-border migration. Migration as a way for people to cope with environmental changes can provide substantial benefits to both origin and destination areas, as well as to the migrants themselves. By taking action today, governments can reduce the likelihood of future humanitarian crises and maximize the possibilities for people to remain and produce food in their own communities. The ADB also posits that if deteriorating environmental conditions make that impractical, governments should ensure that affected people have the option to relocate to a more secure place with livelihood options.[69]

3.10 Hunger and Violence

Famines are as old as humanity, but the numbers are still unclear.[70] The FAO's method of calculating hunger has been the subject of heated debate. After having reached the staggering figure of one billion in 2010, the FAO changed its methodology. In its report for 2015 it estimated that 795 million people are undernourished globally, down 167 million over the last decade, and 216 million less than in 1990–1992.[71] This of course is slow

---

[69] Asian Development Bank (ADB), *Addressing Climate Change and Migration in Asia and the Pacific – Final Report* (ADB 2012).

[70] Joachim De Weerdt *et al.*, *The Challenge of Measuring Hunger through Survey – Discussion Paper Series 365/2015* (LICOS Centre for Institutions and Economic Performance 2015).

[71] FAO, IFAD and WFP, *The State of Food Insecurity in the World 2015. Meeting the 2015 International Hunger Targets: Taking Stock of Uneven Progress* (FAO 2015).

progress compared with progress in reducing extreme poverty. At the same time, the world produces more than enough food relative to the needs of its whole population and is still far from running out of land to sustain agricultural growth, even if productivity growth were to slow down in the coming decades.[72] The question that arises is whether climate change exacerbates global food insecurity in such a way that trade and aid cannot make sufficient food flow towards the poor and hungry. Research informing the debate on violence, hunger and climate change is predominantly carried out by economists and political scientists. Economic literature has been growing along with Amartya Sen's landmark work on famines and his call for 'entitlement' reiterated in 2010.[73]

Ravaillon argues that famines have defied simple explanations as well as geographical boundaries.[74] He points out that episodes of famine occur in both socialist and capitalist economic systems; they come with and without declines in food supply, and with and without wars or unusual political or social instability. He also considers that an objective and holistic view of the determinants of individual survival prospects in economies under stress offers the greatest hope of understanding why famines happen, and of usefully informing domestic and international public action on how best to avoid or relieve them. That is the main insight that economic analysis can offer to replace the earlier reliance on often doubtful 'single-cause' explanations for famines. Market and non-market (including governmental) institutions which in normal times work quite adequately, though not perfectly, can easily turn even a moderate aggregate shock into a devastating blow to poor people. For Ravaillon, a famine can thus be viewed as a tragic magnification of normal market and governmental failures.

While famines are not easily predicted, there are some common elements. Entitlement failure is one. Although famines have happened with and without crop failures, every famine appears to have entailed a collapse in the legal command over food of at least some subgroups within society, either through a loss of endowment or a contraction in the amount of food that can be acquired from given endowments. According to Ravaillon, vulnerability to famine also increases with weak social and physical infrastructure, a weak and unprepared government and a relatively closed political regime.

---

72 Hector Maletta, *Towards the End of Hunger. Lima* (Universidad del Pacífico 2016).

73 United Nations Development Programme (UNDP), *Human Development Report 2010: The Real Wealth of Nations: Pathways to Human Development* (UNDP 2010).

74 Ravaillon (n 57).

A growing literature on the nexus between climate variables and violence confirms that, like migration, 'food riots' often have multiple causes. Contrary to frequently advanced popular beliefs, few if any scientific studies directly correlate rising world market prices to local food riots.[75] A field study in Indonesia, with data from 1990–2003 and using variations in minimum temperature rather than average temperature, indicates that a climate-related reduction in future rice production could increase violence. This study was based on seasonal effects reducing (per capita) the availability of rice when the minimum temperature increased during the core month of the 'wet planting season,' that is, December. The authors accept that it is difficult to correlate such events with climate change or with world food price hikes. They also point out the need to consider the specificities of different crops.[76]

Legal scholars have so far offered little insight on how to regulate agriculture and food security in a climate change context. This is perhaps because such 'best practices' have never been codified at the international level. At any rate, an overview of recent literature shows a remarkable lack of analytical research on the legal side of climate change-related agricultural production and trade policies, with the possible exception of FDI aspects. The following section looks at the regulatory side of the agricultural trade and investment interface with climate change.

4. INTERNATIONAL REGULATORY FRAMEWORK FOR TRADE AND INVESTMENT

4.1 Trade

How do trade rules shape climate action for agriculture? Which policies are mutually reinforcing, and where does the present regulatory framework impede climate focused measures with a trade impact? Answers to these questions must start with a general review of the relevant trade rules.

4.1.1 Role of trade regulation

The academic debate on the trade challenges to climate mitigation and adaptation is still ongoing. Because adaptation especially in developing

---

[75] Nicholas Minot, *Food Price Volatility in Africa: Has It Really Increased? – IFPRI Discussion Paper 01239* (IFPRI 2012). Also see Minot (n 55).

[76] Raul Caruso, Ilaria Petrarca and Roberto Ricciuti, *Climate Change, Rice Crops and Violence: Evidence from Indonesia – CESifo Working Paper No 4665* (Center for Economic Studies and Ifo Institute 2014).

countries depends on sustainable growth and trade, this is not merely a question of priorities and sequencing. For instance, tariff reductions for 'green' goods under the (plurilateral) Environmental Goods Agreement in the WTO would be a welcome but clearly insufficient contribution to international climate governance. Trade rules complicate matters as soon as more fundamental questions arise. For instance, the shift to a low carbon economy demands internalization of carbon emission costs. But a credible formula for an 'acceptable interference with international trade' is yet to be found for the WTO rules applying to this case of the polluter-pays principle, as will be shown below for the EU's aviation scheme.[77] This problem also impairs the 'coalitions of the willing' proposed in the form of a 'Climate Club'.[78]

Some scholars still hold that the WTO rules framework is sufficiently flexible to allow for carbon taxation as long as such measures are not protectionist. This chapter tries to show that, contrary to much of the prevailing discourse and literature, WTO's present rules can prevent both autonomous and collective action for mitigating climate change or preserving biodiversity proposed by Hoekman and Mavroidis.[79] Mere exhortations in environmental agreements not to impede trade will not do, even when they are drafted on the basis of the exception clause in GATT Article XX. One example is UNFCCC Article 3.5: 'Measures taken to combat climate change, including unilateral ones, should not constitute a means of arbitrary or unjustifiable discrimination or a disguised restriction on international trade.' Similar references in the new Climate Agreement adopted in Paris (COP21) are equally unhelpful.[80]

Cottier calls for international climate cooperation by way of enhanced market access and the establishment of stable and fair conditions of competition for domestic operators, exporters and investors alike.[81] He also considers that only a departure from the Westphalian system of territorially defined allocations of powers and responsibilities can achieve

---

77 Felicity Deane, *Emissions Trading and WTO Law* (Edward Elgar 2015).

78 David G Victor, *The Case for Climate Clubs* (ICTSD and World Economic Forum 2015).

79 Bernard Hoekman and Petros C Mavroidis, *Regulatory Spillovers and the Trading System: From Coherence to Cooperation* (ICTSD and World Economic Forum 2015).

80 See n 6.

81 Thomas Cottier, 'Confidence-Building for Global Challenges: The Experience of International Economic Law and Relations' in Ruth Greenspan Bell and Micah S Ziegler (eds), *Building International Climate Cooperation: Lessons from the Weapons and Trade Regimes for Achieving International Climate Goals* (World Resources Institute 2012).

a climate-secure future. While international environmental law is now moving from trans-boundary pollution to shared or *common concerns*, this is not yet the case for the trading system enshrined in the WTO, whose objectives are the pursuit of welfare through an open trading system and the containment of unilateral climate measures and pressures. Existing customary law affirms the sovereign right of states to manage their own natural resources and to support effective adaptation. But for Cottier *cum suis*, it also prohibits a state from allowing activities on its territory to inflict serious damage on the environment of other states or on parts of the environment that do not belong to any state.[82]

In her book on border carbon adjustments (BCAs), Holzer finds emerging agreement that this is an appropriate and necessary policy tool to address the carbon leakage and competitiveness concerns associated with emissions pricing systems. Article 17 of the Kyoto Protocol to the UNFCCC specifically allows emissions trading where countries are committed to limitation and reduction of their emissions under Article 3 of that agreement. WTO law may not be directly applicable here, but such measures have an indirect trade impact. For example, the General Agreement on Trade in Services (GATS) does not explicitly mention emissions allowance trading. But Article 5(x)(F) of the GATS Financial Services Annex applies to 'other negotiable instruments and financial assets.' Hence, Holzer concludes that including imports in national systems, and emissions allowance rebates on exports, keeps systems vulnerable to WTO challenges, notwithstanding UNFCCC provisions.[83]

In the absence of a solution negotiated in the WTO, trade rule deficiencies are also likely to hamper climate action for agriculture. The contribution of trade to global food security is now well-researched and generally agreed upon within the scholarship, and has been confirmed at various World Food Summits. But even hard-core free-traders concede that trade liberalization will not automatically increase food security, especially in poor food-importing countries – nor will it directly promote climate adaptation.[84] Fragmentation of trade and environment rules impacts many

---

[82] Thomas Cottier and Sofya Matteotti-Berkutova, 'International Environmental Law and the Evolving Concept of "Common Concern of Mankind"' in Thomas Cottier, Olga Nartova and Sadeq Bigdeli (eds), *International Trade Regulation and the Mitigation of Climate Change* (Cambridge University Press 2009).

[83] Kateryna Holzer, *Carbon-related Border Adjustment and WTO Law* (Edward Elgar 2014).

[84] For a good overview from a NGO perspective see Jennifer Clapp, *Trade Liberalization and Food Security: Examining the Linkages* (Quaker United Nations Office 2014). For the debate between the former UN Special Rapporteur on

climate adaptation instruments, notably: border measures, harmonization of unilateral standards, consumer choice instruments, water management, tax and subsidies for technology, infrastructure and institutional innovations, and global data collection.

The next subsections address more specifically agricultural subsidies, regional and multilateral approaches, autonomous climate measures, and labelling schemes and standards. They basically show that even though the WTO obligation of non-discrimination has its merits in preventing protectionism, it fails to solve several trade-related problems of climate action.

4.1.2 Agricultural trade-related subsidies

As of 2016, the WTO has not even begun to discuss how to deal with footprint differences, including virtual water content, for specific agricultural products. Trade diplomats may argue that the lack of agreed upon definitions and measurements prevents a debate on this particularly sensitive aspect of the so-called 'production and process method' problem, and how to treat otherwise 'like products'. The UNFCCC and the Kyoto Protocol acknowledge the problem but, with the exception of New Zealand, no national emissions trading schemes, carbon taxes or a combination of both address GHG emissions (or carbon sequestration) by agriculture, including that of the EU.

Meanwhile, trade-distorting farm subsidies (sometimes labelled 'food security enhancements') remain limited but not prohibited. The Doha Round negotiations failed to further reduce even those support measures that increase GHGs. In two interesting contributions, Peters (for the EU)[85] and Winebarger (for the US)[86] have even argued that such farm policies imply a violation of legal obligations under the Kyoto Protocol, and of decisions promoting UNFCCC and Kyoto Protocol objectives.

---

the Right to Food, Olivier de Schutter, and then WTO Director General Pascal Lamy, see Christian Häberli, 'What's Wrong with WTO Rules Applying to Food Security?' in Rosemary Rayfuse and Nicole Weisfelt (eds), *The Challenge of Food Security* (Edward Elgar 2012).

[85] Victoria Peters, 'A Legal Obligation to Mitigate Greenhouse Gas Emissions from Agriculture: A Challenge to the European Union's Emissions Trading System and the EU Member States with the Largest Agricultural Impact' (2015) 19 UCLA J Int'l L & For Aff 213.

[86] Lisa Winebarger, 'Standing behind Beastly Emissions: The US Subsidization of Animal Agriculture Violates the United Nations Framework Convention on Climate Change' (2012) 27(4) Am U Int'l L Rev 991.

4.1.3 Multilateral and regional approaches

Cottier also discusses the decarbonization of economies where biofuels and biomass can contribute to an effective abatement of CO_2 emissions and to climate change mitigation.[87] This requires enhanced disciplines applying to anti-dumping measures. He argues that clean energy adds additional arguments for the reform of anti-dumping and for shifting it toward the law of unfair competition, which also could include services currently absent from such disciplines. For Cottier, production subsidies should be substantially limited but allow for R&D by activating the disciplines of non-actionable subsidies. Countervailing duties should be limited to constellations with strict compliance requirements. Structural adjustment problems of the industry should be addressed by means of safeguard measures which do not increase long-term costs to consumers, nor prevent competitive products from reaching the market. Tariff deconsolidation may be indicated where protection is sought for reasons of strategic trade policy. Flanking disciplines should be strengthened in offsetting subsidies for fossil fuels. The rules on non-discrimination and subsidies are able to frame government support and differential taxation appropriately. Policy space may also be developed for local content requirements for least developed countries, either by means of R&D or with the concept of graduation. According to Cottier, implementing an obligated transfer of technology to the least developed countries also requires a review of relevant intellectual property rules.

In the absence of satisfactory multilateral rules, Leal-Arcas looks for progress through regional approaches across a number of international law instruments, such as trade policy tools (permissions, prohibitions, international standards) and financial instruments (taxes and funding programmes).[88] The question is, of course, how to persuade countries to be parties to trade agreements with GHG reduction commitments. Nonetheless, according to Leal-Arcas, regional trade arrangements could be designed to provide an attractive package with which to settle trade-offs and conflicts of interest, as well as to facilitate the ultimate goal of creating a global climate regime. Agricultural economists further consider that a successful conclusion of the Doha Round in the WTO would greatly

---

[87] Thomas Cottier, 'Renewable Energy and WTO Law: More Policy Space or Enhanced Disciplines?' (2014) 1(14) RELP 40.

[88] Rafael Leal-Arcas, 'Climate Change Mitigation from the Bottom Up: Using Preferential Trade Agreements to Promote Climate Change Mitigation' (2013) 7(1) CCLR 34.

improve conditions of competition for agricultural trade and food security in developing countries.[89]

Laborde argues that 50 percent of global gains from the market access and domestic support reforms envisaged in the so-called 'Modalities' (December 2008) will come from the agricultural sector: a 36 percent reduction in average agricultural bound tariffs, a 26 percent reduction in average applied levels of trade-distorting protection, and US$15 billion of border taxes would be removed.[90] He adds that even if (mainly) developing countries will eventually operate less drastic cuts, these modalities would still lead to more trade liberalization than any other WTO, or non-WTO, agreement discussed so far. Hence, food-exporting developing countries presently facing such barriers would benefit from such results and could in turn improve their own climate change resilience.

4.1.4 Autonomous climate measures

Difficulties can arise when countries introduce GHG reduction measures, even when they are authorized to do so under an international climate agreement. If their climate action implies trade measures designed to limit the loss of competitiveness with products and services from countries without such commitments, a number of WTO rules may become an issue. For instance, a proposal by the EU to tax biodiesel met with stiff resistance from the American Soybean Association, arguing the US biodiesel regime was compatible with WTO rules and with the sustainable land use requirements of the EU's Renewable Energy Directive.[91]

Autonomous carbon taxation (for example, 'cap-and-trade' measures) has to take into account competitiveness and carbon leakage issues faced by other domestic sectors subject to a price on carbon. BCA is one already discussed measure. Governments trying to ensure carbon neutrality may apply countervailing (WTO-problematic) border measures to avoid what otherwise is basically a case of (WTO-unproblematic) non-discrimination.

The EU Aviation Directive is a case in point where several countries including China, Malaysia and the USA argued that the implementation measures violated WTO non-discrimination rules. The EU had to

---

[89] Kimberly Ann Elliott, *Food Security in Developing Countries: Is There a Role for the WTO?* (Center for Global Development 2015).

[90] David Laborde, 'Implications of the Draft Market Access Modalities on Bound and Applied Tariffs' in Ricardo Meléndez-Ortiz, Christophe Bellmann and Jonathan Hepburn (eds), *Tackling Agriculture in the Post-Bali Context* (ICTSD 2014).

[91] ASA Seeks Suspension of European Union Energy Policy Over Biodiesel. *Inside U.S. Trade*, 18 March 2011.

suspend this climate-friendly measure. Bartels shows that BCA might not withstand a WTO dispute settlement and, in addition, that the EU's scheme violated its international civil aviation obligations. Hence, a carbon scheme that is administratively feasible and WTO-compatible remains a challenge.[92] As pointed out by Meltzer, it will be crucial to manage the trade and climate change intersection in ways that maintain the integrity of the WTO system, while recognizing the imperative need for action to address climate change.[93] He argues that the type of WTO rules which the EU Aviation Directive might have violated are actually useful disciplines guiding countries in the development and application of climate action, without impeding international trade. Hence, the basic challenge is to find an appropriate balance between policy space of WTO Members for CO_2 reduction measures while maintaining an open and non-discriminatory trading system that supports economic growth and global welfare.

4.1.5 Labels and standards

Governments prescribe or promote labels and standards as trade-directing tools for climate adaptation. Most of them are private, voluntary standards initiated, implemented and maintained by retailers and their upstream suppliers. Retailers would be able to collect the market premium that consumers are willing to pay on low carbon goods, and they would also have an incentive to join supply chain management systems funnelling a part of the premium back to low carbon producers before this becomes a legal requirement. Early adoption of such systems may well create a first-mover advantage over competitors in the food chain. Well-managed and monitored consumer information labels are generally considered a good idea. Some renowned scholars, however, have raised concerns.

In his provocative book *The Bottom Billion*, Collier argues that the fair trade price premium motivated by charity may actually make it harder for poor producers to move into other activities: 'They get charity as long as they stay producing the crops that have locked them into poverty.'[94] In their survey of American consumers, Sachdeva *et al.* reach a more nuanced conclusion on the positive and negative spillovers of 'green consumerism'.[95]

---

[92] Lorand Bartels, 'The WTO Legality of the Application of the EU's Emission Trading System to Aviation' (2012) 23(2) EJIL 429.

[93] Joshua Meltzer, 'Climate Change and Trade: The EU Aviation Directive and the WTO' (2012) 15(1) J Int'l Econ L 111.

[94] Paul Collier, *The Bottom Billion: Why the Poorest Countries are Failing and What Can be Done about It* (Oxford University Press 2007).

[95] Sonya Sachdeva, Jennifer Jordan and Nina Mazar, 'Green Consumerism:

MacGregor extensively discusses various carbon labelling initiatives.[96] He considers that a developing country may have different priorities than the importing market with higher product standards, GHG emission caps and built-in assumed transport cost differences. True carbon cost calculations are correctly based on life cycle analysis and appropriately priced environmental externalities. But according to MacGregor, methods with sufficient scientific robustness have not yet been developed and the cost of verification and monitoring may be higher for developing country producers. Furthermore, there are no agreed upon standards reflecting the differences in historical GHG emissions and the still low per capita output in poor countries. The UNFCCC acknowledges global inequity in responsibility for dealing with climate change. Hence, this idea of 'ecological space' for non-restrictive economic development would logically need to be reflected in a universal standard or a credible carbon footprint label, indicating the total additional emissions that the planet can sustain without serious disruption to its climate. For MacGregor, agricultural trade and a move away from subsistence farming may indeed provide considerable benefits which should not be curtailed by arbitrary climate standards. Thus the roles of private standards and public legislation in addressing carbon concerns in the food system require a development impact assessment. A coordinated and credible scaling up of successful approaches to sustainable food trade would then release the potential for consumers to obtain contractual carbon reductions in the food system.

4.2 Investment

4.2.1 Rules to change

Adaptation of agriculture to climate change is a challenge to the whole food supply chain. The economic and social implications of the global agri-food value chains for developing countries and global poverty reduction are beginning to be understood.[97] Governments use investment tools to prevent or alleviate food price shocks they expect to result from production problems at home or abroad. They offer advantageous

---

Moral Motivations to a Sustainable Future' (2015) 6 *Current Opinion in Psychology* 60.

[96] James MacGregor, *Carbon Concerns: How Standards and Labelling Initiatives Must Not Limit Agricultural Trade From Developing Countries – ICTSD–IPC Platform on Climate Change, Agriculture and Trade Issue Brief No 3* (ICTSD and International Food and Agricultural Trade Policy Council 2010).

[97] Johan FM Swinnen, *Global Agricultural Value Chains, Standards, and Development – EUI Working Paper RSCAS 2014/30* (LICOS 2014).

investment conditions, and even 'stabilization clauses' to foreigners when local investment and technologies fail to produce the desired results. Between 2000 and 2011, large scale land acquisitions increased by 203 million hectares globally, a huge figure by any account.[98] As late as 2008 the World Bank acknowledged decades of underinvestment even in its own programmes, only to note a rapid increase of local and foreign investment in 2010 (often at the cost of environmental and social concerns related to 'land grabbing').[99] The 'global land rush' has since reached more countries in Asia and Africa, with more operators and variations than ever before. Yet, the displacement of local populations remains a serious issue. In 2015, the World Bank identified serious shortcomings in its own projects for the implementation of resettlement policies causing people to lose land or other assets, and even access to resources.[100] Somewhat surprisingly, few investment lawyers seem to recognize a need for specific rules for FDI in agriculture. Only Cotula sees new questions here regarding the role of international law and governance in farmland acquisition, processing, infrastructure, logistics and food waste mitigation.[101]

Esty *et al.* consider that the WTO is a crucial player in the establishment of environmental governance.[102] A Global Environmental Organization of the kind they propose can only avoid significant environmental degradation with support from the WTO in particular, with which a new trade-environment structure should be built so as to internalize the significant externalities resulting from GHG emissions. Failing that, the trading system would (inadvertently) produce economically inefficient results, lower social welfare and create fundamental unfairness across nations and within societies, as well as on an intergenerational basis. They thus call for a multi-tier compliance mechanism, reinforced by the WTO and

---

[98] Christian Häberli, 'Foreign Direct Investment in Agriculture: Land Grab or Food Security Improvement?' in Thomas Eger, Stefan Oeter and Stefan Voigt (eds), *Economic Analysis of International Law* (Mohr Siebeck 2014).

[99] Matias E Margulis, Nora McKeon and Saturnino M Borras (eds), *Land Grabbing and Global Governance* (Routledge 2014).

[100] World Bank, 'World Bank Acknowledges Shortcomings in Resettlement Projects, Announces Action Plan to Fix Problems – IBRD Press Release No 2015/332/ECR' (4 March 2015), accessed 7 November 2016 at http://www.worldbank.org/en/news/press-release/2015/03/04/world-bank-shortcomings-resettlement-projects-plan-fix-problems.

[101] Cotula (n 4).

[102] Daniel C Esty and Anthony LI Moffa, 'Why Climate Change Collective Action has Failed and What Needs to be Done Within and Without the Trade Regime' (2012) 15(3) J Int'l Econ L 777.

designed to track transboundary environmental harms and member states' adherence to treaty commitments.

The question here is under which regulatory conditions agricultural FDI can support climate adaptation. Generally speaking, foreign investors in the agricultural sector are *under-regulated* in the current framework of international investment law, voluntary guidelines and fragmented national investment legislation, and *over-protected* in regional and bilateral investment treaties and domestic regulation. As pointed out by Dessy *et al.*, given that many farmland investments are in countries with weak land governance, host state regulatory capacity must be strengthened.[103]

Adherence to international guidelines for agro-FDI requires more than discretionary reporting by investors. Home states must not remain inactive where their international obligations and reputations are concerned. As a minimum, their support for any agro-FDI project must be subject to both *ex ante* and *ex post* impact assessments, and to some sort of independent monitoring. On a general level, they should review their FDI regulation and policies, their bilateral investment treaties (BITs), their trade agreements and their role in relevant international financial institutions, with a view to safeguarding the public interest concerning global food security. There can be no presumption that foreign landowners and long term lessees will do this automatically. Current global governance structures for agro-FDI distribute unevenly the rights and obligations between the host state, the investor and the investor's home state. All too often there are precious few legally enforceable obligations at the regional and international level to conduct agricultural investment in climate-smart ways. On the contrary, international investment law not only protects the investor from any direct or indirect expropriation, but even condones so-called stabilization clauses in investment contracts whereby a host state commits, for instance, to not enforce international climate adaptation measures such as carbon taxation, or not to reduce water allocation for irrigation. This may prevent a host state from protecting local food security rights and thus from fulfilling its international right-to-food obligations under the International Covenant on Economic and Social Rights of the UN. Moreover, weak host states often lack the institutional capacity to monitor the investor's behaviour, and the capacity to ensure that their food security is not undermined by the investment.[104]

Unless these issues are addressed in international negotiations, climate

---

[103] Sylvain Dessy, Gaston Gohou and Désiré Vencatachellum, *Foreign Direct Investments in Africa's Farmlands: Threat or Opportunity for Local Populations? – Working Paper 12-03* (Centre interuniversitaire sur le Risque, les Politiques Economiques et l'Emploi 2011).

[104] Christian Häberli and Fiona Smith, 'Food Security and Agri-Foreign

change action is likely to further exacerbate the problem. Finally, an internationally coordinated approach may further promote integration of international human rights in the context of an investment dispute.[105]

The remainder of this section looks at the regulatory issues of: specific performance standards and related investment subsidies, food stockpiles, and international water management.

4.2.2 Investment and input subsidies

Generally speaking, WTO rules for environmental subsidies are found in the GATT, the Agreement on Agriculture and the Agreement on Subsidies and Countervailing Measures (SCM). Processes and production methods may be relevant to determining the issue of 'like products' in GATT Articles I, III, XI and XXIII, the SCM Agreement, the Antidumping Agreement, the Agreement on Technical Barriers to Trade (TBT Agreement) and the Import License Agreement. The general exceptions in GATT, Article XX, paragraphs (b) and (g) might protect climate adaptation measures if they were adequately prescribed in multilateral environmental agreements. In such a process, it will be important to minimize adverse impacts on poor developing countries, as prescribed in Articles 2.3 and 3.14 of the Kyoto Protocol. The scholarly debate here has started but there is still no relevant case law showing a legal way out of this conundrum.

Blandford and Josling have studied climate related investment measures in the form of performance standards, best-practice requirements, subsidies, carbon taxes and tax rebates, and public expenditure for research and extension.[106] They consider that climate-smart subsidies could be given for such practices as minimum tillage or the co-generation of on-farm bioenergy. Conservation payments could incorporate incentives for carbon sequestration. These might be consistent with WTO rules if they are part of a comprehensive environmental programme, and limited to the extra costs incurred by farmers. Nonetheless, they also recognize that such policies, particularly on export crops, impact the competitive position of

---

Direct Investment in Weak States: Finding the Governance Gap to Avoid "Land Grab"' (2014) 77(2) MLR 189.

[105] Christine Kaufmann, 'International Law in Recession? The Role of International Law When Crisis Hits: Food, Finance and Climate Change' in U Fastenrath *et al.* (eds), *From Bilateralism to Community Interest, Essays in Honour of Bruno Simma* (Oxford University Press 2011).

[106] David Blandford and Tim Josling, *Greenhouse Gas Reduction Policies and Agriculture: Implications for Production Incentives and International Trade Disciplines - Issue Brief No 1* (ICTSD and International Food and Agricultural Trade Policy Council 2009).

agriculture and the agro-investment climate. Hence, challenges by trade partners remain possible even when the payments are tied to sequestration. Besides, subsidies tied to specific uses of farm commodities have unintended consequences for food supplies and prices, and tying such subsidies to climate change goals is likely to confound an already confused situation. Given these complexities, the aim should be to identify and address the positive and negative impacts of farming on atmospheric GHG concentrations that are uncontroversial and relatively easy to measure. For Blandford and Josling, domestic schemes that contribute to effective climate action should be possible. However, to avoid investment disputes, this requires an international consensus on which measures are likely to be climate-effective. The move towards 'decoupled' direct payments unrelated to price and current output has provided an opportunity for such a consensus. From an economics viewpoint, a constructive first step would be to 'recouple' such payments in a way that climate change mitigation is encouraged without jeopardizing food security.

One problem with the SCM Agreement from a climate adaptation perspective has already been pointed out by Sykes.[107] He notes that import substitution subsidies and export subsidies are automatically 'specific,' and thus prohibited by Article 3. He claims that this prohibition makes little economic sense and recalls that the general non-discrimination provisions already prohibit import substitution subsidies in the form of local content requirements under GATT, Article III:4. Economically, consumer subsidies might make more sense for climate action; in the form of domestic payments (without exports) they would fall under Article III:8(b) and thus not violate Article III:4. At the same time, and if they involve trade restrictions, they could still be challenged as actionable subsidies under the SCM Agreement, and countervailed if there are exports.

Condon notes that in 2002, Marceau and Trachtman had suggested that it would require a 'heroic approach to interpretation' to apply GATT, Article XX as justification for a violation under another WTO agreement. Nonetheless, he considers that:

> environmental subsidies could be non-actionable under the SCM Agreement if differences in carbon footprints can be used to conclude that products are not 'like' as that term is used in the SCM Agreement. However, if the likeness of products depends on factors that affect the product as such, these cases would be less relevant. He thus concludes that multilateral environmental agreements on climate change will probably be necessary in determining the consistency

---

[107] Alan O Sykes, *The Limited Economic Case for Subsidies Regulation* (ICTSD and World Economic Forum 2015).

of climate change measures with the SCM and the TBT Agreements, and the effective protection offered by GATT Article XX.[108]

4.2.3 Stockpiles

Many governments resort to public stockpiles for staple food to mitigate cyclical causes of food insecurity, avoid supply disruptions and reduce price volatility. Such stocks are very costly, prone to corruption, and they may defeat market-based risk management tools such as weather insurance. Whether they can help adaptation to structural developments like climate change is an open question, so far without empirical evidence. But after the last food crisis in 2007–2009 many developing countries saw few alternatives besides heavy producer subsidies. In addition, some such schemes also involve price support for producers and/or consumers. This is the present situation of stockpile policies which economic scholars have to consider when discussing ways and means to improve food security of small farms and of poor urban households in developing countries.

More generally, Gouel and Jean see a role for price stabilization in a small, open, developing country confronted with shocks in both yield and foreign price.[109] They consider that public intervention can mitigate unwanted domestic and international price fluctuations. Optimally designed trade and/or storage policies can thus increase social welfare where consumers are risk-averse and markets are incomplete. Gouel and Jean thus argue that an optimal combination of storage and trade policies can contribute to stabilizing domestic food prices. On the other hand and as shown above by Do *et al.*, export restrictions, even as part of such a policy mix, can be harmful to trade partners.[110] However, to refrain from using them is also costly and entails substantial transfers from consumers to producers. The same is true for other measures isolating domestic markets from the world price spikes, such as lowered import duties and export restrictions. Government intervention would then be required to address the combination of household risk aversion with incomplete markets (for example, the lack of insurance markets for food prices). Gouel and Jean point out that this kind of market imperfection can be addressed by optimal discretionary public interventions through trade and storage policies, as well as by subsidizing imports and taxing exports. Proceeds from export

---

[108] Bradly J Condon, 'Climate Change and Unresolved Issues in WTO Law' (2009) 12(4) J Int'l Econ L 895.

[109] Christophe Gouel and Sébastien Jean, *Optimal Food Price Stabilization in a Small Open Developing Country – Policy Research Working Paper 5943* (Centre d'Études Prospectives et d'Informations Internationales 2012).

[110] Do, Levchenko and Ravaillon (n 56).

taxation can cover the fiscal cost of import subsidies. Import subsidies in turn alleviate the traditional limitation of food storage through their 'non-negativity'. When stocks are exhausted, subsidizing imports prevents price spikes; when stabilization is pursued through storage subsidies only, it does not improve consumers' welfare. Since storage policies are generally thought of as a way to help consumers, these results sound like a warning that storage policies designed without any flanking trade policy might be inconsistent: the limited insulation provided by trade costs, especially when they are relatively small, does not allow any independent food price policy to be pursued. However for Gouel and Jean, a well-designed combination of trade and storage taxes and subsidies can be a cost-efficient, price-stabilizing policy for certain developing countries.

At the Ninth WTO Ministerial Conference in Bali (December 2013), the relevant rules for food stockpiles were examined to facilitate the use of this instrument without breaching their subsidy limits or impairing other developing countries' market access. After a week of negotiations, a decision on Public Stockholding for Food Security Purposes was adopted with a 'peace clause' shielding certain stockpile programmes from subsidy complaints in formal litigation. At the Tenth WTO Ministerial Conference in Nairobi (December 2015) the impasse continued without changes to the present rules in the WTO Agreement on Agriculture.

The need for changing the rules for stockpiles remains to be demonstrated. Such changes might nevertheless be necessary for other, non-trade-distorting forms of regional 'virtual' food reserves or for government mandated but privately owned stockpiles.[111]

4.2.4 International water (and aquaculture) management

Climate adaptation strategies raise particularly complex questions in respect to water. Agriculture as an economic sector is by far the largest consumer of fresh water (irrigation and rainwater). All agricultural imports contain 'virtual' water from the producer country. There is a need to consider the impact of the ongoing expansion of irrigated agriculture and other productivity improvements, and of the growing rainfall irregularities, on demand and welfare both locally and internationally. Changing rainfall patterns (leading to shifts in agricultural productivity) and increasing intensity of hurricanes and typhoons require international governmental cooperation.

The dearth of scientifically informed decision making and the difficulty

---

[111] Christian Häberli, *After Bali: WTO Rules Applying to Public Food Reserves – FAO Commodity and Trade Policy Research Working Paper No 46* (FAO 2014).

of integrating science into the social and economic dimensions of fisheries have, so far, prevented a climate-resilient regulatory solution. Moomaw and Blankenship have counted 65 fisheries treaties plus a dozen additional ocean management and protection agreements.[112] However, even if these treaties were enacted, evolving knowledge of the oceans now shows that an 'irreversible fisheries tipping point' may be closer than anticipated. The authors point out that fisheries are no longer threatened only by overfishing, but by a decline in overall oceanic productivity as well. The myriad compounding threats include: destructive fishing practices, management by species rather than by ecosystems, oceanic and terrestrial-based pollution, the rising aquaculture industry, invasive species, human population growth and development, UV radiation, climate change, and ocean acidification. Therefore, the 'fisheries problem' is no longer just a sector problem, but rather a component of a vast interconnected, multifaceted network of global environmental issues and unsustainable practices that require the inclusive approach of Sustainable Development Diplomacy. Moomaw and Blankenship consider that there is still time to start reversing the destructive trends impacting the oceans, but this must be done soon, and in a comprehensive, holistic, adaptive and innovative manner.

Here again, WTO rules for fish (and shipping) subsidies are part of the problem. Lee sees even a case of 'WTO Minus' in the negotiated but not adopted drafts, arguing that the new rules for regulating and prohibiting fisheries subsidies proposed therein constitute a 'significant departure from the basic framework of the existing subsidies norms and jurisprudence found in the SCM Agreement', because 'some of the major elements of the fisheries subsidies norms fail to reflect the actual practice of subsidy investigations or disputes'.[113] This regulatory failure comes at the expense of climate-exposed small operators in poor countries.

5. CONCLUSION

This chapter has attempted to show, first, that international food security governance largely fails the test of coherence even without the climate change factor; both national policies and the relevant multilateral rules framework are, in reality, farm security policies with little regard for

---

112 Moomaw and Blankenship (n 34).

113 Jaemin Lee, 'Looking for a Panacea in the SCM Agreement? Systemic Challenges for Post-Bali Fisheries Subsidies Discussion and some Food for Thought to Overcome Them' (2014) 9 Asian J WTO Int Heal 477.

poor consumers' interests. Second, the present rules' deficiencies and the lax subsidy constraints continue to allow those policies and practices to negatively impact producers in other countries. Global food security is thus impaired by an under-restrained (yet WTO-legal) misuse of policy space. Actually, even policies which might be compatible with WTO rules may violate the UN-enshrined Right to Food. Third, it appears that even some of the most fundamental non-discrimination rules of the WTO hamper collective action and autonomous climate mitigation and adaptation measures. Consequently, unreformed agricultural trade and investment rules could make climate change mitigation and adaptation even more difficult.

The good news is that this world seems capable, at present projections for global warming, of producing enough food for considerably more than nine billion people. In addition, this seems doable without a significant expansion of present commercial production surfaces and even with lower productivity increases than those recorded in the period 1960–2010. The real problem would then be the development disparities that imply a lack of resilience by weak states, resource-poor farmers, pastoralists and small-scale fishermen, as well as poor consumers at or below the absolute poverty line. Such disparities, not least in the form of gender issues, are likely to be compounded by climate change and by the fragmentation of trade and investment rules supposed to ensure that sufficient food flows where and when it is needed (except by the poorest population segments having to rely on food aid). Agriculture is more affected by climate change and contributes more to GHG emissions, but it also offers more development opportunities in a climate change perspective than other sectors.

The need for regulatory climate action starts with the present rules framework, which currently allows a misuse of policy space to undermine global food security. Many farm support policies fail to address national food security, and they neglect negative spill-overs onto other producers. The latter problem might be solvable with a successful conclusion of the Doha Round negotiations in the WTO. However, climate-smart production, trade and agro-investment policies lack legal security in the multilateral trade and investment regime. Even some of the most fundamental WTO rules on non-discrimination stand in the way of collective climate action and autonomous mitigation measures. Hence, unreformed agricultural trade and investment rules are particularly unhelpful for addressing the challenges of climate change.

The complexity and the urgency of the many issues involved show that climate change is not just another structural challenge to the question of how to feed nine billion people. They also explain why international regulatory action must be comprehensive. The present and apparently non-negotiable deficiencies of several WTO rules, and the lack of stringent disciplines for trade-distorting farm and fish subsidies, jeopardize climate action for agriculture. Agro-FDI is under-regulated and over-protected. UNFCCC and FAO have yet to agree on climate-smart agricultural practices and standards. Such a multi-forum negotiation will be, technically and politically, extremely difficult. But this is also where regulatory action must start. Agricultural policy reforms without international guidance may increase national farm security but not food security facing a climate change stress test. International standards and rules would thus help to define policy space in an international, non-distorting context for both agro-FDI and trade.

Multilaterally agreed upon climate, environmental, and agricultural sustainability standards could pave the way for the changes in trade and investment rules necessary to make this world climate-proof. These are the fundamental non-discrimination rules of the GATT and the TBT Agreements, which now prevent autonomous and even collective climate action, and the rules prohibiting or limiting climate change support countervailing action against environmental dumping under the WTO agreements on subsidies, anti-dumping and agriculture. Changes are also required for international investment treaties, regional trade agreements, and sectorial agreements on energy, aviation, water management, shipping, fishing and migration. Even preferential treatment for climate-friendly products and processing methods from developing countries must be re-evaluated.

The climate challenge for international diplomacy is daunting, and it is not clear that sufficient political will, or even knowledge and inter-agency cooperation, is available for the multifaceted tasks at hand. Yet, as the case of the EU Aviation Directive has shown, climate action that impacts trade requires legal certainty in all relevant domains of the international regulatory framework.

The climate challenge to agriculture and to the whole food value chain appears manageable on a global scale, but not without changes to trade and investment regulation, and not without special consideration of its development dimension. The time to act is now.

11. Agriculture and payments for ecosystem services in the era of climate change

JB Ruhl

1. INTRODUCTION

Agriculture is a form of ecosystem management. Farms and other agricultural working lands are highly managed, human dominated ecosystems, but they are ecosystems nonetheless. Agriculture alters and then manages ecological processes and functions on small and large scales. Like all ecosystems, agricultural lands provide valuable ecosystem services to human populations. Agricultural lands are managed primarily to provide food and fiber resources, but those are as important to humans as any of the ecosystem services provided by, say, a large wilderness area. Agriculture policy, therefore, is largely about what flows of ecosystem services policymakers hope to derive from agricultural working lands.

Ecosystem services are economically valuable benefits which humans derive from ecological resources directly, such as storm surge mitigation provided by coastal dunes and marshes, and indirectly, such as nutrient cycling that supports crop production.[1] Under the widely adopted typology developed in the *Millennium Ecosystem Assessment*, ecosystem services flow to human communities in four streams: (1) *provisioning* services are commodities such as food, wood, fiber and water; (2) *regulating* services moderate or control environmental conditions, such as flood control by wetlands, water purification by aquifers and carbon sequestration by forests; (3) *cultural* services include recreation, education and

---

[1] For descriptions of the ecosystem services science and policy framework, see *Millennium Ecosystem Assessment, Ecosystems and Human Well-Being: Synthesis* (2005), accessed 7 November 2016 at http://www.milleniumassessment.org/documents/document.356.aspx.pdf; Gretchen C. Daily (ed.) *Nature's Services: Societal Dependence on Natural Ecosystems* (Island Press 1997). For coverage of the emergence of the ecosystem services concept in law and policy, see JB Ruhl, Steven E Kraft and Christopher L Lant, *The Law and Policy of Ecosystem Services* (Island Press 2007); James Salzman, 'A Field of Green? The Past and Future of Ecosystem Services' (2006) 21 J. Land Use and Envtl. L. 133; JB Ruhl and James Salzman, 'The Law and Policy Beginnings of Ecosystem Services' (2007) 22 J. Land Use & Envtl. L. 157.

aesthetics; and (4) *supporting* services, such as nutrient cycling, soil formation and primary production, make the other three service streams possible.[2] For example, aquatic resources such as coastal wetlands can provide bountiful supplies of ecosystem services to human populations in all of these categories, including water, groundwater recharge, storm and flood mitigation, sediment control, water purification, climate regulation, water supply and recreation.[3]

Farms and other agricultural working lands also can supply all of these ecosystem service types to local and distant human populations. The difference in this respect between a farm and a preserved wilderness is that the farm is actively managed to provide a desired blend of ecosystem services, whereas wilderness is managed to preserve natural conditions with the ecosystem services flow being incidental to that goal. Farmers make their living mainly by providing provisioning services in the form of food and fiber; that is what private markets and public subsidy programs pay farmers to do. The flow of regulating and supporting services from agricultural lands to human populations, if any, has historically been incidental to the intentionally-managed flow of provisioning services.

Prioritizing the production of food, fiber and other provisioning services from agricultural working lands has taken its toll on the environment. In many areas, for example, farms are the primary cause of aquatic resources impairment.[4] One of the core objectives of agriculture policy, therefore, has been to influence the impact of agricultural working lands on the environment, usually by providing public monetary incentives to change practices. In the United States, for example, paying farmers to alter practices on behalf of environmental quality has been a theme of federal farm policy for decades, embodied in programs such as the Conservation Reserve Program (CRP), which pays farmers to take land out of production for defined periods to enhance its conservation values, and the Environmental Quality Incentives Program (EQIP), which pays farmers to employ better practices on working lands.[5] The main thrust of these

---

[2] *Millennium Ecosystem Assessment*, above note 1, at vi. Although this typology is not universally accepted, I will use it here given its wide adoption.

[3] Stuart Butchart et al., *Millennium Ecosystem Assessment, Ecosystems and Human Well-Being: Wetlands and Water: Synthesis* (2005), accessed 7 November 2016 at http://www.millenniumassessment.org/documents/document.358.aspx.pdf.

[4] JB Ruhl, 'Farms, Their Environmental Harms, and Environmental Law' (2000) 27 *Ecology Law Quarterly* 263.

[5] For thorough reviews of agricultural land retirement and working land conservation subsidy programs in the United States, see Megan Stubbs, Congressional Research Service, *Conservation Provisions in the 2014 Farm Bill* (2014).

'land sparing' and 'land sharing' payment programs, however, has been to improve environmental quality.[6] To the extent they also enhance the flow of regulating and supporting services to human populations, that has historically been largely incidental to their environmental protection purpose.

In recent years, however, greater attention has been given to changing the flow of ecosystem services from agricultural working lands as an explicit policy goal.[7] Ecologists and economists focusing on agriculture have forged a more complete vision of the natural capital capacity of agricultural working lands as capable of providing a stream of diverse goods and services, including regulating and supporting ecosystem services such as increased biodiversity, carbon sequestration, pollination, groundwater recharge and improvement of water quality.[8] Hence there is growing demand for policy-relevant research into the regulating and supporting of ecosystem services available from agriculture, and growing interest in creating incentives for farmers to provide those services.[9]

Consistent with historical agriculture policy, it is likely that the primary form of incentives for that purpose will be payments. Indeed, the expanding body of research focused on the design of 'payments for ecosystem services' (PES) programs has agricultural working lands as one of its primary themes. Broadly framed, a PES program involves the exchange of value for land management practices intended to provide or ensure the provision of ecosystem services.[10] Agricultural PES (AgPES) programs in

---

[6] Ben Phalan *et al.*, 'Reconciling Food Production and Biodiversity Conservation: Land Sharing and Land Sparing Compared' (2011) 333 *Science* 1289.

[7] G Philip Robertson, 'Farming for Ecosystem Services: An Ecological Approach to Production Agriculture' (2014) 64 *BioScience* 404.

[8] Scott M Swinton, 'Reimagining Farms as Managed Ecosystems' (2008) 23(2) *Choices* 28–31, accessed 7 November 2016 at http://www.choicesmagazine.org/magazine/pdf/article_17.pdf.

[9] Scott M Swinton *et al.*, 'Ecosystem Services from Agriculture: Looking Beyond the Usual Suspects' (2006) 88 Amer. J. Agric. Econ. 1160, 1164. An excellent survey of the literature supporting this movement is found at G Philip Robertson and Scott M Swinton, 'Reconciling Agricultural Productivity and Environmental Integrity: A Grand Challenge for Agriculture' (2005) 3 Frontiers in Ecology and the Env't 38.

[10] For general background on PES programs, see B Kelsey Jack, Carolyn Kousky and Katherine RE Sims, 'Designing Payments for Ecosystem Services: Lessons from Previous Experience with Incentive-Based Mechanisms' (2008) 105 Proceedings of the Nat'l Acad. of Sci. 9465. For discussion of specific PES initiatives in a variety of nations, see James Salzman, 'Creating Markets for Ecosystem Services: Notes from the Field' (2005) 80 N.Y.U. L. Rev. 870; Brian C Steed, 'Government Payments for Ecosystem Services – Lessons from Costa Rica' (2007) 23 J. Land Use & Envtl. L. 177.

particular are designed to provide incentives to change on-farm practices and thereby rebalance the ecosystem services profile to increase off-farm flows of regulating and supporting services. By no means, however, are AgPES programs intended to take away from the goal of improving the environmental quality performance of agricultural working lands. To the contrary, improving environmental quality and enhancing the flows of regulating and supporting ecosystem services can go hand in hand with agriculture policy goals. At the very least, conventional environmental quality payment programs such as the CRP and EQIP – even if improving the environmental quality impacts of agriculture remains their primary purpose – can account for their ecosystem services benefits as an additional policy justification. But that is not the same as designing a true AgPES program: one specifically oriented toward paying to yield ecosystem services as the return on investment.

Indeed, AgPES programs are even more likely to become important policy tools within the larger space of climate change adaptation policies. Ironically, because they are highly managed, agricultural working lands hold significant potential to rebalance the off-site flow of regulating and supporting ecosystem services. As preservation and conservation resources experience the stress of climate change, the flow of regulating and ecosystem services from them may become less secure, making agricultural working lands even more important as a source of those services.

Although the case for using AgPES programs seems fairly sound and their role appears likely to grow more important over time, there are a number of policy challenges and concerns. Valuing ecosystem services, particularly many regulating and supporting services for which no functioning economic markets exist, is not a straightforward undertaking. Property rights in ecosystem services are also often poorly defined, as are the baseline expectations of agricultural working lands in terms of off-farm flows of ecosystem services.[11] Knowing how much to pay and calculating the return on investment thus are not self-evident propositions. Designing an AgPES program, therefore, requires close attention to economic, ecological and legal details.

This chapter focuses primarily on the details of AgPES program design with special attention to their potential role in climate change adaptation. The chapter first provides brief overviews of the ecosystem services framework and the design basics of AgPES programs. It then turns to the promise and challenges of AgPES programs in light of climate change adaptation policy. In closing, the chapter suggests some of the legal

---

[11] Salzman, above note 10.

questions that should be asked and answered when using AgPES as a climate change adaptation policy instrument.

2. THE ECOSYSTEM SERVICES FRAMEWORK[12]

The ecosystem services framework went from an academic research theme in the mid-1990s to a dominant policy theme by 2010, because it operationalizes the obvious. By explicitly describing ecosystems as providing economically valuable benefits *to humans* and proposing a rigorous, scientifically-based argument for integrating those values into private and public decisions, the ecosystem services framework put something on the table that had been missing from the conservation side of the negotiation of environmental policy and decision making: money. For too long the argument in support of conservation has depended largely on an appeal to environmental well-being and the intrinsic values of nature; whereas the argument in support of resource development has had human well-being, especially in the form of economic progress, on its side. The ecosystem services framework might not even up the stakes in that sense, but by putting raw economic values and other contributions to human well-being into play on behalf of conservation, it goes a long way to change the dynamics in the ongoing policy negotiation between resource conservation and development.

Framed as such, the ecosystem services approach has powerful implications for private markets and public policy, but it is often difficult to tap into its full potential. For example, knowledge about the value of ecosystem services improves the information available to landowners in deciding what constitutes the most efficient use of the land and its associated resources. To take advantage of that information, however, private landowners need some way of capturing the value of the services in markets, which is difficult for services like pollination from wild pollinators and groundwater recharge of aquifers from wetlands. Many ecosystem services are classic public goods. The natural capital owner of trees cannot prevent others from benefitting from the photosynthesis service (the service is non-excludable), and if you were for some reason to buy photosynthesis from the owner, you could not prevent other people from benefitting from

---

[12] This section is adapted with permission from JB Ruhl, 'In Defense of Ecosystem Services' (2015) 32 Pace Envtl. L. Rev. 306, and JB Ruhl, 'Agriculture and Ecosystem Services – Strategies for State and Local Government' (2008) 17 NYU Environmental Law Journal 424.

it (the service is non-rivalrous).[13] The hardest cases in this respect are regulating and supporting services. We already have functioning markets and long-standing policies for provisioning services like crops, and cultural services like recreation. It is much more difficult to get a handle on how to bring the regulating and supporting services themes into private markets and public policy.

There are two major obstacles in this regard. First, as mentioned previously, regulating and supporting services often have the qualities of public goods. Because the natural capital owner cannot easily market the service, there is no reason to take it into account when considering alternative uses of the resource.[14] And because the service beneficiaries do not have to pay for the service, they do not invest in the natural capital needed to support it. The inevitable consequence of this incentive structure is the depletion of natural capital.[15]

Second, what to do about the first major obstacle is complicated by the lack of clear property rights.[16] Who owns pollination? Who owns photosynthesis? Are these property rights that run with the land? Or is providing these public goods a baseline expectation society demands of property owners? What services must property owners deliver to society versus those they may claim as theirs to control and choose to deplete or, if they can, sell into markets? And if regulation steps in to prevent property owners from making that choice, have property rights been taken without just compensation?

Even if these obstacles could be easily overcome, the ecosystem services framework does not enjoy universal support. The chief concern coming from those who assign policy primacy to environmental conservation is that incorporating ecosystem services into markets and public policy will commodify nature[17] and oversimplify environmental challenges in the public mind.[18] A related theme focused particularly in the Global South

---

13 Christopher L Lant *et al.*, 'The Tragedy of Ecosystem Services' (2008) 58 *BioScience* 969, 970–1.

14 *Ibid.*

15 *Ibid.*

16 *Ibid.*

17 Nicolas Kosoy and Esteve Corbera, 'Payments for Ecosystem Services as Commodity Fetish' (2010) 69 Ecological Econ. 1228; Brett Sylvester Matulis, 'The Economic Valuation of Nature: A Question of Justice?' (2014) 104 Ecological Econ. 155.

18 Roldan Muradian *et al.*, 'Reconciling Theory and Practice: An Alternative Conceptual Framework for Understanding Payments for Environmental Services' (2010) 69 Ecological Econ. 1202; Richard B Norgaard, 'Ecosystem Services: From Eye-Opening Metaphor to Complexity Blinder' (2010) 69 Ecological Econ. 1219;

comes from land tenure advocates who argue that by commoditizing ecosystem services, we open the door to greater state and corporate control of the environment at the expense of poor and other marginalized populations.[19] The fear is that the powerful alignment of state and corporate interests ultimately will push indigenous and other marginalized populations off their lands because of the very ambiguous land tenure systems that many countries have.[20] In general, the concern is that the market efficiency, payments, and property rights themes of the ecosystem services framework, even when purportedly aimed at ameliorating poverty, ignore equity concerns.[21]

Some argue that the ecosystem services framework masks the tradeoffs of managing for particular ecosystem services.[22] Do we want to manage or engineer ecosystems to provide our preferred flows of ecosystem services? If so, what ecosystem structures and processes are relegated to the sidelines? There is also a social justice dimension to this specialization question:[23] if we are going to manage for specific ecosystem services, which services and for which beneficiaries? Particularly with climate change likely to disproportionately impact vulnerable low income populations, how do we ensure that public investments in natural capital (as part of adaptation planning) provide an equitable climate resilience profile?[24]

Finally, there is the so called stacking problem, which raises the concern that landowners and other market participants will game the ecosystem services framework.[25] This concern postulates a day when a property

---

Bhaskar Vira and William M Adams, 'Ecosystem Services and Conservation Strategy: Beware the Silver Bullet' (2009) 2 *Conservation Letters* 158.

[19] For an extensive discussion of this theme, see Sharachchandra Lele, 'Environmentalisms, Justices and the Limits of Ecosystem Services Frameworks' in Thomas Sikor (ed.), *The Justices and Injustices of Ecosystem Services* (Island Press 2013) 119.

[20] Diana Suhardiman et al., 'Payments for Ecosystem Services in Vietnam: Market-Based Incentives or State Control of Resources?' (2013) 6 Ecosystem Serv. 64.

[21] Unai Pascual *et al.*, 'Exploring the Links Between Equity and Efficiency in Payments for Environmental Services: A Conceptual Approach' (2010) 69 Ecological Econ. 1237.

[22] Elena Bennett *et al.*, 'Understanding Relationships Among Multiple Ecosystem Services' (2009) 12 *Ecological Letters* 1394; Robert B Jackson *et al.*, 'Trading Water for Carbon with Biological Carbon Sequestration' (2005) 310 *Science* 1944.

[23] Katie K Arkema *et al.*, 'Coastal Habitats Shield People and Property from Sea-Level Rise and Storms' (2013) 3 *Nature Climate Change* 913.

[24] *Justices and Injustices*, above note 19.

[25] Morgan Robertson *et al.*, 'Stacking Ecosystem Services' (2014) 12 *Frontiers in Ecology & the Environment* 186.

owner, say of a wetland, could split the property into all the discrete services it provides and sell them in different credit markets operated by public agencies and private interests.[26] A credit for the habitat, a credit for the groundwater recharge, a credit for this other service, and so on, until the wetland has been sold 20 times. But it is just one wetland, not 20 discrete streams of services. Or is it? Can we really slice it and dice it so that these credits are going in many different directions and we suffer no net loss of ecological structures and functions?

All of the above concerns merit attention, and they can be particularly acute in the agricultural context, where in many nations the land tenure rights of agricultural land users are tenuous. Even where these concerns can be addressed, however, agricultural ecosystem services policy faces core challenges simply by the nature of agriculture and agriculture policy. We need to know more about the geographic and economic contexts before we can assess the prospects of realigning the ecosystem services profile of agricultural land uses. Agriculture presents a difficult geographic scenario for developing generalized strategies for ecosystem services. Farms are numerous, dispersed, come in all sizes and produce many different commodities under many different climate and landscape conditions. Farms also manage ecological resources for relatively small spatial scales (the farm) and short temporal scales (the next harvest). The focus on optimizing on-site provisioning services also tends to sever farms and larger agricultural landscapes from surrounding ecological resources. Managing ecosystem services sustainably, by contrast, requires multiscale approaches that integrate connected ecological units across space and time.

These geographic disconnects strongly influence the economics of farming and the bias toward provisioning services. The payoff for providing regulating and supporting services, assuming some mechanism for compensation, is likely often to be marginal compared to the return on investment in commodity production or, even more so, selling to urban development interests. In the absence of any compensation, in other words, economically rational farmers will not provide free regulating and supporting services to off-site lands at levels above the required baseline, unless doing so is incidental to optimization of on-site commodity production or is forced by more regulation. To put the policy problem in standard economic terms, farmers view the provision of regulating and supporting

---

Royal C Gardner and Jessica Fox, 'The Legal Status of Environmental Credit Stacking' (2013) 40 Ecology L.Q. 713.

[26] *Justices and Injustices*, above note 19.

services to outside communities as a positive externality; doing so benefits others, but nobody has been willing to pay for these benefits.[27] The market primarily rewards farmers for producing commodities, and crop and livestock production subsidies can further distort the trade-off between commodity production and ecological conservation.

Promoting farm production of regulating and supporting ecosystem services, therefore, is a balancing exercise between providing farmers with the flexibility to continue benefitting from their skill at managing provisioning services on the one hand, and providing them with the impetus to produce more regulating services for society on the other. Ideally the economics of farming, including market distorting subsidy policies, can be altered to change the flow of ecosystem services, rather than forcing the change through command and control regulation. Thus the turn to AgPES in current agriculture policy discourse.

3. DESIGNING AGPES PROGRAMS[28]

Like all PES programs, an AgPES program provides incentives in exchange for provision of desired ecosystem services. AgPES program payments are not subsidies: the value of the incentive should approximate the value of the ecosystem services.[29] Using AgPES programs to enhance delivery of these and other regulating and supporting services from agricultural lands to surrounding communities thus presents a number of threshold design issues. First, the baseline expectations of agricultural land uses must be defined so that the AgPES program can identify appropriate incentives and when to apply them. In general, the regulatory baseline for agriculture has been set quite low, meaning farmers have relatively high expectations

---

[27] Jules Pretty *et al.*, 'Policy Challenges and Priorities for Internalizing the Externalities of Modern Agriculture' (2001) 44 J. Envtl. Planning and Mgmt. 263.

[28] The section is adapted with permission from JB Ruhl, 'Agriculture and Ecosystem Services – Paying Farmers to Do the New Right Thing,' in Mary Jane Angelo, Jason J. Czarnezki, and William S Eubanks II (eds), *Food, Agriculture, and the Environment* (Environmental Law Institute/Island Press, 2013), and JB Ruhl, 'Agriculture and Ecosystem Services – Strategies for State and Local Government' (2008) 17 NYU Environmental Law Journal 424 (2008).

[29] Indeed, some advocates of AgPES programs in agricultural settings emphasize this feature to differentiate PES programs from subsidy programs. See, e.g., John M Antle, *Payments for Ecosystem Services and U.S. Farm Policy* 17 (Am. Enter. Inst. Forum on the 2007 Farm bill and Beyond, Working Paper, 2006), accessed 7 November 2016 at http://aic.ucdavis.edu/research/farmbill07/aeibriefs/20070515_antlefinal.pdf.

for when they deserve incentives to push them toward policy objectives.[30] The AgPES program thus must form a clear understanding of existing agricultural practices, the desired practices (for example, riparian buffers, wetland recharge features, native vegetation open space) and performance levels that trigger incentives.

Next, the goal of enhancing regulating and supporting ecosystem services should be based on a known present or expected future demand that can be assigned a value with reasonable geographic and economic specificity. Where are the expected ecosystem services likely to produce benefits, in what form, and how valuable are they in present and expected future land-use scenarios within the AgPES program's jurisdiction? These are macro level questions that require: assessment of the potential capacity of existing and restored natural capital on the target agricultural lands, the present and future configuration of land uses in the managing jurisdiction, and the geographic match between the two. Valuation of the ecosystem service flows then can be based on expected avoided costs; for example, the cost savings of avoided flood control or recharge capital expenditures and the avoided costs of flood and drought damages.[31] Comprehensive inventories of natural and built capital within the AgPES program area will be needed, as will well-conceived future land use planning projections.

With the demand side analysis in place, a supply side assessment is also necessary to identify the most effective and efficient incentive distribution. In all likelihood, the agricultural lands identified as having the capacity to enhance ecosystem service values will be held by numerous owners. Yet the delivery of ecosystem services from the landscape is unlikely to be linear and proportionate, such that securing 10 percent of the targeted lands will yield 10 percent of the ecosystem services. Rather, the connections between ecological resources and the delivery of ecosystem service benefits to human populations is known in many contexts to operate at landscape scales, to involve an array of ecological attributes and to behave in nonlinear relationships over space and time.[32] This raises the difficult question of how precisely to define the proxy for natural capital the program is

---

30 Ruhl, above note 4.

31 Assigning these values is a major research effort today among agricultural economists. See John M Antle and Jetse J Stoorvogel, 'Predicting the Supply of Ecosystem Services from Agriculture' (2006) 88 Am. J. Agric. Econ. 1174, 1174.

32 Edward B Barbier *et al.*, 'Coastal Ecosystem-Based Management with Nonlinear Ecological Functions and Values' (2008) 319 *Science* 321, 321 (demonstrating that wave attenuation benefits of coastal wetlands do not respond linearly to surface area of the wetlands).

designed to maintain, and for which the incentive is paid or doled out.[33] Moreover, securing 10 percent of the natural capital capacity of the targeted agricultural lands might produce zero improvement in ecological service flows; it might take half of the targeted natural capital enhancements before the AgPES program realizes any measurable ecosystem service benefits. In that event, providing incentives to cover anything less than half the resources, while it would secure ecological resources, would not have the desired payoff in ecosystem service value. Incentive programs thus must be carefully designed to correspond with the ecological properties of the targeted natural capital resources on a landscape level first, from which incentives can be provided to specific parcel owners in a coordinated manner.

Finally, any incentive system designed to enhance targeted ecosystem services must account for the larger physical and political systems in which it is operating. Aligning farmer incentives to provide, say, increased water recharge resources or riparian habitat necessarily imposes some tradeoffs both ecologically and economically. Those tradeoffs must be recognized and considered, along with other policies. Will the retention of agricultural land and open space have an impact on the stock of affordable housing in the jurisdiction? Are national farm policies competing with state and local incentive programs, making it difficult for one or the other set of incentives to gain traction?[34] The point is simply that the AgPES program, even assuming it has been thought through with respect to baseline performance expectations and the demand for and supply of services, must be integrated into the larger picture, and its consequences and full range of tradeoffs fully considered.

Overcoming these challenges has proven difficult in practice. In the United States, for example, the 2008 Farm Bill directed the US Department of Agriculture (USDA) to 'establish technical guidelines that outline science based methods to measure the environmental services benefits from conservation and land management activities in order to facilitate the participation of farmers, ranchers, and forest landowners in emerging environmental services markets'.[35] The USDA now explicitly recognizes ecosystem service values as a basis for payments under traditional

---

[33] Jack *et al.* note that 'when marginal benefits from service provision are not constant, more complex incentive schemes are needed to achieve environmental effectiveness'. Jack, Kousky and Sims, above note 10, at 9466.

[34] Jack, Kousky and Sims, above note 10, at 9467 (discussing the potential for different incentive and subsidy programs to work at cross purposes).

[35] Food, Conservation, and Energy Act of 2008, Pub. L. No. 110–246, § 2709, 122 Stat. 1651 (codified as amended at 16 U.S.C. § 3845 (2012)).

conservation program payments,[36] but so far the Farm Bill provisions have not produced new AgPES programs.

An example of one of the few operating AgPES programs in the United States is the Northern Everglades Payment for Environmental Services (NEPES) program,[37] which developed out of the Florida Ranchlands Ecosystem Services Project (FRESP) pilot program.[38] NEPES provides ranchers with payments for altering onsite water flows and retention to improve water quality and quantity conditions in Lake Okeechobee. With over $45 million in funding administered by the South Florida Water Management District, NEPES demonstrates that AgPES programs are a viable policy instrument. There are also a number of watershed-based PES programs around the United States, some of which include agricultural working lands.[39] Overall, though, AgPES programs designed to enhance the flow of regulating and supporting services from agricultural lands exist more in theory than in practice.

Besides the inherent design challenges, which NEPES shows can be overcome, what accounts for the slow development of AgPES programs? Most likely a complex mix of factors is at play and much depends on the political and economic context in which these programs aim to exist. In developed nations, shifting to AgPES faces the policy inertia built from long histories of agricultural subsidies and conservation-based incentive payments. In less developed nations, ambiguous land tenure rights and lack of institutional capacity surely have a role to play. Nevertheless, there is good reason to believe that interest in AgPES programs will not wane but instead will only increase. The reason is climate change.

---

[36] 7 C.F.R. §§ 625.8(f), 1467.20(b) (2014).

[37] *Water Storage Strategies*, S. Fla. Water Mgmt Dist., accessed 7 November 2016 at http://my.sfwmd.gov/portal/page/portal/xweb%20protecting%20and%20restoring/water%20storage%20programs.

[38] Florida Ranchlands Environmental Services Project (2015), http://www.fresp.org (accessed 7 November 2016). For a thorough review of the FRESP program, see Jacob T Cremer, 'Tractors Versus Bulldozers: Integrating Growth Management and Ecosystem Services to Conserve Agriculture' (2009) 39 Envtl. L. Rep. (Envtl. L. Inst.) 10541; Sarah Lynch and Leonard Shabman, 'The Florida Ranchlands Environmental Services Project: Field Testing a Pay-for-Environmental-Services Program' (2007) *Resources* 17.

[39] Genevieve Bennett *et al.*, *Charting New Waters: State of Watershed Payments 2012* 49–66 (Forest Trends 2013), accessed 7 November 2016 at http://www.forest-trends.org/documents/files/doc_3308.pdf.

4. AGPES AND CLIMATE CHANGE

The flow profile of ecosystem services from agricultural lands and the role of AgPES in rebalancing the profile to meet policy goals, are likely to gain even more policy attention as climate change mitigation and adaptation policies emerge. The connection with mitigation policy is obvious: agricultural working lands could provide space for conversion to vegetative cover designed to enhance carbon sequestration.[40] But there is more to it than that. Agriculture is a source of greenhouse gas emissions, so any change in practice that reduces those emissions, such as soil conservation, advances climate change mitigation policy.[41] AgPES programs thus could be targeted at enhancing carbon sequestration services and greenhouse gas regulation services.

Climate change adaptation policy presents a more complex set of incentives and tradeoffs. Climate change will undoubtedly impair the crop and livestock productivity of some agricultural lands.[42] As their capacity for production of those provisioning services is negatively affected, the agricultural landowners are likely to seek alternative forms of land use income, which may include supplying regulating and supporting services through AgPES programs. Climate change, in other words, will in some cases alter the comparative marginal returns from agricultural lands between supplying provisioning services versus supplying regulating and supporting services.

Of course, for that to happen there must be a demand for the regulating and supporting services sufficient to justify payments that will shift the agricultural land supply curve. That is likely to come about in many areas as non-agricultural lands and resources also suffer from climate change. If a regulating or supporting service becomes highly valuable in general as a regional climate change adaptation strategy (for example, flood control, groundwater recharge), all land types in the region that can supply it could be the target of PES programs. If the capacity for supplying those regulating and supporting services from non-agricultural lands is more negatively affected relative to agricultural lands, the latter will be even more important as the source of the regulating and supporting services.

The other side of the coin, however, is that climate change adaptation policy may also fuel demand for the provisioning services agricultural lands

---

40 Robertson, above note 7.

41 Mary Jane Angelo and Joanna Reilly-Brown, 'Whole-System Agricultural Certification: Using Lessons Learned from LEED to Build a Resilient Agricultural System to Adapt to Climate Change' (2014) 85 *Colorado Law Review* 689.

42 *Ibid.*

supply, particularly crops and livestock, if food security is disrupted. Even so, AgPES programs could include regulating and supporting services as their target to promote climate resilient working lands. For example, increased crop diversity could make farm production more resilient to the pests, diseases, invasive species, and extreme weather events climate change redistributes across the landscape.[43] Hence, even if the primary role of agricultural lands is to produce provisioning services, which is likely, AgPES programs can target regulating and supporting services essential to enhancing resilience to climate change.

5. GETTING THE LAW RIGHT

PES programs have recently been the focus of calls to get the science right and get the economics right before diving in,[44] and surely those are essential prerequisites. But another fundamental of PES design is to get the law right as well, and this will be of particular importance for AgPES programs. Just as farms are highly varied around the globe, presenting complex and different scientific and economics scenarios, so too are the legal contexts within which they operate. This varied legal context raises a number of questions relevant to the design of AgPES:

(1) *What is the existing land tenure status of the agricultural land user?* As discussed above, ambiguous land tenure regimes can present formidable obstacles to AgPES programs and even cause them to undermine other policy goals. First, it is essential that an AgPES program match the payment to the person who has control over the land use practices the program hopes to influence. To the extent an AgPES program makes agricultural land more valuable, ambiguous land tenure rights threaten the existing land user, the one most capable of altering the land use practice and who likely most values the payment, with potential competing claims or ouster by more powerful private and public interests. Designing AgPES programs in many nations thus may require also rethinking the land tenure rights associated with agricultural lands that are the target of the payments.

---

[43] Robert Adler, 'Climate Change Adaptation and Agriculture and Forestry Law' in Jonathan Verschuuren (ed.) *Research Handbook on Climate Change Adaptation Law* (Edward Elgar 2013); Angelo and Reilly-Brown, above note 41.

[44] AP Kinzig, 'Paying for Ecosystem Services – Promise and Peril' (2011) 334 *Science* 603; S Naeem *et al.*, 'Get the Science Right When Paying for Ecosystem Services' (2015) 347 *Science* 1206.

(2) *What are the existing baseline rules of environmental impact from agricultural land uses?*
An AgPES program should focus on paying for additionality: for yield of services the agricultural landowner is not already required by law or contract to provide. In many contexts, however, this baseline rule is also ambiguous, or, as in the case of the United States, is unambiguously set low compared to other land uses.[45] Particularly if significant changes in practices will be needed to yield the desired type and level of ecosystem services, ambiguous or low baseline rules can make the AgPES program prohibitively expensive.

(3) *What are existing property rights protections relevant to changes in the baseline rules?*
Where the baseline rules are ambiguous or set low, using public regulations to clarify or raise them is one way of reducing the cost of the AgPES program. But in many jurisdictions, in particular the United States, regulation imposing new constraints on land use is often challenged as requiring just compensation, making the AgPES program just as expensive as had the regulation not been imposed. While such claims are not usually successful, the threat and cost of litigation can deter the coordination of public entities in support of public and private AgPES programs.

(4) *What conditions can legally be attached to the PES payment?*
AgPES programs should be designed to track their return on investment, which may involve monitoring and enforcement of the terms of the payment. This may require codifying those terms either through public regulation or by contract, and there may be procedural and substantive limits applicable to either mechanism.

(5) *What conditions and requirements can legally be imposed on the beneficiaries of the rebalanced services?*
An AgPES program is supposed to have a human population as its beneficiary, otherwise it is not paying for ecosystem services. It may be appropriate to expect those beneficiaries to support the funding of the payments. A public AgPES program, for example, might rely on taxes to fund the payments, in which case it must comply with applicable taxing authority rules. A private AgPES program could rely on contractually defined payments by beneficiaries to fund payments to the landowners, enforcement of which would require compliance with contract law.

---

45 Ruhl, above note 4.

(6) *How might climate change alter any of this legal context?*
Climate change will be a force not only of environmental and social disruption but also of legal disruption. The answers given to any of the foregoing questions today could be different in the future as climate change puts pressure on the rules of property and contract, and on governance in general. Any AgPES program brought on line in such a dynamic legal environment thus should anticipate the need for legal resilience over time through adaptive regulatory mechanisms and contract terms.

6. CONCLUSION

Agricultural working lands are uniquely positioned to be a 'winner' in general and relative to other land use types as public and private demand for regulating and supporting services rises as part of climate change mitigation and adaptation strategies. Lands currently in preservation status cannot add natural capital, and lands currently in developed status have had their natural capital depleted. Agricultural lands, as highly managed blends of natural and engineered ecosystem resources, are in the best position to pivot towards a rebalanced ecosystem service profile. The climate adaptation policy and market demand for regulating and supporting services from agricultural lands could tip many agricultural landowners in that direction. AgPES programs thus will increasingly make sense not only as agricultural policy and environmental conservation policy, but as climate change adaptation policy.

12. Climate change and agrofuels: Brazilian ethanol and the Cerrado biome

Sérgio Sauer, José Paulo Pietrafesa and Pedro Araujo Pietrafesa

INTRODUCTION

The cultivation of sugarcane has been a part of Brazilian history since the mid-16th century when Portuguese colonists sought to develop an agriculture-based economy reliant upon slave labor. Following the exploitation of Brazilian redwood, sugarcane plantations were considered the main source of agricultural and industrial wealth of the colonies and drove the demand for African slaves in colonial Brazil.[1] Even with the end of the so-called 'sugarcane economic cycle', this age-old crop maintained its socioeconomic importance in the Northeast Region. By the middle of the 20th century, sugarcane had spread throughout the country as a result of the oil crisis and the 1985 establishment of the National Ethanol Program (*Pró-Alcool*).

The *Pró-Alcool,* a program to transition from fossil fuel-based gasoline to ethanol on a national scale, was introduced by the then civil military dictatorship (1964–1984) and gained momentum in the second half of the 1970s, before declining in the mid-1980s due to falling oil prices on the international market.[2] Associated with the establishment of the Sugar and Ethanol Institute (IAA) in 1933 and government incentives for the sector,[3] *Pró-Alcool* was largely responsible for the expansion of sugarcane cultivation in savannah and forest areas in the South-Central Region of the country and transformed the state of São Paulo into the largest national producer of sugar and ethanol from the 1970s onwards.

After the ethanol crisis of the 1980s, in 2003 the development of flex-fuel

---

1 Afrânio Raul Garcia Jr, 'A Sociologia rural no Brasil: entre escravos do passado e parceiros do futuro' (2003) 5 (10) *Revista Sociologias* 154–89.

2 John Wilkinson and Selena Herrera, 'Biofuels in Brazil: Debates and Impacts' (2010) 37 (4) *Journal of Peasant Studies* 749–68.

3 André Novo, K Jansen, M Slingerland and K Giller, 'Biofuel, Dairy Production and Beef in Brazil: Competing Claims on Land Use in São Paulo State' (2010) 37(4) *Journal of Peasant Studies*, 769–92.

vehicles brought about an upturn in the production and consumption of automotive fuel derived from sugarcane[4] on the basis that agrofuels have 'potential for reducing global greenhouse gas emissions'.[5] Environmental concerns and the ongoing search for alternative and renewable energy sources resulted in private, and especially international, investment; and government incentives for the expansion of the sugar-energy sector (biofuel production, electricity and sugar) were applied to already established or potential markets, the result of this environmental narrative, in conjunction with the ever increasing search for alternative energy sources.[6]

The decision of many European countries to transition away from petroleum, for example, created an increased demand for Brazilian sugarcane to be used in the production of ethanol for agrofuels. This increased demand, or possibility of increasing the demand, was part of the driving forces leading to the implementation of government incentives, especially the creation of the National Agro Energy Plan for 2006–2011.[7] Sugarcane plantations began to spread west from São Paulo into the world's most biologically diverse savannah,[8] the Cerrado, which became the great agricultural frontier for the expansion of sugarcane plantations, in addition to soy[9] and livestock.[10] With more land being utilized for sugarcane cultivation and the impending threat to important ecological resources, the government created Sugarcane Agroecological Zoning (ZAE),[11] which

---

[4] José Paulo Pietrafesa and Sérgio Sauer, 'Agrocombustíveis: nova dinâmica na velha fronteira, Terceira marcha a ocupar o bioma Cerrado' in Josie Melissa A Agricola (ed.), *Cerrado, energia e sustentabilidade* (PUC 2012).

[5] See Novo et al., above note 3 at 769.

[6] Ben McKay, Sérgio Sauer, Ben Richardson, Roman Herre, *The Politics of Sugarcane Flexing in Brazil and Beyond.* TNI piece series on flex crops & commodities 2014 (Transnational Institute) 4.

[7] Ministério da Agricultura, Pecuária e Abastecimento, *Plano Nacional de Agroenergia 2006–2011* (Embrapa Informação Tecnológica 2005).

[8] Sued WC Melo, Sérgio Sauer and Janaína DAS Diniz, 'Agroextrativismo como estratégia de desenvolvimento rural no Cerrado' in José Paulo Pietrafesa, Selma Simões de Castro, Josana de Castro Peixoto, Maria Gonçalves da Silva Barbalho (eds), *Recursos naturais: Indicadores de uso e manejo de biotas, solos e águas* (Editora Garamond, 2015) forthcoming.

[9] João dos Santos Vila da Silva, 'Zoneamento Agroecológico da Cana-de-açúcar: Discussion on Decreto nº 6.961 of 17/09/2009' and Projeto de Lei nº 6.077 of 2009, Presentation to Brazilian Senate on 1 June 2010.

[10] Sérgio Sauer and José Paulo Pietrafesa, 'Cana de açúcar, financiamento público e produção de alimentos no Cerrado' (2012) 7 (14) *Campo – território: Revista de Geografia Agrária* 1–29.

[11] Celso V Manzatto, Eduardo D Assad, Jesus FM Bacca, Maria J Zaroni, Sandro EM Pereira, *Zoneamento Agroecológico da Cana-de Açúcar: expandir a produção, preservar a vida, garantir o futuro* (Embrapa Solos 2009), accessed

established restrictions on cultivation and allocation of available land in the Amazon, consequently putting pressure on the Cerrado biome.[12]

In the controversial debate on the possible environmental benefits of agrofuels, Brazil plays a 'pivotal role . . . as the second largest liquid agrofuel producer in the world',[13] not only with regard to the production of sugarcane ethanol but also with the extensive production of biodiesel.[14] The government's support, with the 2003 establishment of the National Program for the Production and Use of Biodiesel (PNPB) and the 2005 establishment of the National Agro Energy Plan (PNA), meant that the country had 'a complete socio-technical biofuel configuration in place, with a full chain from sugarcane and ethanol production . . ., supported by a regulatory system, technical research, and financial arrangements'.[15]

This chapter presents the expansion of sugarcane cultivation into the Cerrado biome and the legal incentives that contributed to this phenomenon, with a focus on the Sugarcane Agroecological Zoning and the National Agro Energy Plan (NAP). The chapter then presents the environmental characteristics of the Cerrado and the government financial incentives that made the expansion of the sugar-energy complex economically viable. The third section examines the investments made by international capital in the agrofuel sector in Brazil and how the internationalization of the sugar-alcohol process took place in the Cerrado. The fourth section presents an overview of greenhouse gas emissions in Brazil, and discusses their influence on the cultivation, production and use of agrofuels and climate change. Finally, this chapter discusses the question of measuring the greenhouse gas emissions created by the production of agrofuels in the Cerrado.

1. EXPANSION OF SUGARCANE CROPS IN BRAZIL AND IN THE CERRADO

From 2005 to 2013 the total area of sugarcane harvested worldwide increased by 26 percent, from 19.7 to 26.5 million hectares.[16] During this

---

8 November 2016 at http://www.infoteca.cnptia.embrapa.br/bitstream/doc/661637/1/ZonCana1.pdf.

[12] See Novo *et al.*, above note 3.

[13] See Novo *et al.*, above note 3 at 770.

[14] See Wilkinson and Herrera, above note 2.

[15] See Novo *et al.*, above note 3 at 770.

[16] Food and Agriculture Organization, FAO, 'FAOSTAT', 2014, accessed 8 November 2016 at http://faostat3.fao.org/browse/Q/QC/E.

period the land mass dedicated to sugarcane was increasing at almost one million hectares per year, an unprecedented rate since the Food and Agriculture Organization's (FAO) records began. Brazil is primarily responsible for the increase of sugarcane cultivation worldwide since it added 4 million hectares, thereby increasing its total cultivated area by 69 percent after 2005. This land area exceeds all the other countries of the world combined and garnered Brazil the title of the 'world's sugar bowl'[17] and its status as the largest agrofuel producer in the global south.[18]

According to data from the FAO, the area used for sugarcane production in Brazil (9,835,169 hectares in 2013) is almost double that of the world's second largest producer. India, as the deputy leader with 5 million hectares, experienced a growth of 38 percent in its area of sugarcane cultivation from 2005 (3,661,500 hectares) to 2013.[19] These two countries together accounted for over 56 percent of the world's sugarcane plantations in 2013 with more than 14 million hectares.[20]

The Brazilian National Agro Energy Plan (2006–2011) outlined strategic action related to the use of agricultural products for the production of agrofuel (sugarcane, soy, castor oil, palm oil, and so on) based on the Guidelines for Agro Energy Policy. At the time, this policy was one of the priorities of the Ministry of Agriculture, Livestock and Food Supply (MAPA) to increase public investments and incentives for the expansion of cropland and the construction of new agro-industrial plants for the production of agrofuel.[21] The sugarcane industry continued to expand during the mid-2000s with the installation of a large number of new production units, which attracted heavy investment (including foreign investment) for the purchase of industrial plants and land.[22] However the global financial crisis in late 2008, along with restrictions in the United States and precautions from European Union countries concerning importing ethanol, eventually led to a decline in investments in the sector.[23] Nevertheless the National Supply Company (CONAB) estimated that the expansion in

---

17 See McKay et al., above note 6.

18 Kristina Dietz, Bettina Engels and Oliver Pye, 'Territory, Scale and Networks: The Spatial Dynamics of Agrofuels' in K Dietz, B Engels, O Pye and A Brunnengräber (eds), *The Political Ecology of Agrofuels* (Routledge 2015) 36–52.

19 See McKay *et al.*, above note 6 at 4.

20 See Food and Agriculture Organization, above note 16.

21 See Ministério da Agricultura, Pecuária e Abastecimento, above note 7.

22 Sérgio Sauer and Sérgio P Leite, 'Agrarian Structure, Foreign Investment in Land, and Land Prices in Brazil' 39 (3–4) *Journal of Peasant Studies* 2012, 873–98.

23 See Sérgio Sauer and José Paulo Pietrafesa, above note 12.

the sector was '. . . creating a new development cycle in this sphere which would last for many years'[24] and consequently lead to expansion into the Cerrado.

According to analysts at the Brazilian Institute of Geography and Statistics (IBGE), after 2008 there was a slowdown in the growth rate of the sugar-ethanol sector, but not a downturn. The annual IBGE and CONAB surveys show continued growth in sugarcane production (tons of cane harvested), resulting mainly from the increased area being cultivated. According to the IBGE analysis:

> With the reduction in oil prices and the international crisis that hit in the mid-second half of 2008, companies in the sugar-ethanol sector were seriously affected. Due to a lack of financial resources, several projects were postponed or suspended, thereby reducing the advance of sugarcane. In 2009, the total area planted was of 9,671,546 hectares, an increase of only 2.7 percent when compared to 2008.[25]

Data from the National Institute for Space Research (INPE) in the Canasat system[26] indicate that more than 9.4 million hectares were planted with sugarcane in the 2013–2014 harvest in the center-south,[27] almost double that of 2005–2006 harvest when sugarcane occupied 4,632,111 hectares in the region (see Table 12.1).[28] While sugarcane is still grown in every geographic region of the country, 87.4 percent of Brazil's planted

---

[24] Companhia Nacional de Abastecimento, *Perfil do Setor de Açúcar e Álcool no Brasil*, Brasília, CONAB, 2008 at 12, accessed 14 November 2016 at http://www.conab.gov.br/conabweb/download/safra/perfil.pdf.

[25] Instituto Brasileiro de Geografia e Estatística, *Quinta estimativa da safra nacional de cereais, leguminosas e oleaginosas para 2009*. Rio de Janeiro, IBGE, 2009, 12, accessed 14 November 2016 at http://www.ibge.gov.br/home/estatistica/indicadores/agropecuaria/lspa/default.shtm.

[26] The Canasat Project is the annual monitoring, via satellite, of the cultivation and production of sugarcane in the Center-South, carried out by INPE – further details on the Canasat website http://www.dsr.inpe.br/canasat/ (accessed 8 November 2016).

[27] There are significant differences between IBGE data (9.6 million hectares, in 2009), and INPE/Canasat (9.4 million hectares, without considering data on the North and Northeast) and FAO (9.8 million hectares cultivated), but this does not mean a reduction in the area under cultivation but rather differences in the methodology (and in the year or period) for collecting information, as all the surveys show expansion of the crop in the period.

[28] In the INPE survey, the region called Center-South covers the states of São Paulo, Goiás, Minas Gerais, Mato Grosso, Mato Grosso do Sul, Paraná, Rio de Janeiro and Espírito Santo.

Table 12.1 Expansion area of sugarcane crops in the Center-South region (2008–2014)

Harvest season	Expansion area* (hectares)	% of total cultivated area
2008/09	1,162,115	16.2%
2009/10	727,948	9.2%
2010/11	390,931	4.7%
2011/12	422,995	4.8%
2012/13	559,739	6.2%
2013/14	777,147	8.2%

Note: * Expansion is the class of sugarcane fields available for a first time harvest, including former sugarcane fields, converted to other uses for a period equal to or more than two harvests, which returned to planting sugarcane.

Source: Canasat website.

area (2014–2015 harvest) is located in this region,[29] with more than half of that in the Cerrado biome.

> São Paulo continues to be the largest producer with 52 percent (4,685,700 hectares) of the planted area, followed by Goiás with 9.5 percent (854,200 hectares) . . . Minas Gerais with 8.9 percent (805,500 hectares), Mato Grosso do Sul with 7.4 percent (668,300 hectares), Paraná with 7.1 percent (635,000 hectares), Alagoas with 4.3 percent (385,300 hectares) and Pernambuco with 2.9 percent (260,100 hectares). These seven states account for 92.1 percent of national production. The other producing states have smaller areas, with representations below 3 percent.[30]

In the 2014/2015 harvest, there was a growth of about 2.2 percent compared to the previous harvest or an increase of 193,100 hectares of planted area (see Figure 12.1). According to CONAB this expansion was concentrated primarily in Cerrado areas of the states of São Paulo, Goiás, Mato Grosso do Sul, Paraná and Minas Gerais because they had 'the largest increase in terms of new units and this corresponds to the consolidation of these areas to new industries'.[31] The sugarcane plantation area used for

---

[29] Companhia Nacional de Abastecimento, *Acompanhamento da safra brasileira de cana-de-açúcar: Safra 2014/15 – 3º levantamento* (CONAB 2014), accessed 8 November 2016 at http://www.conab.gov.br/OlalaCMS/uploads/arquivos/14_12_19_09_02_49_boletim_cana_portugues_-_3o_lev_-_2014-15.pdf.

[30] *Ibid* at 16–17.

[31] *Ibid* at 4.

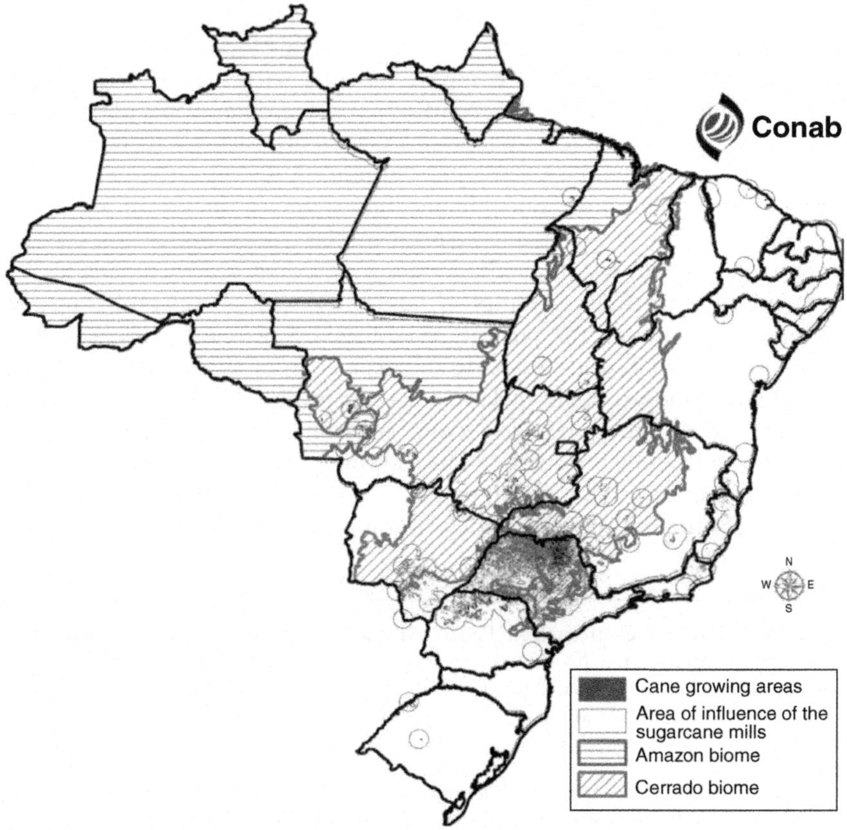

Note: Designed by Ralph de Medeiros Albuquerque (Mader-UnB).

Source: McKay et al.

Figure 12.1 Sugarcane growing areas in Brazil

the production of the 2013/2014 ethanol crop in Cerrado territory was 6,964,450 hectares.[32] The state of Goiás, of which about 90 percent is part of the Cerrado, surpassed Minas Gerais in terms of amount of sugarcane land, ranking second in Brazil at 854,200 hectares in the 2014/2015 harvest.[33]

Between 2000 and 2012, there was a 237 percent increase in sugarcane

32 *Ibid* at 17.

33 See Companhia Nacional de Abastecimento, above note 24 at 12.

plantation area,[34] particularly after 2003 when flex-fuel technology was introduced to the automotive market and transformed the course of the sugar-ethanol industrial complex.[35] Along with the historic government incentives regarding the blend ratio or the ethanol blend mandate, the adoption of flex-fuel vehicles and tax exemptions on their purchase increased the demand for Brazilian ethanol.[36] Despite the increase in private investment and government incentives in preparation for an expected increase in external demand from the United States and the EU, over 90 percent of Brazilian ethanol has been domestically consumed.[37]

In addition to the availability of land, public incentives and transportation infrastructure, the recently established Sugarcane Agroecological Zoning provided another boost to the expansion of sugarcane cultivation in the Cerrado, especially in Goiás. Criticism about possible damage to certain biomes prompted the federal government to set up the Zoning Institute under 2009 Presidential Decree 6961.[38] One of the primary driving forces behind the establishment of the Zoning Institute was a study by the Brazilian Agricultural Research Corporation (EMBRAPA), which aimed

---

[34] José Paulo Pietrafesa and Pedro Araújo Pietrafesa, 'Capital internacional e novas fronteiras na produção de bioenergias: estudo de caso de questões socioambientais' in Sandro Dutra e Silva, José Paulo Pietrafesa, José Luiz de Andrade Franco, José Augusto Drummond and Giovana Galvão Tavares (eds) *Fronteira Cerrado: Sociedade e natureza no Oeste do Brasil* (PUC Goiás 2013), 243–61.

[35] In Brazil, in contrast to Europe and the United States, motor flex fuel technology means adapting engines in cars so that they run on the combustion of either gasoline or ethanol, or a mixture in any proportion. João Humberto Camelini and Ricardo Castilho, 'Etanol e uso corporativo do território' (2012) 11(25) *Mercator* 7–18.

[36] Set up in the 1930s, the obligatory mixture from the 1960s, the amount of ethanol varies between 20 and 25% of the gasoline used in motor vehicles. See McKay et al., above note 6 at 10.

[37] According to 2014 accounts, Brazil should have exported about 1.7 billion liters of ethanol from the 2014/2015 harvest, that is, an amount 35.8% less than the 2.65 billion liters exported from the 2013/2014 harvest. *O Estado de São Paulo*, São Paulo, 04 June 2014, accessed 8 November 2016 at http://g1.globo.com/economia/agronegocios/noticia/2014/06/embarques-de-etanol-devem-perder-receita-em-2014-diz-setor.html.

[38] Based on the EMBRAPA survey, the then President Lula (2003–2010) signed and published Decree number 6.961, in September 2009, establishing the Sugarcane Agroecological Zoning, limiting lands suitable for the cultivation and the National Monetary Council established norms for funding the agrofuel sector in zoning terms. Presidência da República, Decreto nº 6.961, 17 September 2009, accessed 8 November 2016 at http://www.planalto.gov.br/ccivil_03/_Ato2007-2010/2009/Decreto/D6961.htm.

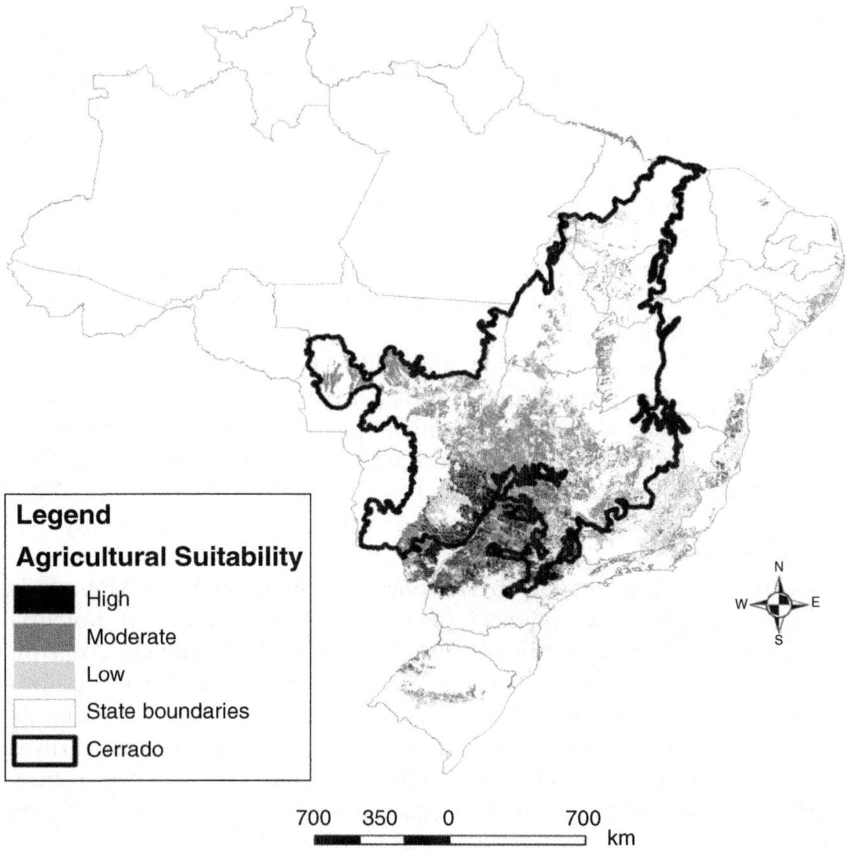

Figure 12.2 ZAE map of land available and appropriate for sugarcane

to 'provide technical support for the formulation of public policies for the expansion and sustainable production of sugarcane in Brazil'. Despite expressing concern about the need to reduce competition between the cultivation of sugarcane and food production, according to the authors, the main focus of this study was clearly to indicate 'areas with agricultural potential for the cultivation of sugarcane with no environmental restrictions'[39] (see Figure 12.2).

The study excluded the Amazon, Pantanal and Upper Paraguay River Basin biomes as well as areas of environmental protection. The survey considered the '. . . areas with agricultural potential (soil

[39] See Manzatto *et al.*, above note 8 at 8–9.

and climate) for the cultivation of sugarcane on land with a slope less than 12 percent, conducive to environmentally sound production with mechanical harvesting'.[40] According to projections made by EMBRAPA, 157.4 million hectares (18.5 percent of the national territory) was the area considered for zoning without restrictions.[41] Of this total, 64.7 million hectares were considered suitable for the cultivation of sugarcane in Brazil based on favorable soil conditions (fertility, slope, and so on) and availability of water.[42] After excluding the Pantanal and the Amazon Rain Forest biomes, the zoning covers the largest area in the Midwest where there are 16.6 million hectares suitable for the expansion of sugarcane farming.[43]

According to the survey data, the midwest has 6,209,273.48 hectares currently used for grazing and another 1,036,777.40 hectares used for agriculture that are classified as highly appropriate. When the amount of highly appropriate land is added to the area of land of average appropriateness, the area that could be cultivated with sugarcane totals 28,119,474.02 hectares,[44] an area three times larger than the current area currently used for sugarcane in the whole country.[45]

Using a method similar to the one used in the EMBRAPA study, Machado, Paglia and Fonseca[46] drew up a map of areas of potential expansion for sugarcane crops focusing on the midwest and the Amazon. The result is very similar to the sugarcane zoning because in addition to the areas already occupied by sugarcane in the Atlantic Forest biome (especially in São Paulo), the potential area is in the Cerrado, (primarily in the states of Goiás, Mato Grosso do Sul and Mato Grosso). The authors then superimposed the areas defined as 'priority' for preservation by the

---

[40] *Ibid* at 9.

[41] *Ibid* at 16.

[42] CEM Silva, *O Cerrado em disputa: Apropriação global e resistências locais* (Confea 2009) at 25. The EMBRAPA study considered that the land presently used for grazing, a total of 34.2 million hectares, was suitable for the expansion of sugarcane – see Manzatto et al., above note 8.

[43] See Manzatto et al., above note 8 at 17.

[44] See Manzatto et al., above note 8 at 17.

[45] See Presidência da República, above note 38.

[46] For further details, see RB Machado, AP Paglia, and RL Fonseca, Áreas e paisagens prioritárias no Cerrado, Pantanal e Amazônia (Conservação Internacional 2007). The authors used three conditions for cultivation as a basis, monthly rainfall in both rainy and dry seasons, and maximum and minimum temperatures, forming a map (areas of rainfall greater than 1200 mm and temperatures between 18°C and 45°C) with potential for the expansion of sugarcane production in Brazil.

Ministry of Environment (MMA) onto this map of potential areas for sugarcane cultivation.[47] The result is quite worrisome because '. . . it indicated that 16 percent of the areas of extreme biological importance in the Amazon coincide with the map of the sugarcane plantation potential. In the Cerrado the level is even higher, as 70 percent of the areas of great biological importance could be at the sugarcane expansion front'.[48]

The Sugarcane Agroecological Zoning excluded the Amazon from the areas of potential expansion, but the Cerrado is still at risk. According to the analysis of the MMA itself, this biome: '. . . has been degraded by the expansion of the Brazilian agricultural frontier, especially in the Western Bahia region, south of Goiás and the Sinop region of Mato Grosso. The Western Bahia region has developed most in terms of the anthropic use of land.'[49]

2. THE CERRADO BIOME AND GOVERNMENTAL SUPPORT FOR SUGARCANE PRODUCTION IN BRAZIL

The Cerrado biome, as a tropical savannah ecoregion, represents the predominant ecosystems found in central Brazil (see Figure 12.3). It is the second largest biome in Brazil in terms of area, covering about 203 million hectares. The biome occupies 25 percent of Brazil's land area[50] 'housing, in its various ecosystems, flora with over 11,000 native plant species'.[51] According to the MMA 'more than 40 percent of the [more than six thousand] woody plant species and 50 percent of the bee species are endemic' meaning that they only exist in the Cerrado.[52] The Cerrado ecosystems play an important role in ecological balance and maintenance of global environmental services. The ecosystems are noted for their heterogeneity and diversity of soils, climates, biota, vegetation and landscapes, which include savanna, forest and pastoral features.

---

47 Ministério do Meio Ambiente, Secretaria de Biodiversidade e Florestas, 'Avaliação e identificação de áreas e ações prioritárias para a conservação, utilização sustentável e repartição dos benefícios da biodiversidade nos biomas brasileiros', MMA/SBF 2002.

48 See Machado, Paglia and Fonseca, above note 46 at 24.

49 Ministério do Meio Ambiente, 'Monitoramento do desmatamento nos biomas brasileiros por satélite: Monitoramento do bioma Cerrado 2009–2010', MMA 2011 at 8.

50 See Melo, Sauer and Diniz, above note 8.

51 See Ministério do Meio Ambiente, above note 49 at 9.

52 *Ibid* at 11.

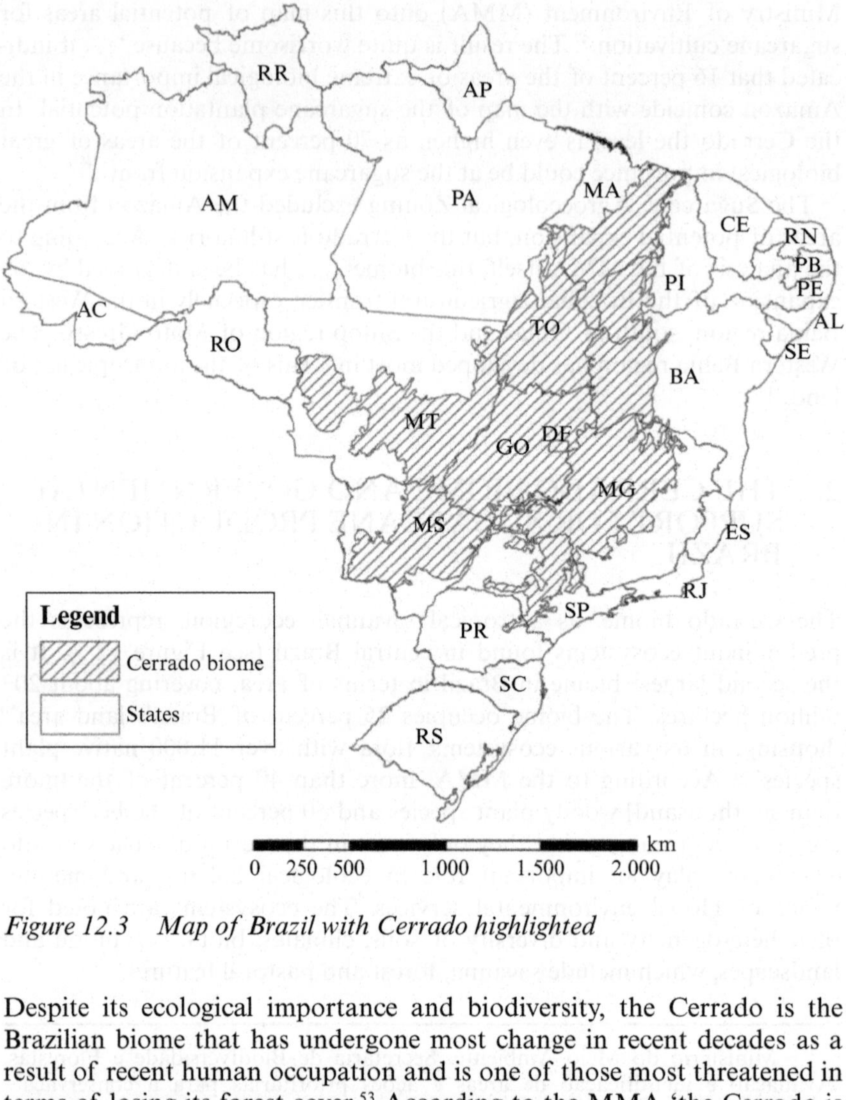

Figure 12.3 Map of Brazil with Cerrado highlighted

Despite its ecological importance and biodiversity, the Cerrado is the Brazilian biome that has undergone most change in recent decades as a result of recent human occupation and is one of those most threatened in terms of losing its forest cover.[53] According to the MMA 'the Cerrado is one of the world's 25 areas in crucial need of conservation, because of its biological richness and the intense anthropogenic pressure to which it is being submitted'.[54]

---

[53] *Ibid.*

[54] See Ministério do Meio Ambiente, above note 47 at 177.

By 2010, approximately half of the original area of the biome had already been transformed into grazing land, farmland (grain, especially soy and corn) and other types of developed land. From 2002 to 2008, 85,074 square kilometers of the Cerrado was deforested, equating to 4.17 percent of the total biome.[55] This represents an average rate of 14,200 km^2 or 0.69 percent per annum. In 2008–2009, 7,637 square kilometers was deforested, equating to 0.37 percent for the year.[56] In the 2009–2010 period, the Cerrado lost 6,469 square kilometers of its forest cover at an annual rate of 0.32 percent, the highest rate recorded for that period for any Brazilian biome.[57]

This systematic devastation of the Cerrado began with agricultural modernization and the adoption of the Green Revolution technological package from 1960 onwards, along with government incentives to occupy and 'develop' the Midwest by creating and implementing massive programs of colonization and monoculture.[58] Two of the most noteworthy of the government programs and incentives are the 1975 Program for the Development of the Savannas (POLOCENTRO) and the 1979 Japanese-Brazilian Program for the Development of the Savannas (PRODECER).[59]

From the 1970s, the degradation and occupation of the Cerrado has intensified and continued through recent decades with the expansion of crops utilized for agro-energy and with the expansion of reliance on soy and sugarcane.[60] Since the 1970s, natural areas with limited anthropic impacts have become synonymous with backwardness and there has been an even greater tendency towards the replacement of native vegetation cover with monocultures, which has resulted in drastically altered landscapes and devastatingly simplified ecosystems.[61] According to the MMA, 'in recent decades, the Cerrado has been seen as an alternative to deforestation in the Amazon, and a more intense exploitation of the region

---

55 See Ministério do Meio Ambiente, above note 49 at 10.

56 *Ibid.*

57 Roseli Senna Ganem, José Augusto Drummond and José Luiz de Andrade Franco, 'Conservação da biodiversidade no bioma Cerrado: ameaças e oportunidades' in Sandro Dutra e Silva, José Paulo Pietrafesa, José Luiz de Andrade Franco, José Augusto Drummond and Giovana Galvão Tavares (eds). *Fronteira Cerrado: Sociedade e natureza no Oeste do Brasil* (PUC Goiás 2013) 331–61.

58 See Melo, Sauer and Diniz, above note 8.

59 RF Ribeiro, '"Largueza" ao "Cercamento": Um balanço dos programas de desenvolvimento do Cerrado' in A Zhouri, K Laschefski and DB Pereira (eds), *A insustentável leveza da política ambiental: desenvolvimento e conflitos socioambientais* (Autêntica 2005) at 172.

60 See Pietrafesa and Pietrafesa, above note 34.

61 Melo, Sauer and Diniz, see above note 8.

has been proposed, either through agricultural expansion, or by planting forests to fix atmospheric carbon'.[62] Due to this agricultural expansion, over 50 percent of the biome has been cleared or transformed by human action and 'more than a hundred endemic Cerrado species are threatened with extinction'.[63]

The MMA itself admits that there has been very little government attention given to the conservation of the Cerrado as 'only 2.2 percent of the biome is legally protected in UCs [conservation areas] of full protection', and estimates indicate that 'at least 20 percent of endemic and endangered species are outside existing parks and reserves'.[64] The MMA has established priority conservation areas to promote the sustainable use of the Cerrado in an effort to protect one of the planet's most threatened biomes.[65] On the other hand, the Federal government, through the National Agro Energy Plan (2006) and Sugarcane Agroecological Zoning (2009), has been removing fiscal and environmental regulations, investing in infrastructure and offering tax incentives and direct funding both for the construction of new industrial plants and for the expansion of sugarcane with the intention to increase agrofuel production in the Cerrado.[66]

Farmland expansion and ethanol and sugar production receive substantial support from public funding, mainly through Petrobrás, the National Bank for Economic and Social Development (BNDES) and the Midwest Constitutional Fund (FCO). BNDES[67] has invested heavily in agriculture

---

[62] See Ministério do Meio Ambiente, above note 47 at 178.

[63] Ministério do Meio Ambiente, *Portaria MMA nº 9*, de 23 de janeiro de 2007 – Áreas prioritárias para conservação, uso sustentável e repartição de benefícios da biodiversidade brasileira (atualização) (MMA 2007) at 56.

[64] See Ministério do Meio Ambiente, above note 63. In another document, the MMA claims that 'in the Cerrado biome, only 7.44 percent of its area is protected by federal, state and municipal conservation units as approximately 2.91 percent of the Cerrado is protected in the form of integral protection conservation units, such as the National Parks'. Ministério do Meio Ambiente, *Monitoramento do bioma Cerrado 2009–2010*, Brasília, 2011 at 8.

[65] Ministério do Meio Ambiente, *Portaria MMA nº 9*, of 23 January 2007.

[66] José Paulo Pietrafesa, Sérgio Sauer and Ana Elizabeth AF Santos, 'Políticas de recursos públicos na expansão dos agrocombustíveis em Goiás: ocupação de novos espaços em áreas de Cerrado', in José Paulo Pietrafesa and Sandro Dutra e Silva, *Transformação do Cerrado: progresso, consumo e natureza* (PUC Goiás 2011).

[67] The BNDES 'continues, in the 21st century, to foster another cycle of sugarcane expansion', since 'the bank grants to ethanol industry increased by 160 percent from 2006 to 2007 . . . The bank's loan portfolio for this industry surpasses US $3 billion' in 2008. Plataforma BNDES, *Sugarcane Industry in Brazil: Air Pollution, Threat to Water Resources, Risk to Foodstuff Production, Outdated*

in the South-Central Region, including the state of Goiás,[68] through the Joint Plan to Support Industrial Technological Innovation in Sugarcane Sectors (Paiss Agrícola).[69] Paiss Agrícola was set up in conjunction with the Funding Authority for Studies and Projects Agency (FINEP) of the Ministry of Science and Technology to support 'agricultural-technological innovation' in the sugar-ethanol sector.[70]

In 2010, BNDES released an additional R$2.3 billion for agrofuel plants which was 11 percent less than the R$2.6 billion released in January–April 2009. Of the R$2.3 billion transferred to the sugar-ethanol sector by May 2010, R$1.1 billion went to ethanol production, R$850 million to sugar, RS$320 million to sugarcane growing and RS$70 million to the cogeneration of electricity.[71] In 2011, BNDES and FINEP supported 35 business plans with R$2.5 billion 'aimed at providing support to industrial-technological innovation in the sugar-based ethanol and chemical sectors'.[72] In 2014, the Paiss Agrícola spent nearly R$2 billion financing 35 other projects for 29 companies. This was the annual total loaned by BNDES for the expansion of the sector.[73]

In 2012, BNDES also launched Prorenova as part of the Paiss Agrícola with a R$4 billion budget. The Prorenova was designed 'to encourage the production of sugarcane by financing the renovation of former sugarcane farms and the expansion of the cultivated area'.[74] Since the launch of

---

Working Relations and Insufficient Protection to Workers Health. Plataforma BNDES: Impactos da indústria canavieira no Brasil – poluição atmosférica, ameaça a recursos hídricos, riscos para a produção de alimentos, relações de trabalho atrasadas e proteção insuficiente à saúde de trabalhadores, accessed 14 November 2016 at http://www.flacsoandes.edu.ec/libros/digital/39821.pdf.

68 See Pietrafesa and Pietrafesa, above note 34.

69 The Paiss Agrícola is made up of different types of funding, including the 'Prorenova' which finances the renewal and establishment of new sugarcane farms. The Paiss will fund projects for technological innovation including the development of new strains of transgenic sugarcane, and the production of second generation ethanol. Three plants have already been funded and two are operating. Alana Gandra, 'Apoio ao setor sucroenergético tem 35 projetos no valor de quase R$ 2 bilhões' *Agência Estado*, São Paulo, 17 December 2014.

70 See McKay et al., above note 6; See Sauer and Pietrafesa, above note 12.

71 Fabiana Batista, 'Desembolso do BNDES a usinas segue firme, mas projetos rareiam' O Valor (São Paulo, 18 May 2010).

72 União da Indústria de Cana-de-Açúcar, 'BNDES, Finep launch program to encourage ethanol innovation' (Unica, 2014), accessed 8 November 2016 at http://www.unica.com.br/in-the-media.

73 See Gandra, above note 69; Sauer and Pietrafesa, above note 4 at 10.

74 Banco Nacional de Desenvolvimento Econômico e Social, 'Prorenova provides supports of R$ 4 billion for renovation and implementation of sugarcane farms', (BNDES, 20 December 2012), accessed 8 November 2016 at http://www.bndes.

Prorenova, interest rates have been reduced, increased funding per hectare has been made available and applicability to foreign-owned companies has been extended to encourage uptake.[75]

In addition to funds from BNDES, the Cerrado sugar-ethanol sector (Goiás, Mato Grosso and Mato Grosso do Sul) benefits from funding from the Midwest Constitutional Fund (FCO). These loans have low interest rates, long grace periods and facilitated deadlines for repayment.[76] The purpose of the FCO is '. . . to contribute to the economic and social development of the Midwest Region by means of funding for productive activities in the industrial, agro-industrial, agricultural, mineral, tourist, trade and service economic sectors', and many of the funded projects supported the sugarcane industry.[77]

A survey carried out on the website of the Goiás Secretariat for Planning (SEPLAN)[78] showed that from 2006 to 2009, the FCO approved funding of approximately R$329 million for the sugar-ethanol sector in the state.[79] During the four years studied (data available on the above-mentioned site), it can be observed that most of the projects and loans were approved in 2006 and 2007, the period during which the National Agro Energy Plan was launched and implemented.[80]

Despite the official argument that the growth of production in Goiás was due to the 'region's vocation for agribusiness', it was likely also the result of numerous government incentives.[81] In addition to the direct investments, there was a series of fiscal incentives such as tax exemptions granted by the state government. According to the newspaper *O Popular*, the state government gave an exemption of approximately R$28.1 billion to 50 tax incentive contracts negotiated with ethanol and sugar plants in 2009.

---

gov.br/SiteBNDES/bndes/bndes_en/Institucional/Press/Noticias/2012/20120111_BNDES_Prorenova.html.

75 See McKay *et al.*, above note 6 at 12.

76 See Sauer and Pietrafesa, above note 12.

77 Ministério da Integração Nacional, 'Fundos Regionais e Incentivos Fiscais', (MIN, May 2012), accessed 8 November 2016 at http://www.integracao.gov.br/web/guest/apresentacao-fco.

78 Data for the Center-West Constitutional fund (FCO) available on SEPLAN (http://www.seplan.go.gov.br/) site for projects approved for Goiás. For further details see Pietrafesa, Sauer and Santos, above note 66.

79 According to data from consultations approved by the FCO Council, these investments should generate more than 7,000 indirect jobs in the State. It must be made clear that this is an estimate usually expanded (or overestimated) to sensitize Council members to approve loans.

80 See Pietrafesa, Sauer and Santos, above note 66.

81 See Sauer and Pietrafesa, above note 12.

Aside from the sugar-ethanol sector, no other sector in the economy received such tax incentives through the Goiás Industrial Development Program (PRODUZIR).[82] According to the National Agro Energy Plan, the main objective of public investment, both in terms of funding and tax exemptions, is to encourage the private expansion of ethanol production. This has been happening in Goiás as ethanol and sugar production plants and sugarcane plantations have continued to expand over the course of recent years.[83]

In addition to the encouragement of agricultural expansion and the lack of government initiatives focused on environmental protection, the topography of the Cerrado biome makes it a prime target for the expansion of the sugarcane industry. The Cerrado's land is flat and other agricultural activities such as soy cultivation and livestock grazing have made soil conditions more conducive to sugarcane cultivation. However, in addition to these favorable conditions, the evolution of the sugar-ethanol sector in the Cerrado requires the existence or construction of necessary infrastructure.[84] A key component of this necessary infrastructure has been the planning of the so-called Petrobrás 'ethanol pipeline'.[85] In 2011, Logum Logistics, a joint venture backed by Petrobrás, Copersucar, Raizen and Odebrecht/ETH, Camargo Corrêa and Uniduto Logistics, received a US$924 million loan from BNDES to begin building an 830-mile ethanol pipeline connecting the states of São Paulo, Minas Gerais and Goiás to consumption sites in the country's Southeast Region.[86] If it is built, once completed, the system will be capable of transporting up to 21 billion liters per year and will take ethanol to the ports of Santos for export and to Brazil's main consumption centers.

In addition to the region's 'vocation', availability of fertile land, plentiful rainfall and proximity to consumer centers, government support has been crucial for the expansion of agrofuel production in the Brazilian Cerrado. The government's investment in infrastructure (especially for shipment of the product) and the direct contribution of public funds have stimulated agrofuel production in the Cerrado biome.[87]

---

[82] The PRODUZIR program was set up by State Law nº 13.591, of 2000, with the main objective of subsidizing the industrial sector (public funding with reduced interest rates) in expanding its production potential.

[83] See Sauer and Pietrafesa, above note 12.

[84] See Pietrafesa and Sauer, above note 4.

[85] See Wilkinson and Herrera, above note 2.

[86] See McKay et al., above note 6 at 9.

[87] See Sauer and Pietrafesa, above note 12.

3. THE NEW CONFIGURATION OF INTERNATIONAL CAPITAL IN THE CONTEXT OF THE SUGARCANE INDUSTRY

After decades of domestic incentives and investment of public funds into the national private sector (mostly family owned companies), international capital has begun to increase in the sector. According to the *Agência Estado,* 58 plants changed hands in Brazil between 2007 and 2009. These transactions involved more than 100 companies. For Fabio Venturelli, the President of the São Martinho group, 'the merger between Cosan, the country's largest sugar and ethanol company, and Shell, one of the world's largest oil companies, is a sign of what's to come'. For him, 'the deal, valued at US$12 billion, was the first . . .' explaining that '. . . the strategic gains are clear and point to a future where ethanol will enjoy global recognition as a renewable fuel'.[88]

Studies by Sauer and Leite,[89] conducted after the World Bank report[90] on the use of land for agricultural production, identified a set of speculative international capital investments which have likely altered aspects such as land prices and the volume of production in the Brazilian farming system. The World Bank emphasized an effect on '. . . the expansion of eight commodities: corn, soy, sugarcane, palm oil, rice, rapeseed, sunflowers and planted forest. In Brazil this [foreign] capital is mainly involved in the first three products',[91] especially sugarcane and soy because of the international potential demand for agrofuels. Foreign investments have resulted in an increasingly concentrated sugar-milling sector since 'seven milling groups control 55 percent of Brazil's sugarcane production with smaller, often family or cooperative-owned mills, disappearing due to indebtedness and takeovers'.[92]

Because Brazilian companies had begun to enter into the sugarcane industry funded by international capital, from the year 2000 onwards there was a steady flow of international funding that exceeded state policy grants through government incentives and loans. The milestone for this change in the financing and control of capital in the industry was the entrance

---

88 Eduardo Magossi and Gustavo Porto, 'BP compra produtora de etanol CNAA, em acordo de US$ 680 milhões,' *O Estado de São Paulo*, São Paulo, 12 March 2011.

89 See Sauer and Leite, above note 22.

90 The World Bank, *Rising Global Interest in Farmland: Can it yield sustainable and equitable benefits?* (Washington DC 2010).

91 See World Bank, above note 90 at 8.

92 See McKay *et al.*, above note 6 at 6.

of the Cosan Group to the São Paulo Bovespa Stock Market in 2003 and the internationalization of its capital. Mendonça, Pitta and Xavier[93] presented an outline of mergers and foreign acquisitions in the agrofuel and petroleum sector, which signaled successive acquisitions:

> The Royal Dutch Shell oil company appears on the scene with the establishment of a joint venture with the Cosan group. This association resulted in the setting up of the Raizen company, one of the five largest economic groups in the country, with an estimated market value of US$20 billion. The corporation is responsible for an annual production of 2.2 billion liters of ethanol and 4 million tons of sugar. Since it started, its target has been to increase its ethanol production to 5 billion liters per annum by 2014. To do so, it has spread its control over 'new' sugarcane-producing regions such as the northwest of São Paulo and the states of Goiás and Mato Grosso do Sul.[94]

It was not only foreign oil companies that were consolidating their position in the agrofuel sector. The Brazilian state company Petrobrás, through Petrobrás Biofuel (PBio), also set-up the Nova Fronteira Bioenergia, a merger with the São Martinho Group in upstate São Paulo and invested approximately R$520 million in expanding the Boa Vista plant in the municipality of Quirinopolis (Goiás). Petrobrás sought to create the largest sugarcane ethanol plant in the world, with a processing capacity increasing from three million tons in 2012 to eight million tons of ground sugarcane in the 2014–2015 harvest.

In addition to the oil companies' investments in agrofuel production, after 2006 other large companies in the food sector also began to enter into the Goiás sugarcane industry. With the creation of the National Agro Energy Plan, there was an increase in authorization for mergers and the expansion of the sugarcane industry in the states of São Paulo, Minas Gerais, Goiás and Mato Grosso do Sul, which attracted investments from food companies like Cargill and Bunge.

The information in Table 12.2 indicates that after decades of using public resources, concentrating land ownership, exploiting degrading forms of labor and increasing environmental liability, this national 'complete biofuel socio-technical configuration'[95] and production chain presents itself as 'ready', 'productive' and 'technologically efficient' to international groups and foreign capital. Approximately 80 percent of the sugarcane planted and processed by these companies was grown on their

---

93 Maria L Mendonça, FT Pitta and CV Xavier, *A agroindústria canavieira e a crise econômica mundial* (Outras Expressões 2012).

94 See Mendonça, Pitta and Xavier, above note 93 at 8.

95 See Novo *et al.*, above note 3 at 770.

Table 12.2 The corporate control of Brazilian sugarcane

Company	Mills/related assets in Brazil	Production/ year	Share of Brazilian market
Copersucar S.A (2014 – merger with Cargill, 50% owned by each company)	Controls sale of sugar and ethanol volumes produced by 47 member mills and 50 non-member units	Sugar: 6.9 million tons Ethanol: 3.7 billion liters	Sugar-ethanol complex (transportation, storage, processing and commercialization): 22% Sugar sales: 19% Ethanol sales: 16.3%
Raizen (2011 merger – Royal Dutch Shell and Brazilian conglomerate Cosan)	24 mills 4,700 Shell service stations 54 airports 60 terminals	Sugar: 4 million tons Ethanol: 2 billion liters	Sugar-ethanol complex: 9.5% Sugar: 11% Ethanol: 8.8%
Biosev (2009 merger – Louis Dreyfus Commodities sister company after merging with Brazilian sugarcane producer and processor Santelisa Vale)	12 mills	Sugar: 2.8 million tons Ethanol: 1.8 billion liters	Sugar-ethanol complex: 7%
Odebrecht Agroindustrial (Brazilian-based engineering, construction, and chemical conglomerate)	9 mills	Sugarcane: 40 million tons Ethanol: 3 billion liters	Sugarcane: 6.8% Ethanol: 13.2%
Guarani (Acquired by French sugar conglomerate Tereos in 2002; Petrobrás buys 46% stake in 2011)	7 mills	Sugarcane: 20 million tons Sugar: 1.9 million tons Ethanol: 8.6 billion liters	Sugarcane: 3.4% Sugar: 5.2% Ethanol: 3.8%

Table 12.2 (continued)

Company	Mills/related assets in Brazil	Production/ year	Share of Brazilian market
Bunge (US-based food processing corporation)	8 mills	Sugarcane: 21 million tons	Sugarcane: 3.6%
Adecoagro (Argentina-based company, heavily backed by US billionaire George Soros)	3 mills	Sugarcane: 17 million tons	Sugarcane: 2.9%

Source: McKay et al.

own property. Thus, in addition to transferring technology to international ownership, large tracts of land are also being handed over[96] resulting in cases of land grabbing[97] and green grabbing.[98]

The Goiás State Department of Agriculture reported that in 2009 more than 80 industrial plant projects had been approved, which indicated the continued growth of sugar and alcohol production in the State. In the industrial sector, there were 77 new units under construction in Brazil in 2005. In early 2007 this number had risen to 90 plants under construction and over 150 projects for new units, 60 of which were in the state of Goiás.[99] Not all of these plans were actually carried out, primarily as a result of the fall in oil prices on the international market, which slowed down investment in new agro-industrial units.

4. BRAZIL, GHG EMISSIONS AND CLIMATE CHANGE

It is essential to analyze the potential climate change impacts arising from the cultivation of sugarcane crops for agrofuel purposes because Brazil has

[96] See Pietrafesa, Sauer and Santos, above note 66.

[97] See Sauer and Leite, above note 22.

[98] Saturnino Borras and Jenny C Franco, 'Global Land Grabbing and Trajectories of Agrarian Change: A Preliminary Analysis' (2012) 12(1) *Journal of Agrarian Change* 34–59.

[99] See Ministério da Integração Nacional, above note 77.

committed itself to the unconditional defense of this 'supposedly green energy'.[100] A study set out to question the claim that the use of ethanol '. . . as fuel reduces CO_2 emissions into the atmosphere by 70 percent when compared to gasoline' and therefore constitutes 'one of the best alternatives for reducing the emission of gases which cause the greenhouse effect'.[101] Although it will not be discussed in depth in this chapter, it is first essential to clarify that the history (and in many cases the current state) of the sugarcane industry in Brazil poses '. . . terrible aggression to the Atlantic rainforest ecosystem, degrading working conditions and relations, and expansion posing a threat to food supply'.[102] Second, it is essential that these social and environmental problems, including the issue of climate change, resulting from the expansion of this monoculture be '. . . framed in terms of justice'.[103]

In this discussion of the relationship between sugarcane and climate change, we have used here data available from the Greenhouse Gas Emission Estimates System (SEEG) organized by the Climate Observatory[104] and from Brazilian government official communications on the monitoring of national emissions of greenhouse gases (GHG).[105] To estimate emissions from the ethanol production cycle in the Cerrado, we have used the methodological procedures and results of a study conducted by Claros Garcia and von Sperling.[106] Also, some context is required to understand the effect that Brazilian agrofuels have on anthropogenic GHG emissions.

---

[100] See Plataforma BNDES, above note 67 at 131.

[101] See Companhia Nacional de Abastecimento, above note 29 at 4.

[102] See Plataforma BNDES, above note 67 at 131.

[103] Markus Wissen. 'The Political Ecology of Agrofuels' in K Dietz, B Engels, O Pye and A Brunnengräber (eds), *The Political Ecology of Agrofuels* (Routledge 2015) 16.

[104] The Climate Observatory is a network of non-governmental organizations and social movements working in the field of climate change. Observatório do Clima, 'Sistema de Estimativas de Emissão de Gases do Efeito Estufa' (SEEG, April 2014), accessed 8 November 2016 at http://seeg.eco.br.

[105] National communications are part of the agreement between the signatories of the United Nations Framework Convention on Climate Change, such as monitoring of the emission of GHGs and policies on the mitigation and adaptation to climatic change. Ministério de Ciência, Tecnologia e Inovação, 'Comunicação Nacional Inicial do Brasil à Convenção-Quadro das Nações Unidas sobre Mudança do Clima' (MCTI 2004).

[106] We used this study on the sugarcane GHG emissions as it was carried out on plants in Minas Gerais in the Cerrado biome. JC Claros Garcia and E Von Sperling, 'Emissão de gases de efeito estufa no ciclo de vida do etanol: estimativa nas fases de agricultura e industrialização em Minas Gerais' (2010) 15(3) *Revista Engenharia Sanitária Ambiental* 217–22.

The Ministry of Science, Technology and Innovation (MCTI) is responsible for the dissemination of the National Inventory of GHG emissions not controlled by the Montreal Protocol.[107] The MCTI followed the directions and guidelines established by the Intergovernmental Panel on Climate Change (IPCC) when measuring Brazilian emissions. These studies observed the emissions resulting from six different human activities: agriculture, change in the use of land and forests, energy, industry, solvents and other products and the treatment of solid waste. The results from 2005 through 2013 are presented in Table 12.3.

Based on the methodology of the Intergovernmental Panel on Climate Change (IPCC), the method adopted by the Brazilian authorities to measure emissions resulting from agriculture and livestock is to calculate the enteric fermentation of ruminant livestock, livestock manure in grazing land, agricultural soils, rice cultivation and the burning of agricultural residue. In calculating GHG emissions resulting from changes in the use of land and forest, the government's report considered the effects of conversion of forest land to agriculture or livestock breeding, changes in the carbon content of soils and planted forests (specifically those used for industrial purposes). The emissions attributed to the energy sector in Table 12.3 were estimated based on the production, processing and consumption of energy.[108]

When emissions are separated by sector, it is apparent that deforestation for all biomes has historically been the main producer of GHG emissions, followed by energy then agriculture and livestock breeding. Industry and solid waste account for a very modest share of Brazilian GHG emissions as illustrated by the following graph (see Figure 12.4).

In recent years, however, there has been a considerable decrease in emissions resulting from deforestation, with a reduction of 963,706,181 tg between 2005 and 2013. During the eight years illustrated in Table 12.3, there was an increase in emissions created by both the energy and agriculture and livestock sectors.[109] Agriculture is a major contributor to atmospheric pollution. In 2010, agriculture was responsible for emitting 35 percent of the GHGs produced in Brazil. Of this 35 percent, 56.4 percent was the result of enteric fermentation. The management of agricultural soils accounted for the second largest percentage of GHG emissions produced by agriculture and livestock breeding. (Table 12.4) The reason

---

107 Ministério de Ciência, Tecnologia e Inovação, Inventário Nacional de Emissões e Remoções Antrópicas de Gases do Efeito Estufa não Controlados pelo Protocolo de Montreal, Brasilia (MCTI 2010).

108 See Ministério de Ciência, Tecnologia e Inovação, above note 105.

109 See Observatório do Clima, above note 104.

Table 12.3 Total emissions of greenhouse gases from Brazil from 2005 to 2013 in Tg CO_2eq

Categories	2005	2007	2009	2011	2013
Agriculture and Livestock	392,045,484	383,263,662	395,621,577	417,640,974	416,694,029
Energy	333,834,060	353,603,767	351,716,949	406,801,978	473,265,687
Use of Land and Forest	1,506,174,138	1,071,686,777	654,589,678	568,352,680	542,467,957
Industry	76,269,914	75,321,066	67,849,323	84,982,871	86,512,911
Solid Waste	41,229,018	43,307,843	46,156,494	48,137,009	48,738,582
Total	**2,349,552,614**	**1,927,183,115**	**1,515,934,021**	**1,525,915,512**	**1,567,679,177**

Note: According to the Ministry of Science, Technology and Innovation (MCTI), all greenhouse gases are part of national estimates. But for methodological reasons of comparison and summing of gases, a measure set up by the IPCC was used to determine a common unit. The unit proposed was the carbon equivalent (CO_2eq). The emissions of 1kg of CH_4 (methane) is equal to 23kg of CO_2eq; 1kg of N_2O (nitrous oxide) equals 296kg of CO_2eq.

Source: Observatório do Clima.

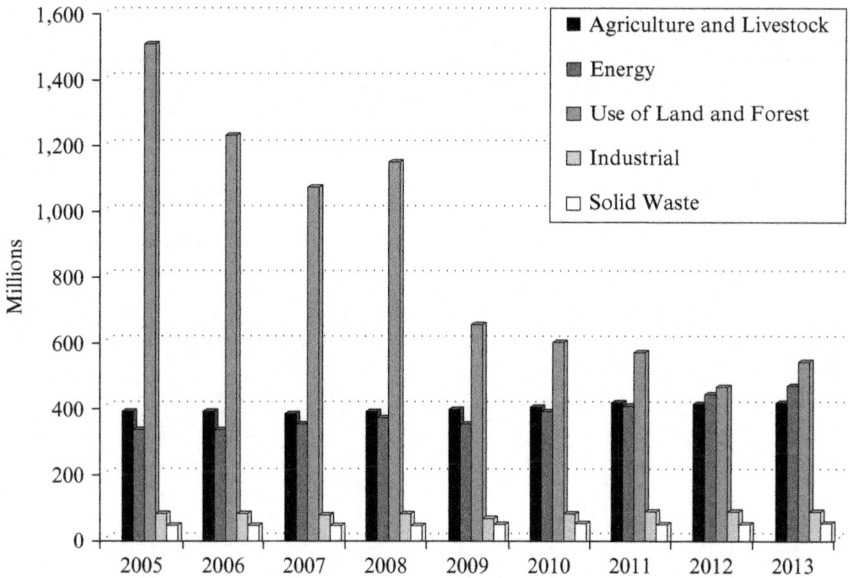

Note: Drawn up by Pedro Araújo Pietrafesa.

Source: Observatório do Clima.

Figure 12.4 Brazilian emissions of greenhouse gases (Tg CO_2eq) per sector

that agricultural soil management produces such a large amount of GHG emissions is that arable and grazing land is treated with fertilizer, crop residue, animal waste, agricultural industrial waste and lime and because the organic matter added to the land is mineralized.[110] Although Table 12.4 does not illustrate disaggregated data, all agricultural operations use the same soil management techniques, so sugarcane fields are included. Since the sugarcane sector occupies over 9 million hectares of land, it is undoubtedly one of the major producers of GHGs.[111]

As depicted in Figure 12.5, there was no overall drastic variation in GHG emissions from agriculture and livestock breeding from 2010 to 2013. However, despite constant technological improvements, land management practices have still resulted in increases in GHG emissions from agricultural soils.

110 See Ministério de Ciência, Tecnologia e Inovação, above note 105.
111 See Ministério de Ciência, Tecnologia e Inovação, above note 107.

Table 12.4 Agricultural emissions by sector in 2010

Agricultural Emissions in 2010	
Enteric Fermentation	56.4%
Agricultural Soils	35.2%
Animal Waste	4.9%
Rice Cultivation	2.0%
Sugarcane Burning	1.5%

Source: Secretaria de Políticas e Programas de Pesquisa e Desenvolvimento, Ministério de Ciência Tecnologia e Inovação, Estimativas anuais de emissões de gases de efeito estufa no Brasil (MCTI 2013).

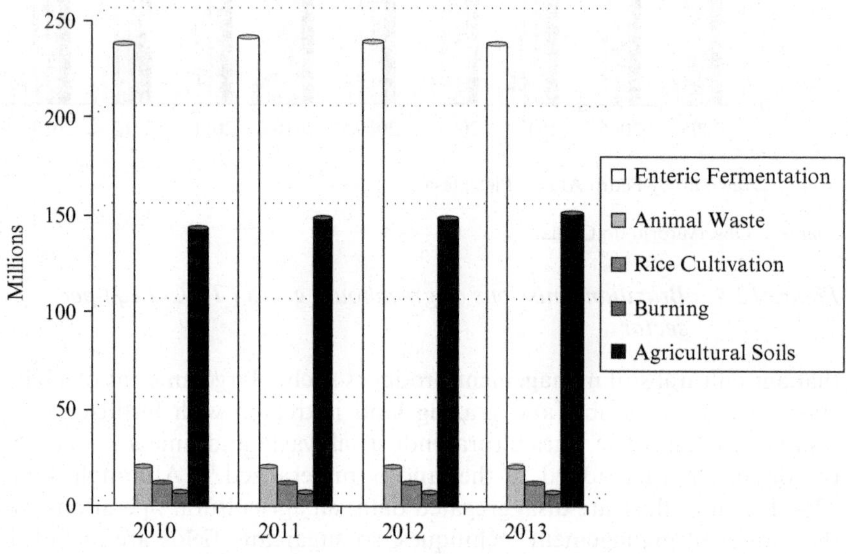

Note: Drawn up by Pedro Araújo Pietrafesa.

Source: Observatório do Clima.

Figure 12.5 Emissions from agriculture and livestock breeding from 2010 to 2013

Enteric fermentation is the largest emitter of GHGs from Brazilian agriculture and livestock breeding. The IBGE estimates that there are at least 211 million head of cattle in Brazil, which explains the high levels of emissions of CO_2eq resulting from this activity. The utilization of fertilizers, lime and machinery for soil management are also contributing factors to the emission of GHGs. Under normal circumstances, the soil releases 60 billion tons of CO_2 into the atmosphere per year. Preparation for cultivation increases that amount by 2.5 billion tons of CO_2 per year. Agricultural soils primarily emit CO_2 during the period in which they are devoid of vegetation and waiting for rain in order to plant.[112]

One of the contributing factors to the reduction of GHG emissions is the decrease of deforestation rates in Brazil, primarily in the Amazon rainforest.[113] In 2014 the rate of deforestation of the Amazon biome was 18 percent lower as compared to the previous year. The total deforested area of the Amazon biome is 20 percent.[114] In the Cerrado biome deforestation reached 50 percent of its territory; most of the deforestation is concentrated in the South-Central Region of the biome where large tracts of agricultural monocultures (soy and sugarcane) and grazing land (for beef cattle)[115] are located. In 2011, 7,637 km^2 of land in the Cerrado was deforested which exceeded deforestation rates in the Amazon. One of the main reasons for the rapid rate of deforestation was the replacement of pastures and soy plantations with sugarcane fields, which resulted in additional land clearing in the Cerrado to compensate for pasture and soy acreage lost to sugarcane. The area of the biome located in the states of Tocantins, Piauí and Maranhão, which had historically been the best preserved, has become the latest Brazilian agricultural frontier.[116]

According to official data (see Table 12.4), sugarcane burning accounted for only 1.5 percent of the total GHG emissions in Brazil in 2010. These data must be considered in context because emissions from sugarcane plantations and ethanol production are not only the result of burning the

---

112 Julio Zanella, 'Manejo do solo pode reduzir gases-estufa' (Portal Unesp, May 2005), accessed 8 November 2016 at http://www.unesp.br/aci/jornal/198/agronomia.php.

113 Ministério de Ciência, Tecnologia e Inovação, see above note 105.

114 Instituto Nacional de Pesquisas Espaciais, 'Monitoramento da Floresta Amazônica Brasileira por Satélite' (OBT, December 2014), accessed 8 November 2016 at www.obt.inpe.br.

115 Instituto Brasileiro de Meio Ambiente e dos Recursos Naturais Renováveis, 'Projeto de Monitoramento de Desmatamento dos Biomas Brasileiros' (PMDBBS, October 2011), accessed 14 November 2016 at www.siscom.ibama.gov.br/monitorabiomas/.

116 These three states still have native Cerrado forest, however they form the main agricultural frontier. According to the Ministry of the Environment, degradation of the biome follows the expansion of such frontier, 'especially in West Bahia, South Goiás and the Sinop region of Mato Grosso. West Bahia is developing more in terms of the anthropic use of soil in the Cerrado'. Ministério do Meio Ambiente, see above note 47.

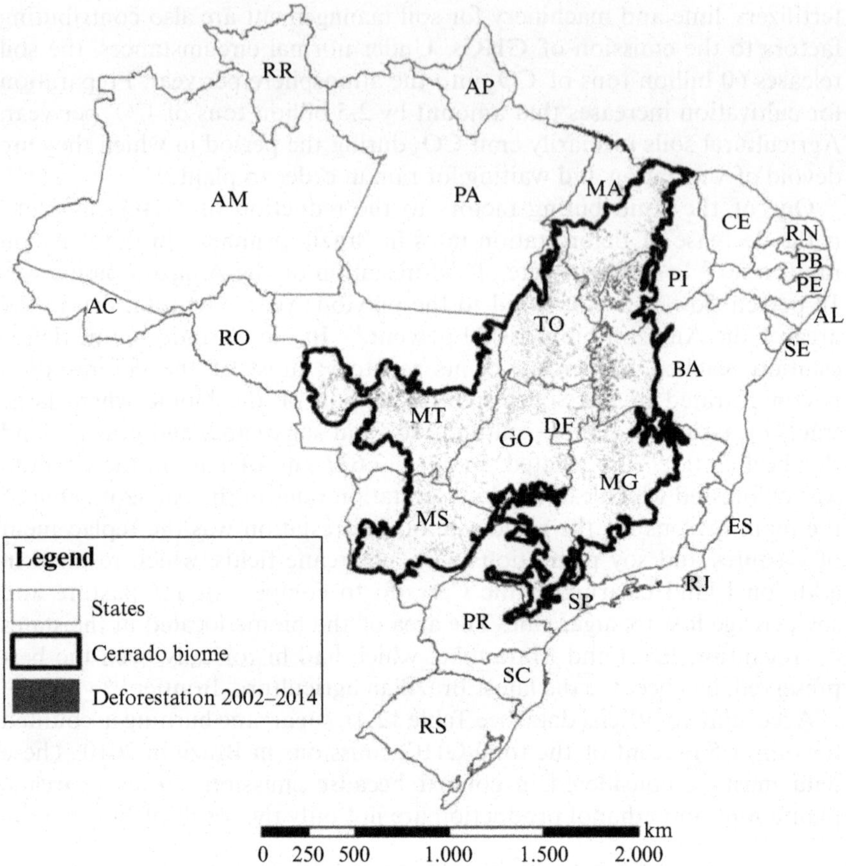

Note: Designed by Ralph de Medeiros Albuquerque (Mader-UnB).

Figure 12.6 Cerrado deforestation 2002–2014

harvest. Furthermore there is disagreement regarding the extent of the use of fire since the Sugar Industry Union (Unica) claims that more than half of sugarcane harvesting is mechanized in São Paulo and therefore burning is not used. Other sources claim that only 25 percent is mechanized in Brazil.[117]

[117] Plataforma BNDES, see above note 67, 171.

5. ETHANOL, CLIMATE CHANGE AND THE CERRADO

According to the Ministry of the Environment, the production of biofuels is one of the industries associated with deforestation of the Cerrado.[118] Based on available data and research, the cultivation of sugarcane accounted for the deforestation of a relatively small section of the biome as compared to soy crops and pasture for livestock. The main expansion of sugarcane cropland has occurred in previously deforested areas of the Cerrado. In 2008, approximately 45 percent of the area cultivated for sugarcane replaced livestock pasture and 53 percent occurred in previously agricultural areas.[119]

Sawyer has called the phenomenon of substituting agriculture and livestock breeding with sugarcane plantations the 'domino effect' because the expansion of sugarcane dislocates livestock breeding and agricultural crops to more distant and less valuable land.[120] This 'domino effect' requires a comprehensive analysis of land use and therefore claims that 'the expansion of sugarcane occurred in areas previously occupied by pastures' must be considered in a broad context. Since agriculture and livestock breeding replaced in one region (areas near the mills) will ultimately occupy new areas, there is no overall reduction but expansion in pastureland or grain acreage. Therefore replacement of agriculture and livestock breeding with sugarcane production still results in changes in soil use and the deforestation of the Cerrado.

Changes in land use for planting sugarcane and soy are one of the indirect but central causes of the deforestation of the Cerrado. Areas that the MMA and IBAMA have labeled as 'very high' or 'extremely high priority' for the promotion and sustainable use of the biome are under threat of being converted to sugarcane farmland for the production of ethanol.[121] According to the surveys, these areas that warranted heightened levels of protection accounted for 2 percent of the sugarcane expansion by 2008.[122]

Consequently the sugarcane sector exerts an indirect influence on the deforestation of sections of the Cerrado biome as a result of the 'domino

---

118 Ministério do Meio Ambiente, see above note 47.

119 Selma Simões Castro *et al.*, 'A expansão da cana-de-açúcar no cerrado e no estado de Goiás: elementos para uma análise espacial do processo' (2010) 30(1), *Boletim Goiano de Geografia* 171–91.

120 Donald Sawyer, 'Fluxos de carbono na Amazônia e no cerrado: um olhar sócio ecossistêmico' (2009) 24(1) *Sociedade e Estado* 149–71.

121 Ministério do Meio Ambiente, see above note 64.

122 Simões Castro et al., see above note 119.

effect' that it causes. It is therefore crucial to use these facts to qualify statements that deforestation or forest cover losses should not be associated with '. . . bioethanol production because the expansion of sugarcane production occurred in areas previously occupied by low productivity pastures or annual crops, largely for export, such as soy, activities with lower carbon storage uptake than those involved in the plantation of sugarcane'.[123]

Furthermore, deforestation in the Cerrado, in addition to directly emitting GHGs and degrading the soil, can contribute to global warming through another process: the jeopardization of hydrological cycles. A study by Oliveira *et al.* has shown that the deforestation of the Cerrado for the cultivation of crops and pasture has the potential to increase the surface water runoff by up to five times. This contributes to erosion and flooding because water is no longer trapped by vegetation or riverbeds.[124] The tributaries of Brazil's main river basins rise and widen in the Cerrado biome. The uncontrolled expansion of monocultures and grazing land has led to flooding in rainy seasons, erosion, silting, reduced infiltration, and accelerated evaporation of dams and reduced evapotranspiration, which generates new rains. All of these effects of the deforestation of the Cerrado jeopardize the water cycle of the Amazon, Tocantins-Araguaia, São Francisco and Paraná River basins.

With regard to direct emissions from the production and use of ethanol, Claros Garcia and von Sperling[125] and Macedo *et al.*[126] investigated the quantity of GHG emissions created by sugar-ethanol plants in Minas Gerais throughout the whole process of the transformation of sugarcane into agrofuel. Claros Garcia and von Sperling warned that certain factors should be taken into account when estimating emissions from ethanol production. First, there are differences between the structures and agro-industrial technologies used in the plants in south-central Brazil, which could lead to either low or high emission estimates. To minimize this problem, we used parameters resulting from the analysis of the environmental licenses of 11 distilleries with the most commonly used practices in

---

123 Banco Nacional de Desenvolvimento Econômico e Social and Centro de Gestão e Estudos Estratégicos, 'Bioetanol de cana-de-açúcar: energia para o desenvolvimento sustentável' (BNDES 2008) 94.

124 Paulo Tarso S. Oliveira *et al.*, 'The Water Balance Components of Undisturbed Tropical Woodlands in Brazilian Cerrado' (2014) 11 *Hydrology and Earth System Science Discussion*, 12987–13018.

125 See Claros Garcia and Von Sperling, above note 106.

126 Isaias C. Macedo *et al.*, 'Greenhouse Gases Emissions in the Production and Use of Ethanol from Sugarcane in Brazil: The 2005/2006 averages and a prediction for 2020' (2008) *Biomass and Bioenergy*.

the production cycle of the plants based on the findings of Claros Garcia and von Sperling.

Second, the period of the sugarcane production cycle used as a basis for calculation was six years, counting the fallow period (six months), the sugarcane planting (one and a half years between planting and the first cutting) and four cycles of sugarcane production (four years with four cuts per year). Third, as previously mentioned, in agricultural operations there is manual harvesting at different levels even in the most modern plants where the straw is removed by burning. Fourth, measurements of GHG emissions are made for an average production of 70–80 tons of sugarcane per hectare. Fifth, ethanol production plants are normally located in areas close to plantations at distances of not more than 100 kilometers, thereby reducing the use of fuel to transport the sugarcane.[127]

The calculation of direct emissions from the production of ethanol in the Cerrado is made based on the results found by Claros Garcia and von Sperling by applying an equation[128] to one hectare, then multiplying by the number of hectares intended for the production of agrofuel for the 2013–2014 harvest as measured by CONAB in 2013.[129]

According to the authors, the calculations showed that the burning of the sugarcane fields, the fuel consumed in the production of ethanol and the use of nitrogen to increase soil fertility are the three main factors for GHG emissions in the production process. As previously mentioned, despite the modernization of harvesting methods, these innovations have not eliminated the use of burning in the management of all sugarcane plantations.[130] Burning is therefore a key element in the measurement of GHG emissions derived from ethanol production.

---

127 See Claros Garcia and Von Sperling, above note 106; Macedo *et al.*, above note 126.

128 Claros Garcia and Von Sperling, see above note 106 at 218. The equation used was: Ei (GHG emissions) = Ij (fuel consumption in agricultural operations, transport of seedlings, filter cake, ashes and sediment, transport of lime and fertilizers, fertirrigation, mechanical harvest, loading and transporting of sugarcane; consumption of lime, fertilizers and pesticides; burning before and after harvesting; release of N_2O in the soil; production of seedlings; consumption of chemicals at the industrial plant) × NC (Number of cycles in which the quantity Ij is consumed) × Fej (GHG emission factor corresponding to input Ij; where j = fuel, lime, fertilizer, pesticides, cane burned in the harvest and chemical products in the industry) / by 6 years (complete cycle).

129 See Companhia Nacional de Abastecimento, above note 29.

130 Since 1998, there have been federal and state legal norms (Decree nº 2.661, of 08 July 1998), to reduce the use of burning in harvesting sugarcane. However there have been various judicial measures, with provisional decisions, prohibiting

Table 12.5 Emission of greenhouse gases in the sugarcane agricultural and industrialization process per hectare

Category	Greenhouse Gas Emission ($TgCO_2eq$/hectare)	Emissions (%)
Fuel Consumption	3,3718	21.9
Agricultural Operations	0,6977	–
Transportation of Feedstocks	0,1728	–
Fertigation	0,302	–
Sugarcane Harvest, loading and transportation	2,1993	–
Lime and fertilizer consumption	2,9838	19.38
Agrochemical pesticides consumption	0,3039	1.97
Sugarcane Burning	4,3431	28.21
N_2O Release	3,3152	21.53
Seedling Production	0,7281	4.73
Chemical Product - Industrial Stage	0,3501	2.27
Total	18,7678	100

Note: Drawn up by Pedro Araújo Pietrafesa.

Source: Juan Carlos Claros Garcia and Eduardo Von Sperling.

Claros Garcia and von Sperling (Table 12.5) estimated that burning accounts for 28.21 percent of GHG emissions in the cycle of ethanol production.[131] The consumption of fossil fuels, especially diesel, in mechanical harvesting, loading and transportation of sugarcane emits another 21.9 percent of the GHGs. If burning were to be replaced by mechanical harvesting, one could predict a reduction in emissions from the use of burning, but even so there would still be emissions from the combustion of fossil fuels in farming operations. The release of N_2O in the soil and the use of fertilizer and lime were the other two factors that heavily contributed to GHG emissions at 21.53 percent and 19.38 percent of the total amount of emissions estimated for ethanol production respectively.

Generalizing the results found by Claros Garcia and von Sperling in Minas Gerais plants, it can be estimated that direct emissions from ethanol production in the Cerrado biome were approximately 130,707,404.71 $TgCO_2eq$ for the 2013–2014 harvest. This approximate amount of total

and limiting this activity, but this is common practice in a large part of productive regions. See Plataforma BNDES, above note 67.

[131] See Claros Garcia and Von Sperling, above note 106.

emissions from the ethanol production cycle in the biome equates to 8.33 percent of total national emissions from the sources of GHGs monitored in Brazil for 2013.

Although emissions from the ethanol production cycle are estimated and generalized under constant conditions, in which all plants in the Cerrado region use the same forms of land management and industrial production of agrofuels, the result presented indicates a considerable contribution considering that they represent only one economic sector in the rural sphere. This finding should at least call into question the unconditional proclamation that ethanol is 'the best solution' for reducing the emission of GHGs into the atmosphere.[132]

On the other hand, as illustrated by Figure 12.7, the energy sector is Brazil's second largest emitter of GHGs. Emissions from fuel combustion accounted for 96 percent of GHGs from the energy sector in 2013.[133] In Figure 12.7 it can be observed that in recent years there has been a tendency towards an increase in emissions from automotive combustion, partially due to an increase in the national fleet in the second half of the 2000s resulting from tax incentives for the purchase of new cars.[134]

Studies confirm that government agencies officially advertise the fact that the GHG emissions from the combustion of ethanol in cars is five times less than the GHG emissions from the combustion of fossil fuels, representing a reduction of over 70 percent of these emissions.[135] Until 2013 the blend ratio was 25 percent ethanol per liter of automotive gasoline consumed in the country. From 2014 this percentage increased to 27.5 percent. This means that Brazilian emissions from fossil fuel combustion are lower as compared to emissions of other countries. However, as discussed above, the contribution of ethanol cannot be stated this simply. While there are reductions in GHG emissions associated with using flex engines, ethanol production creates other social and environmental problems associated with sugarcane monoculture and agrofuel production and it would take decades to offset these effects.[136] With the increase in light vehicles in the Brazilian market due to the lack of public policy promoting

---

[132] See Companhia Nacional de Abastecimento, above note 24; Banco Nacional de Desenvolvimento Econômico e Social, see above note 123.

[133] See Observatório do Clima, above note 104.

[134] See Camelini and Castilho, above note 35; Wilkinson and Herrera, above note 2.

[135] Luís Henrique de Barros Soares *et al.*, 'Mitigação das emissões de gases efeito estufa pelo uso do etanol da cana-de-açúcar produzido no Brasil' (2009) 27 *Embrapa* 1–14.

[136] See Sawyer, above note 120.

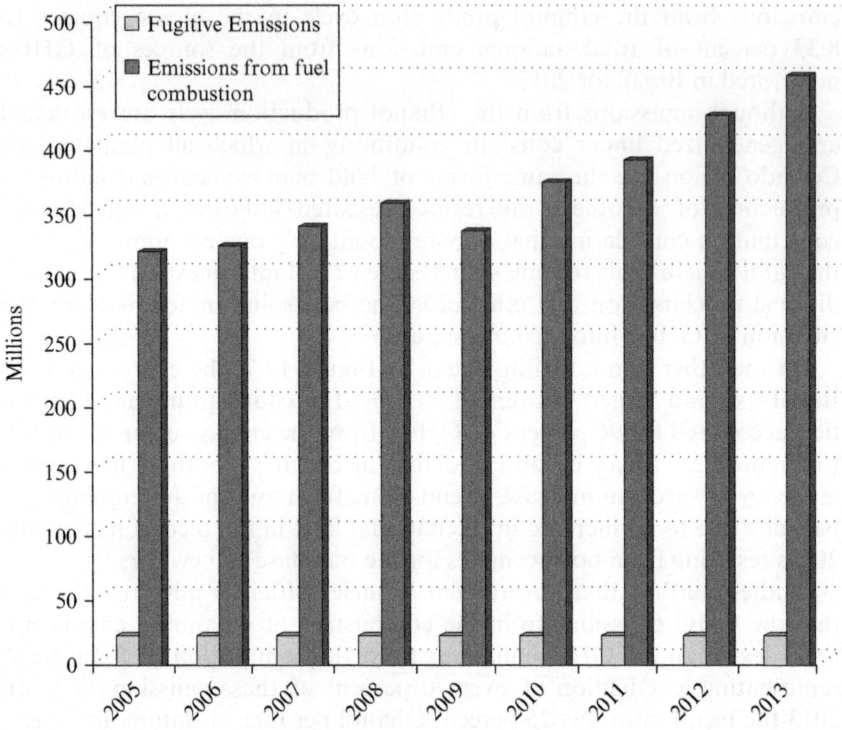

Note: Drawn up by Pedro Araújo Pietrafesa.

Source: Observatório do Clima.

Figure 12.7 GHG emissions from the energy sector in Brazil, 2005–2013

public transport, any decrease in GHG emissions could be jeopardized by the number of new individual vehicles joining the fleet each year.

CONCLUSION

Despite all the technological advances that would indicate modernization, the cultivation of sugarcane is a prime example of extensive monoculture based on land concentration and the over-exploitation of labor. The agrofuel industry is a major contributor to both social and environmental issues facing the Brazilian countryside. These issues include burning without government control or monitoring, excessive use of chemical fertilizers and instances of labor exploitation.

From a strictly environmental point of view, one of the great debates of the moment regarding climate change mitigation involves agro-energy, or more precisely, the possibility of substituting energy derived from petroleum with renewable sources derived from agriculture. When analyzing this phenomenon, if only the consumption of agrofuels in combustion engines of vehicles is considered, there is a false impression of a drastic reduction in the emission of GHGs due exclusively to a change in the consumption of the energy matrix. As discussed in this study, an analysis of such mitigation must be considered in the context of the whole cycle of production and use of agrofuels.

The case study of the Brazilian Cerrado region illustrates the areas that are being impacted by the expansion of sugarcane cultivation for ethanol production and the climate change impacts of the agrofuel production industry. Despite increased energy gains,[137] the production of ethanol is responsible for approximately 8.33 percent of total GHG emissions in Brazil.

In addition to direct emissions, ethanol production is associated with the main source of GHG production in Brazil: use of land and deforestation. Although, deforestation in Brazil is decreasing, it still is one of the principal sources of greenhouse gas emissions in the country. Deforestation is overtaking the Cerrado biome as forests are directly replaced with sugarcane crops or via a 'domino effect' whereby sugarcane replaces other crops and pastureland, which are subsequently relocated and cause deforestation elsewhere. The annual deforestation rate of the Cerrado is greater than that of the Amazon, and the changes in land use resulting from sugarcane expansion undermine all justifications for supporting for this industry, especially the argument that it constitutes a 'clean fuel'. However, this industry continues to grow as the government promotes it by implementing programs like the ZAE. Another major effect of this industry is the disruption of the hydrological cycles of the Cerrado and consequently of Brazil's river basins.

One final characteristic of the ethanol production cycle that is less connected to the climate issue is its funding source. At first funding was provided by national capital with the aid of federal and state government funding bodies. Now the funding process is undergoing internationalization with investments from multinational petrochemical companies and companies from other economic fields. However, the public sector is still playing a strategic role in supporting and financing sugar-energy projects

---

[137] See Banco Nacional de Desenvolvimento Econômico e Social, above note 123 at 83.

by methods such as implementing the mandatory ethanol blend in gasoline and through investments in infrastructure and direct financial support. There is also a lack of government policy stipulating the conditions for granting public funds for environmentally beneficial purposes such as technology to reduce GHG emissions from plants, programs to limit the deforestation of the Cerrado and promotion of environmentally conscious soil management techniques.

13. Adaptation to climate change in smallholder agriculture in Kenya: the role of law

Robert Kibugi

1. INTRODUCTION

The evidence of climate change is now irrefutable. The Fifth Assessment Report (AR5) of the Intergovernmental Panel on Climate Change (IPCC),[1] released in 2014, noted that 'warming projections under medium scenarios indicate that extensive areas of Africa will exceed 2°C by the last 2 decades of this century relative to the late 20th century mean annual temperature and all of Africa under high emission scenarios'.[2] In addition, the AR5 stated that African ecosystems are already being affected by climate change, which will further amplify existing stress on water availability in Africa with strong adverse effects on food security.[3] The 2014 IPCC report further supports the observation made earlier by the 2010 East African Community (EAC) Climate Change Policy that: 'EAC Partner States are generally dependent on climate-sensitive natural resources, have high poverty levels, coupled with low adaptive capacities.'[4] The latter policy also noted that 'a high reliance on rain-fed agriculture makes rural livelihoods and food security highly vulnerable to consequences of climate variability and change'.[5] This background is particularly meaningful for countries situated in Africa.

In Kenya, the impacts of climate change have most recently been investigated by the 2013–2017 National Climate Change Action Plan

---

[1] RK Pachauri and LA Meyer (eds), *Climate Change 2014: Synthesis Report. Contribution of Working Groups I, II and III to the Fifth Assessment Report of the Intergovernmental Panel on Climate Change* (IPCC 2014).

[2] Isabelle Niang *et al.*, 'Africa' in Christopher B Field *et al.* (eds), *Climate Change 2014: Impacts, Adaptation, and Vulnerability, Part B: Regional Aspects, Contribution of Working Group II to the Fifth Assessment Report of the Intergovernmental Panel on Climate Change* (Cambridge University Press 2014) 22.

[3] *Ibid.*

[4] East African Community, *East African Community Climate Change Policy* (EAC Secretariat 2010) 1.

[5] *Ibid* at 15.

(NCCAP),[6] and the 2014 Draft Framework Climate Change Policy,[7] with particular focus on priority economic sectors, including agriculture and land management. Kenya's understanding of the gravity of climate change impacts was summarized by the then President, Mwai Kibaki, in the Preface of the NCCAP. Here, Kibaki noted that Kenya 'continues to take bold measures to secure the country's development against the risks and impacts of climate change', and that the NCCAP 'encourages people-centred development, ensuring that climate change actions support Kenya's achievement of development goals'.[8] As the NCCAP states, the impacts of climate change have indeed been largely adverse: 'Kenya has in recent years had its share of climate-related impacts: prolonged droughts; frost in some of the productive agricultural areas; hailstorms; extreme flooding; receding lake levels; drying of rivers and other wetlands; among others leading to large economic losses and adversely impacting food security.'[9] This is cause for great concern, especially given the consequential impacts on agriculture. Over 70 percent of rural livelihoods depend on rain-fed subsistence agriculture, and are therefore significantly impacted by the declining agricultural production that results from unpredictable rainfall and reduced soil productivity through erosion and increased evapo-transpiration.[10]

This chapter describes some of the climate change-related challenges in Kenya, with specific reference to the vulnerability of smallholder (and rain-fed) agriculture. The chapter examines the role of law in framing appropriate tools for the government, and that of farmers in the development and use of coordinated adaptation strategies, strengthened resilience and enhanced adaptive capacity. Climate change strategies must become mainstream within agriculture law so that adaptation is internalized as an imperative for agricultural decision making, from national policy to land use choices by individual farmers.

Sections 2 and 3 provide the background to agriculture in Kenya, including an assessment of how climate change impacts smallholder

---

[6] Kenya Ministry of Environment and Mineral Resources, *National Climate Change Action Plan 2013–2017* (Ministry of Environment and Mineral Resources 2013).

[7] Kenya Ministry of Environment, Water and Natural Resources, 'Draft Sessional Paper on National Climate Change Framework Policy' (2014), accessed 14 November 2016 at http://iscrc.ilegkenya.org/documents/Draft-Climate-Change-Policy.pdf.

[8] Kenya Ministry of Environment and Mineral Resources (n 6) v.

[9] *Ibid* at 4.

[10] Kenya Ministry of Environment, Water and Natural Resources (n 7) 31.

farming. Section 2 examines the importance of adaptation to agriculture. Section 3 reviews the role of the law, especially the role of the Kenyan Constitution, in setting the basis for climate change adaptation in government functions and decision making and examines how legal tools could be used to mainstream climate change adaptation within the agricultural sector.

2. BACKGROUND TO AGRICULTURE IN KENYA

As is true in many parts of the world, agriculture and the environment (our natural resource base) are critical to the Kenyan economy and society. Notably, the *Second Medium Term Plan to Vision 2030*,[11] Kenya's current national development plan, describes agriculture as central to the economic pillar for guiding Kenya to middle income status.[12] However, according to the NCCAP, the agriculture sector (including livestock) is very sensitive to climate change, meaning that agricultural systems will need to adapt: both to ensure provision of adequate food for a growing population and to increase export crop production that will generate foreign exchange.[13] The urgency of the matter is clear, as agriculture is the mainstay of Kenya's economy and currently contributes 26 percent of the GDP directly, another 25 percent indirectly, and accounts for 65 percent of Kenya's total exports. It also provides more than 18 percent of the population with formal employment.[14]

Kenya's agriculture is mainly rain fed and dependent on the bimodal rainfall in most of the country.[15] In the humid, high altitude areas productivity (as well as predictability) of a good crop is high.[16] However, the population density in these areas has increased and land has been subdivided into such small sizes that it is becoming uneconomical for farm enterprises.[17] Farming is thus primarily small scale, with about 75 percent of total agricultural output coming from rain-fed agricultural lands on farms

---

[11] Kenya, *Sessional Paper No 10 of 2012 on Vision 2030* (Kenya Government Printer 2012).

[12] Kenya Ministry of Planning and Devolution, *Second Medium Term Plan to Vision 2030* (Ministry of Planning and Devolution 2013) 51.

[13] Kenya Ministry of Environment and Mineral Resources (n 6) 31.

[14] Kenya Ministry of Agriculture, *Agriculture Sector Development Strategy 2010–2020* (Ministry of Agriculture 2010) 1.

[15] *Ibid* at 10.

[16] *Ibid.*

[17] Pachauri and Meyer (n 1).

averaging 0.3 to 3 hectares in size,[18] and operated by about three million families.[19] Most agricultural activities are directed towards providing food for domestic consumption[20] and serve as the means of livelihood for most of the rural population.[21] The linkage between agriculture, poverty and food insecurity raises the question as to what extent the agriculture sector is equipped for adapting to, and mitigating, climate change. The Kenyan agricultural sector starts from a poor base, with existing challenges likely to be made more prominent by climate change. This vulnerability is seen through weaknesses in other sectors such as poor extension services that leave farmers without necessary knowledge, the loss of soil fertility and land degradation.[22]

3. UNDERSTANDING THE CHALLENGE POSED BY CLIMATE CHANGE TO KENYAN AGRICULTURE

Rain-fed agriculture, which accounts for 98 percent of the agricultural activities in the country, is very vulnerable to increasing temperatures, droughts and floods, all of which reduce agricultural productivity.[23] A huge demand for agricultural land has forced many people to migrate to arid and semi-arid areas, taking with them unsustainable farming practices that often accelerate land degradation and increase the vulnerability of smallholder farmers.[24] Land degradation is the result of inappropriate land use or poor land management practices, including the use of agrochemicals and tillage methods, which lead to a decline in soil fertility, the erosion of top soils, the silting of flood plains and dams, unstable hydrological conditions and an overall reduction in land productivity.[25] The ten-year Agriculture Sector Development Strategy (ASDS), adopted in 2010, states that 'the rising population density has contributed to the subdivision of land to uneconomically small units . . . with continuous cultivation leading

---

[18] Joab Osumba and Janie Rioux, *Scoping Study on Climate-smart Agriculture in Kenya Smallholder Integrated Crop-livestock Farming Systems* (FAO 2015) 2.

[19] Gordon O Ojwang, Jaspat Agatsiva and Charles Situma, *Analysis of Climate Change and Variability: Risks in the Smallholder Sector Case studies of the Laikipia and Narok Districts Representing Major Agro-ecological Zones in Kenya* (FAO 2010) 9.

[20] *Ibid.*

[21] Kenya Ministry of Agriculture (n 14) 28.

[22] *Ibid* at 24–7.

[23] Osumba and Rioux (n 18) 5.

[24] Ojwang, Agatsiva and Situma (n 19) 10.

[25] *Ibid.*

rapid depletion of soil nutrients, declining yields and environmental degradation'.[26] It follows that poor land use practices lead to degradation, and lower levels of fertility and agriculture productivity exacerbate the vulnerability of smallholder farmers.

Although it does not directly address climate change, Kenya's 2009 National Land Policy[27] identifies sustainable land use practices as being key to the provision of food, and sets out policy priorities such as addressing land deterioration due to population pressure and massive soil erosion.[28] Equally, the Land Act,[29] enacted in 2012, implicitly recognizes these challenges and includes an express requirement for identifying ecologically sensitive areas within public lands as well as for preventing environmental degradation and climate change.[30] Although this power is limited to public land only, this category includes ecologically sensitive areas like rivers, public forests and protected areas that provide critical ecological services to other categories of land use (for example, agriculture). As the guiding agriculture sector policy, the ASDS further acknowledges that 'climate fluctuations have a bearing on the way the environment and natural resources are managed . . . and the effect has been unpredictable weather that in turn has affected agricultural activities'.[31]

As a dominant socio-economic and environmental practice, smallholder agriculture faces significant challenges in developing countries which translate into serious risks for farmers. Smallholder farmers depend directly on agriculture for their livelihoods with limited resources to cope with shocks. Any reduction to agricultural productivity can therefore hamper local food security and well-being.[32]

---

26 Kenya Ministry of Agriculture (n 14) 25.

27 Kenya, *Sessional Paper No.3 of 2009 on National Land Policy* (Kenya Government Printer 2009).

28 *Ibid* para 118.

29 Land Act 6 of 2012.

30 S 11 of the Land Act.

31 Kenya Ministry of Agriculture (n 14) 68.

32 Celia A Harvey *et al.*, 'Extreme Vulnerability of Smallholder Farmers to Agricultural Risks and Climate Change in Madagascar' (2014) 369 Phil Trans R Soc B 1639.

4. AGRICULTURE AND CLIMATE ADAPTATION IN THE KENYAN CONTEXT

According to the Kenyan government, climate change threatens the country and could undermine its pursuit of sustainable development.[33] The risks posed by climate change to national socio-economic activities thus demand a path forward that responds to national needs and can insulate agriculture, forestry and other vulnerable sectors from harmful impacts.[34] This can explain why society perceives a link between sustainable development and climate resilient pathways,[35] a belief so prevalent that the main national climate policy in Kenya focuses on so-called 'low carbon climate resilient development' as a way to respond to climate change.[36]

Low carbon climate resilient development demands national prioritization of both mitigation and adaptation measures, which must be undertaken in a coordinated fashion to address evident and unknown risks. Because of high levels of poverty and underdevelopment, Kenya currently has low climate resilience and low adaptive capacity, particularly in the smallholder agriculture sector,[37] signaling the need for a stronger focus on climate adaptation and the reduction of vulnerability. This point is also clearly conveyed in the country's *Intended Nationally Determined Contribution* (INDC)[38] submitted to the United Nations Framework Convention on Climate Change (UNFCCC) Secretariat in July 2015. In its INDC, Kenya observed that being a minimum emitter, priority will be placed on adaptation to climate change.[39] Adaptation in this context is defined as adjustments in natural or human systems, in response to actual or expected climatic stimuli or their effects, which moderate harm

---

33 Kenya Ministry of Environment, Water and Natural Resources (n 7) paras 1.1.1, 1.3.2.

34 F Denton *et al.*, 'Climate-resilient Pathways: Adaptation, Mitigation, and Sustainable Development' in Christopher B Field *et al.* (eds), *Climate Change 2014: Impacts, Adaptation, and Vulnerability, Part B: Regional Aspects, Contribution of Working Group II to the Fifth Assessment Report of the Intergovernmental Panel on Climate Change* (Cambridge University Press 2014) 1101–31, 1109.

35 *Ibid.*

36 Kenya Ministry of Environment, Water and Natural Resources (n 7) ch 4.

37 Kenya Ministry of Environment, *Intended Nationally Determined Contribution* (Ministry of Environment 2015).

38 *Ibid.*

39 Government of Kenya, *Intended Nationally Determined Contribution (INDC)* (Government of Kenya 2015) 4.

or exploit beneficial opportunities.[40] Adaptation can also be categorized as (i) autonomous and (ii) planned adaptation.[41] Autonomous adaptation is reactive, occurring after impacts of climate change have been observed, for example, when a farmer reacts to changing precipitation patterns by changing crops, using different harvest periods or shifting planting/sowing dates. Planned adaptation measures are conscious policy options or response strategies, often multi-sectoral in nature, aimed at altering the adaptive capacity of the agricultural system or facilitating specific types of adaptation, and enhancing resilience.[42] In order to attain this, a policy of planned or anticipatory adaptation is required, both at the policy level and at the level of land-use decision-making on farms.

5. LEGAL TOOLS FOR THE MAINSTREAMING OF CLIMATE CHANGE IN THE AGRICULTURAL SECTOR

Kenya has a variety of legal and policy tools with which to govern socio-economic development and the environment. The Constitution of Kenya, adopted in August 2010, contains several provisions relevant to the climate mitigation and adaptation discourse. The range of other tools available further shows how thoroughly Kenya's agricultural sector has envisioned their adaptation to climate change.

5.1 Constitutional Foundations

Kenya's Constitution is concerned with stewardship, and the Preamble sets forth the tone of government: 'the people of Kenya' are "respectful of the environment" as a national heritage', and are 'determined to sustain it for the benefit of future generations'. Article 10 sets out several 'principles and values of national governance' which are binding on all persons (including public officials and private persons) when they apply the Constitution, any laws, or make any public policy decisions. One of these principles is that of sustainable development. While the Constitution does not specifically define 'sustainable development', when read together with some other constitutional provision it becomes clear that this term describes a balance

---

40 James McCarthy *et al.*, *Climate Change 2001: Impacts, Adaptation and Vulnerability* (Cambridge University Press 2001) 982.

41 *Ibid.*

42 FAO, *Adaptation to Climate Change in Agriculture, Forestry and Fisheries: Perspective, Framework and Priorities* (FAO 2007) 5.

between socio-economic utilization of natural resources and upholding environmental protection.

The Constitution thus provides the baseline framework for environment and land management in Kenya. Article 69(1) mandates measures the Kenyan government must take to fulfill the environmental right in Article 42. These include: ensuring sustainability in utilization, management and use of natural resources; ensuring a minimum 10 percent tree cover; ensuring public participation in natural resources management; and implementing environmental impact assessment mechanisms, among others. Article 69(2) imposes a mandatory duty on every person in Kenya, in collaboration with the state and other persons, to ensure environmental conservation and ecologically sustainable development and use of natural resources.

Article 60 of the Kenyan Constitution sets out the principles of national land policy for the sustainable management of all land in the nation. These principles reflect the objectives of the 2009 Sessional Paper on National Land Policy,[43] and require implementation of:

(a) equitable access to land;
(b) security of land rights;
(c) sustainable and productive management of land resources;
(d) transparent and cost effective administration of land;
(e) sound conservation and protection of ecologically sensitive areas; and
(f) elimination of gender discrimination in laws and customs.

By implication these principles address land use matters and determine that 'sustainable land use practices are key to the provision of food security . . .'.[44] This is to be read with the 2009 *National Land Policy*, which calls for measures against:

(i) land deterioration due to population pressure, massive soil erosion and climatic variability, which results in abandonment of agriculture activities;
(ii) overstocking of livestock in arid and semi-arid lands; and
(iii) introduction of innovative land uses and diversification of the rural economy.[45]

---

[43] Kenya, *Sessional Paper No. 3 of 2009 on National Land Policy* (Kenya Government Printer 2009).

[44] *Ibid* at 28.

[45] *Ibid* at 43, 28.

5.1.1 A matter of mainstreaming climate change in different sectors and agencies

Article 10 of the Constitution appears to require that sustainable development considerations be internalized or integrated into mechanisms of constitutional interpretation, statutory formulation/implementation, and in the making of public policy decisions. Sustainable development in this context, although a complex construct, refers in essence to the balancing of interests, between socio-economic and environmental considerations. Constitutional objectives ideals and other provisions find more detailed expression in sectoral legislation and policy frameworks in Kenya. Typically, Article 60's principles on land management must be elaborated upon in national land policy and be implemented in accordance with the dictates of land legislation. This amounts to the 'mainstreaming' of constitutional matters into the public administration space. In addition to laws and policies, other available tools which officials have at their disposal are planning, budgets and actual interventions.

The 2014 Draft Climate Change Policy[46] and the 2014 Climate Change Bill[47] (the Draft Policy and the Bill, respectively) promote the 'mainstreaming' of climate change to ensure that any impact on national development is addressed optimally. According to the Draft Policy, climate change mainstreaming is necessary to equip the Kenyan government and its various agencies with effective tools to respond to the complex challenges of climate change.[48] 'Mainstreaming', according to the policy, implies the integration of climate change policy responses into national, county and sectoral planning, budgeting and management processes.[49] The Bill codifies this approach, to the extent that it requires the national response to climate change to focus on 'mainstreaming climate change responses into sector functions', and 'to mainstream climate change disaster risk reduction actions in development programmes' through a National Climate Change Action Plan (NCCAP) that should be reviewed every five years.[50] The Bill also requires regular review of other sectoral laws and policies in order to determine amendments needed to enhance the capability

---

46 Kenya Ministry of Environment, Water and Natural Resources (n 7).

47 At the time of writing, the Bill had been passed by the Senate and National Assembly and was awaiting reconciliation of amendments before presentation to the President for Assent.

48 Kenya Ministry of Environment, Water and Natural Resources (n 7) 6.

49 Kenya Ministry of Environment, Water and Natural Resources (n 7) para 1.3.5.

50 S 13(3) of the Bill.

of these laws to tackle climate change challenges and exploit emerging opportunities.[51]

The notion of mainstreaming climate change interventions, including adaptation, into the work of sectoral line functionaries and legislation suggests that the agriculture sector is also affected. The legal and other tools developed in and for the agricultural sector may be expected to be structured to mainstream adaptation efforts, specifically.

5.2 Mainstreaming Adaptation to Climate Change through the Use of Legal Tools Governing Agriculture

The 2009 *National Land Policy* signaled the government's intention to put all land to productive, sustainable use by implementing key land policy principles such as land quality conservation, effective land use planning, and environmental auditing and assessment.[52] The right to own and use land encompasses these land use principles, and, under Kenya's Constitution, is protected as a fundamental human right.[53]

Property rights to tenure in resources that contribute to agricultural productivity, such as land and water, can influence the patterns of natural resource management – especially by determining *who* can do *what* with a particular resource, such as a parcel of land, and sometimes also *when* and *how* they can do it.[54] These tenure rights are fundamental to the roles of landowners in making decisions on the socio-economic utilization of land, and taking the steps necessary to sustain its environmental quality. Kenya's 2012 Land Registration Act[55] illustrates the breadth of tenure rights on private land, providing that 'the registration of a person as the proprietor of land shall vest in that person the absolute ownership of that land together with all rights and privileges belonging or appurtenant thereto'.[56] Therefore, tenure rights could play a central role by carving out a specific role for tenure holders (such as farmers) to integrate land use practices that build adaptive capacity through adjustments in both land use behavior and deployment of technologies.[57] While neither the Land Act nor the Land

---

[51] Kenya Ministry of Environment, Water and Natural Resources (n 7) para 9.1.4.

[52] *Ibid.*

[53] Art 40 of the Constitution.

[54] Ruth Meinzen-Dick *et al.*, 'Gender, Property Rights and Natural Resources' (1997) 25(8) *World Development* 1303.

[55] Land Registration Act 3 of 2012.

[56] S 24 of the Land Registration Act.

[57] See IPCC, 'Fourth Assessment Report: 17.3 Assessment of Adaptive

Registration Act specifies this possibility, the principles of sustainable land management in the Constitution and in the National Land Policy do support this. It is, however, likely to be done through sectoral land use legislation similar to the provisions in the 2013 Agriculture, Fisheries and Food Authority Act (AFFA),[58] discussed below.

AFFA provides government with the power to prescribe 'rules for the preservation, utilization and development of agricultural land'.[59] Using these powers, the government may typically prescribe the manner in which land owners (including those not in actual occupation) manage their land (that is, in accordance with the rules of good estate management); and further, how those in actual occupation of land should farm on that land (that is, in accordance with the rules of good husbandry). The threshold for fulfilling the prescribed conditions is set out such that:

> (a) an owner of agricultural land shall be deemed to fulfil his or her responsibilities to manage it in accordance with the rules of good estate management if, having regard to the character and situation of the land and other relevant circumstances, it enables an occupier of the land reasonably skilled in husbandry to maintain efficient production as respects both the kind of produce and the quality and quantity thereof;
> (b) the occupier of agricultural land shall be deemed to fulfil his or her responsibilities to farm it in accordance with the rules of good husbandry if the occupier is maintaining a reasonable standard of efficient production, as respects both the kind of produce and the quality and quantity thereof, while keeping the land in a condition to enable such a standard to be maintained in the future.

The above provisions show that compliance with the rules depends on the cooperation of the farmer/land owner. Notably, the statutory standard is that the farmer must be skilled in land husbandry.

Good land husbandry is generally defined as the 'active process of implementing and managing preferred systems of land use in such ways that there will be an increase – or, at worse, no loss – of productivity, stability, and usefulness for the chosen purpose'.[60] The UN Food and Agriculture Organization (FAO) defines land husbandry

---

Capacity, Options and Constraints' (2007), accessed 9 November 2016 at https://www.ipcc.ch/publications_and_data/ar4/wg2/en/ch17s17-3.html.

[58] Agriculture, Fisheries and Food Authority Act 13 of 2013.

[59] S 22 of AFFA.

[60] TF Shaxson, 'Conservation-effectiveness of Farmers' Actions: A Criterion of Good Land Husbandry' in E Baum, P Wolff and MA Zobisch (eds), *Acceptance of Soil and Water Conservation: Strategies and Technologies. Vol.3 in Topics in Applied Resource Management in the Tropics* (German Institute for Tropical and Subtropical Agriculture 1993) 105.

more specifically as a system of 'water, soil fertility and biomass management'[61] that is essential to sustainable land use. The FAO further states that sound land husbandry requires farmers to 'concentrate on the sound management of good, productive land before it becomes degraded, since there is a faster and larger increase in yields on deep soil than on exhausted stony soil'.[62]

It is important to note that at the time of writing, AFFA's anticipated rules on the preservation, utilization and development of agricultural land have not yet been developed. This leaves room for the concept of land husbandry to be mainstreamed into agriculture land use decision making in Kenya, using farmers themselves for land management in a way that facilitates adaptation to climate change. Earlier literature on agriculture land husbandry argues that farmers as the main 'husbanders' of the land are central to successful implementation, as they 'consider' the eventual 'rejection' or 'integration' of husbandry options into land use decision making.[63] It follows that the proposed rules can meaningfully contribute to build the adaptive capacity of farmers through the mainstreaming of adaptation techniques into husbandry of farmland.

AFFA acknowledges that Kenyan farmers have a direct and important role to play in agricultural land use decision making It ensures effective participation of farmers through close consultation with all registered farmers' organizations in the development of policies or regulations and prior to the making of any major decision that has or may have an effect on the agricultural sector. The setting of land preservation rules typically fit the profile of 'any major decision'. It is notable that the AFFA law speaks of engaging with farmers through associations, and even requires the Minister to provide for procedures for internal democracy in the organizations.[64] This is a positive development, because in any event, under Kenya's 2013 Statutory Instruments Act[65] when making any subsidiary legislation such as the rules on land utilization (under the AFFA), the government must consult with those persons who are likely to be affected by the regulation.[66] In addition, where the rules are likely to impose significant costs on the community, the government must prepare

---

[61] Eric Roose, *Land Husbandry: Components and Strategy* (FAO 1996) 24.
[62] *Ibid.*
[63] Francis Shaxson *et al.*, *Better Land Husbandry: Rethinking Approaches to Land Improvement and the Conservation of Water and Soil* (Overseas Development Institute 1997) 6.
[64] S 40(2)(b) of AFFA.
[65] Statutory Instruments Act 23 of 2013.
[66] S

a regulatory impact statement.[67] Such a statement would assess the effects of the proposed measures (regulatory and otherwise), set out practicable means of achieving them and assess the cost-benefit impact of the proposed regulations in the community.[68] The proposal for consultation with farmers, read together with the law governing the making of subsidiary legislation, creates a strong legal basis for compulsory consultation with farmers.

Previously, farmers (especially those operating as smallholders) had no legal mechanism to formally to share their skills and knowledge with one another, or to collectively interact with public officials. By contrast, the Forest Act 2005 created Community Forests Associations (CFA)[69] as the vehicle by means of which forest adjacent communities can get involved in participatory forest management for livelihood and conservation.[70] Borrowing from the structure of the CFA system on community forestry, the concerned forest communities engage with the forest authorities at the lowest level of governance, embracing the principle of subsidiarity. This is arguably an option that agriculture law should opt for, in order to have the farmers associations set up at the lowest possible level, to allow interaction with public officials at farm-level, whether for policy consultation, or for purposes of agriculture extension.

To optimize the legal instruments above, climate change adaptation efforts must become mainstream. The exchange of knowledge is necessary to ensure agricultural lands maintain productivity, are resilient to climate change, and that farmers build their adaptive capacity to reduce vulnerability. The FAO, for instance, has developed the term 'Climate Smart Agriculture' (CSA): agriculture that improves food security and helps communities adapt to and mitigate climate change by adopting appropriate practices, policies and financial support.[71] CSA seeks to support countries in putting in place the necessary policy, technical and financial means to mainstream climate change considerations into agricultural sectors.[72]

In practice, CSA can look very different depending upon the scale at which it is being applied. At the smallholder farm level, it may involve improved land management techniques for higher production such as a

---

67 S 7 of the Statutory Instruments Act.

68 *Ibid.*

69 See s 47 that makes provision for participatory forest management through membership to Community Forests Associations.

70 *Ibid.*

71 FAO, *Climate Smart Source Book* (FAO 2013) ix.

72 *Ibid.*

more targeted use of fertilizers.[73] CSA can also focus on soil and nutrient management, for instance, to enhance the availability of nitrogen and other nutrients essential to increase yields.[74] This can be achieved through composting manure and crop residues, matching nutrients more precisely to plant needs, controlled release and deep placement technologies or using legumes for natural nitrogen fixation – all adaptation practices that increase organic nutrient inputs and retention, and reduce the need of synthetic fertilizers that may be financially out of reach for smallholders.[75] Farmers can also apply conservation agriculture techniques similar to CSA, employing minimal mechanical soil disturbance, a maintained cover of carbon rich organic matter that feeds the soil, and crop rotations that include trees.[76]

The above techniques could be adopted through national laws and policies, to become nationwide practices that truly benefit farmers. Kenya's Sustainable Development Goal 2 makes a strong case for mainstreaming climate change adaptation into agriculture, by focusing on food security and sustainable agriculture.[77] Sub-goal 2.4 urges developed nations to implement resilient agricultural practices that increase productivity and production by 2030. Preferred practices include anything that helps maintain ecosystems, strengthens capacity for adaptation to climate change and extreme weather events, and improves land and soil quality.[78] It is imperative that Kenyans continue to improve knowledge, skills, techniques and technologies to further enhance resilience and build the adaptive capacity of agriculture, especially vulnerable smallholder lands. Interventions like Climate Smart Agriculture suggest the need for further research and inquiry into the generation and application of new and appropriate knowledge. Agriculture research and extension arguably is the mechanism through which this could and should typically be undertaken.

The current Agriculture Sector Development Strategy addresses some of the agricultural challenges faced in Kenya today, highlighting reduced effectiveness of extension services over the last two decades due to the use of inappropriate extension methods.[79] The ASDS notes that

---

[73] World Agroforestry Centre, *Making Climate-Smart Agriculture Work for the Poor – Policy Brief No 11* (World Agroforestry Centre 2012) 2.

[74] FAO, *'Climate Smart' Agriculture: Policies, Practices and Financing for Food Security, Adaptation and Mitigation* (FAO 2010) 1.

[75] *Ibid.*

[76] *Ibid* at 5.

[77] See UNGA Res 70/1 (25 September 2015) UN Doc A/RES/70/1, 13.

[78] *Ibid.*

[79] Kenya Ministry of Agriculture (n 14) 24.

although Kenya has a well-developed agricultural research system, the use of modern science and technology in agriculture production is limited because of inadequate research-extension-farmer relationships.[80] Since the ASDS was adopted in 2010, the 2013–2017 NCCAP has identified priority adaptation actions that can increase climate resilience in agriculture. These include the promotion of drought tolerant crops, water harvesting, integrated soil fertility management, insurance schemes, price stabilization schemes for livestock, strategic food reserves, providing farmers and pastoralists with climate change-related information, and integrating climate change into agricultural extension services.[81] These techniques could all bring climate change mitigation into mainstream use and be implemented through statutory regulations under AFFA. However, this would require further research on how they enhance the adaptive capacity of smallholder farmers in various contexts. As a consequence, neither the role of agricultural research nor the utility of agricultural extension services can be discounted.

5.2.1 The role of agricultural extension in climate change adaptation

Agricultural extension is the function of providing need and demand-based knowledge and skills to rural populations in a non-formal, participatory manner, with the objective of improving their quality of life.[82] The ASDS notes that the collapse of agricultural extension services is a key contributor to a failure in agriculture productivity. Agricultural extension plays a key role in disseminating knowledge, technologies and agricultural information, and links farmers with other actors in the economy.[83]

5.2.2 The role of farmers, researchers and extension works

Conceptually, agricultural extension is intended to help farmers make best use of productive resources at their disposal[84] through provision of important information such as patterns in crop prices, new seed varieties, crop cultivation management and marketing, and training in new

---

80 *Ibid.*

81 Kenya Ministry of Environment and Mineral Resources (n 6) 32.

82 Robert Kibugi, 'Implementing Stewardship in Kenyan Land Use Law: The Case for a Sustainability Extension' in Paul Martin *et al.* (eds), *Environmental Governance and Sustainability* (Edward Elgar 2012) 293.

83 Kenya Ministry of Agriculture (n 14) 19.

84 Milu Muyanga and TS Jayne, *Agricultural Extension in Kenya: Practice and Policy Lessons – Working Paper 26* (Egerton University 2006) 1. See also Milu Muyanga and TS Jayne, 'Private Agricultural Extension System in Kenya: Practice and Policy Lessons' (2008) 14(2) JAEE 111, 114.

technologies.[85] Thus, the ideal agricultural extension structure centers on the relationship between research institutions and extension staff, creating a feedback loop from farmers to researchers and back again, with the extension staff acting as agents. Kenyan agriculture policy, through the ASDS, emphasizes the need for a linkage between agricultural researchers, extension agents and farmers to ensure that agriculture practices overall benefit from appropriate research.[86] The policy indicates that 'research-extension links will be strengthened to ensure demand-driven research and effective application of research technologies on the farm'.[87] This sets the policy basis for establishing a working relationship between agricultural extension staff and end-user farmers. The 2012 National Agriculture Sector Extension Policy (NASEP)[88] identifies this as a priority, proposing strategies to 'ensure that sector ministries, research organizations and training institutions take a leading role in ensuring a workable mechanism for strengthening research-extension-client linkage and feedback'.[89] This is important, because in earlier evaluations by the World Bank of agricultural extension in Kenya, farmers reported that the availability and quality of information services had declined within just a few years.[90] However, World Bank evaluators were quick to highlight that the failure of a particular extension project does not extinguish the rationale or utility of an extension function *per se*.[91] Bearing in mind the challenges facing smallholder agriculture in Kenya, and the need to develop adaptive capacity for farmers, the development of these agricultural linkages, and feedback between farmers, extension workers, and research institutions is imperative.

In 2013, Kenya enacted new legislation, the Kenya Agriculture and Livestock Research Act (KARLA),[92] governing agriculture research and consolidating multiple existing agriculture sector institutions into one entity. Thus, specialized research institutes on coffee, tea, cotton, livestock, and so on, were all merged to form the Kenya Agriculture and Livestock Research Organization (KARLO). KARLO is comprised of the original

---

85 *Ibid.*

86 Kenya Ministry of Environment, Water and Natural Resources (n 7) 33.

87 *Ibid.*

88 Kenya Ministry of Agriculture, *National Agriculture Sector Extension Policy* (Ministry of Agriculture 2012).

89 *Ibid* at 35.

90 Madhur Gautam, *Agricultural Extension, The Kenyan Experience: An Impact Evaluation* (World Bank 2000) xv.

91 *Ibid* at 7.

92 Kenya Agriculture and Livestock Research Act 17 of 2013.

research institutes[93] that, according to the legislation, are required to organize and carry out on-station and on-farm research in their respective sectors of agricultural research. They also must identify value chain constraints in the fields of agriculture, water, livestock, fisheries, forestry and environment, and prepare research programmes.[94] Importantly, for the first time in Kenya, agriculture legislation has created a direct link between research institutions and agricultural extension services. In this context, the KARLA legislation requires research institutes to collaborate with the extension and education services to disseminate research results and technologies.[95]

5.2.3 The utility of agricultural extension in enhancing the role of farmers to build adaptive capacity

Assuming these relationships function optimally, the question arises: how do agricultural extension services aid smallholder farmers in reducing their vulnerability to climate change? Agricultural extension has a dual focus. The first is communication, which involves transferring information about new practices and technology from research to potential users and obtaining feedback from researchers. This transfer of knowledge aims to change the behavior of land users, sometimes necessitating major modifications in production systems. The second focus is education, which works in synergy with communication. Land-users work directly with researchers and other experts to make wise land use choices that will prime their production systems for adaptation to climate change.

Because land use activities like agriculture are inherently influenced by broad cultural, political and socio-economic forces, the role of extension services in changing best practices cannot be overstated.[96] Behavior modification is key to enhancing adaptive capacity and mitigating the effects of climate change. Modification may include, for instance, exploration of new farming techniques, such as Climate Smart Agriculture mentioned earlier, or a collaboration between farmers and the government to create innovative land use rules that enhance both resilience and productivity. Research has established that for adaptation to be sustainable, there must be a somewhat cyclic movement of knowledge, with feedback, between policy makers, researchers, extension workers and farmers – in essence,

---

[93] Part VI of KARLA.

[94] S 32 of KARLA.

[95] S 32(2)(d) of KARLA.

[96] Frank Vanclay and Geoffrey Lawrence, 'Agricultural Extension in the Context of Environmental Degradation: Agricultural Extension as Social Welfare' (1994) 5(1) *Rural Society* 1.

a combination of all local knowledge.[97] Extension services can promote this kind of collaboration through spread of farmer based innovations like farmer-exchange visits, where farmers observe the adaptations practiced by fellow farmers first hand.[98] The farmers, as landowners or occupiers, can thereafter make their own choices about which techniques to use in their own land use practices.

6. CONCLUSION

The challenge posed by climate change to humanity generally cannot be disputed. In developing parts of the world, these challenges are more complex, owing to more vulnerable populations and fragile economies. Smallholder farming in particular faces difficulty because farmers experience lower levels of resilience and restrictions to their adaptive capacity. Increasing levels of land degradation, loss of fertility, and extreme weather events exacerbate already dire circumstances. The role of the law and transformative policies in countries like Kenya facing these challenges is therefore crucial – because the law can be used to provide impetus to new forms of knowledge or practice, for example. Structuring legal intervention around the concept of mainstreaming is important, as it ushers in the optimal exercise of (government) functions, whereby the agriculture sector factors climate change considerations into the planning, design, budgeting and implementation of national policies for the good of the entire nation. The role of farmers is key under these circumstances, as they hold the primary decision-making power over land through tenure rights. In addition, land use regulations can guide farmers into practices that enhance stewardship and maintain productivity, doing what is necessary for climate change adaptation. Thus, the continuous development of knowledge, its dissemination and the synergistic relationships between farmers, researchers, policymakers and extension workers are important cogs in influencing behavior changes favorable to adaptation.

---

97 Chinwe Ifejika Speranza, *Resilient Adaptation to Climate Change in African Agriculture* (German Development Institute 2010) 151.

98 *Ibid* at 156.

14. Agricultural land use conflict in the context of climate change: an Australian case study

Amanda Kennedy and Amy Cosby

1. INTRODUCTION

The rapid expansion of extractive developments in recent years has seen conflict over agricultural land use increase in prevalence and scale in communities around the world. This includes conflict over the expansion of existing fossil fuel operations (for example, coal mining) and the development of newer industries (for example, coal and shale bed methane gas) upon land already used for agricultural purposes.[1] In the context of global climate change and concerns over food security, it is imperative that regulatory arrangements for determining land use are equitable, effective and efficient to avoid widespread social conflict.

A significant portion of current fossil fuel development in Australia takes place upon or within close proximity to agriculturally productive land.[2] Rights to explore and produce subsurface minerals and petroleum are retained by the Crown in Australia, which has created additional challenges for agriculture in the context of climate change. Unprecedented conflicts over agricultural land use have emerged, centred upon environmental and economic costs and benefits. Coexistence has been championed by industry and government,[3] yet regulatory approaches have failed to alleviate concerns or overcome increasingly intractable disputes. Critics have argued that regulatory responses to date have merely prioritized the economic imperatives of fossil fuel production over both agricultural

---

[1] Dimiter Kenarov, 'Dairy Farms Suffer in US Shale Gas Fracking Boom' (*The Ecologist Online*, 2013), accessed 9 November 2016 at http://www.theecologist.org/News/news_analysis/1784330/dairy_farms_suffer_in_us_shale_gas_fracking_boom.html.

[2] Rosemary Lyster, 'Coal Seam Gas in the Context of Global Energy and Climate Change Scenarios' (2012) 29(2) EPLJ 91, 97; Katherine Owens, 'Strategic Regional Land Use Plans: Presenting the Future for Coal Seam Gas Projects in New South Wales?' (2012) 29(2) EPLJ 113, 114.

[3] Owens (n 2) at 114.

development and environmental sustainability,[4] with some noting that Australia's mineral exploration and extraction laws actively 'encourage' the production of these finite resources.[5]

Using a case study from Australia, this chapter explores one example of agricultural land use conflict against the backdrop of global climate change, and the regulatory responses to manage it. The chapter begins with a brief discussion of the Australian dimensions of climate change and increasing global energy demand, placing in perspective the growing interest in domestic fossil fuel production upon agricultural land. It then considers the potential impacts associated with fossil fuel exploration, which, when overlaid with specific concerns from the agricultural sector, accounts for the escalation of conflict over agricultural land in Australia. Attention then turns to an Australian case study of agricultural land use conflict, and the regulatory reforms implemented to manage conflict within that jurisdiction. It emerges that current approaches have yielded little success in resolving agricultural land use disputes, largely because they have resulted in the marginalization of local communities, and the prioritization of development above environmental and social concerns. The chapter concludes by considering an alternative direction for land use decision making grounded in principles of environmental justice, an approach which has application to other types of land use conflicts beyond those considered here.

2. CLIMATE CHANGE AND GLOBAL ENERGY DEMAND: THE AUSTRALIAN DIMENSION

The forecast for future global energy demand is astronomical, with the International Energy Agency (IEA) estimating a 37 percent increase in consumption by 2040.[6] Australia possesses a range of natural resources that are abundant and of high quality, and which already play a major role in satisfying global demands for energy resources as the world's

---

4 Meg Sherval and Nicole Graham, 'Missing the Connection: How the Strategic Regional Land Use Policy Fragments Landscapes and Communities in NSW' (2013) 38(3) Alt LJ 176.

5 Cindy Chen and Alan Randall, 'The Economic Contest Between Coal Seam Gas Mining and Agriculture on Prime Farmland: It May Be Closer than We Thought' (2013) 15(3) JESP 1.

6 International Energy Agency, 'World Energy Outlook 2014 Factsheet' (2014), accessed 9 November 2016 at http://www.worldenergyoutlook.org/media/weowebsite/2014/WEO2014FactSheets.pdf.

ninth-largest energy producer.[7] In the near future, Australia will strengthen this position with increased production of coal and natural gas.

Coal remains a significant component of Australia's energy supply mix, and export demand is rising to meet the increasing global hunger for energy.[8] Currently, black coal is Australia's largest energy export, contributing 27 percent to global export and projected to increase by 54 percent by 2049–50.[9] However, natural gas sources are poised to overshadow this in the near future. Interest in unconventional energy sources has rapidly expanded in Australia, driven largely by export demand for liquefied natural gas (LNG). Typically, these unconventional energy sources involve the extraction of (predominantly) methane gas from underground (for example, from coal seams in the case of coal seam gas (CSG)). As of January 2014, Australia's gas reserves were 1.219 trillion cubic metres (TCM). To put this in perspective, in 2013 Australia produced 61.7 billion cubic metres (BCM) of gas, while the Australian population utilized 35.65 BCM and exported another 32.56 BCM.[10] CSG currently represents 12 percent of gas production in Australia, and 34 percent of eastern Australia, with its export value expected to rise rapidly to over $60 billion by 2017–18, making Australia the largest exporter of gas in the world.[11]

The growth in coal and gas production will inevitably result in increasing greenhouse gas emissions. Coal-fired power generation already accounts for 33 percent of Australia's greenhouse gas emissions, with 0.82 tons of CO_2 produced per megawatt hour of electricity.[12] Australia's domestic

---

[7] Arif Syed, 'Australian Energy Projections to 2049–50' (*Bureau of Resource and Energy Economics*, 2014), accessed 9 November 2016 at http://www.industry.gov.au/Office-of-the-Chief-Economist/Publications/Documents/aep/aep-2014-v2.pdf.

[8] Department of Industry and Science, 2015 Australian energy update' (2015), accessed 14 November 2016 at http://www.industry.gov.au/Office-of-the-Chief-Economist/Publications/Documents/aes/2015-australian-energy-statistics.pdf; Australian Department of Industry, Innovation and Science, 'Resources and Energy Quarterly' (*Office of the Chief Economist*, 2015), accessed 9 November 2016 at http://www.industry.gov.au/Office-of-the-Chief-Economist/Publications/Documents/req/REQ-March15.pdf. Australia is predicted to become the largest exporter of coal, iron ore and liquid natural gas in the world by 2020.

[9] Syed (n 7).

[10] CIA, 'The Word Factbook – Australia' (2015), accessed 9 November 2016 at https://www.cia.gov/library/publications/the-world-factbook/geos/as.html.

[11] Australian Department of Industry, 'Energy Green Paper' (*Department of Industry*, 2014), accessed 9 November 2016 at http://ewp.industry.gov.au/files/egp/energy_green_paper.pdf.

[12] Andrew Stock, 'Australia's Electricity Sector: Ageing, Inefficient and

emissions contribute 1.5 percent to the global total;[13] however, when emissions from exported coal are also taken into account, this rises to 4.8 percent.[14] Unconventional gas extraction will also lead to a significant increase in greenhouse gas emissions. While burning CSG to generate power produces fewer greenhouse gas emissions than conventional energy sources,[15] the manner of its extraction can lead to the escape of fugitive emissions up to 21 times more powerful than CO_2.[16] A naturally occurring product, CSG is mostly made up of methane absorbed into coal found in coal seams. The extraction of CSG involves reducing water pressure in coal seams to enable the gas to 'de-sorb' and release. In some cases, where seams of lower permeability are encountered, fracture stimulation ('hydraulic fracturing') may be necessary to release the gas. This involves injecting a combination of sand, water and some chemicals into the well to stimulate the release of gas.[17] The use of hydraulic fracturing for CSG extraction is not as common as compared to its use for shale gas extraction; moreover, in the case of CSG, fracture stimulation is usually performed at much lower pressures than for shale deposits.[18] There have been no comprehensive studies in Australia related to the amount of fugitive emissions released by CSG extraction;[19] however, research into shale gas extraction in the USA has found the process of 'fracking' releases over 30 percent more methane

Unprepared' (*Climate Council*, 2014), accessed 9 November 2016 at http://www.climatecouncil.org.au/uploads/f9ba30356f697f238d0ae54e913b3faf.pdf.

[13] Australian Bureau of Statistics, '4613.0 – Australia's Environment: Issues and Trends, Jan 2010' (2010), accessed 9 November 2016 at http://www.abs.gov.au/AUSSTATS/abs@.nsf/Lookup/4613.0Feature+Article1Jan+2010.

[14] Peter Christoff, 'Why Australia Must Stop Exporting Coal' (*The Conversation*, 2 November 2012), accessed 9 November 2016 at https://theconversation.com/why-australia-must-stop-exporting-coal-9698.

[15] Lyster (n 2). A study in China determined coal exported from Australia when used for power generation produced between 5–87% less greenhouse gas emissions when compared to burning Australian CSG or LNG for the same purpose, see T Clark, R Hynes and P Mariotti, *Greenhouse Gas Emissions Study of Australian CSG to LNG* (WorleyParsons and APPEA 2011).

[16] NSW Chief Scientist and Engineer, 'Initial Report on the Independent Review of Coal Seam Gas Activities in NSW' (NSW Government 2013), accessed 9 November 2016 at http://www.chiefscientist.nsw.gov.au/_data/assets/pdf_file/0016/31246/130730_1046_CSE-CSG-July-report.pdf.

[17] WorleyParsons, 'Groundwater Risks Associated with Coal Seam Gas Development in the Surat and Southern Bowen Basins – Final Report' (*Australian Department of Natural Resources and Mines* 2013), accessed 9 November 2016 at https://www.dnrm.qld.gov.au/__data/assets/pdf_file/0013/106015/act-5-groundwater-risks-report-text.pdf.

[18] Chief Scientist and Engineer (n 16) 40.

[19] Matt Grudnoff, 'Fracking the Future – Busting Industry Myths about Coal

than conventional gas drilling.[20] It is well known that methane is a powerful greenhouse gas with a greater potential to contribute to global warming than CO_2, particularly in the decades following the initial emissions.[21] Overall, the emissions footprint of shale gas was found to be 50 percent greater over a 20-year period when compared to coal.[22] Overall, the extraction of CSG is certainly not socially or environmentally 'neutral',[23] and has attracted criticism as a result.[24]

Increasing greenhouse gas levels caused by the burning of fossil fuels for energy and transport have been identified as a major cause of climate change in Australia.[25] The average terrestrial temperatures in Australia have increased approximately 0.7°C since the 1950s.[26] Anthropogenically-induced global warming has been recognized as one of the causes of prolonged drought in parts of Australia,[27] leading to predictions of increased incidence and intensity of extreme weather events.[28] Australia's agricultural sector is particularly vulnerable to climate change, with the majority of food production systems in Australia sensitive to changing weather patterns and extreme events.[29] The effects of climate change have

---

Seam Gas' (*Australian Institute*, 2014), accessed 9 November 2016 at http://www.tai.org.au/content/fracking-future.

20 Robert W Howarth, Renee Santoro and Anthony Ingraffea, 'Methane and the Greenhouse-gas Footprint of Natural Gas from Shale Formations – A Letter' (2011) 106 *Climatic Change* 679.

21 *Ibid.*

22 *Ibid.* The extraction process for shale and CSG is similar. See also Tim Poisel, 'Coal Seam Gas Exploration and Production in New South Wales: The Case for Better Strategic Planning and More Stringent Regulation' (2012) 29 EPLJ 129, 132.

23 Rob White, 'Resource Extraction Leaves Something Behind: Environmental Justice and Mining' (2013) 2 IJCJ 50.

24 Howarth, Santoro and Ingraffea (n 20); Lyster (n 2) 91.

25 Bureau of Meteorology, 'State of Climate' (*Australian Government*, 2014), accessed 9 November 2016 at http://www.bom.gov.au/state-of-the-climate/; Australian Greenhouse Gas Emissions Information System, 'Activity Tables: Activity Table1990–2012 –Energy – Stationary Energy' (*Australian Department of the Environment*, 2015), accessed 9 November 2016 at http://ageis.climatechange.gov.au/QueryAppendixTable.aspx.

26 Karl Braganza and John A Church, 'Observations of Global and Australian Climate' in Helen Cleugh *et al.* (eds), *Climate Change: Science and Solutions for Australia* (CSIRO 2011).

27 *Ibid.*

28 Kevin Hennessy, 'Climate Change Impacts' in Helen Cleugh *et al.* (eds), *Climate Change: Science and Solutions for Australia* (CSIRO 2011).

29 Prime Minister's Science, Engineering and Innovation Council, 'Australia and Food Security in a Changing World – Preparing for the Future with Foresight'

been predicted to decrease Australian agricultural production by 9–10 percent by 2030, and a further 13–19 percent by 2050.[30] Australian farms produce over 90 percent of Australia's domestic food supply, and around one percent of global food consumed. Australian agriculture contributes significantly to the domestic economy, predicted to be worth $37.9 billion in 2014–2015.[31] Australia is only one of 11 countries worldwide that is a net food exporter,[32] with its produce estimated to feed around 40 million people across the world.[33] With the global population expected to reach 9.3 billion by 2050,[34] and the Australian population projected to reach 35 million by the same time,[35] the impact of climate change upon Australian agricultural production poses a real risk to food security.

However, with much of Australia's existing and proposed coal and CSG developments located in close proximity to agricultural land, the expansion of fossil fuel exploration and production over the coming years raises a number of concerns in addition to the atmospheric impacts noted above. Within Australia, more immediate concerns around current land use and impact have sparked considerable conflict over agricultural land. Against this broader backdrop of climate change, conflicts have played out across several agricultural communities within Australia in recent years. The following section briefly details the concerns which have underpinned recent conflicts within agricultural communities.

(*Office of the Chief Scientist, Australian Government*, 2010), accessed 9 November 2016 at http://www.chiefscientist.gov.au/wp-content/uploads/FoodSecurity_web.pdf.

30 Ejaz Qureshi *et al.*, 'Impact of Water Scarcity in Australia on Global Food Security in an Era of Climate Change' (2013) 38 *Food Policy* 136.

31 ABARE, 'Agricultural Commodities: September Quarter 2014' (*Australian Department of Agriculture*, 2014), accessed 9 November 2016 at http://www.agriculture.gov.au/abares/publications/display?url=http://143.188.17.20/anrdl/DAFFService/display.php?fid=pb_agcomd9abcc20140916_11a.xml.

32 PMSEIC Expert Working Group, 'Australia and Food Security in a Changing World' (*Prime Minister's Science, Engineering and Innovation Council*, 2010), accessed 9 November 2016 at http://www.chiefscientist.gov.au/wp-content/uploads/FoodSecurity_web.pdf.

33 *Ibid.*

34 UN, 'United Nations Press Release: World Population to Reach 10 Billion by 2100 if Fertility in All Countries Converges to Replacement Level' (2011), accessed 9 November 2016 at http://esa.un.org/unpd/wpp/other-information/press_release_wpp2010.pdf.

35 Australian Government, 'Australia to 2050: Future Challenges – The 2010 Intergenerational Report' (*Commonwealth of Australia*, 2010) 5, accessed 9 November 2016 at http://archive.treasury.gov.au/igr/igr2010/Overview/pdf/IGR_2010_Overview.pdf.

3. EFFECTS OF INCREASED ENERGY PRODUCTION ON AUSTRALIAN AGRICULTURE

The most immediate impact of expanding coal and gas developments is the loss of arable, food producing land.[36] According to the World Bank, 52.8 percent of Australia's land during 2010–14 was classified as agricultural, down from 53.3 percent (409 million hectares) in 2005–09.[37] While only 2 percent of Australian land is currently under mining lease (and only 0.2 percent being actively mined),[38] 57 percent of Australian land is subject to coal and gas exploration applications.[39] In the agriculturally productive North West region of New South Wales alone, 640,000 ha of land are subject to exploration licenses for coal, and 6 million ha for CSG.[40] Between 2001 and 2009, 47 million ha of agricultural land was lost to other purposes, including mining.[41] Figure 14.1[42] illustrates the value of agricultural production ($/km^2) in areas where the exploration or extraction of coal and/or CSG is taking place in Australia.

Of course, not every coal or gas development near agricultural land will result in the direct conversion of the land to other purposes. Industry and government argue that in many cases, agricultural enterprises can coexist with mining and drilling activities, minimizing the loss of agricultural land.[43] Yet agricultural communities have raised many concerns, believing

---

[36] Lyster (n 2) 97; Owens (n 2) 114; Chen and Randall (n 5) 1.

[37] World Bank, 'Agricultural Land (% Land Area)' (2015), accessed 9 November 2016 at http://data.worldbank.org/indicator/AG.LND.AGRI.ZS/countries.

[38] Minerals Council of Australia, 'The Minerals Industry', accessed 9 November 2016 at http://www.minerals.org.au/corporate/about_the_minerals_industry; Australian Collaborative Land Use and Management Program, 'National Scale Land Use Data', accessed 9 November 2016 at http://www.agriculture.gov.au/abares/aclump/land-use/data-download.

[39] Samantha Hepburn, 'FactCheck: is 53% of Australia under exploration licence for unconventional mining?' (The Conversation, 25 March 2015), accessed 14 November 2016 at https://theconversation.com/factcheck-is-53-of-australia-under-exploration-licence-for-unconventional-mining-38702.

[40] Lock the Gate Alliance, 'North West NSW', accessed 24 February 2015 at http://www.lockthegate.org.au/northwestnsw.

[41] Joanne Millar and Jane Roots, 'Changes in Australian Agriculture and Land Use: Implications for Future Food Security' (2012) 10(1) Int J Agri Sustain 25.

[42] Accessed 9 November 2016 at http://apo.org.au/files/Resource/ABARES_TowardsNationalReportingOnAgriculturalLandUseChangeInAustralia_Oct_2013.pdf.

[43] Owens (n 2) 115; see also World Class Mining, 'Mining and Farming Thriving Together' (*NSW Mining*, 2014), accessed 9 November 2016 at http://www.worldclassminers.com.au/news/environment/mining-and-farming-thriving-together.

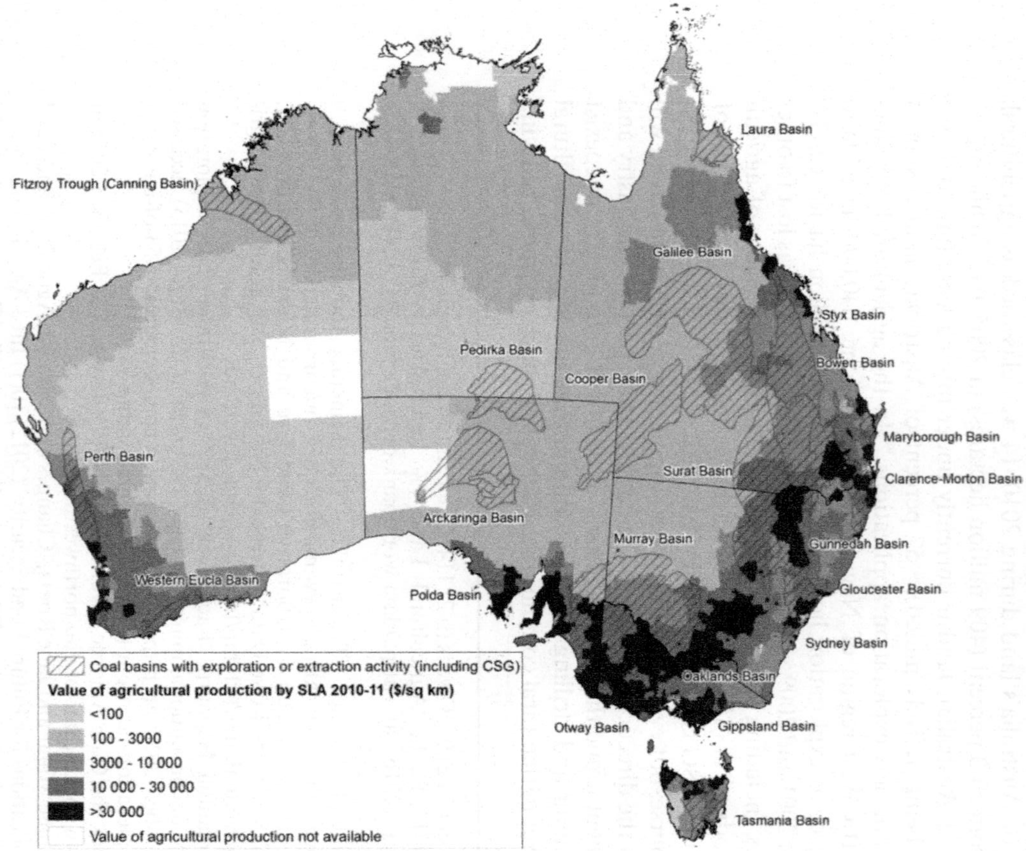

Source: Jodi Mewett *et al.*, 'Towards National Reporting on Agricultural Land Use Change in Australia – Technical Report 13.06' (*Australian Department of Agriculture & ABARES*, 2013).

Figure 14.1 Value of agricultural production ($\$/km^2$) and areas where coal or CSG mining occurs in Australia

there will still be significant impacts upon agricultural production, as well as environmental and social factors.

Water availability is of great concern to agricultural communities. Water is an important and scarce resource with many vital uses including residential water supply and irrigation for agricultural enterprises. The extraction of CSG in particular has the potential to negatively affect the quantity of water resources, with the increased strain on aquifers and surface water potentially decreasing water tables and pressure levels, and potentially displacing existing uses (such as farm access to groundwater).[44] The cumulative effects of extraction across regions are also unknown, and the potential for depressurisation to cause a change in 'equilibrium flow conditions' has been raised.[45]

Water quality may also be affected by coal and gas extraction activities. In the case of coal mining, this can occur through the spilling, leaching or leaking of chemicals and/or heavy metals; erosion on the mine site leading to sediments settling in waterways; and acid mine drainage.[46] The process of extracting CSG entails water being released into wells drilled into coal deposits, causing depressurisation which releases the gas. The wastewater which is pumped out of wells during the extraction process is highly saline, and also contains other chemicals which are present both in and around the coal seam. There is considerable community apprehension as to how this water is disposed of, or treated for re-use. Where gas is tightly held within the coal seam, hydraulic fracturing may be utilized to release the gas.[47] The practice of hydraulic fracturing has been the source of much concern, particularly the risk of groundwater contamination (through changed surface drainage patterns).[48] There has been some evidence of

---

44 Chen and Randall (n 5) 5.

45 Owens (n 2) 115; Eco Logical Australia, 'Proposed Framework for Assessing the Cumulative Risk of Mining on Natural Resource Assets in the Namoi Catchment' (*Namoi Catchment Management Authority*, 2011) 30, accessed 9 November 2016 at http://education.nwlls.com/client/multimedia/namoi_risk_assessment_final_v5_14sept11.pdf; Geoscience Australia and MA Habermehl, 'Summary of Advice in Relation to the Potential Impacts of Coal Seam Gas Extraction in the Surat and Bowen Basins, Queensland, Phase One Report Summary for Australian Government' (*Australian Department of Sustainability, Environment, Water, Population and Communities*, 2010), accessed 9 November 2016 at http://www.environment.gov.au/epbc/notices/pubs/gladstone-ga-report.pdf.

46 Australian Bureau of Statistics, 'Mining and the Environment' (2003), accessed 9 November 2016 at http://www.abs.gov.au/ausstats/abs@.nsf/90a12181d877a6a6ca2568b5007b861c/ce28d7fbe5faa308ca256cae0015da32!OpenDocument.

47 Poisel (n 22) 133.

48 Mary O'Kane, 'Independent Review of Coal Seam Gas Activities in NSW – Managing Environmental and Human Health Risks from CSG Activities' (*NSW*

water source contamination in eight CSG wells in the town of Miles on the Darling Downs in Queensland.[49] The uncertainty of these risks is intensified where little is known about the geology and groundwater conditions for a certain area.[50]

Noise and dust disturbance from drilling and extraction activities are frequently noted,[51] which in the agricultural context can include risks to crops from dust drift across properties. Coal mining is the leading cause of air particle pollution in Australia, with the rate of coarse particle pollution increasing by over 50 percent in the past five years.[52] In 2013–14, 430,000 tons of coarse particle pollution was attributed to the coal industry, accounting for 47 percent of the Australian total.[53] Recent data from Australia's National Pollutant Inventory also reveals that air pollution from unconventional gas extraction has increased sharply over the last five years.[54]

The proliferation of coal and gas operations has also produced growing concern over impacts upon human health. The town of Tara, in the state of Queensland, is the most densely populated region in Australia where CSG infrastructure is present.[55] One family resident on a rural block

---

Government, Chief Scientist and Engineer, 2014) 46, accessed 9 November 2016 at http://www.chiefscientist.nsw.gov.au/_data/assets/pdf_file/0006/56922/140930-Final-Managing-Environmental-and-Human-Health-Risks.pdf.

[49] Lyster (n 2) 98.

[50] Chen and Randall (n 5) 1.

[51] Climate and Health Alliance, 'Coal and Health in the Hunter: Lessons from One Valley for the World' (2015), accessed 9 November 2016 at http://caha.org.au/wp-content/uploads/2015/02/CAHA.CoalHunterValley.Report.FINAL_.Approvedforprint.pdf; Angela Werner *et al.*, 'Environmental Health Impacts of Unconventional Natural Gas Development: A Review of the Current Strength of Evidence' (2015) 505 Sci Total Environ 1127.

[52] Australian Department of Environment National Pollutant Inventory, '2013/2014 Data within Australia - Particulate Matter 10.0 um from Coal Mining', accessed 9 November 2016 at http://www.npi.gov.au/npidata/action/load/summary-result/criteria/industry-source/060/substance/70/destination/ALL/source-type/INDUSTRY/substance-name/Particulate%2BMatter%2B10.0%2Bum/subthreshold-data/Yes/year/2014.

[53] *Ibid.*

[54] Australian Department of Environment National Pollutant Inventory, '2013/2014 Data within Australia – Particulate Matter 10.0 um from Oil and Gas Extraction', accessed 9 November 2016 at http://www.npi.gov.au/npidata/action/load/summary-result/criteria/industry-source/070/substance/70/destination/ALL/source-type/INDUSTRY/substance-name/Particulate%2BMatter%2B10.0%2Bum/subthreshold-data/Yes/year/.

[55] Geralyn McCarron, 'Symptomatology of a Gas Field – An Independent Health Survey in the Tara Rural Residential Estates and Environs' (2013), accessed

within two kilometres from CSG wells reports chronic health issues which have forced them to leave their property.[56] This was not an isolated case, with an independent study of 38 households (a total of 117 people) revealing that 82.58 percent of residents had reported negative impacts on their health from CSG.[57] The potential health harms for communities near coal mines are more widely reported, with coal mining activities linked to increases in lung cancer, heart disease and chronic respiratory conditions.[58]

Coal and gas developments also have the potential to fragment landscapes. The clearing of land for mining and drilling activities, as well as the spatially dispersed network of pipes, wells, water storage and treatment facilities and roads all contribute to the (often irreparable) scarring of landscapes, negatively impacting both agricultural land as well as ecosystems.[59] In some cases, this can result in reduced productivity and increased costs to farmers, as well as threats to habitat, biodiversity and ecological value.[60] The increased industrial activities associated with coal and gas production can also disrupt the 'rural' way of life. Heavy equipment is required for both coal and gas exploration and production, which may need to be brought in via large trucks on rural roads not typically constructed to withstand such traffic.[61] There is also evidence that competition from extractive industries can have a negative impact upon the agricultural workforce.[62] The favorable wages paid in extractive industries may make it difficult for other businesses in the local region to remain competitive, halting the growth and viability of other sectors in an area. There may also be a reliance on mobile skilled 'fly in fly out' labour, which has the potential to drive up the cost of living, and also increase the pressure on local

---

9 November 2016 at http://www.ntn.org.au/wp/wp-content/uploads/2013/05/Symptomatology-of-a-gas-field-An-independent-health-survey-in-the-Tara-rural-residential-estates-and-environs-April-2013.pdf.

56 Michele Jedlicka, 'Dreams Dashed After CSG Experience' (*Inverell Times,* 19 January 2015), accessed 9 November 2016 at http://www.inverell-times.com.au/story/2820934/dreams-dashed-after-csg-experience/?utm_content=buffer4d16e&utm_medium=social&utm_source=twitter.com&utm_campaign=buffer.

57 McCarron (n 55).

58 Ruth Colagiuri, Johanne Cochrane and Seham Girgis, *Health and Social Harms of Coal Mining in Local Communities: Spotlight on the Hunter Region* (Beyond Zero Emissions 2012).

59 Chen and Randall (n 5) 8.

60 *Ibid.*

61 Climate and Health Alliance (n 51); Werner *et al* (n 51).

62 Millar and Roots (n 41).

infrastructure and services to deal with a growing population.[63] Concerns that economic benefits will only flow to a limited few not necessarily located within the community are also prevalent.[64] Communities may lose unity they previously experienced, due to differing opinions of extractive industries, the disproportion of incomes between those who work for a mining or drilling company and those that do not, and also the temporary and transient nature of mine employees. These factors may lead to a less resilient and less varied local economy.

While coal, and even CSG, have been produced in Australia for some time,[65] never before have these industries impacted agricultural land so significantly. The loss of prime agricultural land to coal mining and CSG exploration, alongside the risks posed to scarce water resources, public health, landholder amenity and the agricultural industry, has been a considerable cause of distress to affected communities.[66] Public conflicts have emerged between the need to ensure both economic and energy security, and the need to protect both the environment and food producing capacity. At the heart of this conflict is the legal framework for determining appropriate land uses.

In terms of legal rights to explore and extract subsurface gas and minerals, the law is unequivocally tilted in favor of gas and mineral development. Proponents of gas and mineral developments in Australia possess strong property rights, courtesy of the 'dominial' system of subsurface mineral ownership. In each state, legislation is present which vests ownership of mineral and gas rights in the Crown.[67] This vesting means

---

63 NSW Department of Planning and Infrastructure, *Strategic Regional Land Use Plan – Upper Hunter* (NSW Government 2012) 38–40.

64 Oliver Milman, 'NSW Urged to Ban New Coal Mines in the Hunter Valley on Health and Climate Grounds' (*The Guardian*, 23 February 2015), accessed 9 November 2016 at http://www.theguardian.com/australia-news/2015/feb/23/nsw-urged-ban-new-coal-mines-hunter-valley.

65 Australian Bureau of Statistics, 'The Australian Mining Industry: From Settlement to 2000' (2000), accessed 9 November 2016 at http://www.abs.gov.au/ausstats/abs@.nsf/94713ad445ff1425ca25682000192af2/93136e734ff62aa2ca2569de00271b10!OpenDocument; SBS News, 'Factbox: CSG in Australia' (2013), accessed 9 November 2016 at http://www.sbs.com.au/news/article/2013/03/12/factbox-csg-australia.

66 Geoffrey Lawrence, Carol Richards and Kristen Lyons, 'Food Security in Australia in an Era of Neoliberalism, Productivism and Climate Change' (2013) 29 J Rural Stud 30; Poisel (n 22) 129, 132.

67 See, for example: Mining Act 1992 (NSW); Petroleum (Onshore) Act 1991 (NSW); Mineral Resources (Sustainable Development) Act 1990 (Vic); Mineral Resources Act 1989 (Qld); Petroleum and Gas (Production and Safety) Act 2004 (Qld); Mining Act 1971 (SA), Mining Act 1978 (WA).

that landholders only retain control of the physical surface of their property, while the state controls the mineral and gas reserves underneath. States have the right to license subsurface exploration and extraction to third parties in return for the payment of royalties, while surface owners hold no proprietary interests in the subsurface rights underneath their land. Moreover, while legislation in each jurisdiction provides some protections to landholders regarding land access and compensation, there is no *prima facie* right of veto to prevent gas or mineral companies from accessing private land where they hold exploration or production licenses. Landholders and communities have little capacity to challenge exploration or extraction, and protests and public debate over land use decision making have become commonplace.[68]

The following section explores the regulatory framework for coal and gas development within one Australian jurisdiction, the state of New South Wales. In that state, the regulatory framework for assessment and approval of mining and CSG projects has undergone review, largely in response to conflict over coal and gas development on or in close proximity to agricultural land. Review of the legal landscape, however, has done little to alleviate community concern, serving mainly to prioritize development interests and minimize opportunities for meaningful public participation and objection to land use decisions. Discussion then turns to a specific case study to illustrate the dimensions of conflict under this regulatory regime.

4. THE REGULATORY FRAMEWORK FOR COAL AND GAS DEVELOPMENT IN NEW SOUTH WALES

The Mining Act 1992 (NSW) governs the exploration and production of coal within New South Wales, while the Petroleum (Onshore) Act 1991 (NSW) covers coal seam gas.[69] As noted earlier, these legal instruments

[68] David Shearman, 'Mandatory Federal Guidelines Needed on Fracking, Coal Mining' (*WA Today*, 14 April 2015), accessed 9 November 2016 at http://www.watoday.com.au/comment/mandatory-federal-guidelines-needed-on-fracking-coal-mining-20150414-1mkk4m.html; Thom Mitchell, 'Backing Big Coal: Mike Baird Promises to Up the Ante for Acting on Your Conscience' (*New Matilda*, 12 April 2015), accessed 9 November 2016 at https://newmatilda.com/2015/04/12/backing-big-coal-mike-baird-promises-ante-acting-your-conscience.

[69] Under s 3 of the Petroleum (Onshore) Act 1991 (NSW), petroleum includes naturally-occurring hydrocarbons in gaseous, liquid or solid state. Coal is expressly excluded from this definition, and is considered a mineral under the Mining Act.

vest ownership of subsurface mineral and petroleum resources in the state, enabling it to license both the exploration and production of mineral and petroleum resources by third parties.[70] Under both of these instruments, there are limited opportunities for both landholders and members of the public to participate in the decision-making processes relevant to coal and gas exploration and production licenses.[71] The public comment process for coal and CSG exploration titles, for example, is specifically limited to matters relating to the exploration process only, and not potential production.[72] Moreover, as environmental assessment is not required for exploration or production licenses, there is limited information available for the public to comment on at this stage of the process.

In addition to exploration and production licenses, projects also require planning consent under the Environmental Planning and Assessment Act 1979 (NSW). For exploration activities, this involves a preliminary 'Review of Environmental Factors' (REF), which provides a basic overview of the proposed exploration work.[73] There are no opportunities for affected landholders to challenge a proponent's REF. If a program of exploration is approved, a land access agreement must be entered into with the landholder before the proponent can enter the land.[74] This typically involves a landholder negotiating the terms of access directly with the proponent; however, if no agreement is reached within 28 days, the proponent can seek arbitration.[75] If exploration is approved, the access agreement typically prescribes conditions of access, as well as compensation to be paid for loss or interference.[76] Once an access agreement is entered into, a landholder cannot obstruct the proponent from entering the land.[77]

If an economically recoverable resource is discovered in the exploration process, a proponent may then seek planning approval for production. This first involves the preparation of an environmental impact statement (EIS), lodged with the development application.[78] The relevant regulations

---

70 Petroleum (Onshore) Act 1991 (NSW) ss 29, 41; Mining Act 1992 (NSW) ss 27, 73.

71 Mining Act 1992 (NSW) Sch1 Cl 26.

72 NSW Department of Industry, Resources and Energy, 'Public Comment', accessed 9 November 2016 at http://www.resourcesandenergy.nsw.gov.au/landholders-and-community/minerals-and-coal/public-comment.

73 Environmental Planning and Assessment Act 1979 (NSW) Part 5.

74 Petroleum (Onshore) Act 1991 (NSW) s 69C; Mining Act 1992 (NSW) s 140.

75 Petroleum (Onshore) Act 1991 (NSW) s 69F–G; Mining Act 1992 (NSW) ss 139, 140, 143–153.

76 Petroleum (Onshore) Act 1991 (NSW) s 69D; Mining Act 1992 (NSW) s 141.

77 Petroleum (Onshore) Act 1991 (NSW) s 69C; Mining Act 1992 (NSW) s 140.

78 Environmental Planning and Assessment Act 1979 (NSW) Part 4, Div 4.1.

set out a range of factors which must be addressed within the EIS, including the stated objectives of the proposed development, feasible alternatives to development (including the consequences of not pursuing the development), the likely environmental impacts of the proposed development, and any measures to mitigate or avoid these impacts.[79] Once received, the application is placed on public exhibition for a minimum of 30 days.[80] During this time, the public can make a written submission about the project.[81] It is critical for objectors to make a submission at this point, as the ability to appeal development decisions later rests upon whether a written submission was received during the exhibition period. Proponents then respond to any concerns raised in public submissions.

Development applications are usually then considered by the Planning Assessment Commission (PAC), an independent planning body who has authority to review development applications, as well as act as decision maker under delegation from the Minister for Planning.[82] The PAC considers the likely impacts of the development (including environmental, social and economic impacts), the requirements of any environmental planning instruments (such as State Environmental Planning Policies), and any public submissions.[83] The most relevant State Environmental Planning Policy is the State Environmental Planning Policy (Mining, Petroleum Production and Extractive Industries) 2007 (the 'Mining SEPP'). This policy provides specific criteria for decision makers to consider, including compatibility with nearby land uses, the impact of transport, the efficiency of resource recovery, and post-development land rehabilitation.

As a part of both its review and decision-making functions, the PAC may consult the community. This may be done by way of a public hearing, or a public meeting. Public hearings are usually held as a part of the initial review process, at the request of the Minister.[84] Public meetings are held as a part of the decision-making process, where more than 25 public submissions have been made in objection to a development. However, if a public hearing has already been held as part of a review process, the PAC typically will not hold an additional public meeting.[85] The main distinction between these two types of public consultation is that there are no merits

---

79 Environmental Planning and Assessment Regulations 2000 (NSW) Sch. 2, Cl. 7.

80 Environmental Planning and Assessment Regulations 2000 (NSW) Cl. 83.

81 Environmental Planning and Assessment Act 1979 (NSW) s 89F(3).

82 Environmental Planning and Assessment Act 1979 (NSW) s 23D.

83 Environmental Planning and Assessment Act 1979 (NSW) s 79C.

84 Environmental Planning and Assessment Act 1979 (NSW) s 23D (1)(b)(iii).

85 NSW Government Planning Assessment Commission, 'Guide to

appeal rights following a public hearing.[86] A merits appeal may be made against a determination where there has been a PAC public meeting, but only if there was no public hearing held as a part of the review process. Further, members of the public can only lodge an appeal if they qualify as an 'objector' (by having previously submitted a written submission during the public submission period).[87] Concern has been expressed that referral to a public hearing can be a strategic manoeuvre to prevent the public from lodging merits appeals against decisions.[88]

Finally, while the state regulatory framework primarily drives the approval process for coal and gas developments, federal legislation may also be relevant. In particular, any actions that will have a significant impact on a matter of national environmental significance under the Commonwealth Environment Protection and Biodiversity Conservation Act 1999 (EPBC Act) require approval from the Federal Environment Minister as well. The matters of national environmental significance include whether or not a development will impact upon world heritage properties or national heritage places, Ramsar wetlands, or threatened species or ecological communities.

Following widespread concern surrounding coal and gas developments (particularly on agricultural land), and in recognition of the lack of strategic spatial distribution of these developments, in 2012 the NSW state government proposed a suite of reforms to respond to land use conflict.[89] Central to these reforms was the introduction of the Strategic Regional Land Use Policy (SRLUP), which introduced additional steps relevant to the approval of coal and gas developments situated on agricultural land. The Policy package included a range of initiatives to 'better balance growth in the mining and coal seam gas (CSG) industries with the need to protect important agricultural land and water resources'.[90] The initiatives included: the introduction of exclusion zones for development in residential areas; the establishment of a Land and Water Commissioner to provide independent

---

Commission Meetings' (2013), accessed 9 November 2016 at http://www.planning.nsw.gov.au/~/media/Files/PAC/General/Guide-to-Commission-Meetings.ashx.

86 Environmental Planning and Assessment Act 1979 (NSW) s 23F.

87 Environmental Planning and Assessment Act 1979 (NSW) s 98.

88 Peter Hannam and Leesha McKenny, 'New Planning Minister Rob Stokes Signals Overhaul of CSG, Coal Mining Projects' (*Sydney Morning Herald,* 2 May 2015), accessed 9 November 2016 at http://www.smh.com.au/environment/new-planning-minister-rob-stokes-signals-overhaul-of-csg-coal-mining-projects-20150501-1mxhxk.html.

89 Owens (n 2) 113.

90 NSW Department of Planning and Infrastructure, 'Strategic Regional Land Use Policy' (2012), accessed 9 November 2016 at http://www.nsw.gov.au/sites/default/files/initiatives/strategicregionallandusepolicy-brochure_sd_v01.pdf.

advice on gas and mineral exploration, and to oversee land access agreements; an Aquifer Interference Policy which sets out factors to be assessed relevant to underground water; specific codes of practice relevant to the CSG industry concerning hydraulic fracturing and gas well integrity; and the requirement for Agricultural Impact Statements at exploration and development application stages for projects situated on agricultural land.

A key feature of the SRLUP was the introduction of the Mining and Petroleum Gateway Panel, responsible for the pre-assessment of the impact of significant mining or gas development proposals upon Strategic Agricultural Land (SAL). This assessment takes place prior to the lodgement of a development application pursuant to the Environmental Planning and Assessment Act 1979 (NSW). SAL is defined as highly productive land with unique natural resource characteristics. It is divided into two sub-categories: Critical Industry Clusters (CICs) and Biophysical Strategic Agricultural Land (BSAL). CICs are a concentration of significant industries that may be impacted by coal or gas mining development; thus far, equine and viticulture industry clusters within the Hunter Valley have been identified. The extraction of CSG is prohibited on CIC land; however, coal mining may be permissible following further assessment. If high value agricultural land is not included in a CIC, it may nonetheless be mapped as BSAL land. BSAL land is identified as productive land containing high quality soil and water resources. Strategic Regional Land Use Plans contain the mapped strategic agricultural land, with the first plans released covering the Upper Hunter and New England North West areas. These two areas have experienced the most rapid growth in coal and CSG exploration and development in the state. Mapping for the remainder of the state was released in early 2014.

A coal or gas development proposal on BSAL land will trigger a Gateway Panel Assessment to determine if the development will impact agricultural productivity. The Gateway Panel Assessment is a scientific analysis conducted by a panel of independent experts. There is no public consultation during this process. Following their assessment, the Gateway Panel may issue one of two types of Gateway Certificates: a certificate confirming unconditional approval, or a conditional certificate which can recommend further studies or project modifications. The Panel is not vested with the authority to reject a proposal outright and refuse a Gateway Certificate. For this reason, the Gateway process has been criticized as being a 'gateway without a gate',[91] offering no protection through a right to refuse development on agriculturally productive land. Once the

---

91 Mike Foley, 'No Gate in Gateway: Simson' (*Farmonline*, 3 October 2013),

Gateway Certificate is obtained, the proponent can continue to lodge their development application (as discussed above).

Strategic spatial planning has the potential to identify the capacity of a landscape for a particular land use (for example, agriculture, mining), and may be able to provide the foundation for an integrated, region-wide cumulative assessment of development.[92] However, the suite of SRLUP reforms introduced in NSW were criticized as constituting a 'dangerous shortcut' in planning regulation, and were expected to do little to alleviate land use conflict.[93] Owens argues that the SRLUP was an attempt to 'depoliticize the issues at stake, and provide a reductionist and once-off solution to a complex political question'.[94] She takes issue with the lack of technical knowledge about the functioning of groundwater systems, noting that cumulative regional scale impacts cannot be adequately taken into account by the various decision-making bodies, as there is limited background knowledge about groundwater dynamics and the impact that mining and drilling activities will have on these systems.[95] Further, the rigidity of the approach to spatial planning within the Strategic Regional Land Use Plans is also seen as a limitation, in that 'local aspirations' and landscape potential,[96] as well as a region's 'sense of place',[97] are not accounted for. For example, the Upper Hunter Strategic Regional Land Use Plan failed to recognize land within the Gloucester Shire as BSAL, despite the fact that the area contains productive agricultural land, and locals perceive and place great value upon their agricultural identity.[98]

The trivialization of environmental risks through the SRLUP's reliance on cost-benefit analysis is also problematic.[99] Under the Policy, contests over land are ultimately determined according to the notion of profit, failing to recognize the 'multiple and diverse values of land'.[100] The vernacular of these conflicts reveals that landholders and communities value

---

accessed 9 November 2016 at http://www.theland.com.au/news/agriculture/general/news/no-gate-in-gateway-simson/2673757.aspx.

92 Owens (n 2).

93 *Ibid.*

94 *Ibid.*, 113.

95 *Ibid.*, 118.

96 *Ibid.*, 121.

97 Sherval and Graham (n 4) 177; Meg Sherval and Kristian Hardiman, 'Competing Perceptions of the Rural Idyll' (2014) 45(2) *Australian Geographer* 185.

98 *Ibid.*

99 Owens (n 2) 120.

100 Sherval and Graham (n 4) 178.

landscapes as 'places'; as Sherval and Graham observe, 'the problem is not that land may be profitable; the problem is the land is *not only* profitable' [emphasis added].[101] The Policy package lacks mechanisms to determine the total cost of development projects, particularly social and environmental costs which may transcend spatial and temporal boundaries. In short, the SRLUP 'entrenches imbalance by rendering it impossible to undertake genuine evaluation of the merits of a [coal or CSG] proposal', and by effectively guaranteeing project approval.[102]

The prominence of economic prosperity as a metric for approval is evident not only in the framing of the SRLUP, but in subsequent amendments to the Mining SEPP. As noted earlier, the Mining SEPP is a legal instrument which provides guidance to decision makers as to the assessment of development applications for coal and CSG developments under the Environmental Planning and Assessment Act 1979. In late 2013 amendments to the Mining SEPP saw the inclusion of a specific criterion for assessment – namely, that decision makers must prioritize the 'significance' of a resource as their 'principal consideration' when determining a development application, with regard to the economic benefits of resource development.[103] The economic benefits of a proposed development may be measured by the amount of royalties generated, the number of jobs created, as well as any increases in regional expenditure or capital investment. For some, these amendments confirmed that the regulatory framework is only concerned with the commodification of land:

> . . . the amendments appear to endorse and reinforce the view that the approval of mining projects is 'almost inevitable', even where they have significant negative impacts on human settlements and the environment.[104]

While the overarching legislation, the Environmental Planning and Assessment Act 1979 (NSW), provides discretionary standards to be

---

[101] *Ibid.*

[102] *Ibid.*

[103] Reg 12AA State Environmental Planning Policy (Mining, Petroleum Production and Extractive Industries) 2007.

[104] NSW Environmental Defender's Office, 'Submission on Amendments to the Mining SEPP – State Environmental Planning Policy (Mining, Petroleum Production and Extractive Industries) Amendment (Resource Significance) 2013' (2013) 4, accessed 9 November 2016 at http://d3n8a8pro7vhmx.cloudfront.net/edonsw/pages/335/attachments/original/1380680264/130809MiningSEPPamendments.pdf?1380680264.

taken into account when assessing a development application (such as environmental and social impacts), the Mining SEPP amendments clearly prioritize the economic benefits of development among the factors to be considered.[105] Recent planning determinations by the PAC would seem to indicate that environmental and social factors are trumped by the economic imperatives of the amended Mining SEPP.[106]

Prompted by community concern regarding the NSW regulatory framework, the federal government weighed in on the land use conflict debate by introducing a new matter of national environmental significance to the Environment Protection and Biodiversity Conservation Act 1999 (Cth). Known as the 'water trigger', this matter of national environmental significance was introduced by the former federal government in 2013, and provides the Commonwealth with the ability to approve, refuse or place conditions upon large coal and CSG developments where a development will, or is likely to, have a significant impact on a water resource.[107] The bill that established the 'water trigger' also introduced an Independent Expert Scientific Committee (IESC) to provide advice on CSG and large coal mining proposals.[108] Bioregional assessments have commenced, which aim to better understand the impacts of CSG and large coal mining developments on water resources.[109] These assessments will be used by the IESC in preparing their advice and are expected to be completed by 2016.

While 'not conceptually elegant, nor practically ideal', the water trigger provides in theory an added layer of Commonwealth oversight which appears to diverge from the 'questionable standard of state approval processes'.[110] However, in late 2013 the newly installed Abbott Federal

---

[105] S 79c Environmental Planning and Assessment Act 1979.

[106] Peter Hannam, 'Rio Tinto Wins Planning Nod for Controversial Warkworth Coal Mine' (*Sydney Morning Herald,* 5 March 2015), accessed 9 November 2016 at http://www.smh.com.au/environment/rio-tinto-wins-planning-nod-for-controversial-warkworth-coal-mine-20150305-13vonn.html.

[107] S 24D Environment Protection and Biodiversity Conservation Act 1999.

[108] Independent Expert Scientific Committee on Coal Seam Gas and Large Coal Mining Development (IESC), 'Home Page', accessed 9 November 2016 at http://www.iesc.environment.gov.au/.

[109] Australian Government, Bioregional Assessments, 'Welcome', accessed 9 November 2016 at http://www.bioregionalassessments.gov.au/index.shtml.

[110] Isabelle Whitehead, 'Better Protection or Pure Politics? Evaluating the "Water Trigger" Amendment to the EPBC Act' (2013), accessed 9 November 2016 at http://www.nela.org.au/NELA/Documents/Better_Protection_or_Pure_Politics.pdf.

Government announced that it would seek to create a 'one stop shop' for both the assessment and approval of projects, which would see states accredited through bilateral agreements for assessment and approval of decisions where the 'water trigger' is relevant.[111] Assessment bilateral agreements enable the federal government to rely upon a state assessment when determining whether to approve development; whereas approval bilateral agreements enable the state to actually make the determination in place of the federal government Minister. Assessment bilateral agreements are already in force; however, approval bilateral agreements have not been able to be finalized as the necessary legislation remains opposed in the Senate.[112]

Overall, recent attempts at reform within NSW have done little to alleviate agricultural land use conflicts. As it now stands, the regulatory framework fails to adequately recognize non-economic land values and restricts community capacity to seek review of decisions, while prioritizing short-term economic gain. The law has been used to channel, shape and direct conflict by those in power. As Turton observes, the law is 'a powerful instrument of State authority, with significant capacity to shape the social and legal spaces that extractive industries operate in'.[113] Drawing on Delaney's[114] concept of the 'nomosphere' to explain the ubiquitous interconnection of law, spatiality and power, Jessup explains that the law can be used to displace and reposition conflict by 'construct[ing] and prioritiz[ing] particular spaces and scales'.[115] In his study of conflict over the siting of a waste development facility, he argues that 'the most powerful legal actors are capable of redefining the space and scale of a conflict to ensure a specific resolution'.[116] Ultimately, this redefinition of the legal landscape has the capacity to disempower, displace and ignore those not in power, as well as legitimize (and over time normalize) a power imbalance.[117] Within NSW, proponents are arguably encouraged by a favorable regulatory environment which emphasizes the economic

---

[111] Environment Protection and Biodiversity Conservation Amendment (Bilateral Agreement Implementation) Bill 2014.

[112] *Ibid.*

[113] David Turton, 'Unconventional Gas in Australia: Towards a Legal Geography' (2015) 53(1) Geographical Research 53, 55.

[114] David Delaney, *The Spatial, the Legal and the Pragmatics of World Making: Nomospheric Investigations* (Routledge 2010).

[115] Brad Jessup, 'Environmental Justice as Spatial and Scalar Justice: A Regional Waste Facility or a Local Rubbish Dump out of Place' (2013) 9(2) JSDLP 70, 105.

[116] *Ibid* at 75.

[117] *Ibid* at 106.

benefits of development, and communities have expressed frustration with the lack of recognition of their values and concerns. This has been particularly evident in the North West region, a rich agricultural area where multiple coal and gas interests are at play. To further demonstrate the dimensions of the conflict over agricultural land, and the limitations of the present regulatory framework, a case study of this area is considered below.

5. CASE STUDY: THE NORTH WEST REGION

The North West region spans an area of approximately 82,000 km^2, and is home to around 113,000 people. Approximately 7,500 people are employed in agricultural industry,[118] with mining and extractive industries currently employing approximately 2,000 people.[119]

The region produces a significant amount of Australia's food and fiber, enjoying a climate which is conducive to year round crop and pasture growth.[120] The majority of agricultural income is from cotton production ($656 million),[121] with the North West also producing 100 percent of the value of peanut production, 95 percent of sorghum and 53 percent of cotton in NSW.[122] Two of the sub-regions in the North West, the Southern Plains (Liverpool Plains and Gunnedah Local Government Area (LGA)) and the Northern Plains (Moree Plains and Narrabri LGA) are also home to the majority of coal and CSG developments in the region.[123] The agricultural land in the Southern Plains is the most productive in NSW, with a wide variety of winter and summer crops grown in this area. This is due to the productive combination of volcanic soils, reliable rainfall, adequate

---

[118] NSW Government Local Land Services North West, 'Regional Profile', accessed 9 November 2016 at http://northwest.lls.nsw.gov.au/our-region/region-profile.

[119] ABARES, *Agriculture and Fisheries in the New England and North West Region of New South Wales* (Australian Department of Agriculture 2015). Note that this report overlaps the North West region with the neighbouring New England area; however mining activity in this area is limited.

[120] Environment Protection and Biodiversity Conservation Amendment (Bilateral Agreement Implementation) Bill 2014.

[121] ABARES (n 119) 4.

[122] *Ibid*. Note that this report overlaps the North West region with the neighbouring New England area; however cropping activity in the New England area is limited.

[123] NSW Department of Planning and Infrastructure, *Strategic Regional Land Use Plan – New England North West* (NSW Government 2012) 41.

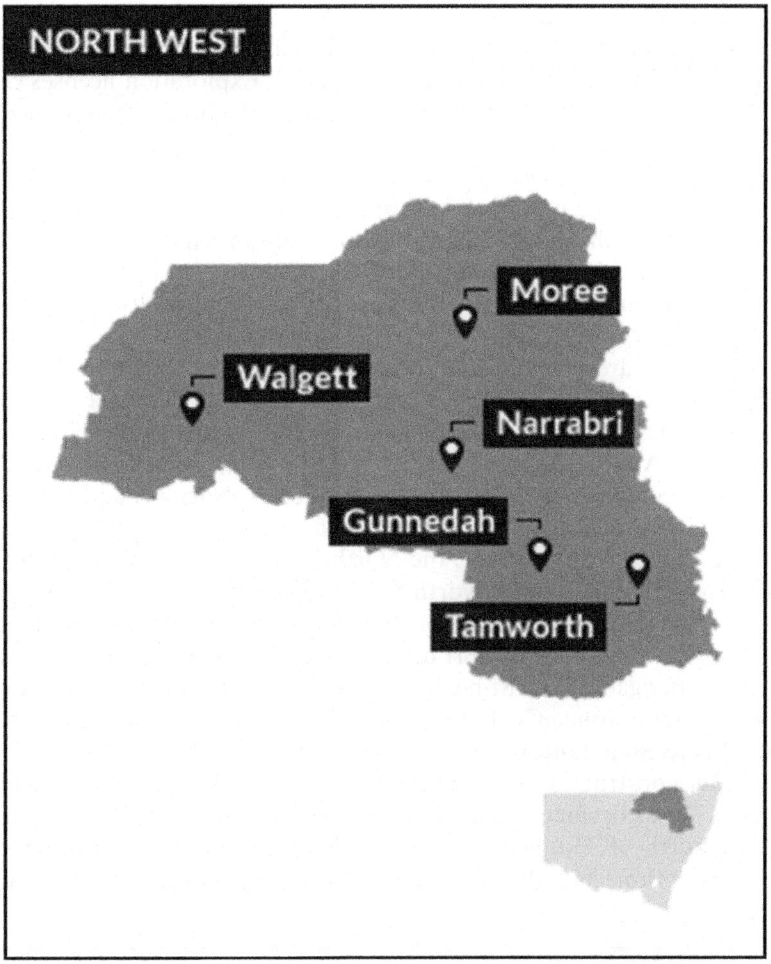

Figure 14.2 Map of the North West region in New South Wales[124]

climatic conditions and surface and groundwater availability.[125] A major volume of the most fertile and rare black earth and chernozem soils in the country are found in this sub-region.[126]

---

124 NSW Government Local Land Services North West, 'Home', accessed 9 November 2016 at http://northwest.lls.nsw.gov.au/home.
125 *Ibid* at 16.
126 *Ibid.*

CSG is still in its early stages of development in the region; however, approximately 3,434,124 hectares (34.4 percent of the region) have been identified as having CSG recovery potential.[127] Exploration licenses cover almost the entire North West region.[128] Pilot production activities for CSG are already taking place in the Pilliga forest and Gunnedah basin, and plans for a major development in Narrabri are underway. The Narrabri Gas Project has the potential to supply 50 percent of NSW's natural gas requirements, with up to 850 production wells anticipated in the 98,000-hectare gas field located in the Pilliga forest and on private land.[129] The project has been forecasted to generate $1.6 billion in royalty payments for the NSW state government over its 20–25 year lifespan.[130]

Coal mines are already established in the region with approximately 3,954,228 hectares of land (39.7 percent of the region) either subject to existing mines or exploration licenses or are known areas of resource potential.[131] A recent survey has shown that mining companies own close to 90,000 hectares of land across the Gunnedah, Narrabri and Liverpool Plains regions, including land identified as having high quality soil and water resources.[132] In 2010–11, the North West produced 5.7 million tons (Mt) of export quality coal worth $585 million, with an expansion of 20 Mt per annum possible in the short to medium term.[133] Existing coal mines in the region include: Narrabri underground coal mine, located south of Narrabri; Boggabri Coal Mine, located within the Leard State Forest near Boggabri; Tarrawonga Coal Mine, located on former grazing land adjacent to the Leard State Forest near Boggabri; the Maules Creek Mine, which is still under construction within the Leard State Forest near Boggabri and which is set to become the largest coal mine in Australia; Rocglen Mine, located north of Gunnedah; and the Werris Creek Mine, located near Tamworth. Projects still under development include the Caroona Coal

---

127 *Ibid* at 17–18.

128 NSW Department of Industry, Resources and Energy, 'MinView', accessed 9 November at http://www.resources.nsw.gov.au/geological/online-services/minview.

129 Santos, 'The Narrabri Gas Project – Overview', accessed 9 November 2016 at https://narrabrigasproject.com.au/uploads/2014/08/FACT-SHEET-Narrabri-Gas1.pdf.

130 *Ibid.*

131 NSW Department of Planning and Infrastructure (n 123) 18.

132 Lock the Gate, 'Flogging the Farm: Coal Mine land Ownership in the Hunter, Mid-West and North West of New South Wales' (2015), accessed 9 November 2016 at https://d3n8a8pro7vhmx.cloudfront.net/lockthegate/pages/1678/attachments/original/1421283067/Flogging_the_farm.pdf?1421283067.

133 NSW Department of Planning and Infrastructure (n 123) 124.

Project at Caroona (owned by BHP) and the Watermark Coal Project near Breeza (owned by Shenhua). Both of these proposed developments are located on farmland near Gunnedah.

The plight of Shenhua's proposed Watermark open-cut coal development, situated 25 kilometres south of Gunnedah adjacent to the highly fertile Liverpool Plains, provides some insight into the frustration the current planning assessment process has caused. The NSW government issued an exploration license for the Watermark project in 2008 in return for a payment of $300 million, with the promise of a further $200 million upon the grant of a mining production lease.[134] In late 2012, the proponent commenced the process of development consent for the mine under the Environmental Planning and Assessment Act 1979. The proposed development is set to become one of Australia's largest coal mines, projected to produce 10 million tons of coal per annum over a 30-year period.[135] It promises vast economic benefits, including royalties of $1.5 billion over the projected 30-year lifespan of the mine.[136] During the construction phase it is anticipated there will be 600 full time equivalent employees, and during the operation of the mine 434 people will be employed.[137] Around 30 farms have already been purchased by the proponent, with half of those leased back to landholders to continue existing agricultural operations.[138]

The EIS for the project was placed on public exhibition in early 2013, with over 130 submissions received – the majority of which were negative.[139] Many expressed concerns that the mine would considerably

---

[134] Anonymous, 'Shenhua's $1.5b Dose of Dismay' (*Stock and Land,* 24 February 2015), accessed 9 November 2016 at http://www.stockandland.com.au/news/agriculture/agribusiness/general-news/shenhuas-15b-dose-of-dismay/2724476.aspx?storypage=0.

[135] Mike Foley, 'Government Backs Shenhua against Backlash' (*The Land*, 2 February 2015), accessed 9 November 2016 at http://www.theland.com.au/news/agriculture/general/news/government-backs-shenhua-against-backlash/2722501.aspx.

[136] Colin Beetles, 'Mine Conundrum in Barnaby's Backyard' (*Stock and Land*, 10 March 2015), accessed 9 November 2016 at http://www.stockandland.com.au/news/agriculture/general/politics/mine-conundrum-in-barnabys-backyard/2725381.aspx?storypage=0.

[137] Shenhua Australia Holdings, 'About the Project', accessed 9 November 2016 at http://www.shenhuawatermark.com/html/Watermark/WatermarkProject/TheProject/.

[138] Lisa Herbert, 'Shenhua Surprised by Lack of EIS Responses' (*ABC Rural,* 29 May 2013), accessed 9 November 2016 at http://www.abc.net.au/news/2013-05-16/shenhua-eis/4692854.

[139] NSW Planning and Environment, 'State Significant Development Assessment Watermark Coal Project (SSD-4975)' (2014), accessed 9 November

impact agricultural production in the area, risking damage to underground aquifers on some of Australia's best cropping country. The community raised concerns about noise, dust pollution and the impact of the mine on the local koala population, as well as the significant disturbance to 29 Aboriginal cultural sites, including two sites of high archaeological significance.[140] Water was the principal concern of most objectors, with concern that pits dug into the ridge country would alter the surface water flow into aquifers used to irrigate the adjacent black soil plains.[141] An independent review of the EIS, commissioned by community groups within the area, revealed that there were 'significant gaps and deficiencies' in water baseline data which made it difficult to assess the true impact of the project.[142] The proponent maintained that the impact upon water would be minimal, and that water resources would be protected through setbacks and other adaptive management plans.

Community concerns persisted throughout the assessment process. While the proponent was not required to prepare a Gateway application (as the application process commenced prior to the release of the SRLUP), the then Minister requested in November 2013 that the Gateway Panel provide advice on the potential impacts of the mine on biophysical strategic agricultural land (BSAL).[143] The Panel expressed concern in their report that the proponent's EIS, and their agricultural assessment in particular, was only undertaken for the disturbance area and not for the entire project area. They noted that the proponent did not consider impacts upon surrounding buffer land which was also classified as BSAL, and urged that the agricultural enterprises and systems of the surrounding locality also be considered. Overall, they found that Shenhua's methodology for considering the impact upon agricultural land was flawed,

---

2016 at https://majorprojects.affinitylive.com/public/5a2cf0939ad5112e7a3a99d4f4f0df82/Watermark%20Coal%20Project%20-%20Assessment%20Report.pdf.

[140] *Ibid.*

[141] Mike Foley, 'Farmer Fury: Shenhua Shakes Foundations' (*The Land,* 5 February 2015), accessed 9 November 2016 at http://www.theland.com.au/news/agriculture/general/news/farmer-fury-shenhua-shakes-foundations/2722967.aspx.

[142] Sarah-Jane Collins, 'Chinese Mine in NSW Fails to Meet Standards: Report' (*Sydney Morning Herald,* 12 May 2013), accessed 9 November 2016 at http://www.smh.com.au/environment/chinese-mine-in-nsw-fails-to-meet-standards-report-20130512-2jgaf.html.

[143] NSW Government Mining and Petroleum Gateway Panel, 'Shenhua "Watermark" Coal Project Advisory Report' (2015), accessed 9 November 2016 at https://majorprojects.affinitylive.com/public/a5c4a793b6ba02545a334e-be30fc9b55/20140110_%20Gateway%20Panel%20Advice%20Shenhua%20Watermark%20Project.pdf.

resulting in a misleading underestimation of agricultural impact. They advised that the project should not proceed until a proper assessment of BSAL was undertaken. However, as the Minister had only sought the *advice* of the Panel (rather than a Gateway Panel assessment), their recommendations were merely guidance that the PAC could consider in their deliberations. Following the Gateway Panel's subsequent findings that the nearby proposed Caroona coal mine would also have 'direct and significant impacts' upon farmland (for which it issued a 'Conditional Certificate'), the Gateway process was heavily criticized as 'powerless,' and a mere 'legislative hoop'.[144]

In May 2014, the NSW Department of Planning and Environment prepared its preliminary assessment report, recommending that the project be approved.[145] The Department considered the Gateway Panel's advice, as well as commissioning an independent review of the project's potential groundwater impacts. The independent review found that the project was unlikely to have a significant impact upon water resources if it were to be restricted to the less fertile slopes above the plains. The PAC, as instructed by the Minister, then began a review of the development application, considering the EIS, public submissions, the Department of Planning and Environment's preliminary assessment report and the advice from the Gateway Panel.[146] They were also instructed by the Minister to conduct a public hearing into the development. The direction to hold a public hearing removed subsequent rights to lodge a merits appeal following the PAC's eventual verdict. The hearing took place in June 2014, at which hundreds of people participated in a public protest to draw attention to their concerns about the proposal.[147] The PAC released their review report a few weeks after the hearing, finding that the mine was approvable, subject to further water modelling to corroborate Shenhua's assertions. In their

---

144 Peter Hannam, 'BHP's Caroona Coal Mine Fails Gateway Tests' (*Sydney Morning Herald,* 11 July 2014), accessed 9 November 2016 http://www.smh.com.au/environment/water-issues/bhps-caroona-coal-mine-fails-gateway-tests-20140711-zt47v.html.

145 NSW Planning and Environment (n 139).

146 Minister for Planning and Infrastructure, 'Request to the Planning Assessment Commission Watermark Coal Project' (2013), accessed 9 November 2016 at https://majorprojects.affinitylive.com/public/3a4b7c4da728a9dc3edcf1616 5e913a9/1.%20Watermark%20Coal%20Project_%20Terms%20of%20Reference% 20PAC%20Review.pdf.

147 Kate Ramien, 'Hundreds Join Shenhua Coal Mine Protest March' (*Namoi Valley Independent,* 26 June 2014), accessed 9 November 2016 at http://www.nvi.com.au/story/2378037/videosgallery-hundreds-join-shenhua-coal-mine-protest-march/?cs=373.

review they recommend that the lease be amended to exclude the black soil plains.

In January 2015, the PAC made their final determination to approve the mine. Despite assurances that mining would be restricted to ridge country, the community maintains that their concerns regarding water and agricultural impact have not been adequately considered in the decision-making process.[148] Agriculture is part of the fabric of the North West, with a long and rich tradition in the area, and landholders feel that this has not been valued in the decision-making process.[149] Objectors have also argued that the PAC did not sufficiently consult experts on the geology and hydrogeology of the local area. Expert advice commissioned by community groups, as well as that previously sought from the Commonwealth IESC, reveals many flaws in the water modelling utilized by the proponent.[150] Each of these studies indicate that knowledge about connectivity in the aquifer structures between the ridges and the plains is limited, and unanimously advise that further research is required to understand the behaviour of ground and surface water systems, particularly where cumulative impacts are to be accounted for. They argue that Shenhua failed to adequately consider this connectivity in their EIS, relying instead upon 'select, non-unique interpretations and overly simplified representations of the available field data'.[151] It has also been contended that the definition of 'floodplain' used in the license is inconsistent with gazetted definitions of the term, resulting in uncertainty as to the protection of the black soil plains.[152]

In February 2015, the Federal Environment Minister referred the development proposal to the Independent Expert Scientific Committee under the 'water trigger' of the Commonwealth Environment Protection

---

148 Foley (n 135).

149 Caroona Coal Action Group, 'PAC Processes Fails to Address Community Concerns' (2015), accessed 9 November 2016 at http://ccag.org.au/water/pac-process-fails-to-address-community-concerns/#sthash.oZXQMvtJ.dpuf.

150 IESC, 'Advice to Decision Maker on Coal Mining Project – Proposed Action: Watermark Coal Project (EPBC 2011/6201) – New Development' (2013), accessed 9 November 2016 at http://www.environment.gov.au/epbc/notices/assessments/2011/6201/2011-6201-iesc-advice.pdf; University of New South Wales (UNSW) Engineering, 'Independent Peer Review of Watermark Coal Project Groundwater Modelling Report' (2014–15), accessed 9 November 2016 at http://www.wrl.unsw.edu.au/independent-peer-review-of-watermark-coal-project-groundwater-modelling-report.

151 Ibid.

152 Eg Water Act – see Mike Foley, 'Shenhua Mine Approval "Fail"' (*The Land,* 30 April 2015), accessed 9 November 2016 at http://www.theland.com.au/news/agriculture/general/news/shenhua-mine-approval-fail/2730820.aspx?storypage=0.

and Biodiversity Conservation Act. This has provided community members with some hope that their concerns over the water impacts of the mine will be considered, as well as the cumulative impact of other developments planned for the area.[153] As of May 2015, the Federal Environment Minister was yet to make a decision on the approval of the mine. Meanwhile, one community group has launched a judicial review challenge in the NSW Land and Environment Court against the approval, arguing that the PAC failed to adequately assess whether the mine would place the local koala population at risk of extinction;[154] the NSW Farmers Association has also called for judicial review on the basis that the PAC based its final decision upon an erroneous definition of floodplain.[155]

Existing developments in the North West area are already considerable. Overlaid with planned coal and gas projects, the cumulative impacts stand to be significant. Many feel that the current regulatory framework has not enabled these cumulative impacts to be properly assessed, generating substantial public furore. Protests and blockades at sites throughout the region have become a frequent occurrence, with local court lists becoming congested with activists facing civil disobedience charges.[156] Many have expressed that such 'non-violent direct action' has been the only way to have their voices heard.[157] The complexity of the planning system in NSW has also been a consistent source of ire, with participation in the land use

---

153 Emma Brown, 'Shenhua Coal Mine Project Halted by Federal Government as "Questions Need to Be Answered" on Water' (*NSW Country Hour,* 2 March 2015), accessed 9 November 2016 at http://www.abc.net.au/news/2015-03-02/shenhua-halted-by-federal-government/6273674.

154 Jake Sturmer and Rebecca Armitage, 'Shenhua Watermark Coal Mine Could Put Local Koala Population at Risk of Extinction, Environmental Lawyers Say' (*ABC News,* 4 May 2015), accessed 9 November 2016 at http://www.abc.net.au/news/2015-05-04/proposed-coal-mine-could-put-local-koalas-at-risk-of-extinction/6441176.

155 Gabrielle Chan, 'Shenhua Faces Legal Action by Farmers over Approvals Process for $1.2bn Mine' (*The Guardian,* 1 May 2015), accessed 9 November 2016 at http://www.theguardian.com/environment/2015/may/01/shenhua-faces-legal-action-by-farmers-over-approvals-process-for-12bn-mine.

156 Jake Sturmer, 'Protesters Chain Themselves to Land Clearing Equipment at Maules Creek Coal Mine Site' (*ABC News,* 15 February 2015), accessed 9 November 2016 at http://www.abc.net.au/news/2015-02-15/activists-disrupt-maules-creek-coal-mine/6109164; Breanna Chillingworth, 'Protesters Getting Stitched up in Narrabri' (*Northern Daily Leader,* 30 April 2014), accessed 9 November 2016 at http://www.northerndailyleader.com.au/story/2248488/protesters-getting-stitched-up-in-narrabri/; Anon, 'Big Fines for Area's Protesters' (*The Courier,* 2 April 2015)..

157 Anon (n 156); Roderick Makim, 'Wallaby Tackles Mine Protest' (*The Land,* 1 December 2014), accessed 9 November 2016 at http://www.theland.com.au/news/agriculture/general/news/wallaby-tackles-mine-protest/2718671.aspx.

decision-making process coming at a significant financial and personal cost to some. One local community organisation, the Caroona Coal Action Group, maintains that they have spent millions of dollars in obtaining independent legal and scientific advice, and that individuals have sacrificed their own income in devoting countless hours to scrutinizing large technical development applications.[158] The newly appointed Planning Minister has recently signalled an intention to overhaul the approval process for large coal and CSG projects, indicating that communities will be given greater opportunities to test decisions, and that the social and environmental impacts of proposals will be weighted equally with economic impacts.[159] Yet the Premier has also recently commented that he intends to reduce the time taken to approve mining applications in the future, as well as enforcing harsher penalties against those who engage in civil disobedience in protesting development.[160]

Overall, the experience in the North West illustrates that the present regulatory regime has done little to alleviate agricultural land use conflict. In the following final section, an alternative direction for agricultural land use decision making is considered.

6. RETHINKING AGRICULTURAL LAND USE DECISION MAKING

The expansion of mining and CSG operations in the context of climate change invariably involves trade-offs between economic gain and environmental and social considerations. The competition between extractive industries and agriculture is just one of the planes of fracture that can be expected: as White notes, there is 'nothing particularly profound about this; nor is the trade-off confined solely to mining'.[161] Arguably though, what catalyzes these contests into intractable conflict is the application of regulatory instruments incapable of responding effectively to conflict. Legal arrangements that disempower or disregard stakeholders through the prioritization of certain interests are unlikely to be effective in overcoming broad scale social disputes. When faced with complicated resource allocation situations, decision makers frequently cite economic

---

[158] Ross Tyson, 'Call for Funds to Fight Miners' (*Northern Daily Leader*, 17 July 2014), accessed 9 November 2016 at http://www.northerndailyleader.com.au/story/2422697/call-for-funds-to-fight-miners/.

[159] Hannam and McKenny (n 88).

[160] Mitchell (n 68).

[161] White (n 23) 53.

efficiencies to justify decisions, but in doing so they avoid 'complex justice arguments'.[162] Principles of fairness and justice are overlooked when preference is given to a particular set of interests, which in turn aggravates social conflict.[163]

Current conflicts over agricultural land use in NSW reveal that power is unequally skewed towards economic interests, favoring extractive industries. Regulatory responses in that state have largely been driven by the desire to facilitate extraction, shaped by the 'considerable economic and political leverage' of industry,[164] and have offered inadequate methods to recognize environmental and social interests. Yet the experience in both the North West and other communities throughout Australia highlights that the expansion of coal and CSG developments upon agricultural land raises a wide variety of environmental and social concerns which are difficult to quantify in economic terms. These uncertainties span risks to water, endangered species and habitats, and the potential for air and noise pollution; to changed landscape aesthetics, threats to cultural values and community identities, and a lack of genuine input into decision-making processes. Communities stand in opposition to these developments for a variety of reasons, yet many of these social and environmental concerns are not adequately accounted for in the decision-making process. Much of the objection from landholders and communities is that they have not had a satisfactory opportunity within the land use decision-making process to express their beliefs and views and have them taken into account, nor to challenge decisions in an independent forum.

Government and industry have continuously championed the notion of co-existence, and even community groups acknowledge that they are not opposed to mining per se.[165] As Poisel remarks, '[t]he reasonable position must fit somewhere between the stances of miners, farmers and other industries, where all industries can coexist'.[166] He suggests that establishing an independent body of accredited consultants to review development proposals would be one way to balance competing interests

---

[162] Geoffrey Syme and Blair Nancarrow, 'Justice and the Allocation of Natural Resources: Current Concepts and Future Directions' in S Clayton, *The Oxford Handbook of Environmental and Conservation Psychology* (Oxford University Press 2012) 109.

[163] Catherine Gross, *Fairness and Justice in Environmental Decision Making: Water under the Bridge* (Routledge 2014).

[164] White (n 23) 58.

[165] Caroona Coal Action Group and SOS Liverpool Plains, 'About Us' (2013), accessed 9 November 2016 at http://ccag.org.au/about-us/.

[166] Poisel (n 22) 149.

going forward. Christie argues further that co-existence requires 'a conflict resolution process where everyone may enjoy the same degree of protection from environmental and health hazards, equal access to the decision-making process and to have a healthy environment in which to live, learn and work. That is, one where environmental justice prevails'.[167]

The concept of environmental justice has much to offer regulatory approaches to land use decision making. Incorporating broad definitions of the 'environment' and 'justice', the environmental justice framework encourages exploration of the concerns raised by individuals and communities within land use conflict situations, and advocates for mechanisms that will meaningfully account for these perspectives in decision-making processes. The environmental justice movement originated in the United States in the late 1970s and 1980s, and initially focused upon the inequitable distribution of environmental harms (such as toxic waste dumps) near communities of color.[168] Since that time, the concept has expanded to include a range of environmental harms and risks, as well as recognition that inequities can be linked to class and even those who are simply less powerful in the political sphere.[169] Schlosberg articulates a comprehensive definition of the concept, noting that environmental justice is concerned not only with the fair distribution of environmental harms and risks, but also how distribution is shaped by the recognition of parties, whether there is parity of participation in environmental decision making, and whether individuals and communities possess the capabilities to function as they desire.[170] Schlosberg argues that these aspects of environmental justice are not mutually exclusive, as the struggles faced by communities span plural constructions of fairness and justice.[171] As Jessup notes, environmental justice is now 'analyzed within and between "communities" (however defined) where one community is more vulnerable to environmental deterioration than the other'.[172] It crosses spatial and temporal boundaries to consider vulnerability which might impact across national borders, and

---

167 Edward Christie, 'Coal Seam Gas Development, Environmental Legislation and Power: Towards a Pathway for Conflict Resolution and Environmental Justice' (2012) 32 Qld Lawyer 225, 225.

168 Robert Bullard, *Dumping in Dixie: Race, Class, and Environmental Quality* Vol 3 (Westview Press 2000).

169 David Schlosberg, *Defining Environmental Justice: Theories, Movements, and Nature* (Oxford University Press 2007); Alice Kaswan, 'Distributive Justice and the Environment' (2003) 81(3) NC L Rev 1031.

170 Schlosberg (n 169).

171 *Ibid.*

172 Jessup (n 115) 77.

across different time frames (for example, climate change).[173] Overall, it is a 'radical' discourse as it 'challenges dominant structures, especially power and legal structures, and promotes political and policy instability'.[174]

Framing land use conflict situations through the lens of environmental justice enables us to see how regulatory and policy responses might be reshaped to more effectively deal with conflict. In the case of NSW, while opportunities for public participation are certainly present within the regulatory framework, they are limited to written submissions and oral presentations at public meetings and hearings. Moreover, opportunities for input during the exploration phase are quite restricted, and environmental assessment information is lacking. As Hunold and Young note, once-off 'public hearing' style consultation is 'sporadic', and falls short of meaningful participation.[175] An environmentally just approach would ensure that all parties have the opportunity to participate equally in decision-making processes, and that no one party is able to claim exclusive decision-making power.[176] Prior to decision-making processes, early and meaningful deliberation should take place, whereby affected communities and individuals are identified and informed of development applications, and are provided timely and affordable access to information.[177] Active consultation should be present at all phases of development assessment, from project formulation through evaluation, and should be multi-directional.[178] Throughout the process, members of the public should also be permitted to participate in informal non-legislative activities as well, including public debate and protest.[179]

---

[173] Ryan Holifield, Michael Porter and Gordon Walker, 'Introduction: Spaces of Environmental Justice – Frameworks for Critical Engagement' in R Holifield, M Porter and G Walker (eds), *Spaces of Environmental Justice* (Wiley-Blackwell 2010).

[174] Brad Jessup, 'The Journey of Environmental Justice through Public and International Law' in Brad Jessup and Kim Rubenstein (eds), *Environmental Discourses in Public and International Law* (Cambridge University Press 2012) 47, 54.

[175] Christian Hunold and Iris Young, 'Justice, Democracy, and Hazardous Siting' (1998) 46(1) Polit Stud 82.

[176] *Ibid.*

[177] Felicity Millner, 'Access to Environmental Justice' (2011) 16(1) Deakin LR 189; Brad Jessup, 'Investing the Law with an Environmental Ethic: Using an Environmental Justice Theory for Change' in E Techera (ed), *Environmental Law, Ethics and Governance* (Inter-disciplinary 2010) 21.

[178] Hunold and Young (n 175).

[179] Millner (n 177) 189; Jessup (n 177); Elisa Arcioni and Glenn Mitchell, 'Environmental Justice in Australia: When the RATS became IRATE' (2005) 14(3) *Environmental Politics* 363.

The principles of environmental justice also require that members of the public have access to an independent and impartial body to review decisions.[180] The distinction in NSW between PAC public hearings and public meetings, and the use of the former to preclude access to merits appeals, raises environmental justice concerns. While the opportunity to mount endless challenges to decisions does not necessarily enhance participation,[181] decisions must be open to impartial review by an independent body. The right to seek merits review as to the impact of a decision is therefore crucial; as is a broad interpretation of standing to ensure that all affected parties (including non-human entities) are able to challenge decisions.[182]

Enhanced opportunities for public participation promote justice by respecting the autonomy and interests of citizens, and are more likely to result in decisions which have fully considered the consequences of, and alternatives to, development.[183] Moreover, a fair process increases the likelihood that a decision will be accepted, even if the outcome is not necessarily the one desired by the community.[184] Yet, as Tyler notes, processes will not be viewed as fair if stakeholders are not treated with dignity or respect, or if authorities are not considered neutral or trustworthy.[185] Thus, while improving procedural obligations is vital to ensuring environmental justice, the capacity to access participatory rights may nonetheless be impeded by a lack of recognition embedded in the economic and political status quo.

In NSW, there is a significant burden placed upon communities to digest large and technical EIS documents, as well as to locate and fund independent scientific data and legal advice to challenge proponent applications. From an environmental justice standpoint, weaker participants in land use decision-making processes should be compensated for any power disparities to remove barriers to effective participation. This includes access to technical and legal information, and the economic support necessary to obtain it, so as to offer alternative perspectives to those who are in positions of power.[186] An environmentally just policy framework could include, for example, access to Legal Aid or reduced rate legal assistance for review or enforcement proceedings, as well as ensuring that non-government

---

180 Millner (n 177) 189; Jessup (n 177); Arcioni and Mitchell (n 179).

181 Jessup (n 174) 54.

182 Millner (n 177) 189.

183 Hunold and Young (n 175).

184 Tom Tyler, 'Social Justice: Outcome and Procedure' (2000) 35(2) Int J Psychol 117.

185 Hunold and Young (n 175).

186 Hunold and Young (n 175); Jessup (n 174) 54.

organizations (such as the Environmental Defender's Office) are funded and recognized as a means for the public to access decision-making processes.[187] Removing barriers created by court practice and procedure against vulnerable parties, such as the threat of adverse costs orders following litigation, or undertakings for damages, is also critical.[188]

The dominance of proponent-led science in the assessment process has also been a strong complaint from communities in NSW, as demonstrated in the North West case study. Positive legal duties to actively consider public submissions, including expert data proffered by the community alongside their personal perspectives on impact, can also ensure that 'due account' is taken of alternative discourses in environmental decision-making processes, and that minority groups are not disempowered or ignored altogether.[189] Sensitivity to geographical barriers to participation, which are particularly heightened in rural areas where people may be located some distance from the location of decision making, can also ensure that communities are not marginalized.[190]

Parity of participation is also precluded when powerful stakeholders influence and shape the regulatory landscape. Taft notes that the 'collision' between extractive industry and other interests (such as environmental and social interests) has led industry to seek control of the democratic process by manipulating and ultimately capturing political and legal institutions.[191] The integrity of these 'bulwarks of democracy' is ultimately compromised as the 'referee . . . takes the side of one of the players, corrupting the competition'.[192] In NSW, for example, claims have been made that the CSG company Santos lobbied the state government heavily to modify the Gateway Panel process, as well as its mapping of BSAL in the SRLUP, which resulted in reductions to land considered as strategic agricultural

---

[187] Millner (n 177) 189. In 2013, the Commonwealth Attorney-General's department withdrew over $10 million AUD of funding from the various Environmental Defender's Offices around Australia.

[188] Millner (n 177) 189; Jessup (n 177).

[189] United Nations Environment Programme (UNEP), 'Guidelines for the Development of National Legislation on Access to Information, Public Participation and Access to Justice in Environmental Matters' (2010), accessed 9 November 2016 at http://www.unep.org/environmentalgovernance/Portals/8/GuidelinesAccesstoJustice2010.pdf.

[190] Millner (n 177) 189.

[191] Kevin Taft, 'Fossil Fuels, Global Warming and Democracy: A Report from a Scene of the Collision' (*Australian Policy Online*, 2014), accessed 9 November 2016 at http://apo.org.au/resource/fossil-fuels-global-warming-and-democracy-report-scene-collision.

[192] *Ibid.*, 16.

land in the North West and the Gateway Panel's authority to refuse projects being withdrawn.[193] The prominence of economic imperatives within the assessment framework, and the lack of mechanisms to genuinely consider cumulative impact, has largely obscured the environmental and social concerns expressed by the community. The North West region's sense of place and identity as the 'food bowl' of Australia, located within a rich landscape of significant environmental value, has simply not been accounted for within the existing industry friendly regulatory framework. As Jessup observes, the approach under current development approval processes seeks to uncover 'how much development can occur . . . before already endangered communities or species become further endangered'.[194] He notes that projects are typically assessed individually without regard to cumulative ecological impacts, or non-endangered species. An environmentally just approach, Jessup argues, would see the inclusion of more positive legislative obligations to maintain species and ecosystem integrity.[195] This would reverse the burden of proof, and place the onus on proponents to demonstrate that they will not degrade species, ecosystems and arguably the community's sense of place. The approval process could be based upon understanding landscape capacity first and foremost, and then approving development only where it does not exceed landscape limits.[196]

7. CONCLUSION

Within the broader context of climate change, the rapid expansion of extractive industries across agricultural communities throughout Australia has catalysed new and complicated land use conflicts, for which existing regulatory frameworks have proven inadequate. These challenges are not unique to Australia, nor are they restricted to the expansion of extractive industries. Where regulatory mechanisms fail to adequately manage such conundrums, widespread detrimental social conflict can ensue.

Attempts at reform in the Australian conflict example presented here,

---

193 Tim Lamacraft, 'Santos Lobbied for "No Go" Area Reduction' (*ABC News*, 17 July 2014), accessed 9 November 2016 at http://www.abc.net.au/news/2014-07-17/santos-lobbied-for-22no-go22-area-reduction/5605030; Turton (n 113) 62.

194 Jessup (n 177) 33.

195 Jessup (n 177) 33.

196 John Williams, Tim Stubbs and Ann Milligan, *An Analysis of Coal Seam Gas Production and Natural Resource Management in Australia* (Australian Council of Environmental Deans and Directors 2012).

while championed as offering protection for the environment and agricultural communities, has done little more than facilitate development in pursuit of short-term economic gains. The prioritization of economic interests has produced a regulatory system which ultimately constrains citizen capacity to contest market hegemony. Diverse community concerns regarding impacts upon agricultural production, and the long term environmental effects of increased coal and gas development, have been well ventilated in a range of public forums – but for the most part they have been discounted within the decision-making process. The case presented here demonstrates that by marginalizing certain actors and values, regulatory frameworks will fail to overcome conflict.

Extractive industries place demands on the landscape like all other land uses, including agriculture and human settlement. The critical challenge for regulatory systems is how to best reconcile the desire to pursue economic growth through energy development with the need to achieve food security objectives and improve environmental sustainability in an era of climate change. Current regulatory approaches in Australia, while underpinned by the concept of ecologically sustainable development, are nonetheless geared towards approving development and are resolutely anthropocentric in their focus. Based upon economic efficiency and proponent-led science, the land use regulatory framework has sanctioned projects which have resulted in habitat loss and species threats, as well as others that will contribute significantly to greenhouse gas emissions.[197] This is in addition to permitting developments that will potentially decimate communities.[198] Reform of decision-making structures is vital, not only to overcome broad scale conflict, but also to ensure that the risks and benefits associated with development are more equitably distributed.

Agricultural land use conflict is complex, and is characterized by diverse and competing values of the landscape and its resources. It is impossible to predict in precise terms what issues will emerge within a conflict, and who will contest them. As such, land use regulatory systems must anticipate that conflict will be complicated and value-laden, and not capable of linear analysis. The exclusion of certain values or interests from decision-making processes raises social justice concerns, and also greatly increases the likelihood that a conflict will remain firmly intractable. The concept of environmental justice provides a more constructive framework for approaching

---

[197] Jessup (n 177).

[198] Miles Godfrey, 'Historic Camberwell Stands in the Way of Chinese Mining Giant Yancoal' (*Daily Telegraph,* 28 August 2014), accessed 9 November 2016 at http://www.dailytelegraph.com.au/news/nsw/historic-camberwell-stands-in-the-way-of-chinese-mining-giant-yancoal/story-fni0cx12-1227039963097.

agricultural land use conflict. It brings broader social justice concerns to the foreground and encourages their consideration in regulatory and policy instruments, providing greater recognition of both anthropocentric and ecological interests. Greater incorporation of environmental justice principles within land use regulatory and policy systems would see more equal distribution of environmental risks and harms, and, as a result of just treatment, strengthen the capacity of both individuals and communities to respond to environmental risks and harms, including the threats posed by climate change.

15. Climate change and land grabbing

Akachi Odoemene

1. INTRODUCTION

The past decade has seen wealthy foreign governments, agencies and private investors accelerate pressure on land resources of developing countries.[1] This revaluation of land by powerful economic and political actors has led to a dramatic rise in the number and extent of land transactions.[2] Developing countries are the most targeted, as they have once more become very important in global calculations for land resources.[3] Aggressive commercial land transactions, popularly dubbed 'land grabs,' have been triggered largely by events directly tied to climate change.[4] Both phenomena (climate change and land grabbing) and their critical nexus are today 'political' realities, and remain some of the prototypical challenges of global development in modern times. Both phenomena have impacted each other in significant ways. Importantly, the challenges they pose have prioritized forcefully the issue of responsibility in climate change mitigation and land governance in global discourses.

This chapter offers a nuanced understanding of the critical intersections of and interactions between climate change and land grabbing, and further examines their effects and consequences. For this purpose, the

[1] Cecilie Friis and Anette Reenberg, 'Land Grab in Africa: Emerging Land System Drivers in a Teleconnected World' (2010) 1 GLP Report 1; Lorenzo Cotula *et al.*, *Land Grab or Development Opportunity? Agricultural Investment and International Land Deals in Africa* (FAO, IIED and IFAD 2009); S Daniel and A Mittal, *The Great Land Grab: Rush for World's Farmland Threatens Food Security For the Poor* (Oakland Institute 2009); GRAIN, 'Seized: The 2008 Land Grab for Food and Financial Security' (*GRAIN Briefings*, 24 October 2008), accessed 10 November 2016 at http://www.grain.org/briefings/?id=212.

[2] Osita Agbu, 'The New Scramble for Africa: Re-enactment of History and Africa's Response in the 21st Century', (13th CODESRIA General Assembly, Rabat, Morocco, December 2011).

[3] It will be recalled that the landed resources of Africa became prized by European powers leading to the 1884/1885 Berlin Africa conference. This development eventually triggered what is today known as the scramble for and partition of Africa. Again, the entire colonial period witnessed diverse forms of land dispossessions of locals by the colonial powers.

[4] Friis and Reenberg (n 1); Cotula *et al.* (n 1); Daniel and Mittal (n 1).

chapter is divided into seven sections and proceeds in the following order. First, clarification of the foundational concepts is provided, which will help readers understand the subsequent discussions on the relationship between the two. The next section takes a look at the factors that drive land grabbing, and the connection with the incidence of climate change. A crucial perspective to the climate change and land grab discourse (the link to agriculture) is explored in the fourth section, while the fifth examines the susceptibility of societies to these phenomena, especially the disparate distributional effects on diverse global settings. The focus then shifts to the consequences of these phenomena on affected societies in the sixth section, and conclusions are drawn in the final section by reflecting on the lessons drawn from the study. The scope of this chapter is limited to these noted themes.

2. CONCEPTUAL AND CONTEXTUAL CLARIFICATIONS

There is need for a proper understanding of the two primary phenomena in this chapter: climate change and land grabbing. Climate change has been examined in earlier chapters of this book. Land grabbing, however, both as a notion and political reality, needs to be discussed in somewhat more detail, particularly because of its prominence and consequent overuse (and probably misuse in contemporary literature and discourses). Indeed there is a reluctance and sometimes outright refusal by many scholars to accept the term 'land grab(bing)' as it seems to be biased and 'judgemental'.[5] Indeed, not all land deals being made (especially in the global South) are 'land grabs'. In other words, there is great need for clarification and explication of this term for proper understanding and usage in our context, by delineating what factors make a deal 'bad', and thus, a grab.

One general criticism of most definitions of land grabbing is their inability to highlight the abusive practices that characterize[6] this type of land acquisitions, despite the ubiquity of, and local resistance to, such

---

[5] The contention often held by these scholars is that the 'land grab' notion presupposes a sinister and greedy scenario, whereas this might not be the case in many instances of land transactions. In other words, with the 'land grab' notion there is already an assumption of 'guilt' well before the facts are laid out.

[6] For instance, while cases of such abuses of rural landholders – lack of consent, coercion, violent dispossessions and removal, killings, etc. – are often not unknown to such authors, their definitions of land grabbing often fall short in taking such abuses into cognizance.

abusive behavior. Indeed, with such a negative tag, definitions that presage a 'nothing really wrong' situation (ethically/morally, procedurally or otherwise) do not just add up. The failure of most definitions to engage with the local population's diverse rights to dignity, food, livelihoods, survival and natural resources for both the present and foreseeable future is equally a shortcoming. Of course, something is only defined after it has existed and been critically observed and evaluated, not just in a vacuum. A critical analysis of many land transactions reveals tales of rights abuse or denial, and the consequent outcries from the victims.[7] These have not been included in most definitions, but are quite indispensable in understanding the land grab.

So, what exactly is 'land grabbing'? Under what conditions can it be observed? Is it associated with a particular kind of person's involvement? Is it historical or wholly contemporary? This chapter defines 'land grab' as 'the forced acquisition of land without valid consent and reasonable commitment to the future survival of the dispossessed'. Clearly, this definition does not presuppose the identity of the perpetrator, the scale or extent of the act, nor even the purpose for which the land is or was acquired. Instead, it points to the *process* of acquiring the land as the key to understanding a land grab. To further strengthen this definition, I would argue that three basic, pre-conditional principles of good conduct on the side of the investor must be present in any land acquisition deal so that it is 'legitimate' and, thus, not a 'grab'. These principles,[8] which are premised on internationally recognized rights of local/indigenous peoples and are well grounded in international law, include: (1) primacy of the indigenous peoples,[9] (2) valid consent and (3) non-coercion. Without these principles being met, the deal could be considered a grab.

The first principle, primacy of the indigenous peoples, places the locals

---

7 See Akachi Odoemene, 'Socio-Political Economy and Dynamics of Government-Driven Land Grabbing in Nigeria since 2000' Global Economic Governance Working Paper 2015/103 (University of Oxford 2015).

8 These three broad principles were identified and adopted for this study after a careful analysis of more than a dozen cases of land grabbing incidents in Africa and Southeast Asia, and the dynamics of their processes and effects. The three principles appeared to be the most important factors which generally distinguished a good land deal from a bad one.

9 The term, 'indigenous peoples' is being used here in its broadest sense, that is, as it is used in international law, to include peoples locally referred to as 'aboriginal', 'native', and 'tribal'; people who are still transferring and inheriting lands through the application of customary law. Many local communities in a majority of the developing world are inhabited by such 'indigenous peoples'. They are also often interchangeably referred to as 'locals' in this chapter.

at the centre of a potential transaction and underscores the thinking that the overall interests and welfare of the locals whose lands are sought after are pre-eminent throughout the land acquisition process. The United Nations Declaration on the Rights of Indigenous Peoples (UNDRIP), adopted by the UN General Assembly in September 2007, outlines some of these internationally recognized rights of indigenous peoples. The Convention's Article 45 indicates that the rights in the Declaration are 'the minimum standards for the survival, dignity and well-being of the indigenous peoples', and do not in any way limit greater rights.[10] The Declaration established the primacy that should be accorded indigenous peoples in matters concerning them. The Convention clearly states that this is 'the obligation of states to protect or fulfil'.[11] Particularly important in this context are issues of survival, food sustainability and livelihood support, which are often adversely affected or even totally destroyed by the outcomes of such land dispossessions and subsequent projects.[12] Yet, 'primacy of the locals' is a standard that is almost never adhered to in land deals in many target countries.

In practical terms, giving primacy to the locals means that both investors and the government must come to a full understanding and realization of the diverse and potentially dire consequences of the land acquisition for local landholders, including all economic, social, cultural, political and spiritual consequences, which have an impact on their survival. The first step would be proper consultation with the local populace, followed by addressing concerns through a reasonable commitment to their future survival. For instance, there must be critical evidence that such concerns are (and will continue to be) dealt with in a straightforward manner. 'Primacy of the locals' is the fundamental and all-encompassing principle from which the other two, 'valid consent' and 'non-coercion', derive. In other words, when the locals are truly given primacy in things that concern them, their valid consent would not only be sought, but the use of force or duress in the process of engagement would not arise.[13]

The notion of 'valid consent' presupposes that before any land is acquired, adequate information would be given to the land owner(s) (a person,

---

[10] United Nations Organizations. *Declaration on the Rights of Indigenous Peoples* (UNO 2007).

[11] *Ibid* Art 22(2).

[12] K Nolte, *Large-Scale Agricultural Investments under Poor Land Governance Systems: Actors and Institutions in the Case of Zambia – GIGA Working Paper No 221* (German Institute of Global and Area Studies 2013).

[13] For some details on the nature of land deals, see Lorenzo Cotula, *Land Deals in Africa: What is in the Contracts?* (IIED 2011).

family or community) and informed approval sought and obtained from the person(s). To be 'valid,' such consent must be prior and *informed* with widespread consultations, as well as free (voluntary) and 'open' (devoid of secrecy). The Indigenous and Tribal Peoples Convention of 1989 (No 169) calls for such consent from locals in all affairs that affect them.[14] Article 6(a) of the Convention provides that governments shall 'consult the peoples concerned, through appropriate procedures and in particular through their representative institutions, whenever consideration is being given to legislative or administrative measures which may affect them directly'. Such consultations 'shall be undertaken in good faith and in a form appropriate to the circumstances, with the objective of achieving agreement or consent to the proposed measures'.[15] Unfortunately most land deals in many developing economies exhibit colonial-style deceptions and dispossessions.[16]

Non-coercion is the third principle of good conduct in any land deal. It presumes that from the periods of negotiations to the eventual execution of the project(s) for which the land is acquired, every engagement with the local land owner(s) must exclude any form of coercion or duress; that is, the use of threatened force, intimidation with harm, or actual violence to induce an adversary to behave differently than it otherwise would.[17] Indeed, this is an accepted legal standard in international law enshrined in Article 31(1)(d) of the Rome Statute of the International Criminal Court (ICC). In other words, even during the process of an agreed acquisition and relocation, no form of coercion should be used. However this principle has been frequently violated in many land acquisitions.[18]

These principles were in practice often denied to locals, and land dispossessions which negate these principles have been a predominant historical experience among humanity tracing back through centuries. Such land dispossessions encompass many episodes and innumerable examples, from pre-colonial land seizures associated with territorial wars,[19] through

---

[14] International Labour Organization (ILO), Indigenous and Tribal Peoples Convention, 72 ILO Official Bull, 59; 28 ILM 1382 (1989) (No 169).

[15] ILO, 'Contribution of the ILO' (International Workshop on Free, Prior and Informed Consent and Indigenous Peoples, New York, January 2005) 2.

[16] Courtney Comstock, 'Kofi Annan Blasts Hedge Funders for Acquiring so Much African Land' (27 June 2011), accessed 10 November 2016 at http://www.oaklandinstitute.org/kofi-annan-blasts-hedge-funders-acquiring-so-much-african-land-0.

[17] Daniel L Byman and Matthew C Waxman, 'Kosovo and the Great Air Power Debate' (2000) 24(4) Int Security 9; Henry Campbell Black, *Black's Law Dictionary* (5th edn, West 1979), 504.

[18] Odoemene (n 7).

[19] B White *et al.*, 'The New Enclosures: Critical Perspectives on Corporate Land Deals' (2012) 39(3–4) J Peasant Stud 6230.

European enclosures and dispossession of native peoples in North America and Australasia,[20] to the imperialist expansions into Africa,[21] Asia and the Americas, as well as post-colonial, government-led land acquisitions through myriads of 'land reforms'.[22] Thus, in virtually every society there is something that people might call a land grab of sorts. The consequences were often abuses of varied dimensions and violations of layers of rights. The argument thus is that 'land grabbing' is defined in the *process*: the procedure through which a land acquisition is made, with the minimum international standards being those that have been laid out here. Thus, any land deal that falls short of any of these principles is a land grab.

While much of the current discourse on land grabbing is framed in ahistorical terms[23] (giving the impression that the phenomenon is something new and 'contemporary'), it is certainly not a new phenomenon. It is also not a fresh theme in scholarly literature. Furthermore, tracing the history of land grabbing reveals much more than just the fact that it is not new. Past land grabbing has affected political processes and precedents throughout history, some of which are still shaping how and where land grabbing happens today.[24] What *is* new, however, is that contemporary land dispossessions seem to be unprecedented in velocity and scale. For instance, independent estimates of the size of land that has been acquired every single year since 2007 was some 40 million hectares,[25] and that figure increased tenfold between 2008 and 2009.[26] More recently, figures from Oxfam showed that an area the size of London was being sold off to investors

---

20 Transnational Institute. *The Global Land Grab: A Primer* (TNI Agrarian Justice Programme 2013) 10.

21 To be sure, a discourse of this trend in Africa must necessarily begin with an insight into the infamous Berlin Conference of 1884/5 – an unprecedented landmark event that radically shaped the historical trajectory of the African continent from the late 19th century and onwards.

22 Lorenzo Cotula, *The Great African Land Grab? Agricultural Investments and the Global Food System* (Zed Books 2013); White *et al.* (n 19) 6230; Pauline E Peters, 'Conflicts over Land and Threats to Customary Tenure in Africa' (2013) 112 (449) *African Affairs* 543. Such reforms often claimed rights of universal land ownership – a legal presumption dating back to colonial times.

23 Cotula (n 22).

24 Transnational Institute (n 20).

25 GRAIN, *The Great Food Robbery: How Corporations Control Food, Grab Land and Destroy the Climate* (GRAIN 2012); World Bank Group, *The Global Land Rush: Can it Yield Sustainable and Equitable Benefits?* (World Bank 2010).

26 J Perez, M Gistelinck and D Karbala, 'Sleeping Lions: International Investment Treaties, State-investor Disputes and Access to Food, Land and Water' (Oxfam International 2011) 8.

every six days.[27] Furthermore, it is pertinent to note that quite unlike the historical 'scrambles' of lands which have occurred throughout history, contemporary land grabbing undeniably has active 'local content'. In other words, most current land grabs take place with apparent complicity and connivances of the local elite and the governments of 'host' countries.[28] Thus, these are 'grabs' agreed to, and most times orchestrated by the elite in the affected societies. Many host countries have encouraged this type of investment and are keen to develop it as a potentially lucrative activity[29] within the framework of an alternative development model.[30]

3. DRIVERS OF LAND GRABBING AND THE CONNECTION WITH CLIMATE CHANGE

What exactly drives land grabbing around the world? In addition, what is the nexus with climate change? Contemporary land acquisition development was triggered by a critical convergence of global human-environmental factors which are equally 'new' in the context of land grabbing. These factors include: the fuel (energy) crisis, the food (agriculture) crisis, and the financial (economic) crisis: the '3F Crises Nexus', or what Stephens and Headey *et al.* have appropriately dubbed 'a perfect storm'.[31] However, the

---

27 Damien McElroy, 'Protest at the Great African Land Grab' (*The Telegraph*, 4 October 2012), accessed 10 November 2016 at http://www.telegraph.co.uk/news/worldnews/africaandindianocean/liberia/9584931/Protest-at-the-great-African-land-grab.html. Often times these were very conservative and under-representations of actual situations, as the scale was often bigger. See Ruth Hall, 'Land Grabbing in Southern Africa: The Many Faces of the Investor Rush' (2012) 38(128) Rev Afr Polit Econ 193; World Bank Group (n 25).

28 Akachi Odoemene, 'White Zimbabwean Farmers in Nigeria: Issues in "New Nigerian" Land Deal and the Implications for Food and Human Security' (2012) 10(1) African Identities 63; H Burley and A Bebb, *Africa: Up for Grabs – The Scale and Impact of Land Grabbing for Agrofuels* (Friends of the Earth Europe 2010).

29 Joachim von Braun and Ruth Meinzen-Dick, *Land Grabbing by Foreign Investors in Developing Countries: Risks and Opportunities – IFPRI Policy Brief 13* (International Food Policy Research Institute 2009).

30 Odoemene (n 7); Oakland Institute, 'Understanding Land Investment Deals in Africa' (2011), accessed 10 November 2016 at http://www.oaklandinstitute.org/sites/oaklandinstitute.org/files/OI_Ethiopa_Land_Investment_report.pdf.

31 Phoebe Stephens, 'The Global Land Grab: An Analysis of Extant Governance Institutions' (2011) XX(1) IAR 1, 3, 5; Derek Headey, Sangeetha Malaiyandi and Shengagn Fan, *Navigating the Perfect Storm: Reflections on the Food, Energy, and Financial Crises* (IFPRI 2007).

fuel and food crises were directly triggered by the incidence and pervasiveness of climate change, as we shall see later. The three crises, as driving factors of land grabbing, and their interconnections with climate change, will be considered next.

3.1 Fuel (Energy) Crisis

The most important factor in the 3F Crises Nexus is the drive for the production of biofuels by many countries desirous of a shift from fossil fuel to biofuel consumption.[32] Three goals are set for the desired energy source migration. The first is a major global effort to contain climate change through the use of 'greener' fuels, particularly pioneered by the global West. This development, through the production of ethanol and biodiesel from agricultural crops and feedstock, has resulted in a migration to a cleaner and cheaper transport fuel supplement in an effort to mitigate climate change. The second goal is directed toward the development of alternative energy sources, thus expanding the fuel energy resource mix that could overcome the 'peak oil' problems, which often result in unstable oil prices and energy crisis. The advocates of this historic switch stressed and argued that it would help achieve 'more reliable' energy sources and higher levels of energy security.[33] Fostering rural development in target societies is the third goal of the biofuel development agenda.[34]

Biofuel production specifically is a central issue in land grabbing, and as such, is a big issue in the climate-energy-food nexus.[35] The Global Land Report 2010 (GLP) noted biofuels production as an important driver of global land investments (see Figure 15.1).[36] The rush towards biofuels has

---

[32] The other part of the strategy – Reducing Emissions from Deforestation and Forest Degradation and enhanced forest carbon stocks (REDD+) – equally results in another form of land grabbing: 'green grabbing,' often for 'environmental ends'. It must be noted that while both strategies hold potentials for a socially-inclusive growth, they can and do lead to land grabs. For further insights, see Institute of Social Studies (ISS) 'The MOSAIC Project: Climate Change Mitigation Policies, Land Grabbing and Conflict in Fragile States: Understanding Intersections, Exploring Transformations in Myanmar and Cambodia' (2014), accessed 10 November 2016 at www.iss.nl/MOSAIC; J Fairhead, M Leach and I Scoones, 'Green Grabbing: A New Appropriation of Nature?' (2012) 39(2) J Peasant Stud 237.

[33] Günther Fischer *et al.*, *Biofuels and Food Security* (OFID and IIASA 2009) 30.

[34] *Ibid.*, 34.

[35] Akachi Odoemene, 'Climate Change, Food (In)security and the Land Grab Syndrome in Africa' (14th CODESRIA General Assembly, Dakar, Senegal, June 2015).

[36] Friis and Reenberg (n 1).

been a direct result of global political factors, such as the European Union's mandatory 10 percent biofuel target for transport fuels by 2020, the United States' domestic energy policy of using 36 billion gallons of renewable biofuel by 2022, and India's 20 percent ethanol mandate in 2017.[37] Meeting these targets requires the sourcing, leasing and cultivation of a large expanse of lands for the production of biofuel stocks. For instance, over 70 percent of the total number of land deals cross checked between 2001 and 2011 were in agriculture,[38] while more than 60 of these deals were directed at growing crops that would be used for biofuels.[39] Indeed, even greater biofuel production and associated land acquisitions will be required in the near future by those countries desiring the energy source shift.[40] The targeted land for acquisition has been in poor, developing countries where land is cheap and access to it quite easy.

Another important dimension to the biofuels development is that such biofuels are predominantly for markets in the global West.[41] The United Kingdom is the biggest investor in biofuel production, followed by the

---

[37] Ward Anseeuw *et al.*, *Transnational Land Deals for Agriculture in the Global South: Analytical Report based on the Land Matrix Database* (CDE, CIRAD and GIGA 2012); Stephens (n 31); L Lonza *et al.*, 'EU Renewable Energy Targets in 2020: Analysis of Scenarios for Transport' (*JEC Biofuels Programme*, 2011), accessed 10 November 2016 at http://iet.jrc.ec.europa.eu/sites/about-jec/files/documents/JECBiofuels_Report_2011_PRINT.pdf; Rabah Arezki, Klaus W Deininger and Harris Selod, *What Drives the Global Land Rush?* (CESifo 2011); M Brüntrup, Detrimental Land Grabbing or Growth Poles? Determinants and Potential Development Effects of Foreign Direct Land Investments (German Development Institute 2011); Klaus W Deininger *et al.*, *Rising Global Interest in Farmland: Can it Yield Sustainable and Equitable Benefits?* (World Bank Group 2011); Friis and Reenberg (n 1); Daniel and Mittal (n 1); Lorenzo Cotula and Sonja Vermeulen, 'Deal or No Deal: The Outlook of Agricultural Land Investment in Africa' (2009) 85(6) Int Aff 1233.

[38] Oxfam International, *Land and Power: The Growing Scandal Surrounding the New Wave of Investments in Land – Oxfam Briefing Paper 151* (Oxfam International, 2011) 5.

[39] John Vidal, 'Land Acquired over Past Decade could have Produced Food for a Billion People' (*The Guardian,* 4 October 2012). accessed 10 November 2016 at http://www.theguardian.com/global-development/2012/oct/04/land-deals-preventing-food-production.

[40] Friends of the Earth International (FoEI), 'Africa: Up for Grabs: The Scale and Impact of Land Grabbing for Agrofuels' (*Friends of the Earth International*, 2010), accessed 10 November 2016 at http://www.foeeurope.org/agrofuels/FoEE_Africa_up_for_grabs_2010.pdf.

[41] *Ibid.* There is also the shift in energy policy among Southern African countries, which is geared towards fulfilling their energy needs with their own natural resources. See, e.g., Emmanuel Sulle and Fred Nelson, *Biofuels, Land Access* and *Rural Livelihoods* in *Tanzania* (IIED 2009).

United States, India, Norway and Germany respectively.[42] The major crops used in this development include jatropha, palm oil, cassava and sugar cane.

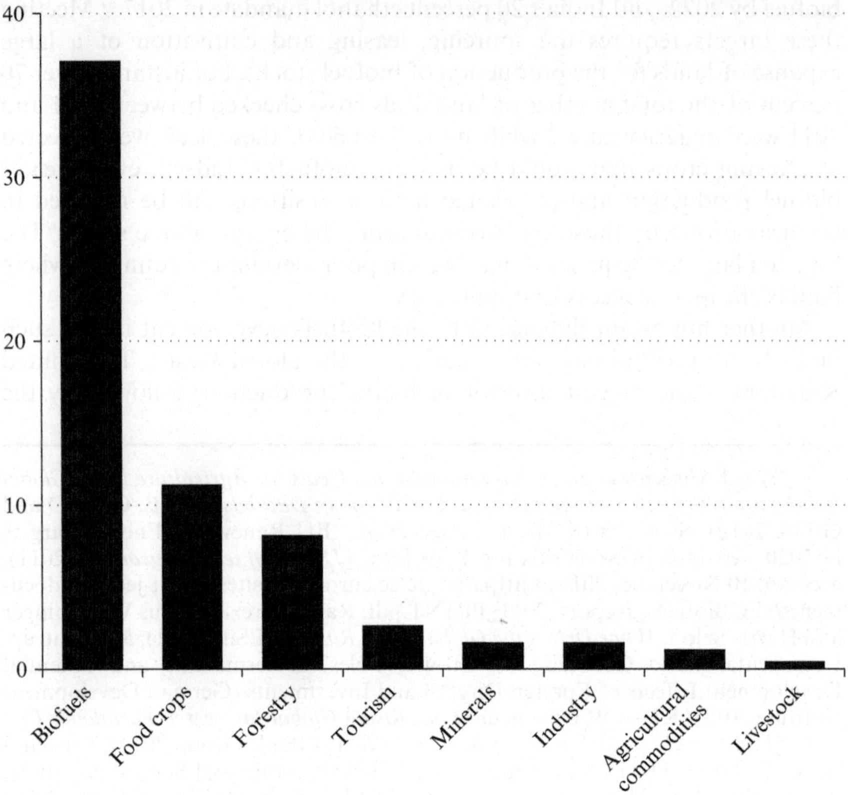

Source: Land Matrix (2012).

Figure 15.1 Global land acquisitions by sectors (in millions of hectares)[43]

[42] Marc Kosciejew, 'The New Scramble for Africa: Land-grabbing Exploitation Presented as Economic Development' (2011), accessed 10 November 2016 at http://www.caaglop.com/robbenisland-blog/economy/the-new-scramble-for-africa-land-grabbing-exploitation-presented-as-economic-development/.

[43] Cited in Saskia Sassen, 'Migration is Expulsion by Another Name in World of Foreign Land Deals' (*The Guardian*, 29 May 2013), accessed 3 December 2015 at http://www.theguardian.com/global-development/poverty-matters/2013/may/29/migration-expulsion-foreign-land-deals.

3.2 Food (Agriculture) Crisis

The second most significant factor for the rise in land acquisitions is the fear of food insecurity, occasioned by the 2007–08 global food crisis. During this crisis world commodity prices rose precipitously, sparking fears among many net-food-importing countries about the security of their food supplies. Many of them sought land and food investment opportunities elsewhere to ensure some level of food security in their home countries. The compromised food production capacity of these countries was a result of either the lack of arable land on which to farm, or the exigencies of climate change including extreme weather events such as droughts and floods, which may have affected agricultural production.[44] For instance, China hosts approximately 20 percent of the world's population but possesses no more than 7 percent of the world's arable land,[45] and for many years was a net exporter of agricultural goods. However, because of its rapid economic growth, higher population income, changes in diets and limited arable land, it has become a net-importer of agricultural goods since the beginning of the 2000s. To guarantee its food security, the Chinese government and private corporations began to invest in land suitable for agriculture outside its national borders.[46] Import-dependent countries, such as Japan and South Korea, also followed this trend.

A slightly different scenario is being experienced among the arid, oil-rich countries from the Gulf States such as Saudi Arabia and the United Arab Emirates. Due to harsh climatic conditions, poor soils and scarce land and water (among other limitations),[47] these Gulf States also began to outsource food supplies through investments in farmlands outside their

---

[44] Anseeuw *et al.* (n 37); Arezki, Deininger and Selod (n 37); Brüntrup (n 37); Deininger *et al.* (n 37); Odoemene (n 7); Friis and Reenberg (n 1); Daniel and Mittal (n 1); Cotula (n 22); Diana Liverman and Kamal Kapadia, 'Food Systems and the Global Environmental Change: An Overview', in John Ingram, Polly Ericksen and Diana Liverman (eds), *Food Security and Global Climate Change* (Earthscan 2010) 3–24.

[45] Duncan Freeman, Jonathan Holslag, and Steffi Weil, *China's Foreign Farming Policy: Can Land Provide Security? – BICCS Asia Paper 3 No 9* (Brussels Institute of Contemporary China Studies 2009).

[46] Kihwan Seo and Natalia Rodriguez, 'Land Grab, Food Security and Climate Change: A Vicious Circle in the Global South' in Netra Chhetri (ed.) *Human and Social Dimensions of Climate Change* (InTech 2012).

[47] S Haralambous, H Liversage and M Romano, 'The Growing Demand for Land Risks and Opportunities for Smallholder Farmers' (32nd Session of the Governing Council of IFAD, Rome, 2009) 17.

borders in an attempt to reduce domestic water use.[48] Thus, by controlling farmlands beyond their national borders, both of these countries are gaining control of the international supply chain of food stuffs.[49] The food produced by these countries or their agents on farmland elsewhere is specifically meant for export or repatriation back to the investor countries.[50] This practice is perceived as an innovative, long-term strategy to ensure the food security of a population at cheap prices.[51] Important investment players in this respect include China, India, Libya, Saudi Arabia, Japan, South Korea and the United Arab Emirates.[52] These countries are currently food self-sufficient,[53] even when much of the host countries' own population lacks sufficient food.[54]

Relevant to the present discussion is the fact that extreme weather events decreased cereal production worldwide by 10 percent in 2005–2006, especially from major cereal producers such as the USA, the EU and China, and was a major cause of the 2007–2008 food crisis.[55] Similarly, the first factor (the fuel crisis) equally interconnects with the second: the food crisis. For instance, it has been argued that fertilizer prices and transportation costs, which increased from 2003 to 2008, were due to the oil price hike.[56] Again, the increased demand for the production of biofuels, owing to the energy crisis, conflicted with food crops as land was diverted for the production of monocultures such as sugar cane, oilseeds, palm oil and cereals for biofuels.[57]

3.3 Financial (Economic) Crisis

In addition to the need for sustainable biofuel energy sources and the fear of food insecurity, a third important factor driving land grabbing is pure

---

48 Food and Water Watch, 'Food and Water Europe: Global Land Grab Undermines Food Security in the Developing World', accessed 2 June 2012 at http://www.foodandwaterwatch.org/factsheet/global-land-grab/.

49 L Cochrane, 'Food Security or Food Sovereignty: The Case of Land Grabs' (*Journal of Humanitarian Assistance,* 2011), accessed 10 November 2016 at http://sites.tufts.edu/jha/archives/1241.

50 Friis and Reenberg (n 1).

51 GRAIN (n 1).

52 Kosciejew (n 42).

53 Seo and Rodriguez (n 46).

54 Annelies Zoomers, 'Globalisation and the Foreignisation of Space: Seven Processes Driving the Current Global Land Grab' (2010) 37(2) J Peasant Stud 429.

55 Liverman and Kapadia (n 44) 3–24.

56 *Ibid.*

57 *Ibid.*

land speculation. To be sure, the aim is neither food nor biofuel, but profits and high returns – either from increasing land values or short-term land exploitation. This factor was heightened after the 2007–2008 financial (and economic) crisis which also had a dramatic impact on food prices. Indeed, Ghosh argues that the two crises (food and financial) should not be treated as separate from each other, but rather seen as well-connected phenomena.[58] Thus, while the global financial crisis is not related to climate change (at least, in no direct manner), it would be necessary to consider its importance here, particularly because of its links to the total land-grabbing dynamic.

When the financial crisis hit the United States, causing a series of collapses in its housing and derivative markets and having a ripple effect on the world, 'investors were left searching for new areas in which to channel their funds'.[59] In this vein, the agricultural sector was 'rediscovered' by multiple actors as a reliable investment destination.[60] George Soros, the US billionaire and investment expert, unequivocally expressed his conviction over this trend in June 2009, noting that 'farmland is going to be one of the best investments of our time'.[61] As the sharp upward trend in commodity prices increased returns on investment in production (as noted by Cotula and Vermeulen),[62] there emerged a great interest by numerous investors in owning land or shares in companies that are involved in the production end of the value chain.[63] Thus, the attention of food corporations, financial investors and other global institutions, including some world class universities, who were hungry for profits in the midst of the deepening financial crises, was drawn to the profits of investment opportunities presented by cheap and 'available' farmlands in the developing world (see Table 15.1).

Without a doubt, any one of the three factors driving the land grab, or the sum of all of them, typically poses significant challenges to the target countries and their populations. Current land-grabbing practices constitute a major threat to land ownership in most target countries, and has led to an all-time surge in farmland purchases. In addition to these global drivers of land grabbing, there are domestic factors that exacerbate the likelihood of land grabbing in each target country.[64] These include ineffectual rural

---

58 Jayati Ghosh, 'The Unnatural Coupling: Food and Global Finance' (2010) 10(1) JOAC 72.

59 Stephens (n 31) 4.

60 Deininger *et al.* (n 37) xxxii.

61 GRAIN (n 1) 121.

62 Cotula and Vermeulen (n 37) 1237.

63 Stephens (n 31).

64 Odoemene (n 7).

Table 15.1 Global estimates (as at 2012) of major land deals

Countries involved	Land purchased/leased (in ha)	Number of deals
United Kingdom	4,941,765	40
USA	4,162,394	42
UAE	3,182,950	19
India	2,101,400	28
China	1,953,527	36
South Korea	1,412,394	16
Saudi Arabia	1,132,945	20
Germany	525,345	22

Source: Kihwan Seo and Natalia Rodriguez, 'Land Grab, Food Security and Climate Change: A Vicious Circle in the Global South' in Netra Chhetri (ed) *Human and Social Dimensions of Climate Change* (InTech 2012).

land claims, contentious land laws, weak governance institutions, corruption, and high poverty levels. Usually governance-related, these domestic factors create together a permissive environment at the national level in which the three principles of good conduct are likely to be abrogated, thus giving rise to a high likelihood that a land acquisition will be a land grab. In other words, the three global crises earlier discussed interact with domestic factors to create environments conducive to grabbing prime agricultural farmlands in poor, developing countries.[65]

At this stage, it would be necessary to consider and have some insight into who the land grabbers and their supporters/sponsors are. Major international development agencies, including the World Bank Group and USAID, are often the architects of these deals.[66] If they are not, the main actors are generally governments of the world's wealthiest and most acquisitive countries, who act directly or through state-owned enterprises such as transnational megacorporations, agribusiness corporations, sovereign wealth funds, private equity firms, pensions/hedge funds and international financial speculators; oligarchs, *sheikhs* and wealthy private investors can also play roles.[67] Such developments are mostly investor-government deals using private capital. Interestingly, many reputable academic institutions,

65 Kosciejew (n 42); Agbu (n 2).

66 Oakland Institute, 'Land Grabs: False Climate Change Solutions Hurt Africa' (*Black Business Quarterly*, 2012), accessed 10 November 2016 at http://www.oaklandinstitute.org/land-grabs-false-climate-change-solutions-hurt-africa.

67 Friis and Reenberg (n 1); Fred Pearce, *The Landgrabbers: The New Fight over Who Owns the Planet* (Eden Project 2012).

such as Harvard and Vanderbilt in the US, have been involved in leasing vast areas of African farmland through hedge fund deals and financial speculators.[68]

4. CRUCIAL AGRICULTURAL DYNAMIC OF THE PHENOMENA

One primary dynamic connects all three global factors that drive land grabbing: agriculture. The factors point to agricultural land investments as a means of solving each of the fuel, food and financial crises, because agriculture (particularly industrialized agriculture) has the unique power to affect climate change.[69] For instance, agriculture (which has always been highly dependent on climate patterns and variations[70]) actually contributes to climate change by anthropogenic emissions of greenhouse gases (GHGs)[71] and the use of fossil fuel-based fertilizers.[72] The conversion of non-agricultural lands, like forests, into agricultural land through deforestation and the burning of biomass has an impact on the climate.[73]

---

[68] John Vidal and Claire Provost, 'US Universities in Africa "Land Grab"' (*The Guardian*, 8 June 2011), accessed 10 November 2016 at http://www.theguardian.com/world/2011/jun/08/us-universities-africa-land-grab.

[69] William Clark *et al.*, *Enhancing Food Security in an Era of Global Climate Change – CID Working Paper No 198* (CID 2010) 5.

[70] For instance, solar radiation, temperature, and precipitation are noted as the main drivers of crop growth.

[71] The Intergovernmental Panel on Climate Change (IPCC) aptly noted that the three main causes of the increase in greenhouse gases observed over the past 250 years have been fossil fuels, land use, and agriculture. About 25% of carbon dioxide emissions are produced by agricultural sources, while most of the methane in the atmosphere comes from domestic ruminants, forest fires, wetland rice cultivation and waste products, and conventional tillage and fertilizer use account for 70% of the nitrous oxides. Furthermore, it has been suggested that agriculture, forestry and land-use change contributed around 20–25% to global annual emissions in 2010 alone (see Smith *et al.*, 'Agriculture, Forestry and Other Land Use (AFOLU)' in O Edenhofer *et al.* (eds), *Climate Change 2014: Mitigation of Climate Change – Contribution of Working Group III to the Fifth Assessment Report of the Intergovernmental Panel on Climate Change* (Cambridge University Press 2014) 383.

[72] See Tommy Boland, 'The Contribution of Global Agriculture to Greenhouse Gas Emissions' (Feeding the World in 2050 – A Policy Symposium, University College Dublin, January 2013).

[73] CE Rosenzweig *et al.*, 'Increased Crop Damage in the US from Excess Precipitation under Climate Change' (2002) 12(3) *Global Environmental Change* 197.

These forces combined have been estimated to be responsible for one-third of climate change.[74] The competition for land, water and energy will only intensify, along with the need to reduce the many negative impacts of agriculture on the environment.[75] Thus, the agricultural sector has become one of the main driving forces behind gas emissions and land use effects.[76]

By contrast, climate change hinders agriculture in a number of biophysical, ecological and socio-economic ways. Changes in the average temperature and rainfall, as well as climate extremes like heatwaves, disrupt seasonal agricultural processes, and disrupt the cycles of pests and diseases. Even the nutritional quality of some foods could be hampered. Atmospheric carbon dioxide and ground-level ozone concentrations can fluctuate and changes in sea level can be observed.[77] These effects are unevenly distributed across the world[78] and create increased vulnerability for the landless and the poor.[79] For instance, the Food and Agricultural Organization of the United Nations (FAO) has noted that developing countries are more vulnerable to climate change than developed countries.[80] This is crucial, as a strong consensus recognizes climate change as posing a threat to already-vulnerable populations.[81] Developing countries are at present bearing the

---

[74] Climate Institute, 'Agriculture', accessed 15 August 2014 at http://www.climate.org/topics/agriculture.html.

[75] SJ Scherr and JA McNeely, 'Biodiversity Conservation and Agricultural Sustainability: Towards a New Paradigm of "Ecoagriculture" Landscapes' (2008) 363 Philos Trans Royal Soc B 477; Millennium Ecosystem Assessment, *Ecosystems and Human Well-being: Synthesis* (Washington DC 2005).

[76] See Climate Institute (n 74).

[77] Ulrich Hoffmann, Agriculture at the Crossroads: Assuring Food Security in Developing Countries under the Challenge of Global Warming (UNCTAD 2013) 3, 5.

[78] JR Porter *et al.*, 'Food Security and Food Production Systems' in Christopher B Field *et al.* (eds), *Climate Change 2014: Impacts, Adaptation, and Vulnerability - Working Group II Contribution to the Fifth Assessment Report of the Intergovernmental Panel on Climate Change* (Cambridge University Press 2014) 488–489.

[79] Climate Institute (n 74).

[80] Food and Agriculture Organization (FAO), 'Climate Change, Bioenergy and Food Security: Options for Decision Makers Identified by Expert Meeting' (High-level Conference on World Food Security: The Challenges of Climate Change and Bioenergy, Rome, June 2008) 3–5; FAO-IWGCC, *Climate Change and Food Security: A Framework Document* (FAO 2008).

[81] This is because of the predominance of agriculture in their weak economies, high poverty rates, the scarcity of capital for adaptation measures, high vulnerability levels due to their warmer baseline climates and their heightened exposure to extreme events. See Heather Thompson, Lea Berrang-Ford and James Ford, 'Climate Change and Food Security in Sub-Saharan Africa: A Systematic

brunt of its adverse effects of climate change, and will continue to do so in the future. The risk of food insecurity will likely increase for some vulnerable groups, namely the poor, in developing countries.[82] It is clear that climate change is having, and will continue to have, particularly serious consequences for developing societies.[83]

For example, sub-Saharan Africa is commonly identified as a region highly vulnerable to climate change,[84] as their food security and agricultural livelihoods are under threat of dangerous anthropogenic interference from climate change.[85] Climate change puts additional pressure on already overexploited natural resources and also negatively affects crop yields, stability of food supplies and the ability of people to access and utilize food.[86] This is grossly problematic for about two-thirds of the work force in sub-Saharan Africa who make a living from agriculture.[87] In fact, it has been projected that by 2080, agricultural output in developing countries may decline by 20 percent because of the cumulative effects of climate change.[88]

Literature Review' (2010) 2 *Sustainability* 2720; C Ringler *et al.*, *Climate Change Impacts on Food Security in Sub-Saharan Africa: Insights from Comprehensive Climate Change Scenarios – IFPRI Discussion Paper 01042* (IFPRI 2010); FAO (n 80); FAO-IWGCC (n 80); PJM Cooper *et al.*, 'Coping Better with Current Climatic Variability in the Rain-fed Farming Systems of Sub-Saharan Africa: An Essential First Step in Adapting to Future Climate Change?' (2008) 126(1–2) *Agric Ecosyst Environ* 24.

[82] Climate Institute (n 74).

[83] Ringler *et al.* (n 81).

[84] Commission on Sustainable Agriculture and Climate Change (CSACC), *Achieving Food Security in the Face of Climate Change* (CGIAR Research Program on Climate Change, Agriculture and Food Security, 2012); Thompson, Berrang-Ford and Ford (n 81); Jens Hesselbjerg Christensen *et al.*, 'Regional Climate Projections' in S Solomon *et al.* (eds), *Climate Change 2007: The Physical Science Basis: Contribution of Working Group I to the Fourth Assessment Report of the Intergovernmental Panel on Climate Change* (Cambridge University Press 2007).

[85] RT Watson *et al.* (eds), *Land Use, Land-use Change, and Forestry: A Special Report of the IPCC* (Cambridge University Press 2000); James J McCarthy, *Climate Change 2001: Impacts, Adaptation, and Vulnerability: Contribution of Working Group II to the Third Assessment Report of the Intergovernmental Panel on Climate Change* (Cambridge University Press 2001).

[86] Joachim von Braun, 'Impact of Climate Change on Food Security in Times of High Energy Prices' (Future of Agriculture: A Global Dialogue amongst Stakeholders, Barcelona, May 2008); FAO-IWGCC (n 80).

[87] ILO, 'Employment by Sector' in International Labour Office, *Key Indicators of the Labour Market (KILM)* 5th edn (ILO 2007).

[88] William R Cline, *Global Warming and Agriculture: Impact Estimates by Country* (Center for Global Development and Peterson Institute for International Economics 2007).

5. SUSCEPTIBILITY TO LAND GRABBING

The increased global demand for food production and alternative energy development led to the deliberate targeting of the global South as an appropriate reservoir of arable land, capable of satisfying international needs.[89] Examining the evidence, one can see that developing economies in the southern hemisphere (Africa, Asia and Latin America) are the main targets for investors seeking farmlands.[90] This recalls one of the goals of biofuels advancement: to foster 'rural development' in target societies. Without a doubt, the targets for 'rural development' are countries where more than 70 percent of the poor reside in rural areas and, thus, are in need of such a 'development' model. Following this train of thought, it would be fair to argue that, as 40 of the 65 'least-developed countries' are in Africa,[91] the continent is the foremost target for arable land.[92] Indeed, the scale and intensity of land grabbing on the continent, particularly in comparison to others, is proof of this fact.[93]

For instance, the International Food Policy Research Institute (IFPRI) estimated that deals of 15 to 20 million hectares of farmland in developing countries were under negotiation between 2006 and 2009,[94] while out of all such land deals counted in the 2008–2009 period, just under half (48 percent) are located in sub-Saharan Africa.[95] In 2009 alone, the World Bank estimated that 56 million hectares of farmland, an area seven times the size of Sierra Leone, were acquired around the world. Two-thirds of these, it noted, were in Africa.[96] Vidal and Provost[97] reported that between 2009 and 2011, nearly 60 million hectares of land were bought or leased in Africa by foreign companies – an area the size of France. It was also estimated that more than 250 million of hectares have been 'grabbed' (sold,

---

89 Seo and Rodriguez (n 46).

90 Oliver De Schutter, 'Large-scale Land Acquisitions and Leases: A Set of Core Principles and Measures to Address the Human Rights Challenge' UN Doc A/HRC/13/33/Add.2 (June 2009).

91 Fischer *et al.* (n 33).

92 shiftN, '2025 – Fields for Food or Fuel? Scenarios for a New Biomass Regime – A Summary Brochure of Key Findings of the "Fields for Food or Fuel" (FFF) Project' (2009), accessed 10 November 2016 at http://issuu.com/shiftn/docs/fields_for_food_or_fuel_brochure/7.

93 Odoemene (n 7).

94 $$

95 $$

96 Deininger *et al.* (n 37); Friis and Reenberg (n 1).

97 Deininger *et al.* (n 37); Friis and Reenberg (n 1).

leased or licensed) by foreign interests in the developing world between 2001 and 2012, with two-thirds of these existing in Africa.[98]

A critical question here is: why the rush for Africa's farmland? There are three reasons for this: beliefs about land availability, land affordability and the vulnerability of African states due to poverty. First, the notion being portrayed is that Africa's arable land is 'empty', 'un-used', 'marginal',

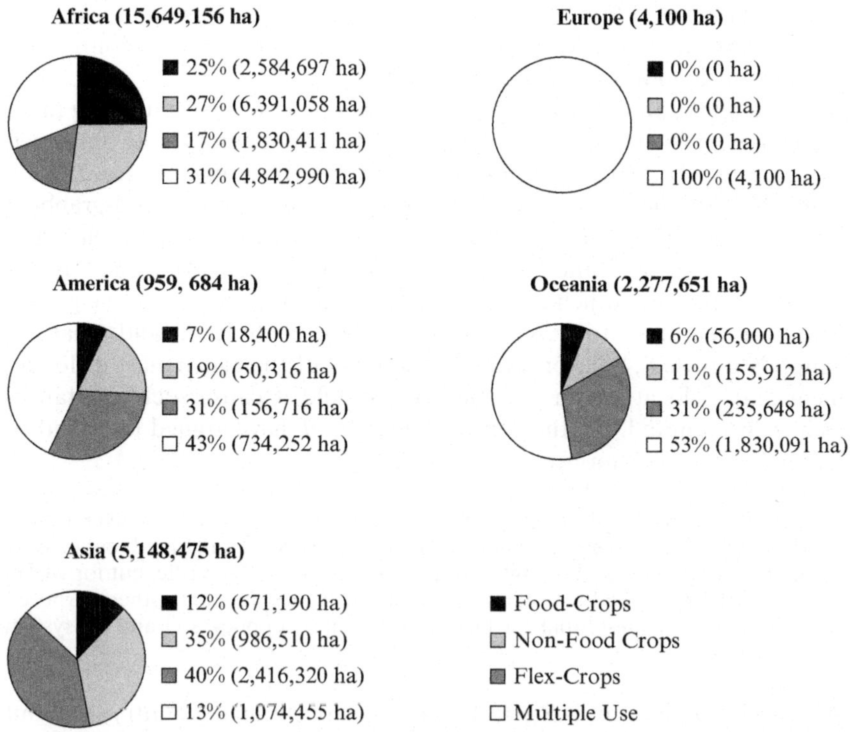

Source: Land Matrix.

Figure 15.2 Top target regions for foreign farmland investment (since 2000)[99]

---

[98] GRAIN (n 25). Instructively though, all these represented only land deals with foreign concerns; almost nothing was reported on those involving the local elite, which in some instances could have been equally pervasive.

[99] Cited in farmlandgrab.org 'Top Target Regions for Foreign Farmland Investment (Since the Year 2000)' (2013), accessed 10 November 2016 at http://farmlandgrab.org/22205.

'idle', 'available in abundance' and unlikely to compete with local food production.[100] This idea was originally espoused by the World Bank, which presented Africa as an uncrowded space of opportunities and thus suggested that as the world's largest excessive expanse of land reserves amidst low population densities, it would be key to meeting the world's growing food and biofuels demands.[101] This appears to have sent a positive signal to potential investors that Africa's land resources are, indeed, 'available' and up for grabs. Without a doubt, the sustenance of this 'land availability' impression by the World Bank has served to justify the continuous grabbing of land in the African continent.[102]

This contention, however, is quite faulty and amounts to nothing more than a myth. The reality is that land in Africa is not empty, idle, or unused as suggested. In fact, lands with such characterizations are rare. If such lands existed, one wonders why in almost all the recorded land-grabbing incidents in Africa thousands of smallholder farmers, their families and several rural communities were often forcibly evicted from and dispossessed of lands on which they have lived, cultivated crops and/or grazed cattle for centuries, to make way for new users? Indeed, lands that are 'empty', 'un-used', 'idle' or 'available' should not elicit such mass dislodgement and dislocations in societies, nor should it cause such resistances as have been noted. On the contrary, Paul *et al.* have argued that land so labelled as 'empty,' 'un-used,' 'marginal,' 'idle':

> . . . is often land used by marginalized people, by economically weaker sectors of communities, especially women. Much of it is communal land, collectively used by local people who might not have an individual land title, but for whom it is a vital resource for water, feed, food, medicines, fuel and other purposes. Such land is also essential for biodiversity, water supplies, soil and ecosystem regeneration.[103]

A typical case in point is Sierra Leone, a relatively small country of about 7.2 million hectares. Of this, about 5.4 million hectares (74 percent) of the country's land is considered suitable for agriculture.[104] From the foregoing,

---

[100] Friis and Reenberg (n 1); World Bank. *Awakening Africa's Sleeping Giant: Prospects for Commercial Agriculture in the Guinea Savannah Zone and Beyond* (World Bank 2009).

[101] *Ibid.*

[102] Odoemene (n 7).

[103] Helena Paul *et al.*, 'Agriculture and Climate Change – Real Problems, False Solutions' (Conference of the Parties, COP15, of the United Nations Framework Convention on Climate Change, Copenhagen, December 2009) 23.

[104] Kingsley Ighobor, 'Sierra Leone: Nursing Agriculture Back to Health' (2014) *Africa Renewal* (Special Edition on Agriculture) 18.

it would be quite normal for observers to speculate that since only about 15 percent of the country's arable land is currently under cultivation, the rest is therefore 'un-used' or otherwise 'available', probably for a long-term lease to investors. However, this would be misleading given the reality on the ground. Indeed, a study on *Rural Agricultural Finance in Sierra Leone* commissioned by the Bank of Sierra Leone and the German Ministry of Economic Development and Cooperation revealed that 'there is no remaining potential to significantly enlarge the area under cultivation anywhere in Sierra Leone'.[105] This is one reason why the country is currently witnessing a great land-grabbing crisis.

With regards to land affordability, there is the knowledge that Africa's land is cheap,[106] undervalued or even free.[107] Really, this appears to be a good business call and an excellent investment opportunity for many cash-rich and wealthy countries, corporations and persons. For example, it has been noted that land in Zambia, the most expensive in sub-Saharan Africa, is just about one-eighth (12.5 percent) the price of similarly sized land in Argentina or Brazil, and less than one-twentieth (5 percent) of that in Germany.[108] Likewise, in Sierra Leone, official regulation requires investors to pay $5 (US) per acre, or $12 (US) per hectare for the leasing of land per year, while the same size land goes for between $1.25 and $6.75 per hectare in Ethiopia. Similarly in Liberia, a comparable plot of land has an agreed upon price of $5–$7 per hectare. In contrast, land prices in Brazil or Argentina are $5,000–$6,000 per hectare.[109] These extremely low

---

105 Daniel Adebowale Bello, 'Land Grabbing in Sierra Leone: Who Benefits – Farmers or Investors?' (*Future Challenges*, 11 November 2011), accessed 10 November 2016 at http://futurechallenges.org/local/land-grabbing-in-sierra-leone-who-is-benefiting-the-investors-or-the-farmers/.

106 Seo and Rodriguez (n 46) 2.

107 Future Agricultures, 'Land Grabbing in Africa and the New Politics of Food' (*Policy Brief 041*, June 2011) 2, accessed 10 November 2016 at http://www.future-agricultures.org/policy-engagement/policy-briefs/1427-land-grabbing-in-africa-and-the-new-politics-of-food/file.

108 Oxfam International (n 38) 5; Evans S Osabuohien, A Ogundipe and UR Efobi, 'The Land Rush in Africa – Implications and Institutional Panacea' (United Nations University: Institute for Natural Resources in Africa (UNU-INRA) International Conference on Sustainable Development in Africa, University of Ghana, Accra, December 2011).

109 Cotula (n 13); Bruce Krasting, 'Land Grab: African Farmland Purchases have Surged 1500 percent since 2008' (*Business Insider Online*, 12 June 2011), accessed 10 November 2016 http://www.businessinsider.com/african-land-grab-acres-for-a-bottle-of-scotch-2011-6; Mats Hårsmar, 'Land Grabbing in Africa' (DevNet International Workshop: 'Land Grabbing in Africa: Global Resource Scarcity and Competition for Survival', Uppsala, October 2011).

land sale or leasehold prices have significantly contributed to the surging interest in Africa's land.

The last reason for the land rush in Africa has to do with the vulnerability of states in Africa. Undeniably, Africa has been 'the sick man of the world' since the late 1970s. It has been a continent of largely ineffectual and unsuccessful states mostly plagued by decades of corruption, misrule and maladministration, poverty and hunger, diseases, and general underdevelopment. Because of these issues, most African states have witnessed debilitating socio-economic, political and cultural breakdowns that have led to unacceptable underdevelopment and gross impoverishment of the majority of its peoples. The grim condition of most African states is being taken advantage of by global players, acting in conjunction with corrupt, rent-seeking leaderships and 'predatory' local elite. They only see an excellent investment opportunity in such circumstances. This situation is made worse by weak and broken governing institutions, as well as ineffectual land policies.[110]

Murphy has pointed out that most of the target countries for such investments have problematic governments (weak, corrupt or both), in which 'accountability, transparency and the enforcement of law can hardly be expected'.[111] This fact is corroborated by Phil Heilberg, a United States investor with an allocation of one million acres in Southern Sudan, who aptly notes: 'Listen, I want to control that ground . . . I want a country that's weaker. There's a cost to dealing with strong countries: resource nationalism'.[112] Furthermore, the deep financial indebtedness of most African states to many of the investor countries is equally an important factor. In the face of such glaring weakness, many of the African states would be unable to insist even on best practices – a minimum that is the benchmark elsewhere. The local people, who lack a voice and political power, face a level of impoverishment that constitutes yet another factor. With such status, Africans often lack 'real powers' to face a land grabbing onslaught on their natural resources by powerful global actors. With the often lopsided power relations and divergence of interests associated with land deals, it is clear that the vulnerability of the African continent was and is being exploited, and that transacting parties will always leverage such weaknesses to secure their own interests.

---

[110] Odoemene (n 7).

[111] Sophia Murphy, *Land Grabs and Fragile Food Systems: The Role of Globalization* (Institute for Agriculture and Trade Policy 2013) 5.

[112] Cited in: McKenzie Funk, 'Will Global Warming, Overpopulation, Floods, Droughts and Food Riots Make This Man Rich? Meet the New Capitalists of Chaos' (27 May 2010) Rolling Stone 62.

6. CONSEQUENCES OF THE LAND GRABBING PHENOMENON

While some have seen great socio-economic potential and benefits in the land grab movement,[113] many others are of the opinion that it poses grave dangers to the existence and livelihoods of the rural poor.[114] For the former, the injection of much-needed capital to revitalize agriculture (through the introduction of commercial and mechanised farming)[115] could be seen as a benefit, also creating on- and off-farm jobs, the development of rural infrastructure such as the provision of amenities (electricity, pipe-borne water, irrigation canals, the construction of schools and health clinics) that will improve local lives and livelihoods. The latter group would attest that land deals have historically resulted in the forceful and violent dispossession, displacement, dislocation and disenfranchisement of local smallholders and communities.[116] Clearly, this movement spells the end of small-scale farming and rural livelihoods in affected societies[117] and women, who make up 70 percent of farmers in the developing world, are often the most vulnerable to this development.

An important consideration in this discourse is the distinctiveness of *land* in the affected societies, and thus the crucial importance of the destructive nature of land alienation. For the local people who have witnessed enormous waves of land grabbing, the land itself lies at the heart of their social, spiritual, economic and political life. Indeed, it is often the only source of wealth, livelihood and survival for most families. This is not merely for its economics, but because of the social recognition and status it confers, as well as its spiritual implications for the owner(s), and the dire cultural challenges of landlessness in such contexts.[118] To be sure, much of

---

113 Von Braun and Meinzen-Dick (n 29); World Bank (n 104).

114 Odoemene (n 7); Akachi Odoemene, 'Land Grab Conflicts in Africa: Engaging Landscapes of Resistance and Alternatives' (Research Proposal Presented for the African Peacebuilding Network (APN) Fellowship Competition of the Social Science Research Council (SSRC), New York, June 2012); Daniel and Mittal (n 1); Oxfam International (n 38) 5; Oxfam International, *Bio-fuelling Poverty: Why the EU Renewable-Fuel Target May be Disastrous for Poor People* (Oxfam International 2007).

115 Cotula (n 13).

116 Friis and Reenberg (n 1); Prosper Bvumiranayi Matondi, Kjell J Havnevik and Atakilte Beyene (eds), *Biofuels, Land Grabbing and Food Security in Africa* (Zed Books 2011).

117 GRAIN (n 1).

118 For instance, in certain societies where the people still bury their dead on their ancestral lands this issue becomes a very emotional one, as attempts at forcing

the rural population lives on lands that have been in the family for generations, transferred from ages to the present. For these rural peoples, land is everything and 'all that matters'; it is a matter of identity and dignity, a factor of production and socio-economics.

With the grabbing of arable land also comes 'water grabbing', as investors' main targets are lands with access to irrigation for better potential production.[119] Thus, the acquisition of water rights is often the ultimate goal of the purchase or long-term lease.[120] This new pressure on water resources will adversely impact small farmers, pastoralists and fisherfolk, who rely on water resources for their livelihoods. In this respect, the Oakland Institute warns that jeopardizing Africa's fragile river systems, which are already under serious stress, will have both political and ecological consequences.[121] Politically, these grabbing transactions have been described as a 'neo-colonial approach' by wealthy countries to take over the key natural resources of poor countries.[122] Most global farmlands (80 percent of which are located in Africa and South America[123]) are either tropical rainforests, protected natural regions, or are already used for cultivation or animal grazing.[124] Thus they represent the most suitable regions for land deal investments. The conversion of tropical forests to crop land (mostly monocrops) is an inevitable ecological threat to the region's biodiversity, carbon stocks and water resources, and ultimately affects the climate.[125] Besides, as the conversion of tropical landscapes to agricultural and pasture areas occur, soil productivity decreases due to less rainfall and drier lands, occasioned by solar radiation partitioning.[126] This change also makes livestock farming impossible.

---

them off the lands through grabbing directly implies cutting them off from their ancestry. This partly explains the spiritual dimensions of land to such peoples.

119 Food and Water Watch, 'Global Land Grab Undermines Food Security in the Developing World' (2009), accessed 5 May 2013 at http://documents.foodandwaterwatch.org/doc/GlobalLandGrab.pdf; Oakland Institute (n 66).

120 Carin Smaller and Howard Mann, 'A Thirst for Distant Lands: Foreign Investment in Agricultural Land and Water' (*IISD*, 2009), accessed 10 November 2016 at http://www.iisd.org/pdf/2009/thirst_for_distant_lands.pdf.

121 Oakland Institute (n 66).

122 Hall (n 27).

123 Cotula *et al.* (n 1).

124 Navin Ramankutty, Jonathan A Foley and Nicholas J Olejniczak, 'People on the Land: Changes in Global Population and Croplands during the 20th Century' (2002) 31(3) *Ambio* 251.

125 Von Braun and Meinzen-Dick (n 29).

126 Roger A Pielke *et al.*, 'The Influence of Land-use Change and Landscape Dynamics on the Climate System: Relevance to Climate-Change Policy beyond the Radiative Effect of Greenhouse Gases' (2002) Phil Trans Royal Soc A 360. As

Land grabbing also increases local food insufficiency and insecurity. As the rural agricultural work force (local smallholder farmers, fishermen and pastoralists) is dislocated and displaced, it becomes unable to produce sufficient crops for societal survival. Furthermore, the export of locally produced agricultural products on sequestered farmlands forces locals to purchase agricultural goods elsewhere, as opposed to benefiting from the harvest of their own lands.[127] Both events working in tandem create food insecurity and increase a person's vulnerability in the affected society. Furthermore, the land grabbing arrangement means that crops are grown on prime agricultural lands, not for the plates of their largely impoverished and hungry citizens, but for the fuel tanks of rich, developed economies elsewhere. Indeed, such trajectories merely reinforce poverty, hunger and destitution amongst the rural poor in affected developing economies.

Land remains the most prized asset in most affected societies and the local attachment to it is not just a myth. Land dynamics (rights, ownership, access and usage) form a critical part of their existence and any attempt to 'squeeze' land from the people would be strongly resisted. Indeed, land has been a predisposing factor or source of provocation and escalation of violent conflicts. As is being witnessed in many affected societies, land grabbing is often, and increasingly, resisted and contested. Such contestations have been on two main fronts. The first is by the locals who genuinely fear a threat to their livelihoods, food systems and survival. In this respect, contestations have taken the form of protests and sometimes even legal challenges to acquisitions.[128] On the second front are international NGOs and a network of civil society and social movements, many of which consider these land acquisitions to be grossly unethical. This is especially true with acquisitions that involve sophisticated forms of land seizures, dispossession of local smallholders and the export of crops harvested from such expropriated lands.[129]

7. CONCLUSION

Climate change is tightly interconnected with land grabbing, with implications that are both diverse and complex. Three global crises – in fuel, food,

abundant as water may seem, predictions from the Intergovernmental Panel on Climate Change (IPCC) suggest that fresh water supplies are likely to be depleted, especially in some parts of Africa.

127 Seo and Rodriguez (n 46).

128 Odoemene (n 7).

129 Seo and Rodriguez (n 46).

and finance –in large part drive wealthy but resource-poor economies to acquire large-scale lands in poor but resource-rich economies, for their own interests. Obviously, this development forges one-sided land deals in which, capitalizing on these crises and the lucrative investment opportunities they create, wealthy outsiders connive with local elites to exploit the land resources of the rural poor. The 'urgency' occasioned by the three crises creates desperation in which the principles of good land deals are violated, thus creating the land grab. This negatively affects populations: not only is it a major threat to local livelihoods, food security and human capital development, but it is a big blow to peasant agriculture and smallholder farming in affected societies. Ironically, these are the same effects climate change has on the same populations.

The main challenge is the obvious lack of political will to resolve critical issues associated with the phenomena, especially from 'big global players' in both climate change and land grabbing.[130] The political controversy surrounding biofuels[131] exposes the agrofuel efforts at climate change mitigation as a falsely upheld solution.[132] Instead the resulting biomass regime has helped trigger massive land grabs that continue to threaten the same vulnerable populations in fragile economies. Again, these phenomena and their critical intersections have produced diverse socio-environmental spillover effects that changed both the nature of the contested resources and the social dynamics of resultant conflicts. Thus, these have not been 'one-way political dynamics'. While widespread resistance to these developments is yet to be experienced, many affected populations have reacted to these factors in varied and significant ways. Even though active resistance by subordinate parties continues to grow, land grabbing has not stopped and meaningful reduction of climate change effects has not occurred. To be sure, these will continue to produce chain reactions in the future.

From all indications, these phenomena and the effects they could very well usher in, have assumed postures of yet another set of global tragic episodes which, similar to the trans-Atlantic slave trade and colonialism, will only be denounced and 'regretted' years after they have wreaked severe havoc in all facets of society.

---

[130] These include: United States, United Kingdom, Canada, China, India, continental Europe, the Gulf states and some other emerging economies.

[131] Jules Pretty *et al.*, 'The Top 100 Questions of Importance to the Future of Global Agriculture' (2010) 8(4) Int J Agr Sustain 219.

[132] Oakland Institute (n 66).

Index

Aberman, Noora 294
absolute ownership doctrine 119
acid rain 50
Adaptation Committee 43
Adaptation Fund 40, 41
adaptation to climate change 34–43, 36, 83, 139, 212
  barriers to 115
  cooperation on 36–8
  financial resources, transfer for 40–3
  genetically modified food crops, as adaptation 176–9
  and increasing temperatures 274, 278, 281, 285
  National Adaption Programmes of Action 35–6
  obligations under UNFCCC and Kyoto Protocol 35–43
  as private and collective 283–4
  and resilience 17–20, 69–71
  strategies 15–20
  technology transfer for 38–9
  trade and investment rules *see* agricultural trade and investment rules, adaptation to climate change
adaptive capacity 70
adaptive cycles 59–60, 60
Adler, Robert W 30, 33
administrative permits 114–15
Advanced Nitrification/De-nitrification (AND) 251
AFOLU (Agriculture, Forestry and Other Land Use) 23, 31
Africa 31, 56, 106, 111
  see also Kenya; sub-Saharan Africa (SSA)
  agricultural trade and investment rules, adaptation to climate change 284, 286
  East African Community Climate Change Policy 367
  export restrictions 291–2
  food security issues 31, 439
  gender issues 294–5
  land grabbing 9, 423, 442, 443, 444
  poverty in 291, 367, 370, 372, 441, 444
  South Africa 278
  Water Efficient Maize for Africa (WEMA) project 183–4
  West Africa 294–5
African Agricultural Technology Foundation (AATF) 184
African Development Bank 285
AgPES (agricultural payments for ecosystem services) programmes 317–19, 323–8
  see also ecosystem services
  baseline rules of environmental impact 329
  and climate change 327–8, 330
  conditions and requirements legally imposed on beneficiaries of services 329
  conditions legally attached to PES payment 329
  designing 323–6, 328–9
  property rights issues 329
Agreement on Agriculture (WTO) 281, 291, 308, 311
Agreement on Subsidies and Countervailing Measures (SCM Agreement) 309, 312
Agreement on Trade Related Aspects of Intellectual Property Rights (TRIPS) 30
agricultural PES programmes *see* AgPES (agricultural payments for ecosystem services) programmes
Agricultural Research Corporation (EMBRAPA), Brazil 338–9, 340
agricultural trade and investment rules, adaptation to climate change 274–314

see also agriculture; agriculture and climate change; bioenergy/ biofuels; investment; trade; water; water law and climate change
autonomous climate measures 303–4
biofuels 282–3
biotechnology 289–90
carbon taxes 292
climate change challenges to agriculture 276–86
country comparison 288–9
development challenges/ opportunities 279–81
export restrictions 291–2
fish 284–5
gender issues 294–5
hunger and violence 296–8
insurance 293
international regulatory framework 298–312
international water management 311–12
interpretation of differing but not contradictory figures 277–9
investment 305–12
irrigation 264, 284, 288, 289, 307
labels and standards 304–5
migration 295–6
mitigation and adaptation 286–98
multilateral and regional approaches 302–3
production and productivity 281–2
regulation of trade 298–301
stockpiles 310–12
subsidies 301, 308–10
trade and investment measures 290–1
urban agriculture 295
vulnerabilities and resilience 285–6
water 283–4

agriculture
see also Climate Smart Agriculture
Agreement on Agriculture (WTO) 281, 291, 308, 311
agricultural wastes versus other biofuel stocks 271–2
and climate change *see* agriculture and climate change
Everglades region, transformations *see* Everglades region, South Florida
land transformation for purposes of 49
in Paris Agreement 22, 43–5
self-perpetuating problems in 98–9
subsistence 274, 281–2
threats to 174–6
UNFCCC, marginal role in 21
urban 217–20

Agriculture, Fisheries and Food Authority Act (AFFA), Kenya 377, 378

agriculture and climate change
see also agricultural trade and investment rules, adaptation to climate change
climate change challenges to agriculture 276–86
climate resilient agriculture, as a mitigation and adaption strategy 17–20
contribution of agriculture to climate change 6–9
impacts of climate change on agriculture 9–10, 79, 106, 179, 180
see also water law and climate change

Agriculture Sector Development Strategy (ASDS), Kenya 370, 371, 380–1, 382

agriculture-dependent communities 99

agro-ecosystems 47–73
see also Everglades region, South Florida
adaptive cycles 59, 60
Anthropocene, agriculture in 48–52
biodiversity 71–2
climate change 67–9
complexity 68, 69, 72, 73
definitions 47–8
ecological components 59
human components 59
and resilience 69–72
ecological 52–8
general 71–2
specific 72
transformative change in 58, 59–67

agrofuels 331–66
sugarcane crops in Brazil and the Carrado, expansion of 333–41
AgSTAR program 267
Air Compliance Consent Agreement 257
air pollution 84, 253, 254, 353, 394
see also pollution
definition of air pollutant 252
Hazardous Air Pollutants (HAPs) 253
regulation of pollutants from CAFOs (CAA) 256–60
American Planning Association 211
American Soybean Association 303
ammonia 235, 242, 250, 253, 256, 257, 258, 259
Anderson, Kym 278
animal agriculture
see also greenhouse gases (GHGs)/greenhouse gas emissions
background 230–8
changing production systems 248–50
in China 236–7
concentrated animal feeding operations (CAFOs) 233–5, 246–51
enteric fermentation 239–40
feed stocks, changing 247–8
global production trends 235–8
history of American livestock systems 230–2
intensive livestock production and GHGs 238–46
land application 242–3
manure storage and management 234, 235, 240–2, 250–1
animal husbandry, small scale 236
Anthropocene, agriculture in 48–52
archetypes, and rural problems 91–2
Argentina 293
Arizona Water Bank 119
ASDS (Agriculture Sector Development Strategy), Kenya 370, 371, 380–1, 382
Ashford, Connecticut 217
Asia 105, 106, 107, 111
see also Bangladesh; China; India; Vietnam
South/Southeast 176, 279, 280, 289, 425
Asian Development Bank (ADB) 296
Atlanta, Georgia 219
atmosphere, systems thinking 76
Austin, Texas 219, 221
Australia
see also invasive species law
Biophysical Strategic Agricultural Land (BSAL) 401, 402, 410, 411, 419
Bureau of Statistics 151
Carbon Farming Initiative Act (2011) 27
climate change and global energy demand 386–90
coal mining 385, 393, 394, 395, 396, 401, 404, 408
coal seam gas (CSG) 387, 388, 389, 394, 395, 396, 397
Commonwealth Environment Protection and Biodiversity Conservation Act 412–13
Commonwealth Scientific and Industrial Research Organization (CSIRO) 145
Constitution 147
Desert Knowledge Cooperative Research Centre 143
environmental impact assessment (EIS) 398, 409, 410, 411, 412
Environmental Planning and Assessment Act (1979), NSW 398, 401, 403–4, 409
Environmental Protection Biodiversity Act (EPBA) 147, 404
fossil fuels in 385, 390
Gateway Panel Assessment 401, 402, 410, 411, 419, 420
Great Barrier Reef region, coral reef ecosystems in 55
Independent Expert Scientific Committee (IESC) 404
Intergovernmental Agreement on National Drought Program Reform 36
invasive species law 147–51
challenge of invasive species, lack

of systemic approach to 158–9
community engagement, facilitating 159–62
effectiveness 151–4
examples of invasive species 140, 142, 143–4
Feral Deer Management Strategy 2013–2018 144
financial resources 156
implementation issues 154–7
legislative instruments 148–9
model 148–9, 156
National Policies 149
Senate enquiry 158–9
transactions costs, effects on invasive species management 157–8
land use conflict case study 385–422
coal and gas development in New South Wales 397–406
decision making, rethinking 414–20
energy production, increased 391–7
North West region 406–14
Mining Act (1992), NSW 397
National Pollutant Inventory 394
Natural Resource Management 149
New South Wales, coal and gas development 397–406
Petroleum (Onshore) Act (1991) 397, 398
Planning Assessment Commission (PAC) 399, 400, 404, 413, 418
Review of Environmental Factors (REF) 398
Sixth National Communication report 36
Southern Plains 406
State Environmental Planning Policy (Mining, Petroleum Production and Extractive Industries), 2007 (Mining SEPP) 399, 403, 404
Strategic Agricultural Land (SAL) 401
Strategic Regional Land Use Plans 402
Strategic Regional Land Use Policy (SRLUP) 400, 401, 402, 403
temperature 389
autopoiesis 97–9
Aviation Directive, EU 303–4

Baffes, John 290
Bald Eagle, and DDT use 92
Bali, WTO Ministerial Conference (2013) 311
Bali Action Plan (2007) 29
Baltimore County, Maryland 214–15
Bangladesh 279
Bartel, Robyn 161
Bayer 195
BCAs (border carbon adjustments) 300, 304
Beale report (2008), invasive species 150
Beekman, New York 216–17
Bemesia genus 57, 59
Bentham, Jeremy 153
Berlin Rules on Water Resources (2004) 127
Best Available Control Technology 254
'best practices,' mitigation 13
beta-carotene 182
bilateral investment treaties (BITs) 307
biodiversity 71–2, 81, 89
bioenergy/biofuels 9, 31, 93, 271–2, 282–3, 359
biofuel weed pathway 140, 141
crop based 282–3
fuel crisis 430–2
second generation biofuels 143, 282
and United Kingdom 431–2
bioethanol 282
Biogas Opportunities Roadmap 268–9
Biomass Research and Development Initiative 266
biomass/system structure 59
Biophysical Strategic Agricultural Land (BSAL), Australia 401, 402, 410, 411, 419
biosecurity 75, 95, 98
invasive species law 136, 140, 147, 149, 150, 151, 154, 158, 160, 164
national system, Australia 147, 150, 154
officers 157, 165

Biosecurity Act (2014), Australia 150, 151–2
biosphere (planet) 48, 54, 67–8
biotechnology 171, 174, 179, 287, 289–90
Blandford, David 308, 309
Blankenship, Sara 285, 312
blindness, vitamin A deficiency 181, 182, 183
BNDES (National Bank for Economic and Social Development), Brazil 344–6
Bonaparte Broward, Napoleon 63
Boone County, Illinois 215
border carbon adjustments (BCAs) 300, 304
Bormann v. Board of Supervisors (1998) 206
Brazil and Cerrado (tropical savanna ecoregion)
  Agricultural Research Corporation (EMBRAPA) 338–9, 340
  animal agriculture 237
  Cerrado biome and governmental support for sugarcane production 341–7
  deforestation in 9, 357, 358, 359, 360, 365, 366
  ethanol production 331, 359–64
  expansion of sugarcane crops 333–41
  and genetically modified food crops 194
  GHG emissions and climate change 351–8
  Goiás Industrial Development Program (PRODUZIR) 347
  Goiás Secretariat for Planning (SEPLAN) 346
  Goiás State Department of Agriculture 351
  Guidelines for Agro Energy Policy 334
  Institute of Geography and Statistics (IBGE) 335
  and insurance 293
  international capital, sugarcane industry 348–51
  Midwest Constitutional Fund (FCO) 344, 346
  Ministry of Agriculture, Livestock and Food Supply (MAPA) 334
  Ministry of Environment (MMA) 341, 342, 343, 344, 359
  Ministry of Science, Technology and Innovation (MCTI) 353
  National Agro Energy Plan (2006–2011) 332, 334
  National Bank for Economic and Social Development (BNDES) 344–6
  National Ethanol Program (*Pró Alcool*) 331
  National Supply Company (CONAB) 334–5, 336, 361
  Petrobrás ethanol pipeline 347
  Sugar and Ethanol Institute (IAA) 331
  sugarcane, corporate control 350–1
  Sugarcane Agroecological Zoning (ZAE) 332–3, 338, 339, 341, 344
  Zoning Institute 338–9
Brecht, Bertold 293
Brookhaven, New York 205–6
Bt corn 172, 193, 194
'business as usual' approach 185, 277

CAA *see* Clean Air Act (CAA)
CAFOs *see* concentrated animal feeding operations (CAFOs)
California
  Cap-and Trade Regulation 27
  drought measures/water use 123–4, 130–1, 132
  farming regions 222
California Air Resources Board (CARB) 256
camels, feral 143
Campbell, Bruce 30, 33
Canada 194
Canasat Project 335
Cancun Agreements (2010) 43
CAP *see* Common Agriculture Policy (CAP)
Cap-and Trade Regulation, California 27
capture, rule of 119
carbon
  border carbon adjustments (BCAs) 300, 304

carbon balances 282
carbon fertilization effect 278
carbon footprint 170, 185, 186, 192, 305
carbon taxes 292, 303
emissions 185, 187
carbon dioxide (CO2)/carbon dioxide emissions 22, 23, 51, 292
Carbon Farming Initiative Act (2011), Australia 27
Caribbean region, coral reef ecosystems in 55
Caroona Coal Action Group, Australia 414
Caroona Coal Project, Australia 408–9, 411
Carson, Rachel 92
Castine, Maine 223, 224
Castro, Fidel 65
cattle, dairy and beef 231, 232
Center for Global Development, Washington 287
Central and Southern Florida Flood Control District (FCD) 64
CERCLA (Comprehensive Environmental Response, Compensation, and Liability Act) 258
chaos theory 77
Cheung, W 284, 285
China
animal agriculture and GHGs 235, 236–7, 239
food security in 285, 433
GHG emissions 288
population 433
urban agriculture 295
Christie, Edward 416
CITES (Convention on International Trade in Endangered Species of Wild Fauna and Flora) 140, 150
city planning 210
Claros Garcia, JC 352, 360, 361, 362
Clean Air Act (CAA) 12, 252–62
overview 252–4
past violations of 258
Prevention of Significant Degradation (PSD) program 254
regulation of air pollutants from CAFOs under 256–60
regulation of GHGs under 254–6
Renewable Fuel Program 271
reporting CAFO GHG emissions under 260–2
Clean Development Mechanism (CDM) 12, 28
Clean Water Act (CWA), US 113, 223, 233
climate, defined 68
Climate Action Plans 211, 224–5
of 2013 267–70
climate change
and AgPES program 327–8, 330
agro-ecosystems and resilience 67–9
assessment of local hazards and vulnerabilities 212
autonomous climate measures 303–4
and Brazil, GHG emissions in 351–8
challenges to agriculture 276–86
complexity 17, 20, 196, 280, 313
contribution of agriculture to 6–9
crisis as anthropogenic environmental problem 2
farming 79
and food production 31, 40, 169, 200
and food security 4–10
and global energy demand 386–90
impacts on agriculture 9–10, 79, 105, 179, 180
and invasive species 145–6
and land use controls 198–201
livestock production, effects on 243–6
phenomenon of 138–9
predictions 69
regulatory deficits, current 275–6
and structural developments 274
as threat to agriculture 174–6
and water law *see* water law and climate change
Climate Change Technology Transfer Centres 39
Climate Observatory 352
Climate Smart Agriculture (CSA) 31–2, 289
animal agriculture and GHGs 269, 270
and Kenya 379, 380, 383

climate systems 76–7
see also governance and systems relationships
Cline, William 277–8
clockwork dolls dance analogy, climate systems 76–7
coal mining, Australia 385, 393, 394, 395, 396, 401, 404, 408
coal seam gas (CSG), Australia 387, 388, 389, 393, 394, 395, 396, 397
Coase, Ronald H 157
Collier, Paul 304
Colorado River 119, 129, 130
Colorado River Compact 121–2
Commission on Sustainable Agriculture and Climate Change 281
Common Agriculture Policy (CAP) 25, 26, 32
common but differentiated responsibilities 36–7
Common Reporting Format (CRF) 25
Commonwealth Scientific and Industrial Research Organization (CSIRO) 145
community engagement, invasive species law 159–62
complexity 18, 88, 89
agro-ecosystems 68, 69, 72, 73
climate change 17, 20, 196, 280, 313
defined 83
institutional 75, 88, 94
Comprehensive Environmental Response, Compensation, and Liability Act (CERCLA) 258
concentrated animal feeding operations (CAFOs)
see also greenhouse gases (GHGs)/ greenhouse gas emissions
definitions 233–5
intensive feedlot approach of 249
reducing GHGs from 246–51
regulation of air pollutants from, under the CAA 256–60
reporting GHG emissions under 260–2
Condon, Bradly J 309
Conference of the Parties (COP 21), Paris *see* Paris Agreement (2015), adopted at COP21
conservation practice standards (CPSs) 264
Conservation Reserve Program (CRP) 316
conservation tillage 189–90, 192
contaminants, runoff 111
Convention on Biological Diversity (1992) 150
Convention on International Trade in Endangered Species of Wild Fauna and Flora (CITES) 140, 150
cooperation, on adaptation 36–8
coral reef ecosystems 55
corn 177, 231, 282, 343
Bt corn 172, 193, 194
corn-fed beef 231
drought resistant 184
ethanol production 7–8
field corn 178
GM 193
transgenic 172
Cottier, Thomas 299–300, 302
cows, lactating 235
CPSs (conservation practice standards) 264
creative destruction 59–60
crisis
ethanol 331
financial and economic 434–7
food 433–4
fuel 430–2
Critical Industry Clusters (CICs) 401
CRP (Conservation Reserve Program) 316
Crystal Peak Fertilizer (CPF) Process, Missouri 251
CSA *see* Climate Smart Agriculture (CSA)
CSG (coal seam gas), Australia 387, 388, 389, 393, 394, 395, 396, 397
Cuba 65, 67

Dahlonega, Georgia 220
DDT, widespread use of 92–3
de Soto, Hernando 230
deer, Australia 143–4
deforestation 13, 21, 51, 167, 343, 353
in Brazil 9, 357, 358, 359, 360, 365, 366

and genetically modified food crops 175, 176, 186
Delaney, David 405
denitrification 243
Dennis, Allen 290
Desert Knowledge Cooperative Research Centre, Australia 143
desert locust system 55–7
desertification 109
Dessy, Sylvain 307
developed/developing countries 30, 32, 274, 278
see also specific countries
Developed Country Parties/ Developing Country Parties 39
Di Falco, Salvatore 288
dingoes, Australia 142
Do, Levchenko 310
Doha Development Round/ Development Agenda (DDA) 275, 302–3, 313
drought and aridity 109–10, 129
drought law, and climate change 124–5
resilience across nations 130–2
DuPont 184, 195

ecolabels 89
ecological resilience, agro-ecosystems 52–8
ecosystem services 16, 315–30
AgPES program *see* AgPES (agricultural payments for ecosystem services) programmes
cultural services 315–16
and farming 316, 322
framework 319–23
and land tenure 321, 322, 328–9
land use controls 203
legal matters 328–30
obstacles 320–1
opponents of 320–1
payments for ecosystem services (PES) programmes 317
property rights issues 320, 329
provisioning services 315
public goods concerns 320
regulating services 315
stacking problem 321–2
supporting services 316
EEMs (emissions-estimating methodologies) 257, 259
Egypt 126
Eisenhower, Dwight David 65
EMBRAPA (Agricultural Research Corporation), Brazil 338–9, 340
Emergency Planning and Community Right-to- Know Act (EPCRA) 258
emissions *see* greenhouse gases (GHGs)/greenhouse gas emissions
Emissions Gap Report, UNEP 190
emissions trading system (ETS) 27, 28
emissions-estimating methodologies (EEMs) 257, 259
Endangered Species Act (ESA) 113
energy, systems thinking 76
Energy Policy Act (2005) 271
engineering resilience 53
enteric fermentation 239–40
environmental impact assessment (EIS) 398, 409, 410, 411, 412
environmental justice 416–17
Environmental Planning and Assessment Act (1979), Australia 398, 401, 403–4, 409
Environmental Policy Instruments, OECD 101
Environmental Protection Agency (EPA), US 198, 226
Endangerment Finding, greenhouse gases 254–5
and greenhouse gas emissions 233, 242, 252–61, 262, 264, 266, 267, 268
National Air Emissions Monitoring Study (NAEMS) 257, 258, 259, 260
National Ambient Air Quality Standards (NAAQS) 253, 254
Science Advisory Board 259
Environmental Protection Biodiversity Act (EPBA), Australia 147, 404
Environmental Quality Incentives Program (EQIP) 263–4, 269, 316
EPA *see* Environmental Protection Agency (EPA), US
EPCRA (Emergency Planning and Community Right-to- Know Act) 258

EQIP (Environmental Quality Incentives Program) 263–4, 269, 316
Esty, Daniel C 306
ethanol production 7–8
  in Brazil 359–64
  ethanol crisis (1980s) 331
ETS (emissions trading system) 27, 28
Euclidean zoning 224
  and community threats 207–9
European Commission 32–3, 44
European Union (EU)
  emissions trading system (ETS) 27
  National Inventory Report, 2014 25
eutrophication 50
Everglades region, South Florida
  see also agro-ecosystems
  Early Drainage era 62–3
  eras of agriculture, 20th century 62
  Everglades Agricultural Area (EAA) 55, 64, 65, 67
  Everglades Land Company 63
  Everglades National Park 64
  flooding 63, 64, 65, 66
  history of transformations in agriculture 61–7
  Internal Improvement Fund 62–3
  Lake Okeechobee 63, 64, 65, 326
  Northern Everglades Payment for Environmental Services (NEPES) program 326
  partitioning of 63–4
  sugarcane production 64–6
  Swamp Lands Act 62
  water quality, decline of 66–7
export restrictions 291–2

fair apportionment principle 120, 128
famine 132, 182, 296, 297
FAO *see* Food and Agriculture Organization (FAO), UN
Farm Bill programs (USDA) 262, 263–6, 325
  Energy Title 265, 266
farming
  and climate change 79
  and concentrated animal feeding operations (CAFOs) 234
  and ecosystem services 316, 322
  Everglades region, South Florida 63
  farmers' markets 220–1
  farmland preservation 201, 209
    incentive programs for 216–17
  hog farms 237, 241
  household strategies 288–9
  integration of farming practices 202
  in Kenya 381–4
  protecting farms from displacement 204–7
  'Right-to-Farm' laws 205, 206
  small farms, protection from urbanization 208
  and suburban development 205
  systems thinking 79
  USDA Farm Bill programs 262, 263–6, 325
  water use 121
FDI *see* foreign direct investment (FDI)
feed stocks, changing 247–8
fertilizers 6, 22, 50, 54, 241, 266, 269, 281, 357
  carbon fertilization effect 278
  chemical 223, 364
  Crystal Peak Fertilizer (CPF) Process, Missouri 251
  fossil-fuel based 437
  Organic Certification for 269
  synthetic 185, 380
financial resources, transfer for adaptation 40–3
FINEP (Funding Authority for Studies and Projects Agency), Brazil 345
fish 284–5
flex fuel cars 282
Flood Control Act (1948), Florida 64
flooding 2, 4, 10, 33, 63, 133, 179, 300, 360, 368
  in Everglades region 63, 64, 65, 66
  and water quality impacts 110–12
Florida (US), transformation of ecosystems in *see* Everglades region, South Florida
Florida Ranchlands Ecosystem Services Project (FRESP) pilot program 326
Fonseca, RL 340
Food and Agriculture Organization (FAO), UN 56, 228, 248, 296, 334, 437

on adaptation strategies 15–16
'climate smart' sourcebook 19
definition of foods security 5
and genetically modified food crops 180, 195
International Plant Protection Convention 150
on land husbandry 377–8
food crises 310, 433–4
food policy councils 221–3
food production 16, 135, 245, 275, 279, 339, 389, 433
and climate change 31, 40, 169, 200
deficiencies 5, 6, 11, 15, 30, 31, 35, 44, 106, 185–6, 283–4
global 5, 105, 194, 274, 440
land use controls 196, 201, 204, 211, 215, 218, 220, 224
local/regional 10, 209, 222, 442
meat 235, 236
mixed crop and livestock farms 280
sustainability 31, 221
food riots 298
food security and climate change 46, 182, 195, 380, 434
achieving food security 4–10, 16, 19, 20
adaptation of agricultural trade and investment rules to climate change 300, 309
adaptation strategies 15–17
in Africa 31, 368, 379, 439
in China 285, 433
contribution of agriculture to climate change 6–9
definitions/dimensions 5, 10
global food security 3, 10, 12, 14, 105, 168, 278, 279, 284, 291, 300, 307, 312, 313
and greenhouse gas emissions 31, 274
impact of climate change on agriculture 9–10
local/regional food security 282, 290, 292, 307, 371
mitigation strategies 12–14
objectives 3, 31, 44, 421
protecting/safeguarding 44, 135, 166, 201, 217, 301
Public Stockholding for Food Security Purposes 311
threats to food security 1, 3, 6, 10, 31, 146, 328, 367, 390, 448
and UNFCCC 21, 30–2, 34, 36
food security enhancements 301
foot-and-mouth outbreak, United Kingdom 136
forecasts 78, 79, 279, 291, 293
weather 15, 79, 293
foreign direct investment (FDI) 275, 277, 298, 306
agro-FDI 307, 314
Forest Act (2005), Kenya 379
Forestry Strategy program, USDA 269, 270
fossil fuels 50, 51, 282
in Australia 385, 390
fracking 388–9, 393, 401
freshwater lakes/wetlands 54–5
freshwater pathways, in hydrological cycle 50–1
FRESP (Florida Ranchlands Ecosystem Services Project) pilot program 326
fuel crisis 430–2
see also bioenergy/biofuels
Funding Authority for Studies and Projects Agency (FINEP), Brazil 345

Gambia 41
Garrido, Alberto 293
Gates Foundation 184
Gateway Panel Assessment, Australia 401, 402, 410, 411, 419, 420
GCF (Green Climate Fund) 29, 42–3
gender issues 294–5
General Agreement on Tariffs and Trade (GATT) 291, 309, 310
General Agreement on Trade in Services (GATS) 300
genetic engineering
as the answer 179–94
as climate mitigation 192
defined 171–4
genetically modified food crops 167–95
as adaptation 176–9
and deforestation 175, 176, 186
golden rice 180–1, 182, 183

mitigation claims 186
soil carbon sequestration, whether promoting 189–92
unintended consequences 192–4
yield, whether increasing 186–8
Global Change Research Program, US 12, 15
Global Environment Facility 41
Global Land Report (GLP) 430
global warming 174, 238, 360, 389
see also temperature
agricultural trade and investment rules, adaptation to climate change 275, 282, 284, 296, 313
glyphosate 193
GMCs *see* genetically modified food crops
Goiás Industrial Development Program (PRODUZIR), Brazil 347
Goiás Secretariat for Planning (SEPLAN), Brazil 346
Goiás State Department of Agriculture, Brazil 351
golden rice 180–1, 182, 183
Gouel, Christophe 310, 311
governance and systems thinking 78–84
see also systems thinking/relationships
agricultural governance 88–9
autopoiesis 97–9
biodiversity 89
climate systems 76–8
clockwork dolls dance analogy, climate systems 76–7
difficult environmental governance challenges 92–4
environmental and social governance 88–9
institutions, role of 87–90
interconnection between natural and social systems in agriculture 99–102
legal governance 84–7, 88, 95–6
natural resource governance 80
objective of governance 80
private governance 88–9
reliable governance 89
rural environments 91–2, 94–7
government programs, voluntary and incentive-based 262–72
USDA Farm Bill programs 262, 263–6
Graham, Nicole 386
grazing systems 107–8
Great Barrier Reef region, coral reef ecosystems in 55
Green Climate Fund (GCF) 29, 42–3
Greenhouse Gas Emission Estimates System (SEEG), Brazil 352
greenhouse gases (GHGs)/greenhouse gas emissions 5, 6, 7, 11, 14, 228, 228–73
see also ammonia; hydrogen sulphide; particulate matter; Volatile Organic Compounds (VOCs)
agricultural 23, 24–34
and animal agriculture
background 230–8
changing production systems 248–50
concentrated animal feeding operations (CAFOs) 233–5, 246–51
enteric fermentation 239–40
feed stocks, changing 247–8
global production trends 235–8
history of American livestock systems 230–2
intensive livestock production and GHGs 238–46
land application 242–3
manure storage and management 240–2, 250–1
reporting CAFO GHG emissions under CAA 260–2
anthropogenic 24
in Brazil 351–8
CAFO GHG emissions, reporting under CCA 260–2
caps, emission 305
carbon 185, 187
carbon dioxide (CO2) 22, 23, 51
case law 254
Clean Air Act (CAA) 252–62
overview 252–4
regulation of air pollutants from CAFOs under 256–60

regulation of GHGs under 254–6
reporting CAFO GHG emissions under 260–2
Emissions Gap Report, UNEP 190
energy generation designed to reduce 112
EPA Endangerment Finding 254–5
global 31
increases in concentration 51
IPCC on 185
land grabbing 437
Mandatory Greenhouse Gas Reporting Rule 260, 261
methane 22–3, 26, 51
nitrous oxide 22–3, 26, 50, 51, 241, 242, 243
non-land 23
planning 211
regulation under the CAA 254–6
reporting under the CAA 260–2
stabilization of 185
targets 185
uniform factors 23
USDA Farm Bill programs 262, 263–6
voluntary and incentive-based government programs 262–72
greenhouse tomato production 57, 58
groundwater contamination 393
groundwater law, and climate change 118–20
Gunderson, Lance H 142

Hazardous Air Pollutants (HAPs) 253
Headey, Derek 429
heat stress 245
Heilberg, Phil 444
Hellmann, Jessica J 146
Helsinki Rules on the Use of International Rivers (1966) 127
Herbert Hoover Dike, Everglades 63
Herrero, Mario 280
Hertel, Thomas W 280
Hochman, Gal 282
hogs, in United States 230, 234
Holling, CS 51, 53
Holzer, Kateryna 300
hothouses 57
Huang, Jikun 288
hunger and violence 296–8
Hunold, Christian 417
hydraulic fracturing 388–9, 393, 401
hydrogen sulphide 235, 256, 257, 258, 259
hydrological cycle, freshwater pathways in 50–1

Idaho, land use controls 206
IFPRI *see* International Food Policy Research Institute (IFPRI)
IGP (Indo-Gangetic Plain) 289
incentive zoning 216–17
Independent Expert Scientific Committee (IESC), Australia 404
India 106, 237, 239, 249, 279, 293
Indigenous and Tribal People's Convention (1989) 427
Indo-Gangetic Plain (IGP) 289
information gathering, land use 213
input subsidies 281, 308–10
Institute of Geography and Statistics (IBGE), Brazil 335
institutions, role of 87–90
'Integrating Climate Resilience into the Agricultural Sector for Food Security in Rural Areas of Mali,' project 41–2
Intended Nationally Determined Contribution (INDC) 372
Intergovernmental Panel On Climate Change (IPCC) 23, 28, 139, 183, 285, 353, 437
Fifth Assessment Report 34, 235–6, 367
and livestock production 245–6
and water 104–5
International Association for Public Participation (IAP2) 160
International Covenant on Economic and Social Rights 307
International Energy Agency (IEA) 386
International Food Policy Research Institute (IFPRI) 190, 276, 280–1, 440
International Fund for Agricultural Development, UN 236
International Law Association (ILA) 127

International Network of Resource Centres on Urban Agriculture and Food Security (RUAF Foundation) 295
International Plant Protection Convention, FAO 150
international regulatory framework 298–312
  international water management 311–12
  investment
    investment and input subsidies 281, 308–10
    rules to change 305–8
    stockpiles 310–12
  trade
    autonomous climate measures 303–4
    labels and standards 304–5
    multilateral and regional approaches 302–3
    role of regulation 298–301
    subsidies, agricultural trade-related 301
International Union for Conservation of Nature (IUCN) 139
invasive species law 18, 135–66
  see also under Australia
  biological infestations of threat 163–4
  biosecurity 136, 140, 147, 149, 150, 151, 154, 158, 160, 164
  community engagement, facilitating 159–62
  distribution of invasive species 146
  economic impacts 144–5
  elusive concept of invasive species 139–45
  examples of invasive species 137, 140, 142, 143–4
  expert-driven approach 160–1
  foot-and-mouth outbreak, United Kingdom 136
  government regulation 152
  importation, illegal 139–40
  improving with legal and institutional innovations 164–5
  institutional arrangements 150
  interactions between climate change and invasive species 145–6
  knowledge types 163
  lack of systemic approach to challenge 158–9
  legal scholarship 163
  meaning of 'law' 137
  motivational postures 153
  ontological views 143, 144
  scientific uncertainty 142
  terminology 142–3
  transactions costs, effects on invasive species management 157–8
  transdisciplinarity 162–4
  weeds 140, 141, 142, 146
  'wicked' problems 146, 159
  wild dogs and sheep slaughter, Australia 136
  wild pig invasion, United States 136
investment
  see also agricultural trade and investment rules, adaptation to climate change
  foreign direct investment (FDI) 275, 277, 298, 306, 307, 314
  and input subsidies 281, 308–10
  research and development (R&D), agricultural 278, 302
  rules to change 305–8
  stockpiles 310–12
IPCC *see* Intergovernmental Panel On Climate Change (IPCC)
irrigation 13, 41, 50, 203, 211, 393, 446
  see also precipitation; rainfall; water; water law and climate change
  agricultural trade and investment rules, adaptation to climate change 264, 284, 288, 289, 307
  artificial 110
  demand 34, 107
  drip 36
  groundwater pumping 118
  improved 15, 34, 117, 130, 284
  large-scale 116, 118
  pasture 123
  and water law 107, 110, 112, 113, 115

Jean, Sébastien 310, 311
Jessup, Brad 405, 416
Jones, Peter G 280
Josling, Tim 308, 309

Kansas, land use controls 208
KARLO (Kenya Agriculture and Livestock Research Organization) 382–3
Kenya
see also Africa
agricultural extension policy 381, 382, 383–4
Agriculture, Fisheries and Food Authority Act (AFFA) 377, 378
agriculture and climate adaptation 372–3
Agriculture Sector Development Strategy (ASDS) 370, 371, 380–1, 382
background 369–70
climate change challenge 370–1
Climate Smart Agriculture (CSA) 379, 380, 383
constitutional foundations 373–6
Draft Climate Change Policy 375
enhancing role of farmers to building adaptive capacity 383–4
Forest Act (2005) 379
Intended Nationally Determined Contribution (INDC) 372
Land Act (2012) 371, 376
Land Registration Act (2012) 376–7
legal tools for the mainstreaming of climate change in the agricultural sector 373–84
mainstreaming of climate change in the agricultural sector
legal tools, use of 376–84
sectors and agencies 375–6
National Agriculture Sector Extension Policy (NASEP) 382
National Climate Change Action Plan (NCCAP) 367–8, 375, 381
National Land Policy 370, 374, 376, 377
role of agricultural extension in climate change adaptation 381
role of farmers, researchers and extension works 381–3
smallholder agriculture in 367–84
Statutory Instruments Act (2013) 378
Sustainable Development Goals 380
Vision 2030, Second Medium Term Plan 369
Kenya Agriculture and Livestock Research Act (KARLA) 382
Kenya Agriculture and Livestock Research Organization (KARLO) 382–3
Kibaki, Mwai 368
King County, Washington 211
Knoxville, Tennessee 224–5
Köppen Climate Classification System 68
Kueffer, Christoph 163
Kyoto Protocol to the United Nations Framework Convention on Climate Change 12
see also United Nations Framework on Climate Change (UNFCCC) and Kyoto Protocol
Adaptation Fund 40
adaption technologies, transfer 39
Annex A 26
Annex I Parties 24–5, 27, 32, 36
subsidies, agricultural trade-related 301
trade and investment, regulatory framework 300

labels and standards 304–5
Laborde, David 303
lagoons 234, 242, 251
land degradation 370, 384
land grabbing 9, 275, 277, 423–45
Africa 9, 423, 442, 443, 444
agricultural dynamic 437–9
conceptual and contextual clarifications 424–9
consequences 445–7
criticism of definitions 424–5
drivers of, and connection with climate change 429–37
financial and economic crisis 434–7
food crisis 433–4
fuel crisis 430–2
historical context 428
non-coercion 425, 427
primacy of the indigenous peoples 425–6

susceptibility to 440–4
valid consent 425, 426–7
Land Registration Act (2012), Kenya 376–7
land tenure, and ecosystem services 321, 322, 328–9
land use charges (LUCs) 282–3
land use conflict (Australian case study) 16–17, 385–422
climate change and global energy demand 386–90
coal and gas development in New South Wales 397–406
decision making, rethinking 414–20
energy production, increased 391–7
North West region 406–14
land use controls 196–227
case law 204–5, 206, 207–8
Climate Action Plans 211, 224–5
and climate change 198–201
ecosystem services 203
farmers' markets 220–1
farms, protecting from displacement 204–7
food policy councils 221–3
food production 196, 201, 204, 211, 215, 218, 220, 224
incentive programs for farmland preservation 216–17
local government action 197, 199–200, 203–4, 208, 210–13
planning, promise of 209–13
purchase of agricultural conservation easements (PACE) 214–15
repurposed land use tools as climate change responses 213–25
resilience, improving 201–4
trade-offs, environmental 199, 201
traditional challenges and land use tools 204–13
transferable development rights (TDR) programs 215–16
United States 205–6
Ashford, Connecticut 217
Atlanta, Georgia 219
Austin, Texas 219, 221
Baltimore County, Maryland 214–15
Beekman, New York 216–17
Berthoud, Colorado 216
Boone County, Illinois 215
Brookhaven, New York 205–6
Castine, Maine 223, 224
Dahlonega, Georgia 220
Idaho 206
Kansas 208
King County, Washington 211
Knoxville, Tennessee 224–5
Malta, New York 206
Milwaukie, Oregon 218–19
Minnesota 206
Oregon 209
San Francisco 220
Southampton, New York 216
Suffield, Connecticut 208–9
Troy, Wisconsin 215
Wichita, Kansas 221
urban agriculture 217–20
zoning (Euclidean), and community threats 207–9
laser land levelling (LLL) 289
Leal-Arcas, Rafael 302
Least Developed Countries Fund (LDCF) 41
Lee, Jaemin 312
legal governance 88
problems challenging traditional approaches 95–6
and systems thinking 84–7
legumes 50
Leite, Sérgio 334, 348
life science companies 194–5
linear and non-linear systems 53
liquefied natural gas (LNG) 387
livestock 22, 135, 185
see also animal agriculture
climate change, effects on production 243–6
history of American livestock systems 230–2
intensive livestock production and GHGs 238–46
mixed crop and livestock farms 280
and water 107–8
LLL (laser land levelling) 289
locusts 55–7
Ludwig, D 52–3
Luhmann, Niklas 97

LULUCF sector (Land Use, Land use Change and Forestry) 23, 25, 26
Lybbert, Travis 280, 289–90

MacGregor, James 305
Machado, RB 340
Malawi 35
malnutrition 35, 127, 200, 275
  in children 181, 281
Malta, New York 206
Mandatory Greenhouse Gas Reporting Rule 260, 261
manure storage and management 234, 235, 240–2
  changes to 250–1
Marceau, Gabrielle 309
Martin, Paul 137, 148, 155, 162
Massachusetts v. EPA (2007) 254
Maximum Achievable Control Technology (MACT) standards 253
meat production 235, 236
Meltzer, Joshua 304
methane 7, 22–3, 26, 51
  animal agriculture and GHGs 239–40, 241
  and fracking 388–9
methanogens (microorganisms) 239
Mexico 293
Midwest Constitutional Fund (FCO), Brazil 344, 346
migration 295–6
Millennium Development Goals 276–7
Millennium Ecosystem Assessment (MEA) 135, 315
Miller, Marc L 142
Milwaukie, Oregon 218–19
Ministry of Agriculture, Livestock and Food Supply (MAPA), Brazil 334
Ministry of Environment (MMA), Brazil 341, 342, 343, 344, 359
Ministry of Science, Technology and Innovation (MCTI), Brazil 353
Minnesota, land use controls 206
mitigation, climate change 31, 275
  agricultural trade and investment rules, adaptation to climate change 286–98
  'best practices' 13
  climate resilient agriculture 17–20
  genetically modified food crops 185–6
  invasive species law 138–9
  obligations for agricultural emissions under UNFCCC and Kyoto Protocol 24–34
    and food security 30–2
    national communication reports (NCs) 26, 36
    national inventory submissions 25
    and trade 32–4
  policy and reporting requirements 24–6
  sector specific commitments 26–9
  strategies 12–14
  technology transfer for 29–30
Monsanto 179, 195
  Roundup Ready Soybeans 172, 193
Moomaw, William 285, 312
Murphy, Sophia 444

NAAQS (National Ambient Air Quality Standards) 253, 254
NAEMs (National Air Emissions Monitoring Study) 257, 258, 259, 260
Nairn Review (1995), invasive species 150
Nairobi, WTO Ministerial Conference (2015) 311
Narrabri Gas Project, Australia 408
National Academies, US 70
National Adaption Programmes of Action (NAPAs) 35–6, 41, 42
National Agriculture Sector Extension Policy (NASEP), Kenya 382
National Agro Energy Plan (2006–2011), Brazil 332, 334
National Air Emissions Monitoring Study (NAEMS) 257, 258, 259, 260
National Ambient Air Quality Standards (NAAQS) 253, 254
National Bank for Economic and Social Development (BNDES), Brazil 344–6
National Climate Change Action Plan (NCCAP), Kenya 367–8, 375, 381
National Communication reports (NCs) 26, 36

National Ethanol Program (*Pró Alcool*), Brazil 331
National Institute for Space Research (INPE), Canasat system 335
National Inventory Report (NIR) 25
National Land Policy, Kenya 370, 374, 376, 377
National Plant Biosecurity Strategy, Australia 149
National Program for the Production and Use of Biodiesel (PNPB), Brazil 333
National Resource Management (NRM) 149, 160
National Resources Conservation Service (NRCS) 263, 264, 268, 270
National Supply Company (CONAB), Brazil 334–5, 336, 361
nationally appropriate mitigation actions (NAMAs) 28–9
natural disasters 78
see also drought and aridity; flooding; storms
natural resource governance 80
Natural Resource Management (NRM), Australia 149, 154, 155
natural selection, Darwinian 98
NCCAP (National Climate Change Action Plan), Kenya 367–8, 375, 381
NCs *see* national communication reports (NCs)
Nepal 35
NEPES (Northern Everglades Payment for Environmental Services) program 326
net food-importing developing countries (NFIDC) 276
Netherlands 239
New Zealand, emissions trading system 28
Nile River Basin 126
Nin-Pratt, Alejandro 281
nitrogen 50, 54
nitrous oxide emissions 22, 26, 50, 51, 241, 242, 243
Nolon, John 210
nomosphere concept 405
North America 106, 110, 244
Northern Everglades Payment for Environmental Services (NEPES) program 326
Nova Fronteira Bioenergia, Brazil 349
NRCS (National Resources Conservation Service) 263, 264, 268, 270
NRM (National Resource Management) 149, 160
nutrient induced shifts 54–5
nutritional deficiencies 181–2

Oak Park zoning ordinance, US 207
Oakland City Council, US 221–3
Oakland Food Policy Council (Oakland FPC) 222
Obama, Barack 267–8
O'Connell, D 72
OECD *see* Organisation for Economic Cooperation and Development (OECD)
Okeechobee, Lake (Everglades) 63, 64, 65, 326
Olmstead, Alan L 286
Oregon, land use controls 209
Organisation for Economic Cooperation and Development (OECD) 80, 86, 278, 283
Environmental Policy Instruments 101
Owens, Katherine 385, 391, 393, 402

PACE (purchase of agricultural conservation easements) 214–15
Paglia, AP 340
Paiss Agrícola (Joint Plan to Support Industrial Technological Innovation in Sugarcane Sectors), Brazil 345
panarchy theory 59–61
parasitoids/predators 58
Paris Agreement on Climate Change (2015), adopted at COP21 1, 12, 276, 299
agriculture in 22, 43–5
Negotiating Texts 43, 44
particulate matter 235, 253, 256, 257, 259
pasture composition 244
Paul, Helena 442
Pauly, D 284, 285

payments for ecosystem services (PES) programmes 317
Pendoley v. Ferreira (1963) 204–5
pest outbreaks 55–8, 144
pesticide restrictions and water use 223–4
Peters, Victoria 301
Petrobrás Biofuel (PBio), Brazil 349
Petrobrás ethanol pipeline, Brazil 347
phosphorous 50, 54, 111
photosynthesis 49–50
Pimentel, D 145
Pittock, Barrie 137
planning, land use controls 16, 209–13
Planning Assessment Commission (PAC), Australia 399, 400, 404, 411, 413, 418
Poisel, Tim 389, 415
Pollen, Michael 181
'polluter pays' principles 99
pollution 76, 101, 127, 252, 312
  air *see* air pollution
  ammonia *see* ammonia
  dust 410
  National Pollutant Inventory, Australia 394
  noise 415
  non-point source 198
  trans-boundary 300
  water 25, 91, 111, 150, 202, 223, 229, 233, 242, 271, 393
POLOCENTRO (Program for the Development of the Savannas) 343
population
  of China 433
  global 5–6, 103
Posner, Richard 163
poverty 11, 195, 275, 291
  absolute or relative 99, 313
  in Africa 291, 367, 370, 372, 441, 444
  alleviation/eradication of 182, 183, 280, 284, 305, 321
  extreme 100, 277, 297
  high levels 367, 436, 438
  'hotspots' 279
  and vitamin A blindness 182, 183
precipitation 106, 107, 109–10
  see also flooding; rainfall; water; water law and climate change; weather
  changes in 9–10, 108, 111, 178, 373
  global 109
  heavy/intense 2, 110, 111, 132, 169
  high precipitation events 178, 179
  patterns 10, 15, 104, 106, 274, 373
  and temperature 10, 68
Prevention of Significant Degradation (PSD) program 254
prior appropriation doctrine, and water law 115–17, 118, 124
private governance 88–9
PRODECER (Japanese-Brazilian Program for the Development of the Savannas) 343
PRODUZIR (Goiás Industrial Development Program), Brazil 347
Program for the Development of the Savannas (POLOCENTRO), Brazil 343
Prorenova 345
PSD (Prevention of Significant Degradation) 254
public participation, planning 212
public private partnerships (PPPs) 270
Public Stockholding for Food Security Purposes 311
purchase of agricultural conservation easements (PACE) 214–15

Quaye, Wilhelmina 290
Queensland, Australia 143–4, 148

rainfall 34, 57, 68, 69, 72, 110, 138, 347, 406, 446
  see also flooding
  abundant 64, 66, 67, 340
  altered patterns 6, 10, 17, 48, 73, 78, 311, 438
  insufficient 287
  variable 69, 279, 289, 311, 368
rational actor model 153
Ravaillon, Martin 297
refugees, climate 296
regime shifts and agricultural systems 53–8
  definition of 'regime' 52
  nutrient induced shifts 54–5

pest outbreaks 55–8
and shifting controls 53–4
Renewable Energy Directive, EU 303
Renewable Fuel Program, CAA 271
reporting requirements
adaptation 36
mitigation 24–6
research and development (R&D), agricultural 278, 302
resilience 18–19
and adaptation 69–71
agricultural, description of 201
and agro-ecosystems 47–73
climate resilient agriculture as adaptation and mitigation strategy 17–20
concept/definitions 51, 52, 53
decreasing 73, 275
ecological 52–8
engineering 53
general, in agro-ecosystems 71–2
increasing 71–2
international, and drought 130–2
land use controls 201–4
specific, in agro-ecosystems 72
theory 58, 59
and vulnerabilities 285–6
Rhode, Paul W 286
'Right-to-Farm' laws 205, 206
riparian rights, and climate change 113, 114–15, 117, 124–5
regulated riparianism 125
Roesch, Stephanie D 280
Rome Statute of the International Criminal Court 427
Rowhani, Pedram 291
Rozelle, Scott 288
ruminant animals 239
runoff 50, 54, 179
agricultural 66, 67, 202, 208, 223
of contaminants 111
nutrient 55, 66
surface water 360
Rural Energy for America Program (REAP) 269
rural environments
archetypes, and rural problems 91–2
collective action solutions 92
coordination, frustration of 96
problem-solving 91–2
spatial feature of rurality, acknowledging 94–7
Russia 293
Ryan, Erin 197

Sachdev, Sonya 304
Salkin, Patricia 210, 226
Salt, D 52
San Francisco 220
Sauer, Sérgio 334, 348
Sawyer, Donald 359
scarcity, water 10, 113, 123, 124, 125, 129
Schindler, Sarah 219
Schlosberg, David 416
SCM Agreement (Agreement on Subsidies and Countervailing Measures) 309, 312
sea level rise 111–12
second generation biofuels 143, 282
SEEG (Greenhouse Gas Emission Estimates System), Brazil 352
Sen, Amartya 297
Senegal 285
shale gas 388
Sherval, Meg 386
Sierra Leone 442–3
signals, and behaviour 79
Silent Spring (Carson) 92
Sindel, Brian 140
Smith, Buckingham 62, 64
Smithfield Agreement, North Carolina 251
socio-ecological system 82
soil, systems thinking 76, 78
soil carbon sequestration 189–92
solar energy 49–50
Soros, George 435
South Africa 278
South America 111
Southampton, New York 216
sovereignty 95, 96
limited territorial 126–7
soy biodiesel 282
Special Climate Change Fund (SCCF) 41
Speranza, Ifejika 286
SRLUP (Strategic Regional Land Use Policy), Australia 400, 401, 402, 403

starvation *see* famine
State Implementation Plans (SIPs) 254
stationarity 108
Stephens, Phoebe 429
Stockholm International Water Institute 284
stockpiles 310–12
stocks 78, 79, 83, 311
storms 63, 78, 98, 110, 133
see also rainfall; weather
land use controls 212, 213
Strategic Agricultural Land (SAL), Australia 401
see also Biophysical Strategic Agricultural Land (BSAL), Australia
Strategic Regional Land Use Plans, Australia 402
Strategic Regional Land Use Policy (SRLUP), Australia 400, 401, 402, 403
sub-Saharan Africa (SSA) 31, 105, 280, 290, 439
Subsidiary Body for Scientific and Technological Advice (SBSTA) 37–8
subsidies
agricultural trade-related 301
energy 282
farm 301
import 311
investment and input subsidies 281, 308–10
subsistence agriculture 274, 281–2
Suffield, Connecticut 208–9
Sugar Industry Union (Unica), Brazil 358
Sugarcane Agroecological Zoning (ZAE), Brazil 332–3, 338, 339, 341, 344
sugarcane production 8
in Brazil and in the Cerrado region, expansion of 333–41
Cerrado biome and governmental support for sugarcane production 341–7
corporate control of, in Brazil 350–1
'domino effect' 359
in Everglades region 64–6
international capital 348–51
sugarcane economic cycle 331
Sumner, Daniel 280, 289–90
Sykes, Alan O 309
systems thinking/relationships 17–18, 78–84
see also governance and systems relationships
agricultural and environmental systems 82–3
building blocks of natural environment 76
chaos theory 77
climate systems 76–8
clockwork dolls dance analogy, climate systems 76–7
coupling of natural and social systems 80
legal governance 84–7
socio-ecological systems 82
stocks 78, 79, 83
'whole-of-system' approaches, need for 101

Tailoring Rule 255
Tanzania, export restrictions 291–2
tariff reductions, 'green' goods 299
taxation, carbon 292, 303
TDR (transferable development rights) 215–16
Technology Executive Committee (TEC), UNFCCC 39
Technology Mechanism 30
technology transfer
for adaptation 38–9
for mitigation 29–30
temperature 48, 52, 58, 68, 69, 73, 79, 104, 111, 177, 389
see also global warming
ambient air 242, 243, 250, 252, 254
annual 286–7
average 298
global 2, 12, 31, 175, 185, 228
global average 2, 45, 106, 108, 110, 174
increasing/rising 36, 57, 72, 242, 370
agricultural trade and investment rules, adaptation to climate change 274, 278, 281, 285

food security issues 2, 4, 6, 10, 12, 15
genetically modified food crops 167, 169, 178
water law and climate change 107, 111
local 34
minimum 298, 340
surface 2, 285
terrestrial ecosystems 50
Thornthwaite Climate Classification System 68
Thornton, Philip K 280
tomato production 57, 58
Tomato Yellow Leaf Curl Virus (TYLC) 57–8
Trachtman, Joel P 309
trade
see also agricultural trade and investment rules, adaptation to climate change
autonomous climate measures 303–4
labels and standards 304–5
liberalization 275
and mitigation measures 32–4
multilateral and regional approaches 302–3
role of regulation 298–301
subsidies, agricultural trade-related 301
trade-offs, environmental 199, 201–3
transactions costs, effects on invasive species management 157–8
transdisciplinarity, invasive species law 162–4
transferable development rights (TDR) programs 215–16
transformative change 58, 59–67
Troy, Wisconsin 215
Tyler, Tom 418

UNEP *see* United Nations Environment Programme (UNEP)
UNFCCC *see* United Nations Framework on Climate Change (UNFCCC); United Nations Framework on Climate Change (UNFCCC) and Kyoto Protocol
Union for Concerned Scientists 188
United Kingdom 431–2
United Nations Declaration on the Rights of Indigenous People's (UNDRIP) 426
United Nations Environment Programme (UNEP) 131, 186, 279
Emissions Gap Report 190
United Nations Framework on Climate Change (UNFCCC) 21–46, 279, 372
see also Kyoto Protocol to the United Nations Framework Convention on Climate Change
adaptation
Adaptation Committee 43
cooperation on 36–8
financial resources, transfer for 40–3
National Adaption Programmes of Action 35–6
need for 34
obligations 35–43
reporting on 36
technology transfer for 38–9
Ad-Hoc Working Group on Long-term Cooperative Action 28
agriculture in 22–3
mitigation obligations for agricultural emissions under 24–34
and food security 30–2
policy and reporting requirements 24–6
sector specific 26–9
technology transfer, for mitigation 29–30
and trade 32–4
Secretariat 28
Subsidiary Body for Implementation 29
subsidies, agricultural trade-related 301
Technology Executive Committee (TEC) 39
trade and investment, regulatory framework 299, 300
and World Trade Organization 33
United Nations (UN) *see* Food and Agriculture Organization (FAO), UN; Kyoto Protocol to the United Nations Framework

Convention on Climate Change; United Nations Declaration on the Rights of Indigenous People's (UNDRIP); United Nations Environment Programme (UNEP); United Nations Framework on Climate Change (UNFCCC)
United States
see also California
Agricultural Act (2014) 266
agriculture, lack of regulation 229
American Planning Association 211
Arizona Water Bank 119
Clean Air Act (CAA) 12, 252–62
overview 252–4
regulation of air pollutants from CAFOs under 256–60
regulation of GHGs under 254–6
reporting CAFO GHG emissions under 260–2
Clean Water Act (CWA) 113, 223, 233
Colorado River Compact 121–2
corn ethanol production 8
crop destruction 136
Crystal Peak Fertilizer (CPF) Process, Missouri 251
DDT, ban on 92
Department of Agriculture *see* US Department of Agriculture (USDA)
disease distribution changes in 246
Endangered Species Act (ESA) 113
Everglades region, transformation of agro-systems in *see* Everglades region, South Florida
and genetically modified food crops 194
Global Change Research Program 12, 15
groundwater law, and climate change 118
history of livestock systems 230–2
interstate apportionment of water 120–2
land use controls 205–6
Ashford, Connecticut 217
Atlanta, Georgia 219
Austin, Texas 219, 221
Baltimore County, Maryland 214–15
Beekman, New York 216–17
Berthoud, Colorado 216
Boone County, Illinois 215
Brookhaven, New York 205–6
Castine, Maine 223, 224
Dahlonega, Georgia 220
Idaho 206
Kansas 208
King County, Washington 211
Knoxville, Tennessee 224–5
Malta, New York 206
Milwaukie, Oregon 218–19
Minnesota 206
Oregon 209
San Francisco 220
Southampton, New York 216
Suffield, Connecticut 208–9
Troy, Wisconsin 215
Wichita, Kansas 221
New Deal era 65
Rural Energy for America Program (REAP) 269
swine in 230, 231, 234
US Sugar Corporation 65
water law 107, 113–25
interstate apportionment 120–2
water markets 123
urban agriculture 217–20
Urban Land Institute 218
US Department of Agriculture (USDA) 230
see also United States
Climate Smart Agriculture 31–2, 269, 270
and ecosystem services 325–6
Farm Bill programs 262, 263–6, 325
Forestry Strategy program 269, 270
Organic Certification for fertilizers 269

Valenzuela, Ernesto 278
Venturelli, Fabio 348
Vietnam 288
Village of Euclid v. Ambler Realty Co. (1926) 207, 208
vitamin A deficiency 181, 182, 183
Volatile Organic Compounds (VOCs) 235, 253, 256, 257, 259

Von Bertalanffy, Ludwig 77
Von Sperling, E 352, 360, 361, 362

Walker, B 52, 53
Wang, Jinxia 288
Warner, Ethan 282–3
wastewater 13, 34, 125, 133, 200, 393
water
  see also precipitation; water law and climate change
  and agricultural productivity 103–4
  availability 393
  decline of quality, in Everglades region 66–7
  efficiency 284
  markets 122–4
  quality 111, 223–4, 393–4
  quantity 106–9
  risks to agricultural supplies and infrastructure 105–12
  scarcity 10, 113, 123, 124, 125, 129
  and systems thinking 76, 78
  trigger for 404, 412–13
Water Efficient Maize for Africa (WEMA) project 183–4
water law and climate change 112–32
  aquifer storage and recovery 119–20
  Clean Water Act (CWA) 113, 223
  and climate change impacts on agriculture *see* water law and climate change
  domestic law (United States) 113–25
  drought and aridity 109–10, 129
    drought law 124–5
    and resilience across nations 130–2
  evolution of law in response to climate change 133–4
  flooding and water quality impacts 110–12
  groundwater law 118–20
  improvements in water use efficiency 123–4
  international water law 125–32, 311–12
    customary 126–8
    drought and resilience, across nations 130–2
    treaties and conventions 128–30
  international water management 311–12
  international waterways, use of 128
  interstate apportionment, in US 120–2
  pesticide restrictions and water use 223–4
  prior appropriation doctrine 115–17, 118, 124
  riparian rights 113, 114–15, 117, 124–5
  water markets 122–4
water pollution 25, 91, 111, 150, 229, 233, 242, 271
  groundwater contamination 393
  land use controls 202, 223
watercourses, international 113, 132
Watermark Coal Project, Australia 409–10
Watson, R 284, 285
weather 23, 68, 71, 73, 286, 293, 371
  see also flooding; precipitation; rainfall; storms
  extreme/severe events 246, 287, 328, 380, 384, 389, 433
    food security issues 3, 4, 6, 9, 10
    and land use controls 200, 212
    and UNFCCC 37, 40
    and water law 110, 111
  forecasts 15, 79, 293
  insurance schemes 293, 310
  shocks 291, 292
weeds 142, 146
  biofuel weed pathway 140, 141
Wessels, K J 282
Westphalian system 299–300
White Fly/Tomato/TYLC system 58
Wichita, Kansas 221
willingness to pay (WTP) 289
Winebarger, Lisa 301
World Bank 391, 436
  agricultural trade and investment rules, adaptation 276–7, 278, 284, 291, 292, 306
  and export restrictions 291, 292
World Food Summits 300
World Trade Organization (WTO) 30, 33, 276
  Agreement on Agriculture 281, 291, 308, 311